MICRO HOUSE®

PC Hardware Library Volume I: Hard Drives

by

Micro House International, Inc.

Scott Mueller

QUE®

A Division of Macmillan Computer Publishing
201 West 103rd Street,
Indianapolis, Indiana 46290

Micro House PC Hardware Library

Volume I: Hard Drives

International Standard Book Number: 0-7897-1663-1

Library of Congress Catalog Card Number: 98-84620

Printed in the United States of America

First Printing: June 1998

00 99 98 4 3 2 1

Trademarks

Credits

Executive Editor:	Jim Minatel
Managing Editors	Thomas Hayes Patrick Kanouse
Senior Editor	Elizabeth A. Bruns
Copy Editor	Patricia Kinyon
Indexers	Christine Nelsen Ginny Bess
Production	Mandie Rowell

Contents at a Glance

Table of Contents

About the Authors

Micro House International, Inc.

Micro House International, a privately held company headquartered in Boulder, Colorado, is the leading provider of PC hardware technical information libraries and hard drive setup and configuration software. Micro House information products and utilities are used by network administrators, help desk and IT professionals, resellers, and system integrators throughout the world for installation, maintenance, and upgrades of multivendor PCs and networks. Among the company's key information products and utilities are SupportSource, Support On Site for Hardware, the Micro House® Technical Library, DrivePro®, and EZ-Drive®. For more information about any Micro House products, please see the Micro House web site at http://www.microhouse.com.

Scott Mueller

Scott Mueller is the Series Editor for the **Scott Mueller Library** series of technical PC books all published by QUE. Scott is president of Mueller Technical Research, an international personal computer research and corporate training firm. Since 1982, MTR has specialized in the industry's longest running, most in-depth, accurate and effective corporate technical training seminars, maintaining a client list that includes Fortune 500 companies, the U.S. and foreign governments, major software and hardware corporations, as well as PC enthusiasts and entrepreneurs. His seminars have been presented to thousands of PC professionals throughout the world.

Scott Mueller has developed and presented personal computer training courses in all areas of PC hardware and software. He is an expert in PC hardware, operating systems, data-recovery techniques, and local area networks. MTR seminars are available on an on-site contract basis or publicly through the American Research Group (ARG). For more information about a custom computer training seminar for your organization, contact:

Mueller Technical Research

21718 Mayfield Lane

Barrington, IL 60010-9733

(847) 726-0709

(847) 726-0710 Fax

Internet: 73145.1566@compuserve.com

For more information about attending a public seminar, contact:

American Research Group

114 Edinburgh South, Suite 200

Cary, NC 27512

(919)461-8600

(919)461-8846 Fax

http://www.arg.com

email: questions@arg.com

Scott has many popular books, articles, and course materials to his credit, including *Upgrading and Repairing PCs*, which has sold more than 1 million copies, making it by far the most popular PC hardware book on the market today. His 2 hour video titled "*Your PC-The Inside Story" is available through* LearnKey, Inc. For ordering information, contact:

LearnKey, Inc.

1845 West Sunset Boulevard

St. George, UT 84770

(800) 937-3279

(801) 674-9733

(801) 674-9734 Fax

When he is not working on PC related books or teaching seminars, Scott can usually be found in the garage working on his LT4 powered '94 Impala SS, LT4 powered '95 Caprice 9C1 (police package), or Buick Turbo V6 powered '89 Trans Am, as well as various other performance car related projects. He can also be found testing the vehicles at the local drag strip, or showing them off at car shows or the local cruise/drive-in scene.

We'd Like to Hear from You!

Que Corporation has a long-standing reputation for high-quality books and products. To ensure your continued satisfaction, we also understand the importance of customer service and support.

Tech Support If you need assistance with the information in this book or you have feedback for us about the book, please contact Macmillan Technical Support by phone at 317-581-3833 or via email at support@mcp.com.

Orders, Catalogs, and Customer Service To order other Que or Macmillan Computer Publishing books, catalogs, or products, please contact our Customer Service Department

Phone: 1-800-428-5331

Fax: 1-800-835-3202

Internation Fax: 1-317-228-4400

Or visit our online bookstore at http://www.mcp.com/.

Introduction

The purpose of this text is to be a practical and easy reference for today's most popular hard disk drive formats used by desktop computers. While PC technicians and support staff will get the most use from this book in their daily dealings with PC hard drives and controllers, PC hobbiests and enthusiasts will also find it a valuable reference to keep handy whenever they need to work on a hard drive or controller.

The *Micro House PC Hardware Library Volume I: Hard Drives* contains complete set-up information and procedures for the industry's most common hard drives and controllers. It focuses on the two most prevalent hard drive interface formats in use today: IDE and SCSI.

The CD-ROM included with the *Micro House PC Hardware Library* contains some popular Micro House utilities to use with your hard drives. A fully-functional copy of ImageCast LE is included multicasting hard drive images. Other utilities are included for identifying a drive model, saving BIOS drive configurations and more. You'll also find the content of the tables from the book on the CD in searchable electronic format.

About This Book

This book is arranged into seven chapters. Each chapter has been designed to help you with every aspect of hard drive and hard drive controller installation.

Chapter 1 covers basic hard drive components and concepts common to many hard drives.

Chapter 2 contains comprehensive information on the IDE and SCSI interface types.

Chapter 3 discusses installation procedures for IDE and SCSI interface drives, including partitioning and high-level formatting.

Chapter 4 is a listing of many popular drive controller cards. This chapter includes a diagram for each card listed showing jumper and connector locations and tables of the relevant connections, pins, jumper defaults, and jumper uses.

Chapter 5 lists over two hundred popular and common IDE and SCSI hard drives. Note that many of the manufacturers have several similar models that are all described in the same figures and tables. For example, the Conner Peripherals CFA850S, CFA1275S, CFP1080S all share the same diagram and configuration tables so these are all listed together. So, be sure to look through all of the models listed on each page to see if the model you need is listed. For each drive (or set or drives) there is a diagram showing the orientation of the connectors and relevant jumpers. The tables with each drive show jumper settings for items such as master/slave configuration on IDE drives, SCSI IDs on SCSI drives, and any other user settable jumpers.

Chapter 6 lists physical and logical drive parameters for over 3600 drives from dozens of major manufacturers. Use these tables to identify drive capacity and to enter logical drive settings in the BIOS where needed.

Chapter 7 is a quick reference to IDE drive capacities and legal BIOS logical parameters for IDE drives. if you have an IDE drive that you cannot locate the manufacturer's recommended logical parameters for in either chapter 6 or in manufacturer drive settings, these generic settings will work for your BIOS as long as you have the correct drive capacity.

Appendix A lists manufacturer contact information for most major hard drive and drive controller vendors. You will also find other drive related vendors for items such as removable media and drive configuration and maintenance utilities.

Chapter 1

Hard Disk Drives

To most users, the hard disk drive is the most important, yet most mysterious, part of a computer system. A *hard disk drive* is a sealed unit that holds the data in a system. When the hard disk fails, the consequences usually are very serious. To maintain, service, and expand a PC system properly, you must fully understand the hard disk unit.

Data Encoding Schemes

Magnetic media essentially is an analog storage medium. The data that we store on it, however, is digital information—that is, ones and zeros. When digital information is applied to a magnetic recording head, the head creates magnetic domains on the disk media with specific polarities. When a positive current is applied to the write head, the magnetic domains are polarized in one direction; when negative voltage is applied, the magnetic domains are polarized in the opposite direction. When the digital waveform that is recorded switches from a positive to a negative voltage, the polarity of the magnetic domains is reversed.

During a readback, the head actually generates no voltage signal when it encounters a group of magnetic domains with the same polarity, but it generates a voltage pulse every time it detects a switch in polarity. Each *flux reversal* generates a voltage pulse in the read head; it is these pulses that the drive detects when reading data. A read head does not generate the same waveform that was written; instead, it generates a series of pulses, each pulse appearing where a magnetic flux transition has occurred.

To optimize the placement of pulses during magnetic storage, the raw digital input data is passed through a device called an *encoder/decoder* (*endec*), which converts the raw binary information to a waveform that is more concerned with the optimum placement of the *flux transitions* (*pulses*). During a read operation, the endec reverses the process and decodes the pulse train back into the original binary data.

Because the number of flux transitions that can be recorded on a particular medium is limited by the disk media and head technology, disk drive engineers have been trying various ways of encoding the data into a minimum number of flux reversals, taking into consideration the fact that some flux reversals, used solely for proper timing, are required. This method permits maximum use of a given drive hardware technology.

Although various encoding schemes have been tried, only a few are popular today. Over the years, the following three basic types have been the most popular:

- Frequency Modulation (FM)
- Modified Frequency Modulation (MFM)
- Run Length Limited (RLL)

The following sections examines these codes, discusses how they work, where they have been used, and any advantages or disadvantages that apply to them.

FM Encoding

One of the earliest techniques for encoding data for magnetic storage is called *Frequency Modulation* (*FM*) encoding. This encoding scheme, sometimes called *Single Density encoding*, was used in the earliest floppy disk drives that were installed in PC systems. The original Osborne portable computer, for example, used these Single Density floppy drives, which stored about 80KB of data on a single disk. Although it was popular until the late 1970s, FM encoding no longer is used today.

MFM Encoding

Modified Frequency Modulation (*MFM*) encoding was devised to reduce the number of flux reversals used in the original FM encoding scheme and, therefore, to pack more data onto the disk. In MFM encoding, the use of the clock transition cells is minimized, leaving more room for the data. Clock transitions are recorded only if a stored 0 bit is preceded by another 0 bit. In all other cases, a clock transition is not required. Because the use of the clock transitions has been minimized, the actual clock frequency can be doubled from FM encoding, resulting in twice as many data bits being stored in the same number of flux transitions as in FM.

Because it is twice as efficient as FM encoding, MFM encoding also has been called *Double Density recording*. MFM is used in virtually all PC floppy drives today and was used in nearly all PC hard disks for a number of years. Today, most hard disks use Run Length Limited encoding, which provides even greater efficiency than MFM.

Because MFM encoding places twice as many data bits in the same number of flux reversals as FM, the clock speed of the data is doubled so that the drive actually sees the same number of total flux reversals as with FM. This means that data is read and written at twice the speed in MFM encoding, even though the drive sees the flux reversals arriving at the same frequency as in FM. This method allows existing drive technology to store twice the data and deliver it twice as fast.

The only caveat is that MFM encoding requires improved disk controller and drive circuitry because the timing of the flux reversals must be more precise than in FM. As it turned out, these improvements were not difficult to achieve, and MFM encoding became the most popular encoding scheme for many years.

RLL Encoding

Today's most popular encoding scheme for hard disks, called *Run Length Limited* (RLL), packs up to 50% more information on a given disk than even MFM does and three times as much information as FM. In RLL encoding, groups of bits are taken as a unit and combined to generate specific patterns of flux reversals. By combining the clock and data in these patterns, the clock rate can be further increased while maintaining the same basic distance between the flux transitions on the disk.

During the late 1980s, the PC hard disk industry began using RLL encoding schemes to increase the storage capabilities of PC hard disks. Today, virtually every drive on the market uses some form of RLL encoding.

Instead of encoding a single bit, RLL normally encodes a group of data bits at a time. The term *Run Length Limited* is derived from the two primary specifications of these codes, which is the minimum number (the run length) and maximum number (the run limit) of transition cells allowed between two actual flux transitions. Several schemes can be achieved by changing the length and limit parameters, but only two have achieved any real popularity: RLL 2,7 and RLL 1,7.

Even FM and MFM encoding can be expressed as a form of RLL. FM can be called RLL 0,1 because there can be as few as zero and as many as one transition cell separating two flux transitions. MFM can be called RLL 1,3 because as few as one and as many as three transition cells can separate two flux transitions. Although these codes can be expressed in RLL form, it is not common to do so.

RLL 2,7 initially was the most popular RLL variation because it offers a high density ratio with a transition detection window the same relative size as that in MFM. This method allows for high storage density with fairly good reliability. In very high-capacity drives, however, RLL 2,7 did not prove to be reliable enough. Most of today's highest-capacity drives use RLL 1,7 encoding, which offers a density ratio 1.27 times that of MFM and a larger transition detection window relative to MFM. Because of the larger relative window size within which a transition can be detected, RLL 1,7 is a more forgiving and more reliable code, which is required when media and head technology are being pushed to their limits.

Another little-used RLL variation called RLL 3,9—sometimes called *Advanced RLL* (ARLL)—allowed an even higher density ratio than RLL 2,7. Unfortunately, reliability suffered too greatly under the RLL 3,9 scheme. The method was used by only a few controller companies that have all but disappeared.

Sectors

A disk track is too large to manage effectively as a single storage unit. Many disk tracks can store 50,000 or more bytes of data, which is very inefficient for storing small files. For that reason, a disk track is divided into several numbered divisions known as *sectors*. These sectors represent slices of the track.

Different types of disk drives and disks split tracks into different numbers of sectors, depending on the density of the tracks. Hard disks usually can use 17 to 100 or more sectors per track. Sectors created by standard formatting procedures on PC systems have a capacity of 512 bytes, but this capacity may change in the future. Sectors are numbered on a track starting with 1, unlike the heads or cylinders which are numbered starting with 0.

When a disk is formatted, additional ID areas are created on the disk for the disk controller to use for sector numbering and identifying the start and end of each sector. These areas precede and follow each sector's data area, which accounts for the difference between a disk's unformatted and formatted capacities. These sector headers, inter-sector gaps, and so on, are independent of the operating system, file system, or files stored on the drive. For example, an older 38MB hard disk has a capacity of only 32MB when it is formatted. Modern IDE (Integrated Drive Electronics) and SCSI (Small Computer System Interface) hard drives are preformatted, so the manufacturers now only advertise formatted capacity. Even so, nearly all drives use some reserved space for managing the data that can be stored on the drive.

Although I have stated that each disk sector is 512 bytes in size, this statement technically is false. Each sector does allow for the storage of 512 bytes of data, but the data area is only a portion of the sector. Each sector on a disk typically occupies 571 bytes of the disk, of which only 512 bytes are usable for user data. The actual number of bytes required for the sector header and trailer can vary from drive to drive, but this figure is typical. A few modern drives now use an ID-less recording that virtually eliminates the storage overhead of the sector header information. In an ID-less recording, virtually all of the space on the track is occupied by data.

You may find it helpful to think of each sector as being a page in a book. In a book, each page contains text, but the entire page is not filled with text; rather, each page has top, bottom, left, and right margins. Information such as chapter titles (track and cylinder numbers) and page numbers (sector numbers) is placed in the margins. The "margin" areas of a sector are created and written to during the disk-formatting process. Formatting also fills the data area of each sector with dummy values. After the disk is formatted, the data area can be altered by normal writing to the disk. The sector header and trailer information cannot be altered during normal write operations unless you reformat the disk.

Each sector on a disk has a *prefix portion*, or header, that identifies the start of the sector and a sector number, as well as a *suffix portion*, or trailer, that contains a *checksum*, which helps ensure the integrity of the data contents. Each sector also contains 512 bytes of data. The data bytes normally are set to some specific value, such as F6h (hex), when the disk is physically (or low-level) formatted. (The following section explains low-level formatting.)

In many cases, a specific pattern of bytes that are considered to be difficult to write are written so as to flush out any marginal sectors. In addition to the gaps within the sectors, gaps exist between sectors on each track and also between tracks; none of these gaps contains usable data space. The prefix, suffix, and gaps account for the lost space between the unformatted capacity of a disk and the formatted capacity.

Table 1.1 shows the format for each track and sector on a typical hard disk with 17 sectors per track.

Table 1.1 Typical 17-Sector/17-Track Disk Sector Format.

Bytes	Name	Description
16	POST INDEX GAP	All 4Eh, at the track beginning after the Index mark.
The following sector data (shown between the lines in this table) is repeated 17 times for an MFM encoded track.		
13	ID VFO LOCK	All 00h; synchronizes the VFO for the sector ID.
1	SYNC BYTE	A1h; notifies the controller that data follows.
1	ADDRESS MARK	FEh; defines that ID field data follows.
2	CYLINDER NUMBER	A value that defines the actuator position.
1	HEAD NUMBER	A value that defines the head selected.
1	SECTOR NUMBER	A value that defines the sector.
2	CRC	Cyclic Redundancy Check to verify ID data.

Table 1.1 Typical 17-Sector/17-Track Disk Sector Format (continued).

Bytes	Name	Description
3	WRITE TURN-ON GAP	00h written by format to isolate the ID from DATA.
13	DATA SYNC VFO LOCK	All 00h; synchronizes the VFO for the DATA.
1	SYNC BYTE	A1h; notifies the controller that data follows.
1	ADDRESS MARK	F8h; defines that user DATA field follows.
512	DATA	The area for user DATA.
2	CRC	Cyclic Redundancy Check to verify DATA.
3	WRITE TURN-OFF GAP	00h; written by DATA update to isolate DATA.
15	INTER-RECORD GAP	All 00h; a buffer for spindle speed variation.
693	PRE-INDEX GAP	All 4Eh, at track end before Index mark.
571	Total bytes per sector	
512	Data bytes per sector	
10,416	Total bytes per track	
8,704	Data bytes per track	

This table refers to a hard disk track with 17 sectors. Although this capacity was typical during the mid 1980s, modern hard disks place as many as 150 or more sectors per track, and the specific formats of those sectors may vary slightly from the example.

As you can see, the usable space on each track is about 16% less than the unformatted capacity. This example is true for most disks although some may vary slightly.

The POST INDEX GAP provides a head-switching recovery period so that when switching from one track to another, sequential sectors can be read without waiting for an additional revolution of the disk. In some drives this time is not enough. Additional time can be added by skewing the sectors on different tracks so that the arrival of the first sector is delayed.

The Sector ID data consists of the CYLINDER NUMBER, HEAD NUMBER, and SECTOR NUMBER fields, as well as a CRC field to allow for verification of the ID data. Most controllers use bit 7 of the HEAD NUMBER field to mark the sector as bad during a low-level format or surface analysis. This system, however, is not absolute; some controllers use other methods to indicate a marked bad sector. Usually, though, the mark involves one of the ID fields.

WRITE TURN-ON GAP follows the ID field CRC bytes and provides a pad to ensure a proper recording of the following user data area as well as to allow full recovery of the ID CRC.

The user DATA field consists of all 512 bytes of data stored in the sector. This field is followed by a CRC field to verify the data. Although many controllers use two bytes of CRC here, the controller may implement a longer Error Correction Code (ECC) that requires more than two CRC bytes to store. The ECC data stored here provides the possibility of DATA-field read correction as well as read error detection. The correction/detection capabilities depend on the ECC code chosen and on the controller implementation. A WRITE TURN-OFF GAP is a pad to allow the ECC (CRC) bytes to be fully recovered.

The INTER-RECORD GAP provides a means to accommodate variances in drive spindle speeds. A track may have been formatted while the disk was running slower than normal and then write-updated while the disk was running faster than normal. In such cases, this gap prevents accidental overwriting of any information in the next sector. The actual size of this padding varies, depending on the speed of disk rotation when the track was formatted and each time the DATA field is updated.

The PRE-INDEX GAP allows for speed tolerance over the entire track. This gap varies in size, depending on the variances in disk-rotation speed and write-frequency tolerance at the time of formatting.

This sector prefix information is extremely important because it contains the numbering information that defines the cylinder, head, and sector. All this information except the DATA field, DATA CRC bytes, and WRITE TURN-OFF GAP is written only during a low-level format. On a typical non-servo–guided (stepper-motor actuator) hard disk on which thermal gradients cause mistracking, the data updates that rewrite the 512-byte DATA area and the CRC that follows may not be placed exactly in line with the sector header information. This situation eventually causes read or write failures of the Abort, Retry, Fail, and Ignore variety. You can often correct this problem by redoing the Low Level Formatting (LLF) of the disk; this process rewrites the header and data information together at the current track positions. Then when you restore the data to the disk, the DATA areas are written in alignment with the new sector headers.

Multiple Zoned Recording

One way to increase the capacity of a hard drive is to format more sectors on the outer cylinders than on the inner ones. Because they have a larger circumference, the outer cylinders can hold more data. Drives without zoned recording store the same amount of data on every cylinder, even though the outer cylinders may be twice as long as the inner cylinders. The result is wasted storage capacity, because the disk media must be capable of storing data reliably at the same density as on the inner cylinders. With older ST-506/412 and ESDI (Enhanced Small Device Interface) controllers, unfortunately, the number of sectors per track was fixed. Drive capacity, therefore, was limited by the density capability of the innermost (shortest) track.

In a zoned recording, the cylinders are split into groups called *zones*, with each successive zone having more and more sectors per track as you move out from the inner radius of the disk. All the cylinders in a particular zone have the same number of sectors per track. The number of zones varies with specific drives, but most drives have 10 or more zones.

Another effect of zoned recording is that transfer speeds vary depending on what zone the heads are in. Because there are more sectors in the outer zones, but the rotational speed is always the same, the transfer rate will be highest.

Drives with separate controllers could not handle zoned recordings because there was no standard way to communicate information about the zones from the drive to the controller. With SCSI and IDE disks, it became possible to format individual tracks with different numbers of sectors due to the fact that these drives have the disk controller built in. The built-in controllers on these drives can be made fully aware of the zoning that is used. These built-in controllers must then also translate the physical cylinder, head, and sector numbers to logical cylinder, head, and sector numbers so that the drive has the appearance of having the same number of sectors on each track. The PC BIOS was designed to only handle a single number of specific sectors per track throughout the entire drive, meaning that zoned drives always must run under a sector translation scheme.

The use of zoned recording has allowed drive manufacturers to increase the capacity of their hard drives by between 20 percent and 50 percent compared with a fixed-sector-per-track arrangement.

Virtually all IDE and SCSI drives today use zoned recording. Because of limitations in the PC BIOS, these drives still have to act as though they have a fixed number of sectors per track. This situation is handled by translation algorithms that are implemented in the controller.

Basic Hard Disk Drive Components

Many types of hard disks are on the market, but nearly all drives share the same basic physical components. Some differences may exist in the implementation of these components (and in the quality of materials used to make them), but the operational characteristics of most drives are similar. Following are the components of a typical hard disk drive (see Figure 1-1):

- Disk platters
- Read/write heads
- Head actuator mechanism
- Spindle motor
- Logic board
- Cables and connectors
- Configuration items (such as jumpers or switches)
- Bezel (optional)

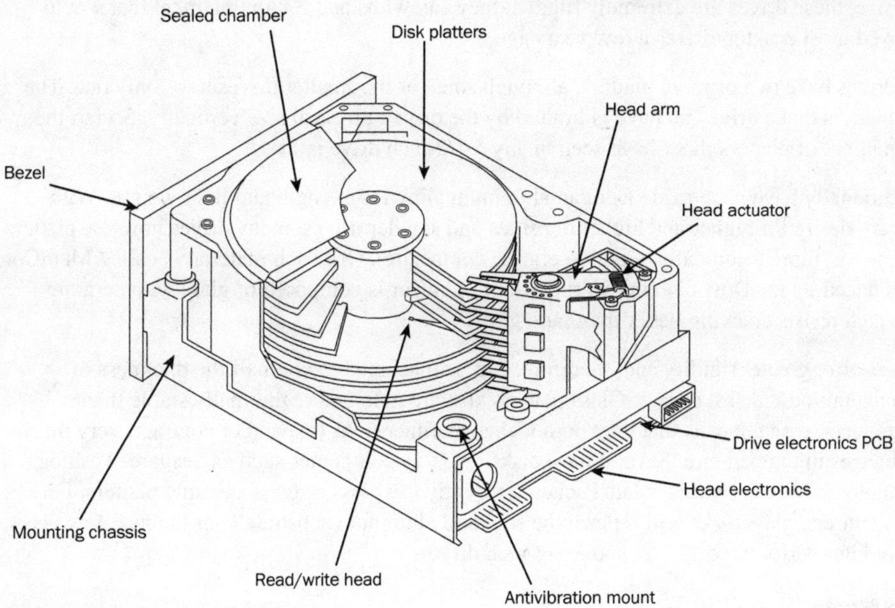

FIG. 1-1 Hard disk drive components.

The platters, spindle motor, heads, and head actuator mechanisms usually are contained in a sealed chamber called the *Head Disk Assembly* (*HDA*). The HDA usually is treated as a single component; it rarely is opened. Other parts external to the drive's HDA—such as the logic boards, bezel, and other configuration or mounting hardware—can be disassembled from the drive.

Hard Disk Platters (Disks)

A typical hard disk has one or more platters, or *disks*. Hard disks for PC systems have been available in a number of form factors over the years. Normally, the physical size of a drive is expressed as the size of the platters. Following are the most common platter sizes used in PC hard disks today:

- 5 1/4 inch (actually 130mm or 5.12 inches)
- 3 1/2 inch (actually 95mm or 3.74 inches)
- 2 1/2 inch
- 1.8 inch

Larger hard drives that have 8-inch, 14-inch, or even larger platters are available, but these drives typically have not been associated with PC systems. Currently, the 3 1/2-inch drives are the most popular for desktop and some portable systems, whereas the 2 1/2-inch and smaller drives are very popular in portable or notebook systems. These little drives are fairly amazing, with current capacities of up to 1GB or more, and capacities of 20GB are expected by the year 2000. Imagine carrying a notebook computer around with a built-in 20GB drive. It will happen sooner than you think! Due to their small size, these drives are extremely rugged; they can withstand rough treatment that would have destroyed most desktop drives a few years ago.

Most hard drives have two or more platters, although some of the smaller drives have only one. The number of platters that a drive can have is limited by the drive's physical size vertically. So far, the maximum number of platters that I have seen in any 3 1/2-inch drive is 11.

Platters traditionally have been made from an aluminum alloy for strength and light weight. With manufacturers' desire for higher and higher densities and smaller drives, many drives now use platters made of glass (or, more technically, a glass-ceramic composite). One such material is called MemCor, which is produced by the Dow Corning Corporation. MemCor is composed of glass with ceramic implants, which resists cracking better than pure glass.

Glass platters offer greater rigidity and, therefore, can be machined to one-half the thickness of conventional aluminum disks, or less. Glass platters also are much more thermally stable than aluminum platters, which means that they do not change dimensions (expand or contract) very much with any changes in temperature. Several hard disks made by companies such as Seagate, Toshiba, Areal Technology, Maxtor, and Hewlett-Packard currently use glass or glass-ceramic platters. For most manufacturers, glass disks will replace the standard aluminum substrate over the next few years, especially in high-performance 2 1/2- and 3 1/2-inch drives.

Recording Media

No matter what substrate is used, the platters are covered with a thin layer of a magnetically retentive substance called *media* in which magnetic information is stored. Two popular types of media are used on hard disk platters:

- Oxide media
- Thin-film media

Oxide media is made of various compounds, containing iron oxide as the active ingredient. A magnetic layer is created by coating the aluminum platter with a syrup containing iron-oxide particles. This media is spread across the disk by spinning the platters at high speed. Centrifugal force causes the material to flow from the center of the platter to the outside, creating an even coating of media material on the platter. The surface then is cured and polished. Finally, a layer of material that protects and lubricates the surface is added and burnished smooth. The oxide media coating normally is about 30 millionths of an inch thick. If you could peer into a drive with oxide-media–coated platters, you would see that the platters are brownish or amber.

As drive density increases, the media needs to be thinner and more perfectly formed. The capabilities of oxide coatings have been exceeded by most higher-capacity drives. Because oxide media is very soft, disks that use this type of media are subject to head-crash damage if the drive is jolted during operation. Most older drives, especially those sold as low-end models, have oxide media on the drive platters. Oxide media, which has been used since 1955, remained popular because of its relatively low cost and ease of application. Today, however, very few drives use oxide media.

Thin-film media is thinner, harder, and more perfectly formed than oxide media. Thin film was developed as a high-performance media that enabled a new generation of drives to have lower head floating heights, which in turn made possible increases in drive density. Originally, thin-film media was used only in higher-capacity or higher-quality drive systems, but today, virtually all drives have thin-film media.

Thin-film media is aptly named. The coating is much thinner than can be achieved by the oxide-coating method. Thin-film media also is known as *plated*, or *sputtered*, media because of the various processes used to place the thin film of media on the platters.

Thin-film plated media is manufactured by placing the media material on the disk with an electroplating mechanism, much the way chrome plating is placed on the bumper of a car. The aluminum platter then is immersed in a series of chemical baths that coat the platter with several layers of metallic film. The media layer is a cobalt alloy about 3 u-in thick.

Thin-film sputtered media is created by first coating the aluminum platters with a layer of nickel phosphorus and then applying the cobalt-alloy magnetic material in a continuous vacuum-deposition process called *sputtering*. During this process, magnetic layers as thin as 1 or 2 u-in are deposited on the disk, in a fashion similar to the way that silicon wafers are coated with metallic films in the semiconductor industry. The sputtering technique then is used again to lay down an extremely hard, 1 u-in protective carbon coating. The need for a near-perfect vacuum makes sputtering the most expensive of the processes described here.

The surface of a sputtered platter contains magnetic layers as thin as 1 u-in. Because this surface also is very smooth, the head can float closer to the disk surface than was possible previously; floating heights as small as 3 u-in above the surface are possible. When the head is closer to the platter, the density of the magnetic flux transitions can be increased to provide greater storage capacity. Additionally, the increased intensity of the magnetic field during a closer-proximity read provides the higher signal amplitudes needed for good signal-to-noise performance.

Both the sputtering and plating processes result in a very thin, very hard film of media on the platters. Because the thin-film media is so hard, it has a better chance of surviving contact with the heads at high speed. In fact, modern thin-film media is virtually uncrashable. Oxide coatings can be scratched or damaged much more easily. If you could open a drive to peek at the platters, you would see that the thin-film media platters look like the silver surfaces of mirrors.

The sputtering process results in the most perfect, thinnest, and hardest disk surface that can be produced commercially. The sputtering process has largely replaced plating as the method of creating thin-film media. Having a thin-film media surface on a drive results in increased storage capacity in a smaller area with fewer head crashes—and in a drive that will provide many years of trouble-free use.

Read/Write Heads

A hard disk drive usually has one read/write head for each platter side, and these heads are connected, or *ganged*, on a single movement mechanism. The heads, therefore, move across the platters in unison.

Mechanically, read/write heads are simple. Each head is on an actuator arm that is spring-loaded to force the head into a platter. Few people realize that each platter actually is "squeezed" by the heads above and below it. If you could open a drive safely and lift the top head with your finger, the head would snap back into the platter when you released it. If you could pull down on one of the heads below a platter, the spring tension would cause it to snap back up into the platter when you released it.

Figure 1-2 shows a typical hard disk head-actuator assembly from a voice-coil drive.

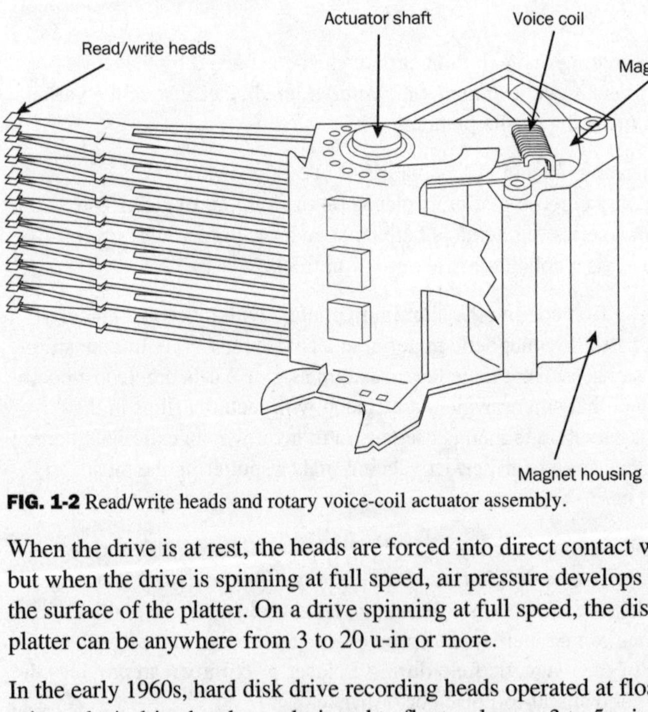

FIG. 1-2 Read/write heads and rotary voice-coil actuator assembly.

When the drive is at rest, the heads are forced into direct contact with the platters by spring tension, but when the drive is spinning at full speed, air pressure develops below the heads and lifts them off the surface of the platter. On a drive spinning at full speed, the distance between the heads and the platter can be anywhere from 3 to 20 u-in or more.

In the early 1960s, hard disk drive recording heads operated at floating heights as large as 200 to 300 u-in; today's drive heads are designed to float as low as 3 to 5 u-in above the surface of the disk. To support higher densities in future drives, the physical separation between the head and disk is expected to be as little as 0.5 u-in by the end of the century.

> **Caution**
>
> The small size of this gap is why the disk drive's HDA should never be opened except in a clean-room environment: Any particle of dust or dirt that gets into this mechanism could cause the heads to read improperly or possibly even to strike the platters while the drive is running at full speed. The latter event could scratch the platter or the head.

To ensure the cleanliness of the interior of the drive, the HDA is assembled in a class-100 or better clean room. This specification is such that a cubic foot of air cannot contain more than 100 particles that measure up to 0.5 micron (19.7 u-in). A single person breathing while standing motionless spews out 500 such particles in a single minute! These rooms contain special air-filtration systems that continuously evacuate and refresh the air. A drive's HDA should not be opened unless it is inside such a room.

Although maintaining such an environment may seem to be expensive, many companies manufacture tabletop or bench-size clean rooms that sell for only a few thousand dollars. Some of these devices operate like a glove box; the operator first inserts the drive and any tools required, and then closes the box and turns on the filtration system. Inside the box, a clean-room environment is maintained, and a technician can use the built-in gloves to work on the drive.

In other clean-room variations, the operator stands at a bench where a forced-air curtain is used to maintain a clean environment on the bench top. The technician can walk in and out of the clean-room field simply by walking through the air curtain. This air curtain is much like the curtain of air used in some stores and warehouses to prevent heat from escaping in the winter while leaving a passage wide open.

Because the clean environment is expensive to produce, few companies except those that manufacture the drives are prepared to service hard disk drives.

Read/Write Head Designs

As disk drive technology has evolved so has the design of the read/write head. The earliest heads were simple iron cores with coil windings (electromagnets). By today's standards, the original head designs were enormous in physical size and operated at very low recording densities. Over the years, many different head designs have evolved from the first simple ferrite core designs into several types and technologies available today. This section discusses the different types of heads found in PC-type hard disk drives, including the applications and relative strengths and weaknesses of each.

Four types of heads have been used in hard disk drives over the years:

- Ferrite
- Metal-in-gap (MIG)
- Thin-film (TF)
- Magneto-resistive (MR)

Ferrite Ferrite heads, the traditional type of magnetic-head design, evolved from the original IBM Winchester drive. These heads have an iron-oxide core wrapped with electromagnetic coils. A magnetic field is produced by energizing the coils. A field also can be induced by passing a magnetic field near the coils. This process gives the heads full read and write capability. Ferrite heads are larger and heavier than thin-film heads and, therefore, require a larger floating height to prevent contact with the disk.

Many refinements have been made in the original (monolithic) ferrite head design. A type of ferrite head called a *composite ferrite head* has a smaller ferrite core bonded with glass in a ceramic housing. This design permits a smaller head gap, which allows higher track densities. These heads are less susceptible to stray magnetic fields than are heads in the older monolithic design.

During the 1980s, composite ferrite heads were popular in many low-end drives, such as the popular Seagate ST-225. As density demands grew, the competing MIG and TF head designs were used in place of ferrite heads, which are virtually obsolete today. Ferrite heads cannot write to the higher coercivity media needed for high-density designs and have poor frequency response with higher noise levels. The main advantage of ferrite heads is that they are the cheapest type available.

Metal-In-Gap *Metal-in-gap (MIG) heads* basically are a specially enhanced version of the composite ferrite design. In MIG heads, a metal substance is sputtered into the recording gap on the trailing edge of the head. This material offers increased resistance to magnetic saturation, allowing higher-density recording. MIG heads also produce a sharper gradient in the magnetic field for a better-defined magnetic pulse. These heads permit the use of higher-coercivity thin-film disks and can operate at lower floating heights.

Two versions of MIG heads are available: single-sided and double-sided. Single-sided MIG heads are designed with a layer of magnetic alloy placed along the trailing edge of the gap. Double-sided MIG designs apply the layer to both sides of the gap. The metal alloy is applied through a vacuum-deposition process called *sputtering*. This alloy has twice the magnetization capability of raw ferrite and allows writing to the higher-coercivity thin-film media needed at the higher densities. Double-sided MIG heads offer even higher coercivity capability than the single-sided designs do.

Because of these increases in capabilities through improved designs, MIG heads for a time were the most popular head used in all but very high-capacity drives. Due to market pressures that have demanded higher and higher densities, however, MIG heads have been largely displaced in favor of TF heads.

Thin Film *Thin-film (TF) heads* are produced in much the same manner as a semiconductor chip—that is, through a photolithographic process. In this manner, many thousands of heads can be created on a single circular wafer. This manufacturing process also results in a very small high-quality product.

TF heads offer an extremely narrow and controlled head gap created by sputtering a hard aluminum material. Because this material completely encloses the gap, this area is very well protected, minimizing the chance of damage from contact with the media. The core is a combination of iron and nickel alloy that is two to four times more powerful magnetically than a ferrite head core.

TF heads produce a sharply defined magnetic pulse that allows extremely high densities to be written. Because they do not have a conventional coil, TF heads are more immune to variations in coil impedance. The small, lightweight heads can float at a much lower height than the ferrite and MIG heads; floating height has been reduced to 2 u-in or less in some designs. Because the reduced height enables a much stronger signal to be picked up and transmitted between the head and platters, the signal-to-noise ratio increases, which improves accuracy. At the high track and linear densities of some drives, a standard ferrite head would not be able to pick out the data signal from the background noise. When TF heads are used, their small size enables more platters to be stacked in a drive.

Until the past few years, TF heads were relatively expensive compared with older technologies, such as ferrite and MIG. Better manufacturing techniques and the need for higher densities, however, have driven the market to TF heads. The widespread use of these heads also has made them cost-competitive, if not cheaper, than MIG heads.

TF heads currently are used in most high-capacity drives, especially in the smaller form factors. They have displaced MIG heads as the most popular head design being used in drives today. The industry is working on ways to improve TF head efficiency, so TF heads are likely to remain popular for some time, especially in mainstream drives.

Magneto-Resistive *Magneto-resistive (MR) heads* are the latest technology. Invented and pioneered by IBM, MR heads currently are the superior head design, offering the highest performance available. Most 3 1/2-inch drives with capacities in excess of 1GB currently use MR heads. As areal densities continue to increase, the MR head eventually will become the head of choice for nearly all hard drives, displacing the popular MIG and TF head designs.

MR heads rely on the fact that the resistance of a conductor changes slightly when an external magnetic field is present. Rather than put out a voltage by passing through a magnetic-field flux reversal, as a normal head would, the MR head senses the flux reversal and changes resistance. A small current flows through the heads, and the change in resistance is measured by this sense current. This design enables the output to be three or more times more powerful than a TF head during a read. In effect, MR heads are power-read heads, acting more like sensors than generators.

MR heads are more costly and complex to manufacture than other types of heads because several special features or steps must be added. Among them are

- Additional wires must be run to and from the head to carry the sense current.
- Four to six more masking steps are required.
- Because MR heads are so sensitive, they are very susceptible to stray magnetic fields and must be shielded.

Because the MR principle can only read data and is not used for writing, MR heads really are two heads in one. A standard inductive TF head is used for writing, and an MR head is used for reading. Because two separate heads are built into one assembly, each head can be optimized for its task. Ferrite, MIG, and TF heads are known as *single-gap heads* because the same gap is used for both reading and writing, whereas the MR head uses a separate gap for each operation.

The problem with single-gap heads is that the gap length is always a compromise between what is best for reading and what is best for writing. The read function needs a thin gap for higher resolution; the write function needs a thicker gap for deeper flux penetration to switch the media. In a dual-gap MR head, the read and write gaps can be optimized for both functions independently. The write (TF) gap writes a wider track than the read (MR) gap does. The read head then is less likely to pick up stray magnetic information from adjacent tracks.

Drives with MR heads require better shielding from stray magnetic fields, which can affect these heads more easily than they do the other head designs. All in all, however, the drawback is minor compared with the advantages that the MR heads offer.

Head Sliders

The term *slider* is used to describe the body of material that supports the actual drive head itself. The slider is what actually floats or slides over the surface of the disk, carrying the head at the correct distance from the media for reading and writing. Most sliders resemble a catamaran, with two outboard pods that float along the surface of the disk media and a central "rudder" portion that actually carries the head and read/write gap.

The trend toward smaller and smaller form-factor drives has forced a requirement for smaller and smaller sliders as well. The typical mini-Winchester slider design is about .160×.126×.034 inch in size. Most head manufacturers now are shifting to 50 percent smaller nanosliders, which have dimensions of about .08×.063×.017 inch. The nanoslider is being used in both high-capacity and small–form-factor drives. Smaller sliders reduce the mass carried at the end of the head actuator arms, allowing for increased acceleration and deceleration and leading to faster seek times. The smaller sliders also require less area for a landing zone, thus increasing the usable area of the disk platters. Further, the smaller slider contact area reduces the slight wear on the media surface that occurs during normal startup and spindown of the drive platters.

The newer nanoslider designs also have specially modified surface patterns that are designed to maintain the same floating height above the disk surface, whether the slider is above the inner or outer cylinders. Conventional sliders increase or decrease their floating height considerably, according to the velocity of the disk surface traveling below them. Above the outer cylinders, the velocity and floating height are higher. This arrangement is undesirable in newer drives that use zoned recording, in which the same bit density is achieved on all the cylinders. Because the same bit density is maintained throughout the drive, the head floating height should be relatively constant as well for maximum performance. Special textured surface patterns and manufacturing techniques allow the nanosliders to float at a much more consistent height, making them ideal for zoned recording drives.

Head Actuator Mechanisms

Possibly more important than the heads themselves is the mechanical system that moves them: the *head actuator*. This mechanism moves the heads across the disk and positions them accurately above the desired cylinder. Many variations on head actuator mechanisms are in use, but all of them can be categorized as being one of two basic types:

- Stepper motor actuators
- Voice-coil actuators

The use of one or the other type of positioner has profound effects on a drive's performance and reliability. The effect is not limited to speed; it also includes accuracy, sensitivity to temperature, position, vibration, and overall reliability. To put it bluntly, a drive equipped with a stepper motor actuator is much less reliable (by a large factor) than a drive equipped with a voice-coil actuator.

The head actuator is the single most important specification in the drive. The type of head actuator mechanism in a drive tells you a great deal about the drive's performance and reliability characteristics. Table 1.2 shows the two types of hard disk drive head actuators and the affected performance parameters.

Table 1.2 Characteristics of Stepper Motor Versus Voice-coil Drives.

Characteristic	Stepper Motor	Voice Coil
Relative access speed	Slow	Fast
Temperature sensitive	Yes (very)	No
Positionally sensitive	Yes	No
Automatic head parking	Not usually	Yes
Preventive maintenance	Periodic format	None required
Relative reliability	Poor	Excellent

Generally, a stepper motor drive has a slow average access rating, is temperature-sensitive during read and write operations, is sensitive to physical orientation during read and write operations, does not automatically park its heads above a save zone during power-down, and usually requires annual or biannual reformats to realign the sector data with the sector header information due to mistracking. Overall, stepper motor drives are vastly inferior to drives that use voice-coil actuators.

Some stepper motor drives feature automatic head parking at power-down. If you have a newer stepper motor drive, refer to the drive's technical reference manual to determine whether your drive has this feature. Sometimes, you hear a noise after power-down, but that can be deceptive. Some drives use a solenoid-activated spindle brake, which makes a noise as the drive is powered down and does not involve head parking.

Many of the older, less expensive, low-capacity hard disks use a stepper motor system. Most hard disks with capacities of more than 40MB have voice-coil actuators, as do all drives with capacities of more than 100MB, which means all drives being manufactured today.

Voice Coil A *voice-coil actuator* is found in all higher-quality hard disk drives, including most drives with capacities greater than 40MB and virtually all drives with capacities exceeding 80MB. Unlike the blind stepper motor positioning system, a voice-coil actuator uses a feedback signal from the drive to accurately determine the head positions and to adjust them, if necessary. This system allows for significantly greater performance, accuracy, and reliability than traditional stepper motor actuators offer.

A voice-coil actuator works by pure electromagnetic force. The construction of this mechanism is similar to that of a typical audio speaker, from which the term *voice coil* is derived. An audio speaker uses a stationary magnet surrounded by a voice coil connected to the speaker's paper cone. Energizing the coil causes the coil to move relative to the stationary magnet, which produces sound from the speaker cone. In a typical hard disk voice-coil system, the electromagnetic coil is attached to the end of the head rack and placed near a stationary magnet. No contact is made between the coil and the magnet; instead, the coil moves by pure magnetic force. As the electromagnetic coils are energized, they attract or repulse the stationary magnet and move the head rack. Such systems are extremely quick and efficient and usually much quieter than systems driven by stepper motors.

Unlike a stepper motor, a voice-coil actuator has no click-stops, or detent positions; rather, a special guidance system stops the head rack above a particular cylinder. Because it has no detents, the voice-coil actuator can slide the heads in and out smoothly to any position desired, much like the slide of a trombone. Voice-coil actuators use a guidance mechanism called a *servo* to tell the actuator where the heads are in relation to the cylinders and to place the heads accurately at the desired positions. This positioning system often is called a *closed-loop, servo-controlled mechanism. Closed loop* indicates that the index (or servo) signal is sent to the positioning electronics in a closed-loop system. This loop sometimes is called a *feedback loop* because the feedback from this information is used to position the heads accurately. *Servo controlled* refers to this index or the servo information that is used to dictate or control head-positioning accuracy.

A voice-coil actuator with servo control is not affected by temperature changes, as a stepper motor is. When the temperature is cold and the platters have shrunk (or when the temperature is hot and the platters have expanded), the voice-coil system compensates because it never positions the heads in predetermined track positions. Rather, the voice-coil system searches for the specific track, guided by the prewritten servo information and can position the head rack precisely above the desired track at that track's current position, regardless of the temperature. Because of the continuous feedback of servo information, the heads adjust to the current position of the track at all times. For example, as a drive warms up and the platters expand, the servo information allows the heads to "follow" the track. As a result, a voice-coil actuator often is called a *track following system*.

Two main types of voice-coil positioner mechanisms are available:

- Linear voice-coil actuators
- Rotary voice-coil actuators

The types differ only in the physical arrangement of the magnets and coils.

A *linear actuator* (see Figure 1-3) moves the heads in and out over the platters in a straight line, much like a tangential-tracking turntable. The coil moves in and out on a track surrounded by the stationary magnets. The primary advantage of the linear design is that it eliminates the head azimuth variations that occur with rotary positioning systems. (*Azimuth* refers to the angular measurement of the head position relative to the tangent of a given cylinder.) A linear actuator does not rotate the head as it moves from one cylinder to another, thus eliminating this problem.

Although the linear actuator seems to be a good design, it has one fatal flaw: The devices are much too heavy. As drive performance has increased, the desire for lightweight actuator mechanisms has become very important. The lighter the mechanism, the faster it can be accelerated and decelerated from one cylinder to another. Because they are much heavier than rotary actuators, linear actuators were popular only for a short time; they are virtually nonexistent in drives manufactured today.

Rotary actuators also use stationary magnets and a movable coil, but the coil is attached to the end of an actuator arm, much like that of a turntable's tone arm. As the coil is forced to move relative to the stationary magnet, it swings the head arms in and out over the surface of the disk. The primary advantage of this mechanism is its light weight, which means that the heads can be accelerated and decelerated very quickly, resulting in very fast average seek times. Because of the lever effect on the head arm, the heads move faster than the actuator, which also helps to improve access times.

The disadvantage with a rotary system is that as the heads move from the outer to inner cylinders, they are rotated slightly with respect to the tangent of the cylinders. This rotation results in an azimuth error and is one reason why the area of the platter in which the cylinders are located is somewhat limited. By limiting the total motion of the actuator, the azimuth error can be contained to within reasonable specifications. Virtually all voice-coil drives today use rotary actuator systems.

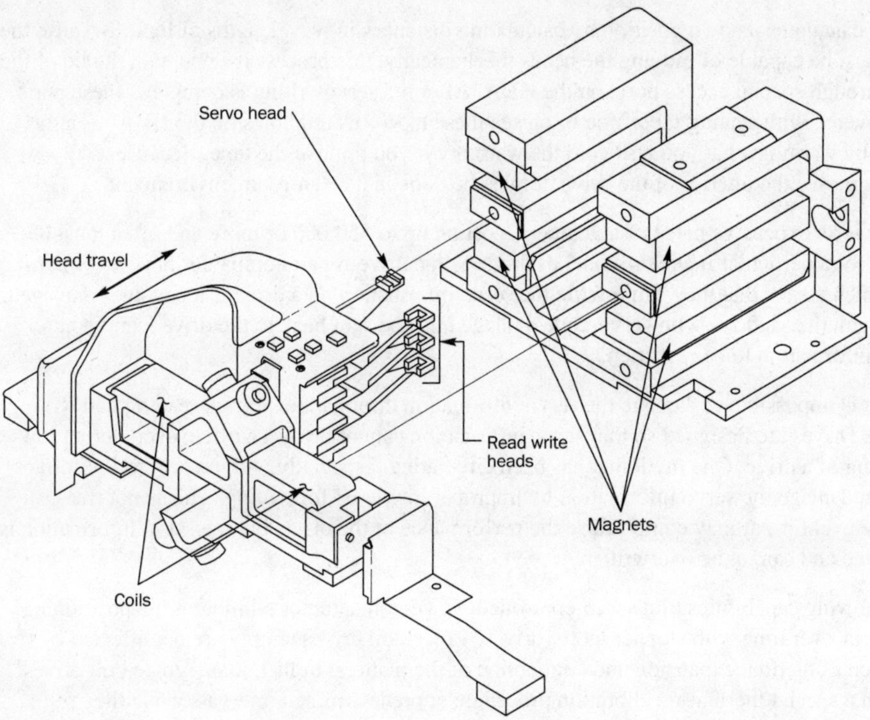

FIG. 1-3 A linear voice-coil actuator.

Servo Mechanisms Three servo mechanism designs have been used to control voice-coil positioners over the years:

- Wedge servo
- Embedded servo
- Dedicated servo

These designs are slightly different, but they accomplish the same basic task: They enable the head positioner to adjust continuously so that it is precisely placed above a given cylinder in the drive. The main difference among these servo designs is where the gray code information is actually written on the drive.

All servo mechanisms rely on special information that is only written to the disk when the disk is manufactured. This information usually is in the form of a special code called a gray code. A *gray code* is a special binary notational system in which any two adjacent numbers are represented by a code that differs in only one bit place or column position. This system makes it easy for the head to read the information and quickly determine its precise position. This guidance code can be written only when the drive is manufactured; the code is used over the life of the drive for accurate positional information.

The servo gray code is written at the time of manufacture by a special machine called a *servowriter*: Basically, a jig that mechanically moves the heads to a given reference position and then writes the servo information for that position. Many servowriters are themselves guided by a laser-beam reference that calculates its own position by calculating distances in wavelengths of light. Because the servowriter must be capable of moving the heads mechanically, this process is done with the lid of the drive off or through special access ports on the HDA. After the servowriting is complete, these ports usually are covered with sealing tape. You often see these tape-covered holes on the HDA, usually accompanied by warnings that you will void the warranty if you remove the tape. Because servowriting exposes the interior of the drive, it must be done in a clean-room environment.

A servowriter is an expensive piece of machinery, costing up to $50,000 or more and often must be custom-made for a particular make or model of drive. Some drive-repair companies have *servowriting capability*, which means that they can rewrite the servo information on a drive if it becomes damaged. Lacking a servowriter, a drive with servo-code damage must be sent back to the drive manufacturer for the servo information to be rewritten.

Fortunately, it is impossible to damage the servo information through any normal reading and writing to a hard disk. Drives are designed so that servo information cannot be overwritten, even during low-level formatting of a drive. One myth that has been circulating (especially with respect to IDE drives) is that you can damage the servo information by improper low-level formatting. This is not true. An improper low-level format may compromise the performance of the drive, but the servo information is totally protected and cannot be overwritten.

The track-following capabilities of a servo-controlled, voice-coil actuator eliminates the positioning errors that occur over time with stepper motor drives. Voice-coil drives simply are not affected by conditions such as thermal expansion and contraction of the platters. In fact, many voice-coil drives today perform a special thermal-recalibration procedure at predetermined intervals while they run. This procedure usually involves seeking the heads from cylinder 0 to some other cylinder one time for every head on the drive. As this sequence occurs, the control circuitry in the drive monitors how much the track positions have moved since the last time the sequence was performed, and a thermal calibration adjustment is calculated and stored in the drive's memory. This information then is used every time the drive positions to ensure the most accurate positioning possible.

Most drives perform the thermal-recalibration sequence every five minutes for the first half-hour that the drive is powered on and then once every 25 minutes after that. With some drives (such as Quantum, for example), this thermal-calibration sequence is very noticeable; the drive essentially stops what it is doing, and you hear rapid ticking for a second or so. At this time, some people think that their drive is having a problem reading something and perhaps is conducting a read retry, but this is not true. Most of the newer intelligent drives (IDE and SCSI) employ this thermal-recalibration procedure for ultimate positioning accuracy.

As multimedia applications grew, thermal recalibration became a problem with some manufacturer's drives. The thermal recalibration sequence could interrupt a data transfer, which would make audio and video playback jitter. These companies released special A/V (Audio Visual) drives that would hide the thermal recalibration sequences and not let them ever interrupt a transfer. Most of the newer IDE and SCSI drives are A/V capable, which means the thermal recalibration sequences will not interrupt a transfer such as a video playback.

While we are on the subject of automatic drive functions, most of the drives that perform thermal-recalibration sequences also automatically perform a function called a *disk sweep*. This procedure is an automatic head seek that occurs after the drive has been idle for a period of time (for example, nine minutes). The disk-sweep function moves the heads to a random cylinder in the outer portion of the platters, which is considered to be the high float-height area because the head-to-platter velocity is highest. Then, if the drive continues to remain idle for another period, the heads move to another cylinder in this area, and the process continues indefinitely as long as the drive is powered on.

The disk-sweep function is designed to prevent the head from remaining stationary above one cylinder in the drive, where friction between the head and platter eventually would dig a trench in the media. Although the heads are not in direct contact with the media, they are so close that the constant air pressure from the head floating above a single cylinder causes friction and excessive wear.

Wedge Servo Some early servo-controlled drives used a technique called a *wedge servo*. In these drives, the gray-code guidance information is contained in a "wedge" slice of the drive in each cylinder immediately preceding the index mark. The index mark indicates the beginning of each track, so the wedge-servo information was written in the PRE-INDEX GAP, which is at the end of each track. This area is provided for speed tolerance and normally is not used by the controller. Figure 1-4 shows the servo-wedge information on a drive.

Some controllers, such as the Xebec 1210 that IBM used in the XT, had to be notified that the drive was using a wedge servo so that they could shorten the sector timing to allow for the wedge-servo area. If they were not correctly configured, these controllers would not work properly with such drives. Many people believed—erroneously—that the wedge-servo information could be overwritten in such cases by an improper low-level format. This, however, is not the case; all drives using a wedge servo disable any write commands and take control of the head select lines whenever the heads are above the wedge area. This procedure protects the servo from any possibility of being overwritten, no matter how hard you try. If the controller tried to write over this area, the drive would prevent the write, and the controller would be unable to complete the format. Most controllers simply do not write to the PRE-INDEX GAP area and do not need to be configured specially for wedge-servo drives.

The only way that the servo information normally could be damaged is by a powerful external magnetic field (or perhaps by a head crash or some other catastrophe). In such a case, the drive would have to be sent in for repair and re-servoing.

One problem is that the servo information appears only one time every revolution, which means that the drive often needs several revolutions before it can accurately determine and adjust the head position. Because of these problems, the wedge servo never was a popular design; it no longer is used in drives.

Embedded Servo An *embedded servo* (see Figure 1-5) is an enhancement of the wedge servo. Instead of placing the servo code before the beginning of each cylinder, an embedded servo design writes the servo information before the start of each sector. This arrangement enables the positioner circuits to receive feedback many times in a single revolution, making the head positioning much faster and more precise. Another advantage is that every track on the drive has this positioning information, so each head can quickly and efficiently adjust position to compensate for any changes in the platter or head dimensions, especially for changes due to thermal expansion or physical stress.

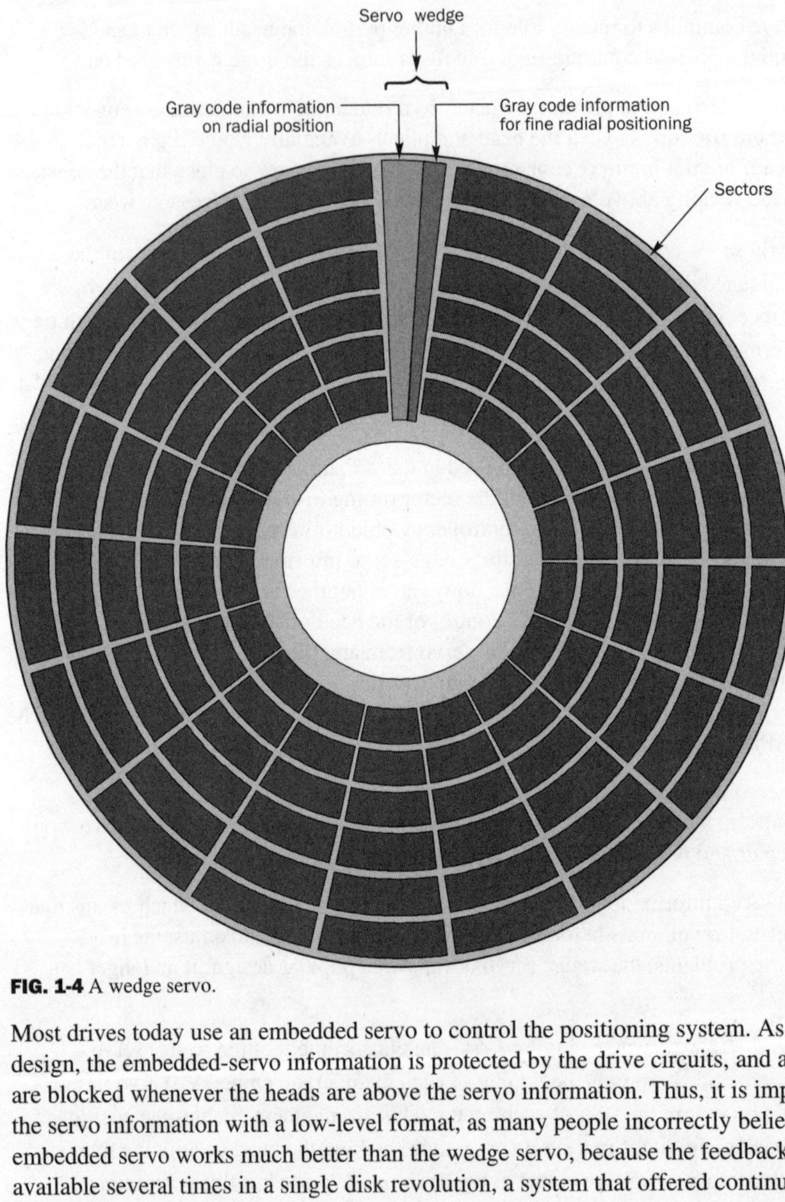

FIG. 1-4 A wedge servo.

Most drives today use an embedded servo to control the positioning system. As in the wedge servo design, the embedded-servo information is protected by the drive circuits, and any write operations are blocked whenever the heads are above the servo information. Thus, it is impossible to overwrite the servo information with a low-level format, as many people incorrectly believed. Although the embedded servo works much better than the wedge servo, because the feedback servo information is available several times in a single disk revolution, a system that offered continuous servo feedback information would be better.

Dedicated Servo A *dedicated servo* is a design in which the servo information is written continuously throughout the entire track, rather than just one time per track or at the beginning of each sector. Unfortunately, if this procedure were performed on the entire drive, no room would be left for data. For this reason, a dedicated servo uses one side of one of the platters exclusively for the servo-positioning information. The term *dedicated* comes from the fact that this platter side is completely dedicated to the servo information and cannot contain any data. Although the dedicated-servo design may seem to be wasteful, none of the other platter sides carry any servo information, and you end up losing about the same amount of total disk real estate as with the embedded servo.

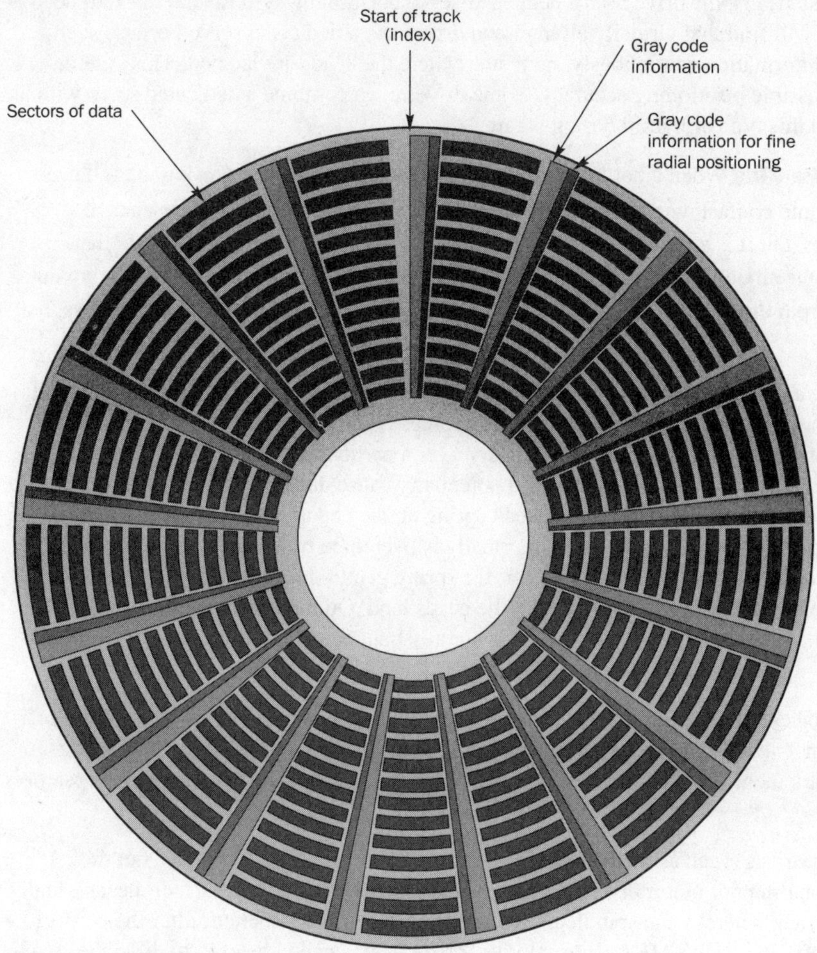

FIG. 1-5 An embedded servo.

When a dedicated-servo drive is manufactured, one side of one platter is deducted from normal read/write usage; on this platter is recorded a special set of gray-code data that indicates proper track positions. Because the head that rests above this surface cannot be used for normal reading and writing, these marks can never be erased, and the servo information is protected, as in the other servo designs. No low-level format or other procedure can possibly overwrite the servo information.

When the drive is commanded to move the heads to a specific cylinder, the internal drive electronics use the signals received by the servo head to determine the position of the heads. As the heads are moved, the track counters are read from the dedicated servo surface. When the requested track is detected below the servo head, the actuator is stopped. The servo electronics then fine-tune the position so that, before writing is allowed, the heads are positioned precisely above the desired cylinder. Although only one head is used for servo tracking, the other heads are attached to the same rigid rack so that if one head is above the desired cylinder, all the others will be as well.

One noticeable trait of dedicated servo drives is that they usually have an odd number of heads. For example, the Toshiba MK-538FB 1.2GB drive on which I am saving this chapter has eight platters but only 15 read/write heads; the drive uses a dedicated-servo positioning system, and the 16th head is the servo head. You will find that virtually all high-end drives use a dedicated servo because such a design offers servo information continuously, no matter where the heads are located. This system offers the greatest possible positioning accuracy. Some drives even combine a dedicated servo with an embedded servo, but this type of hybrid design is rare.

Automatic Head Parking When a hard disk drive is powered off, the spring tension in each head arm pulls the heads into contact with the platters. The drive is designed to sustain thousands of takeoffs and landings, but it is wise to ensure that the landing occurs at a spot on the platter that contains no data. Some amount of abrasion occurs during the landing and takeoff process, removing just a "micro puff" from the media; but if the drive is jarred during the landing or takeoff process, real damage can occur.

One benefit of using a voice-coil actuator is *automatic head parking*. In a drive that has a voice-coil actuator, the heads are positioned and held by magnetic force. When power is removed from the drive, the magnetic field that holds the heads stationary over a particular cylinder dissipates, enabling the head rack to skitter across the drive surface and potentially cause damage. In the voice-coil design, therefore, the head rack is attached to a weak spring at one end and a head stop at the other end. When the system is powered on, the spring normally is overcome by the magnetic force of the positioner. When the drive is powered off, however, the spring gently drags the head rack to a park-and-lock position before the drive slows down and the heads land. On many drives, you can actually hear the "ting…ting…ting…ting" sound as the heads literally bounce-park themselves, driven by this spring.

On a drive with a voice-coil actuator, you can activate the parking mechanism simply by turning off the system; you do not need to run a program to park or retract the heads. In the event of a power outage, the heads park themselves automatically. (The drives unpark automatically when the system is powered on.)

Some stepper motor drives (such as the Seagate ST-251 series drives) park their heads, but this function is rare among stepper motor drives. The stepper motor drives that do park their heads usually use an ingenious system whereby the spindle motor actually is used as a generator after the power to the drive is turned off. The back *EMF* (electro motive force), as it is called, is used to drive the stepper motor to park the heads.

Air Filters

Nearly all hard disk drives have two air filters. One filter is called the *recirculating filter*, and the other is called either a *barometric* or *breather filter*. These filters are permanently sealed inside the drive and are designed never to be changed for the life of the drive, unlike many older mainframe hard disks that had changeable filters. Many mainframe drives circulate air from outside the drive through a filter that must be changed periodically.

A hard disk on a PC system does not circulate air from inside to outside the HDA, or vice versa. The recirculating filter that is permanently installed inside the HDA is designed to filter only the small particles of media scraped off the platters during head takeoffs and landings (and possibly any other small particles dislodged inside the drive). Because PC hard disk drives are permanently sealed and do not circulate outside air, they can run in extremely dirty environments (see Figure 1-6).

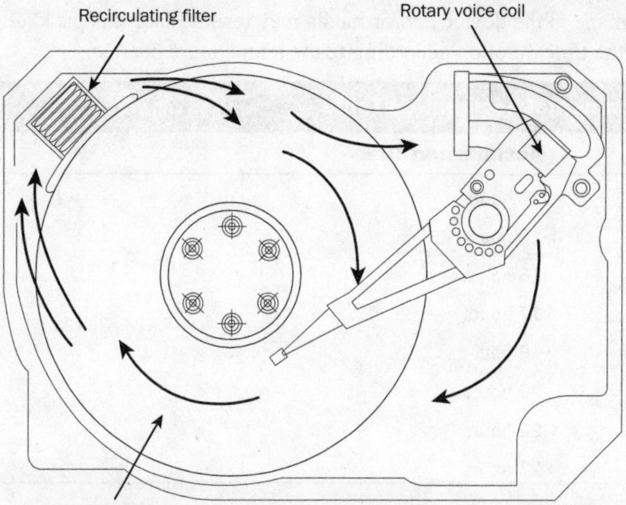

FIG. 1-6 Air circulation in a hard disk.

The HDA in a hard disk is sealed but not airtight. The HDA is vented through a barometric or breather filter element that allows for pressure equalization (breathing) between the inside and outside of the drive. For this reason, most hard drives are rated by the drive's manufacturer to run in a specific range of altitudes, usually from –1,000 to +10,000 feet above sea level. In fact, some hard drives are not rated to exceed 7,000 feet while operating because the air pressure would be too low inside the drive to float the heads properly. As the environmental air pressure changes, air bleeds into or out of the drive so that internal and external pressures are identical. Although air does bleed through a vent, contamination usually is not a concern, because the barometric filter on this vent is designed to filter out all particles larger than 0.3 micron (about 12 u-in) to meet the specifications for cleanliness inside the drive. You can see the vent holes on most drives, which are covered internally by this breather filter. Some drives use even finer-grade filter elements to keep out even smaller particles.

Hard Disk Temperature Acclimation

To allow for pressure equalization, hard drives have a filtered port to bleed air into or out of the HDA as necessary. This breathing also enables moisture to enter the drive, and after some period of time, it must be assumed that the humidity inside any hard disk is similar to that outside the drive. Humidity can become a serious problem if it is allowed to condense—and especially if the drive is powered up while this condensation is present. Most hard disk manufacturers have specified procedures for acclimating a hard drive to a new environment with different temperature and humidity ranges, especially for bringing a drive into a warmer environment in which condensation can form. This situation should be of special concern to users of laptop or portable systems with hard disks. If you leave a portable system in an automobile trunk during the winter, for example, it could be catastrophic to bring the machine inside and power it up without allowing it to acclimate to the temperature indoors.

The following text and Table 1.3 are taken from the factory packaging that Control Data Corporation (later Imprimis and eventually Seagate) used to ship its hard drives:

> If you have just received or removed this unit from a climate with temperatures at or below 50°F (10°C) do not open this container until the following conditions are met, otherwise

condensation could occur and damage to the device and/or media may result. Place this package in the operating environment for the time duration according to the temperature chart.

Table 1.3 Hard Disk Drive Environmental Acclimation Table.

Previous Climate Temp.	Acclimation Time
+40°F (+4°C)	13 hours
+30°F (1°C)	15 hours
+20°F (7°C)	16 hours
+10°F (12°C)	17 hours
0°F (18°C)	18 hours
10°F (23°C)	20 hours
20°F (29°C)	22 hours
30°F (34°C) or less	27 hours

As you can see from this table, a hard disk that has been stored in a colder-than-normal environment must be placed in the normal operating environment for a specified amount of time to allow for acclimation before it is powered on.

Spindle Motors

The motor that spins the platters is called the *spindle motor* because it is connected to the spindle around which the platters revolve. Spindle motors in hard disks always are connected directly; no belts or gears are used. The motors must be free of noise and vibration; otherwise, they transmit to the platters a rumble that could disrupt reading and writing operations.

The motors also must be precisely controlled for speed. The platters on hard disks revolve at speeds ranging from 3,600 to 7,200 RPM or more, and the motor has a control circuit with a feedback loop to monitor and control this speed precisely. Because this speed control must be automatic, hard drives do not have a motor-speed adjustment. Some diagnostics programs claim to measure hard drive rotation speed, but all that these programs do is estimate the rotational speed by the timing at which sectors arrive.

There actually is no way for a program to measure hard disk rotational speed; this measurement can be made only with sophisticated test equipment. Don't be alarmed if some diagnostic program tells you that your drive is spinning at an incorrect speed; most likely the program is wrong, not the drive. Platter rotation and timing information is simply not provided through the hard disk controller interface. In the past, software could give approximate rotational speed estimates by performing multiple sector read requests and timing them, but this was valid only when all drives had the same number of sectors per track (17) and they all spun at 3,600 RPM. Zone recording—combined with a variety of different rotational speeds found in modern drives, not to mention built-in buffers and caches—means that these calculation estimates cannot be performed accurately.

On most drives, the spindle motor is on the bottom of the drive, just below the sealed HDA. Many drives today, however, have the spindle motor built directly into the platter hub inside the HDA. By using an internal hub spindle motor, the manufacturer can stack more platters in the drive because the spindle motor takes up no vertical space. This method allows for more platters than would be possible if the motor were outside the HDA.

> **Note**
>
> Spindle motors, particularly on the larger form-factor drives, can consume a great deal of 12-volt power. Most drives require two to three times the normal operating power when the motor first spins the platters. This heavy draw lasts only a few seconds or until the drive platters reach operating speed. If you have more than one drive, you should try to sequence the start of the spindle motors so that the power supply does not receive such a large load from all the drives at the same time. Most SCSI and IDE drives have a delayed spindle-motor start feature.

Spindle Ground Strap

Some drives have a special grounding strap attached to a ground on the drive and resting on the center spindle of the platter spindle motor. This device is the single most likely cause of excessive drive noise.

The *grounding strap* usually is made of copper and often has a carbon or graphite button that contacts the motor or platter spindle. The grounding strap dissipates static generated by the platters as they spin through the air inside the HDA. If the platters generate static due to friction with the air, and if no place exists for this electrical potential to bleed off, static may discharge through the heads or the internal bearings in the motor. When static discharges through the motor bearings, it can burn the lubricants inside the sealed bearings. If the static charge discharges through the read/write heads, the heads can be damaged or data can be corrupted. The grounding strap bleeds off this static buildup to prevent these problems.

Where the spindle of the motor contacts the carbon contact button (at the end of the ground strap) spinning at full speed, the button often wears, creating a flat spot. The flat spot causes the strap to vibrate and produce a high-pitched squeal or whine. The noise may come and go, depending on temperature and humidity. Sometimes, banging the side of the machine can jar the strap so that the noise changes or goes away, but this is not the way to fix the problem. *I am not suggesting that you bang on your system!* (Most people mistake this noise for something much more serious, such as a total drive-motor failure or bearing failure, which rarely occurs.)

If the spindle grounding strap vibrates and causes noise, you can remedy the situation in several ways:

- Dampen the vibration of the strap by attaching some foam tape or rubber to it.
- Lubricate the contact point.
- Tear off the strap (not recommended!).

On some drives, the spindle motor strap is easily accessible. On other drives, you have to partially disassemble the drive by removing the logic board or other external items to get to the strap.

Of these suggested solutions, the first one is the best. The best way to correct this problem is to glue (or otherwise affix) some rubber or foam to the strap. This procedure changes the harmonics of the strap and usually dampens vibrations. Most manufacturers now use this technique on newly manufactured drives. An easy way to do this is to place some foam tape on the back side of the ground strap.

You also can use a dab of silicone RTV (room-temperature vulcanizing) rubber or caulk on the back of the strap. If you try this method, be sure to use low-volatile (noncorrosive) silicone RTV sealer, which commonly is sold at auto-parts stores. The noncorrosive silicone will be listed on the label as being safe for automotive oxygen sensors. This low-volatile silicone also is free from corrosive acids that can damage the copper strap and is described as low-odor because it does not have the vinegar odor usually associated with silicone RTV. Dab a small amount on the back side of the copper strap (do not interfere with the contact location), and the problem should be solved permanently.

Lubricating the strap is an acceptable, but often temporary, solution. You will want to use some sort of conducting lube, such as a graphite-based compound (the kind used on frozen car locks). Any conductive lubricant (such as moly or lithium) will work as long as it is conductive, but do not use standard oil or grease. Simply dab a small amount of lubricant onto the end of a toothpick, and place a small drop directly on the point of contact.

The last solution is not acceptable. Tearing off the strap eliminates the noise, but it has several possible ramifications. Although the drive will work (silently) without the strap, an engineer placed it there for a reason. Imagine those ungrounded static charges leaving the platters through the heads, perhaps in the form of a spark—possibly even damaging the TF heads. You should choose one of the other solutions.

I mention this last solution only because several people have told me that members of the tech-support staff of some of the hard drive vendors, and even of some manufacturers, told them to remove the strap, which—of course—I do not recommend.

Logic Boards

A disk drive, including a hard disk drive, has one or more logic boards mounted on it. The *logic boards* contain the electronics that control the drive's spindle and head actuator systems and that present data to the controller in some agreed-upon form. In some drives, the controller is located on the drive, which can save on a system's total chip count.

Many disk drive failures occur in the logic board, not in the mechanical assembly. (This statement does not seem logical, but it is true.) Therefore, you can repair many failed drives by replacing the logic board, not the entire drive. Replacing the logic board, moreover, enables you to regain access to the data on the failed drive—something that replacing the entire drive precludes.

Logic boards can be removed or replaced because they simply plug into the drive. These boards usually are mounted with standard screw hardware. If a drive is failing and you have a spare, you may be able to verify a logic-board failure by taking the board off the known good drive and mounting it on the bad one. If your suspicions are confirmed, you can order a new logic board from the drive manufacturer, but unless you have data you need to recover, it makes more sense to just buy a new drive, considering today's cost.

To reduce costs further, many third-party vendors also can supply replacement logic-board assemblies. These companies often charge much less than the drive manufacturers for the same components.

Cables and Connectors

Most hard disk drives have several connectors for interfacing to the system, receiving power, and sometimes grounding to the system chassis. Most drives have at least these three types of connectors:

- Interface connector(s)
- Power connector

- Optional ground connector (tab)

Of these, the *interface connectors* are the most important because they carry the data and command signals from the system to and from the drive. In many drive interfaces, the drive interface cables can be connected in a *daisy chain* or bus-type configuration. Most interfaces support at least two drives, and SCSI supports up to seven in the chain. Some interfaces, such as ST-506/412 or ESDI, use a separate cable for data and control signals. These drives have two cables from the controller interface to the drive. SCSI and IDE drives usually have a single data and control connector. With these interfaces, the disk controller is built into the drive (see Figure 1-7).

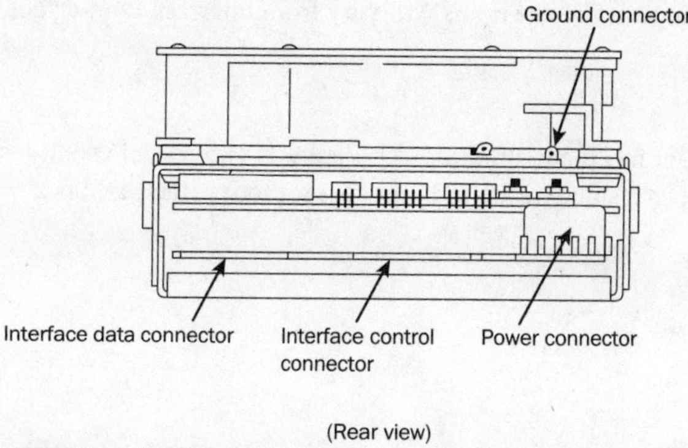

FIG. 1-7 Typical hard disk connections (ST-506/412 or ESDI shown).

The different interfaces and cable specifications are covered in the sections on drive interfaces later in this chapter.

The *power connector* usually is the same type that is used in floppy drives, and the same power-supply connector plugs into it. Most hard disk drives use both 5- and 12-volt power, although some of the smaller drives designed for portable applications use only 5-volt power. In most cases, the 12-volt power runs the spindle motor and head actuator, and the 5-volt power runs the circuitry. Make sure that your power supply can supply adequate power for the hard disk drives installed in your system.

The 12-volt–power consumption of a drive usually varies with the physical size of the unit. The larger the drive is and the more platters there are to spin, the more power is required. Also, the faster the drive spins, the more power will be required. For example, most of the 3 1/2-inch drives on the market today use roughly one-half to one-fourth the power (in watts) of the full-size 5 1/4-inch drives. Some of the very small (2 1/2- or 1.8-inch) hard disks barely sip electrical power and actually use 1 watt or less!

Ensuring an adequate power supply is particularly important with some systems, such as the original IBM AT. These systems have a power supply with three disk drive power connectors, labeled P10, P11, and P12. The three power connectors may seem to be equal, but the technical-reference manual for these systems indicates that 2.8 amps of 12-volt current is available on P10 and P11, and that only 1 amp of 12-volt current is available on P12. Because most full-height hard drives draw much more power than 1 amp, especially at startup, the P12 connector can be used only by floppy drives or half-height hard drives. Some 5 1/2-inch drives draw as much as 4 amps of current during the first few seconds of startup. These drives also can draw as much as 2.5 amps during normal operation. Most PC-compatible systems have a power supply with four or more disk drive power connectors that provide equal power.

A *grounding tab* provides a positive ground connection between the drive and the system's chassis. In most systems, the hard disk drive is mounted directly to the chassis using screws so the ground wire is unnecessary. On some systems, the drives are installed on plastic or fiberglass rails, which do not provide proper grounding. These systems must provide a grounding wire plugged into the drive at this grounding tab. Failure to ground the drive may result in improper operation, intermittent failure, or general read and write errors.

Configuration Items

To configure a hard disk drive for installation in a system, several jumpers (and, possibly, terminating resistors) usually must be set or configured properly. These items will vary from interface to interface and often from drive to drive as well.

The Faceplate or Bezel

Many hard disk drives offer as an option a front faceplate, or *bezel* (see Figure 1-8). A bezel usually is supplied as an option for the drive rather than as a standard item. In most cases today, the bezel is a part of the case and not the drive itself.

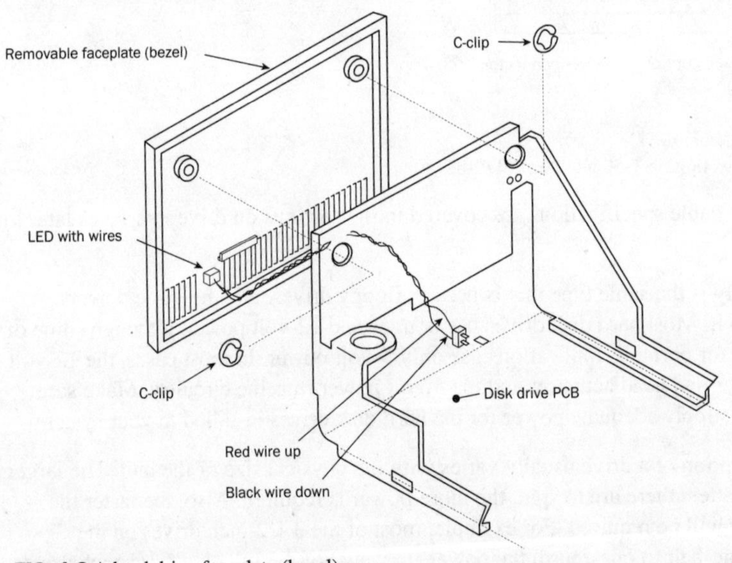

FIG. 1-8 A hard drive faceplate (bezel).

Older systems had the drive installed so that it was visible outside the system case. Drives could have an optional bezel or faceplate to cover the holes. Bezels often come in several sizes and colors to match various PC systems. Many faceplate configurations for 3 1/2-inch drives are available, including bezels that fit 3 1/2-inch drive bays as well as 5 1/4-inch drive bays. You even have a choice of colors (usually, black, cream, or white).

Some bezels feature a light-emitting diode (LED) that flickers when your hard disk is in use. The LED is mounted in the bezel; the wire hanging off the back of the LED plugs into the drive or perhaps the controller. In some drives, the LED is permanently mounted on the drive, and the bezel has a clear or colored window so that you can see the LED flicker while the drive is accessed.

One type of LED problem occurs with some older hard disk installations: If the drive has an LED, the LED may remain on continuously, as though it were a "power-on" light rather than an access light.

This problem happens because the controller in those systems have a direct connection for the LED, thus altering the drive LED function. Some controllers have a jumper that enables the controller to run the drive in what is called latched or unlatched mode. *Latched mode* means that the drive is selected continuously and that the drive LED remains lighted; in *unlatched mode* (to which we are more accustomed), the LED lights only when the drive is accessed. Check to see whether your controller has a jumper for changing this function; if so, you may be able to control the way that the LED operates.

In systems in which the hard disk is hidden by the unit's cover, a bezel is not needed. In fact, using a bezel may prevent the cover from resting on the chassis properly, in which case the bezel will have to be removed.

Hard Disk Features

To make the best decision in purchasing a hard disk for your system, or to understand what differentiates one brand of hard disk from another, you must consider many features. This section examines the issues that you should consider when you evaluate drives:

- Reliability
- Speed
- Shock mounting
- Cost

Reliability

When you shop for a drive, you may notice a feature called the *Mean Time Between Failures* (*MTBF*) described in the brochures. MTBF figures usually range from 20,000 hours to 500,000 hours or more. I usually ignore these figures, because they usually are just theoretical, statistical values. Most drives that boast these figures have not even been manufactured for that length of time. One year of five-day work weeks with eight-hour days equals 2,080 hours of operation. If you never turn off your system for 365 days and run the full 24 hours per day, you operate your system 8,760 hours each year; a drive with a 500,000-hour MTBF rating is supposed to last (on average) 57 years before failing! Obviously, that figure cannot be derived from actual statistics because the particular drive probably has been on the market for less than a year.

Statistically, for the MTBF figures to have real weight, you must take a sample of drives, measure the failure rate for at least twice the rated figure, and measure how many drives fail in that time. To be really accurate, you would have to wait until all the drives fail and record the operating hours at each failure. Then you would average the running time for all the test samples to arrive at the average time before a drive failure. For a reported MTBF of 500,000 hours (common today), the test sample should be run for at least 1 million hours (114 years) to be truly accurate, yet the drive carries this specification on the day that it is introduced.

The bottom line is that I do not really place much emphasis on MTBF figures. Some of the worst drives that I have used boasted high MTBF figures, and some of the best drives have lower ones. These figures do not necessarily translate to reliability in the field, and that is why I generally place no importance on them.

Performance

When you select a hard disk, an important feature to consider is the performance (speed) of the drive. Hard disks come in a wide range of performance capabilities. As is true of many things, one of the best indicators of a drive's relative performance is its price. An old saying from the automobile-racing industry is appropriate here: "Speed costs money. How fast do you want to go?"

You can measure the speed of a disk drive in two ways:

- Average seek time
- Transfer rate

Average seek time, normally measured in milliseconds (ms), is the average amount of time it takes to move the heads from one cylinder to another cylinder a random distance away. One way to measure this specification is to run many random track-seek operations and then divide the timed results by the number of seeks performed. This method provides an average time for a single seek.

The standard way to measure average seek time used by many drive manufacturers involves measuring the time that it takes the heads to move across one-third of the total cylinders. Average seek time depends only on the drive; the type of interface or controller has little effect on this specification. The rating is a gauge of the capabilities of the head actuator.

> **Note**
>
> Be wary of benchmarks that claim to measure drive seek performance. Most IDE and SCSI drives use a scheme called *sector translation*, so any commands to the drive to move the heads to a specific cylinder do not actually cause the intended physical movement. This situation renders some benchmarks meaningless for those types of drives. SCSI drives also require an additional command, because the commands first must be sent to the drive over the SCSI bus. Even though these drives can have the fastest access times, because the command overhead is not factored in by most benchmarks, the benchmark programs produce poor performance figures for these drives.

A slightly different measurement, called *average access time*, involves another element, called latency. *Latency* is the average time (in milliseconds) that it takes for a sector to be available after the heads have reached a track. On average, this figure is half the time that it takes for the disk to rotate one time, which is 8.33 ms at 3,600 RPM. A drive that spins twice as fast would have half the latency. A measurement of average access time is the sum of the average seek time and latency. This number provides the average amount of time required before a randomly requested sector can be accessed.

Latency is a factor in disk read and write performance. Decreasing the latency increases the speed of access to data or files, accomplished only by spinning the drive platters faster. I have a drive that spins at 4,318 RPM, for a latency of 6.95 ms. Some drives spin at 7,200 RPM or faster, resulting in an even shorter latency time of only 4.17 ms. In addition to increasing performance where real-world access to data is concerned, spinning the platters faster also increases the data-transfer rate after the heads arrive at the desired sectors.

The transfer rate probably is more important to overall system performance than any other specification. *Transfer rate* is the rate at which the drive and controller can send data to the system. The transfer rate depends primarily on the drive's HDA and secondarily on the controller. Transfer rate used to be more bound to the limits of the controller, meaning that drives that were connected to newer controllers often outperformed those connected to older controllers. This situation is where the concept of interleaving sectors came from. *Interleaving* refers to the ordering of the sectors so that they are not sequential, enabling a slow controller to keep up without missing the next sector.

Modern drives with integrated controllers are fully capable of keeping up with the raw drive transfer rate. In other words, they no longer have to interleave the sectors to slow the data for the controller.

Another performance issue is the raw interface performance, which, in IDE or SCSI drives, usually is far higher than any of the drives themselves are able to sustain. Be wary of quoted transfer specifications for the interface, because they may have little effect on what the drive can actually put out. The drive interface simply limits the maximum theoretical transfer rate. The actual drive and controller place the real limits on performance.

In older ST-506/412 interface drives, you can sometimes double or triple the transfer rate by changing the controller, because many of the older controllers could not support a 1:1 interleave. When you change the controller to one that does support this interleave, the transfer rate will be equal to the drive's true capability.

To calculate the true transfer rate of a drive, you need to know several important specifications. The two most important specifications are the true rotational speed of the drive (in RPM) and the average number of physical sectors on each track. I say "average" because most drives today use a zoned recording technique that places different numbers of sectors on the inner and outer cylinders. The transfer rate on zoned recording drives always is fastest in the outermost zone, where the sector per track count is highest. Also be aware that many drives (especially zoned recording drives) are configured with sector translation so that the BIOS reported number of sectors per track has little to do with physical reality. You need to know the true physical parameters, rather than what the BIOS thinks.

When you know these figures, you can use the following formula to determine the maximum transfer rate in millions of bits per second (Mbps):

$$\text{Maximum Data Transfer Rate (Mbps)} = \text{SPT} \times 512 \text{ bytes} \times \text{RPM} / 60 \text{ seconds} / 1{,}000{,}000 \text{ bits}$$

For example, the ST-12551N 2G 3 1/2-inch drive spins at 7,200 RPM and has an average of 81 sectors per track. The maximum transfer rate for this drive is figured as follows:

$$81 \times 512 \times 7{,}200 / 60 / 1{,}000{,}000 = 4.98\text{Mbps}$$

Using this formula, you can calculate the true maximum sustained transfer rate of any drive.

Cache Programs and Caching Controllers At the software level, disk cache programs such as SMARTDRV (DOS) or VCACHE (Windows 95) can have a major effect on disk drive performance. These cache programs hook into the BIOS hard drive interrupt and then intercept the read and write calls to the disk BIOS from application programs and the device drivers of DOS. This method speeds access tremendously and can greatly affect disk drive performance measurements.

Most controllers now have some form of built-in hardware buffer or cache that doesn't intercept or use any BIOS interrupts. Instead, the caching is performed at the hardware level and is invisible to normal performance-measurement software. Track read-ahead buffers originally were included in controllers to allow for 1:1 interleave performance. Some controllers have simply increased the sizes of these read-ahead buffers; others have added intelligence by making them a cache instead of a simple buffer.

Many IDE and SCSI drives have cache memory built directly into the drive. These integral caches are part of the reason why most IDE and SCSI drives perform so well. Although software and hardware caches can make a drive faster for routine transfer operations, a cache will not affect the true maximum transfer rate that the drive can sustain.

Interleave Selection In a discussion of disk performance, the issue of interleave always comes up. Although traditionally this was more a controller performance issue than a drive issue, most modern hard disks now have built-in controllers (IDE and SCSI) that are fully capable of taking the drive data as fast as the drive can send it. In other words, virtually all modern IDE and SCSI drives are formatted with no interleave (sometimes expressed as a 1:1 interleave ratio).

Head and Cylinder Skewing Most controllers today are capable of transferring data at a 1:1 sector interleave. This is especially true of controllers that are built in to IDE and SCSI drives. With a 1:1 interleave controller, the maximum data transfer rate can be maintained when reading and writing sectors to the disk. Although it would seem that there is no other way to further improve efficiency and the transfer rate, many people overlook two important factors that are similar to interleave: *head* and *cylinder skewing*.

When a drive is reading (or writing) data sequentially, first all of the sectors on a given track are read; then the drive must electronically switch to the next head in the cylinder to continue the operation. If the sectors are not skewed from head to head within the cylinder, no delay occurs after the last sector on one track and before the arrival of the first sector on the next track. Because all drives require some time (although a small amount) to switch from one head to another, and because the controller also adds some overhead to the operation, it is likely that by the time the drive is ready to read the sectors on the newly selected track, the first sector will already have passed by. By skewing the sectors from one head to another—that is, rotating their arrangement on the track so that the arrival of the first sector is delayed relative to the preceding track—you can ensure that no extra disk revolutions will be required when switching heads. This method provides the highest possible transfer rate when head switching is involved.

In a similar fashion, it takes considerable time for the heads to move from one cylinder to another. If the sectors on one cylinder were not skewed from those on the preceding adjacent cylinder, it is likely that by the time the heads arrive, the first sector will already have passed below them, requiring an additional revolution of the disk before reading of the new cylinder can begin. By skewing the sectors from one cylinder to the next, you can account for the cylinder-to-cylinder head-movement time and prevent any additional revolutions of the drive.

Head Skew *Head skew* is the offset in logical sector numbering between the same physical sectors on two tracks below adjacent heads of the same cylinder. The number of sectors skewed when switching from head to head within a single cylinder is to compensate for head switching and controller overhead time. Think of it as the surface of each platter being rotated as you traverse from head to head. This method permits continuous read or write operation across head boundaries without missing any disk revolutions, thus maximizing system performance.

To understand head skew, you first need to know the order in which tracks and sectors are read from a disk. If you imagine a single-platter (two-head) drive with 10 cylinders and 17 sectors per track, the first sector that will be read on the entire drive is Cylinder 0, Head 0, Sector 1. Following that, all the remaining sectors on that first track (Cylinder 0, Head 0) will be read until Sector 17 is reached. After that, two things could take place:

- The heads could be moved so that the drive could continue reading the next track on the same side of the platter.
- The second head could be selected; therefore, another entire track could be read with no head movement.

Because head movement takes much longer than electronically selecting another head, all disk drives will select the subsequent heads on a cylinder before physically moving the heads to the next cylinder. Thus, the next sector to be read would be Cylinder 0, Head 1, Sector 1. Next, all the remaining sectors on that track are read (2 through 17), and then in our single platter example it is time to switch heads. This sequence continues until the last sector on the last track is read—in this example, Cylinder 9, Head 1, Sector 17.

If you could take the tracks off a cylinder in this example and lay them on top of one another, the tracks might look like this:

Cyl. 0, Head 0:	1 2 3 4 5 6 7 8 9 10 11 12 13 14 15 16 17
Cyl. 0, Head 1:	1 2 3 4 5 6 7 8 9 10 11 12 13 14 15 16 17

After reading all the sectors on head 0, the controller switches heads to head 1 and continues the read (looping around to the beginning of the track). In this example, the sectors were not skewed at all between the heads, which means that the sectors are directly above and below one another in a given cylinder.

Now the platters in this example are spinning at 3,600 RPM, so one sector is passing below a head once every 980 millionths of a second! This obviously is a very small timing window. It takes some time for the head switch to occur (usually, 15 millionths of a second), plus some overhead time for the controller to pass the head-switch command. By the time the head switch is complete and you are ready to read the new track, sector 1 has already gone by! This problem is similar to interleaving when the interleave is too low. The drive is forced to wait while the platter spins around another revolution so that it can begin to pick up the track, starting with Sector 1.

This problem is easy to solve: Simply offset the sector numbering on subsequent tracks from those that precede them sufficiently to account for the head-switching and controller overhead time. That way, when Head 0, Sector 17 finishes and the head switches, Head 1, Sector 1 arrives right on time. The result looks something like this:

Cyl. 0, Head 0:	1 2 3 4 5 6 7 8 9 10 11 12 13 14 15 16 17
Cyl. 0, Head 1:	16 17 1 2 3 4 5 6 7 8 9 10 11 12 13 14 15

Shifting the second track by two sectors provides time to allow for the head-switching overhead and is the equivalent to a head-skew factor of 2. In normal use, a drive switches heads much more often than it switches physical cylinders, which makes head skew more important than cylinder skew. Throughput can rise dramatically when a proper head skew is in place. Different head skews can account for different transfer rates among drives that have the same number of sectors per track and the same interleave.

A nonskewed MFM drive, for example, may have a transfer rate of 380KBps, whereas the transfer rate of a drive with a head skew of 2 could rise to 425KBps. Notice that different controllers and drives have different amounts of overhead, so real-world results will be different in each case. In most cases, the head-switch time is very small compared with the controller overhead. As with interleaving, it is better to be on the conservative side to avoid additional disk revolutions.

Cylinder Skew *Cylinder skew* is the offset in logical sector numbering between the same physical sectors on two adjacent tracks on two adjacent cylinders.

The number of sectors skewed when switching tracks from one cylinder to the next is to compensate for track to track seek time. In essence, all of the sectors on adjacent tracks are rotated with respect to each other. This method permits continuous read or write operations across cylinder boundaries without missing any disk revolutions, thus maximizing system performance.

Cylinder skew is a larger numerical factor than head skew because more overhead exists. It takes much longer to move the heads from one cylinder to another than simply to switch heads. Also, the controller overhead in changing cylinders is higher as well.

Following is a depiction of our example drive with a head-skew factor of 2 but no cylinder skew.

Cyl. 0, Head 0:	1 2 3 4 5 6 7 8 9 10 11 12 13 14 15 16 17
Cyl. 0, Head 1:	16 17 1 2 3 4 5 6 7 8 9 10 11 12 13 14 15
Cyl. 1, Head 0:	8 9 10 11 12 13 14 15 16 17 1 2 3 4 5 6 7

In this example, the cylinder-skew factor is 8. Shifting the sectors on the subsequent cylinder by eight sectors gives the drive and controller time to be ready for sector 1 on the next cylinder and eliminates an extra revolution of the disk.

Calculating Skew Factors You can derive the correct head-skew factor from the following information and formula:

Head skew = (head-switch time/rotational period) × SPT + 2

In other words, the head-switching time of a drive is divided by the time required for a single rotation. The result is multiplied by the number of sectors per track, and 2 is added for controller overhead. The result should then be rounded up to the next whole integer (for example, 2.3 = 2, 2.5 = 3).

You can derive the correct cylinder-skew factor from the following information and formula:

Cylinder skew = (track-to-track seek time/rotational period) × SPT + 4

In other words, the track-to-track seek time of a drive is divided by the time required for a single rotation. The result is multiplied by the number of sectors per track, and 4 is added for controller overhead. Round the result up to a whole integer (for example, 2.3 = 2, 2.5 = 3).

The following example uses typical figures for an ESDI drive and controller. If the head-switching time is 15 us (microseconds), the track-to-track seek is 3 ms, the rotational period is 16.67 ms (3,600 RPM), and the drive has 53 physical sectors per track:

Head skew = (0.015/16.67) × 53 +2 = 2 (rounded up)

Cylinder Skew = (3/16.67) × 53 + 4 = 14 (rounded up)

If you do not have the necessary information to make the calculations, contact the drive manufacturer for recommendations. Otherwise, you can make the calculations by using conservative figures for head-switch and track-to-track access times. If you are unsure, just as with interleaving, it is better to be on the conservative side, which minimizes the possibility of additional rotations when reading sequential information on the drive. In most cases, a default head skew of 2 and a cylinder skew of 16 work well.

Because factors such as controller overhead can vary from model to model, sometimes the only way to figure out the best value is to experiment. You can try different skew values and then run data-transfer rate tests to see which value results in the highest performance. Be careful with these tests, however; many disk benchmark programs will only read or write data from one track or one cylinder during testing, which totally eliminates the effect of skewing on the results. The best type of benchmark to use for this testing is one that reads and writes large files on the disk.

Most real (controller register level) low-level format programs are capable of setting skew factors. Those programs that are supplied by a particular controller or drive manufacturer usually are already optimized for their particular drives and controllers and may not allow you to change the skew.

Notice that most IDE and SCSI drives have their interleave and skew factors set to their optimum values by the manufacturer. In most cases, you cannot even change these values; in the cases in which you can, the most likely result is a slower drive. For this reason, most IDE drive manufacturers recommend against low-level formatting their drives. With some IDE drives, unless you use the right software, you might alter the optimum skew settings and slow the drive. IDE drives that use zoned recording cannot ever have the interleave or skew factors changed, and as such, they are fully protected. No matter how you try to format these drives, the interleave and skew factors cannot be altered. The same can be said for SCSI drives.

Shock Mounting

Most hard disks manufactured today have a *shock-mounted* HDA, which means that a rubber cushion is placed between the disk drive body and the mounting chassis. Some drives use more rubber than others, but for the most part, a shock mount is a shock mount. Some drives do not have a shock-isolated HDA due to physical or cost constraints. Be sure that the drive you are using has adequate shock-isolation mounts for the HDA, especially if you are using the drive in a portable PC system or in a system in which environmental conditions are less favorable than in a normal office. I usually never recommend a drive that lacks at least some form of shock mounting.

Cost

The cost of hard disk storage recently has fallen to 4 cents per megabyte or less. You can purchase 8GB drives for under $300. That places the value of the 10MB drive that I bought in 1983 at about $0.40. (Too bad—I paid $1,800 for it at the time!)

Of course, the cost of drives continues to fall, and eventually, even 4 cents per megabyte will seem expensive. Because of the low costs of disk storage today, not many drives that are less than 1GB are even being manufactured.

Capacity

Four figures commonly are used in advertising drive capacity:

- Unformatted capacity, in millions of bytes (M)

- Formatted capacity, in millions of bytes (M)
- Unformatted capacity, in megabytes (Meg)
- Formatted capacity, in megabytes (Meg)

Most manufacturers of IDE and SCSI drives now report only the formatted capacities because these drives are delivered preformatted. Most of the time, advertisements refer to the unformatted or formatted capacity in millions of bytes because these figures are larger than the same capacity expressed in megabytes. This situation generates a great deal of confusion when the user runs FDISK (which reports total drive capacity in megabytes) and wonders where the missing space is. This question ranks as one of the most common questions that I hear during my seminars. Fortunately, the answer is easy. It only involves a little math to figure it out.

Perhaps the most common questions I get are concerning "missing" drive capacity. Consider the following example: "I just installed a new Western Digital AC2200 drive, billed as 212M. When I entered the drive parameters (989 cylinders, 12 heads, 35 sectors per track), both the BIOS Setup routine and FDISK report the drive as only 203MB! What happened to the other 9MB?"

The answer is only a few calculations away. By multiplying the drive specification parameters, you get this result:

Cylinders:	989
Heads:	12
Sectors per track:	35
Bytes per sector:	512
Total bytes:	212.67M
Total megabytes:	202.82MB

The result figures to a capacity of 212.67M or 202.82MB. Drive manufacturers usually report drive capacity in millions of bytes, whereas your BIOS and FDISK usually report the capacity in megabytes. One megabyte equals 1,048,576 bytes (or 1,024KB, wherein each kilobyte is 1,024 bytes). So the bottom line is that this 212.67M drive also is a 202.82MB drive! What is really confusing is that there is no industry-wide accepted way of differentiating binary megabytes from decimal ones. Usually drive manufacturers will always report metric megabytes, since they result in larger, more impressive sounding numbers! One additional item to note about this particular drive is that it is a zoned recording drive and that the actual physical parameters are different. Physically, this drive has 1,971 cylinders and four heads; however, the total number of sectors on the drive (and, therefore, the capacity) is the same no matter how you translate the parameters.

Although Western Digital does not report the unformatted capacity of this particular drive, unformatted capacity usually works out to be about 19% larger than a drive's formatted capacity. The Seagate ST-12550N Barracuda 2GB drive, for example, is advertised as having the following capacities:

Unformatted capacity:	2,572.00M
Unformatted capacity:	2,452.85MB
Formatted capacity:	2,139.00M

Formatted capacity: 2,039.91MB

Each of these four figures is a correct answer to the question "What is the storage capacity of the drive?" As you can see, however, the numbers are very different. In fact, yet another number could be used. Divide the 2,039.91MB by 1,024, and the drive's capacity is 1.99GB! So when you are comparing or discussing drive capacities, make sure that you are working with a consistent unit of measure, or your comparisons will be meaningless.

The true industry standard abbreviations for these figures are shown in Table 1.4.

Table 1.4 Standard Abbreviations and Meanings.

Abbreviation	Description	Decimal Meaning	Binary Meaning
Kbit	Kilobit	1,000	1,024
K	Kilobyte	1,000	1,024
Mbit	Megabit	1,000,000	1,048,576
M	Megabyte	1,000,000	1,048,576
Gbit	Gigabit	1,000,000,000	1,073,741,824
G	Gigabyte	1,000,000,000	1,073,741,824
Tbit	Terabit	1,000,000,000,000	1,099,511,627,776
T	Terabyte	1,000,000,000,000	1,099,511,627,776

Unfortunately, there are no differences in the abbreviations when used to indicate metric verses binary values. In other words, M can be used to indicate both “millions of bytes” and megabytes. In general, memory values are always computed using the binary derived meanings, while disk capacity goes either way. Unfortunately, this often leads to confusion in reporting disk capacities. Note that bits and bytes are distinguished by a lower- or uppercase *B*. For example, millions of bits are indicated by using a lowercase *b*, resulting in the abbreviation of Mbps for million bits per second, while MBps indicates million bytes per second.

Chapter 2

Hard Disk Interfaces

This chapter describes the hard disk interface, from the drives to the cables and controllers that run them. You learn about the various disk interfaces you can select, and the shortcomings and strengths of each type.

Interfaces Choices

The primary job of the hard disk controller or interface is to transmit and receive data to and from the drive.

Several types of hard disk interfaces have been used in PC systems over the years:

- ST-506/412
- ESDI
- IDE
- SCSI

Of these interfaces, only ST-506/412 and ESDI are what you could call true disk-controller-to-drive interfaces. SCSI and IDE are system-level interfaces that usually incorporate a chipset-based variation of one of the other two types of disk controller interfaces internally. For example, most SCSI and IDE drives incorporate the same basic controller circuitry used in separate ESDI controllers. The SCSI interface adds another layer of interface that attaches thc controller to the system bus, whereas IDE is a direct bus-attachment interface.

In data recovery, it helps to know the disk interface you are working with, because many data-recovery problems involve drive setup and installation problems. Each interface requires a slightly different method of installation and drive configuration. If the installation or configuration is incorrect or accidentally altered by the system user, it may prevent access to data on a drive. Accordingly, anyone who wants to become proficient in data recovery must be an expert on installing and configuring various types of hard disks and controllers.

The ST-506/412 Interface

The ST-506/412 interface was developed by Seagate Technologies around 1980. Most drive manufacturers that made hard disks for PC systems adopted the Seagate ST-506/412 standard, a situation that helped make this interface popular.

The ST-506/412 interface does not quite make the grade in today's high-performance PC systems. This interface was designed for a 5MB drive, and I have not seen any drives larger than 152MB (Modified Frequency Modulation encoding) or 233MB (Run-Length Limited encoding) available for this type of interface. Because the capacity, performance, and expandability of ST-506/412s are so limited, this interface is obsolete and generally unavailable in new systems. However, many older systems still use drives that have this interface.

The ESDI Interface

ESDI, or *Enhanced Small Device Interface*, is a specialized hard disk interface established as a standard in 1983, primarily by Maxtor Corporation.

Compared with ST-506/412, ESDI has provisions for increased reliability, such as building the endec into the drive. ESDI is a very-high-speed interface, capable of a maximum 24Mbps transfer rate. Most drives running ESDI, however, are limited to a maximum 10Mbps or 15Mbps. Unfortunately, compatibility problems between different ESDI implementations, combined with pressure from low-cost, high-performance IDE interface drives, have served to make the ESDI interface obsolete. Few, if any, new systems today include ESDI drives, although ESDI became somewhat popular in high-end systems during the late 1980s.

I no longer recommend installing ESDI drives unless you are upgrading a system that already has an ESDI controller.

The IDE Interface

Integrated Drive Electronics (IDE) is a generic term applied to any drive with an integrated (built-in) disk controller. The IDE interface as we know it is officially called *ATA* (AT Attachment) and is an ANSI standard; however, IDE can roughly apply to any disk drive with a built-in controller.

In a drive with IDE, the disk controller is integrated into the drive, and this combination drive/controller assembly usually plugs into a bus connector on the motherboard or bus adapter card. Combining the drive and controller greatly simplifies installation, because there are no separate power or signal cables from the controller to the drive. Also, when the controller and the drive are assembled as a unit, the number of total components is reduced, signal paths are shorter, and the electrical connections are more noise-resistant, resulting in a more reliable design than is possible when a separate controller, connected to the drive by cables, is used.

Integrating the controller and drive also frees the controller and drive engineers from having to adhere to the strict standards imposed by the earlier interface standards. The resulting drive and controller combinations can offer higher performance than earlier stand-alone controller and drive setups. IDE drives sometimes are called drives with embedded controllers.

The IDE connector on motherboards in many systems is nothing more than a stripped-down bus slot. In ATA IDE installations, these connectors normally contain a 40-pin subset of the 98 pins that would be available in a standard 16-bit ISA bus slot. The pins used are only the signal pins required by a standard-type XT or AT hard disk controller. For example, because an AT-style disk controller uses only interrupt line 14, the motherboard AT IDE connector supplies only that interrupt line. No other interrupt lines are needed. The XT IDE motherboard connector supplies interrupt line 5 because that is what an XT controller would use.

When IDE drives are discussed, the ATA IDE variety usually is the only kind mentioned because it is so popular. But other forms of IDE drives exist, based on other buses. For example, several PS/2 systems came with Micro-Channel (MCA) IDE drives which plug directly into a Micro-Channel Bus slot (through an angle adapter or Interposer card). An 8-bit ISA form of IDE also existed but was never very popular. Most IBM-compatible systems with the ISA or EISA Bus use AT-Bus (16-bit) IDE drives. The ATA IDE interface is by far the most popular type of drive interface available.

The primary advantage of IDE drives is cost. Because the separate controller or host adapter is eliminated and the cable connections are simplified, IDE drives cost much less than a standard controller-and-drive combination. These drives also are more reliable, because the controller is built into the drive. Therefore, the endec or data separator (the converter between the digital and analog signals on the drive) stays close to the media. Because the drive has a short analog-signal path, it is less susceptible to external noise and interference.

Another advantage is performance. IDE drives are some of the highest-performance drives available—but they also are among the lowest-performance drives. This apparent contradiction is a result of the fact that all IDE drives are different. You cannot make a blanket statement about the performance of IDE drives because each drive is unique. The high-end models, however, offer performance that is equal or superior to that of any other type of drive on the market for a single-user, single-tasking operating system.

IDE Bus Versions

Three main types of IDE interfaces are available, with the differences based on three different bus standards:

- AT Attachment (ATA) IDE (16-bit ISA)
- XT IDE (8-bit ISA)
- MCA IDE (16-bit Micro Channel)

Note

Many people are confused about 16- versus 32-bit bus connections and 16- versus 32-bit hard drive connections. A PCI connection allows for a 32-bit (and 64-bit in the future) bandwidth from the bus to the drive controller only. In an IDE (or EIDE) drive configuration, you are still getting only 16-bit bandwidth between the drive and the controller. This usually does not create a bottleneck, however, because one or two hard drives cannot supply the controller enough data to saturate even a 16-bit channel. Fast Wide SCSI-3 is the only device/controller combination that gives you 32 bits from the controller to the drive, primarily because you can hang 15 devices off a SCSI Wide chain and there is a good chance that many devices will saturate a 16-bit channel at some time.

The XT and ATA versions have standardized on 40-pin connectors and cables, but the connectors have slightly different pinouts, rendering them incompatible with one another. MCA IDE uses a completely different 72-pin connector and is designed for MCA bus systems only.

In most cases, you must use the type of IDE drive that matches your system bus. This situation means that XT IDE drives work only in XT-class 8-bit ISA slot systems, AT IDE drives work only in AT-class 16-bit ISA or EISA slot systems, and MCA IDE drives work only in Micro-Channel systems (such as the IBM PS/2 Model 50 or higher). A company called Silicon Valley offers adapter cards for

XT systems that will run ATA IDE drives. Other companies, such as Arco Electronics and Sigma Data, have IDE adapters for Micro-Channel systems that allow ATA IDE drives to be used on these systems. These adapters are very useful for XT or PS/2 systems, because there is a very limited selection of XT or MCA IDE drives, whereas the selection of ATA drives is virtually unlimited.

In most modern ISA and EISA systems, you will find an ATA connector on the motherboard. If your motherboard does not have one of these connectors and you want to attach an AT IDE drive to your system, you can purchase an adapter card that changes your 98-pin slot connector to the 40-pin IDE connector. These adapter cards are nothing more than buffered cables; they are not really controllers. The controller is built into the drive. Some of the cards offer additional features, such as an on-board ROM BIOS or cache memory.

Because of the limited availability and use of XT IDE and MCA IDE drives, only ATA IDE drive types will be discussed here.

ATA IDE CDC, Western Digital, and Compaq actually created what could be called the first ATA-type IDE interface drive and were the first to establish the 40-pin IDE connector pinout.

Eventually, the 40-pin IDE connector and drive interface method was placed before one of the ANSI standards committees which, in conjunction with drive manufacturers, ironed out some deficiencies, tied up some loose ends, and published what is known as the *CAM ATA (Common Access Method AT Attachment) interface*. The CAM Committee was formed in October 1988, and the first working document of the AT Attachment interface was introduced in March 1989. Before the CAM ATA standard, many companies that followed CDC, such as Conner Peripherals, made proprietary changes to what had been done by CDC. As a result, many older ATA drives are very difficult to integrate into a dual-drive setup that has newer drives.

Some areas of the ATA standard have been left open for vendor-specific commands and functions. These vendor-specific commands and functions are the main reason why it is so difficult to low-level format IDE drives. To work properly, the formatter you are using usually must know the specific vendor-unique commands for rewriting sector headers and remapping defects. Unfortunately, these and other specific drive commands differ from OEM to OEM, clouding the "standard" somewhat.

Note

It is important to note that only the ATA IDE interface has been standardized by the industry. The XT IDE and MCA IDE never were adopted as industry-wide standards and never became very popular. These interfaces no longer are in production, and no new systems of which I am aware come with these nonstandard IDE interfaces.

The ATA Specification The ATA specification was introduced in March 1989 as an ANSI standard. ATA-1 was finally approved in 1994, and ATA-2 (also called *Enhanced IDE*) was approved in 1995. ATA-3 is currently in the works. The ATA standards have gone a long way toward eliminating incompatibilities and problems with interfacing IDE drives to ISA and EISA systems. The ATA specifications define the signals on the 40-pin connector, the functions and timings of these signals, cable specifications, and so on. The following section lists some of the elements and functions defined by the ATA specification.

Dual-Drive Configurations Dual-drive ATA installations can be problematic because each drive has its own controller, and both controllers must function while being connected to the same bus. There has to be a way to ensure that only one of the two controllers will respond to a command at a time.

The ATA standard provides the option of operating on the AT Bus with two drives in a daisy-chained configuration. The primary drive (drive 0) is called the *master*, and the secondary drive (drive 1) is the *slave*. You designate a drive as being master or slave by setting a jumper or switch on the drive or by using a special line in the interface called the *Cable Select (CSEL) pin*.

When only one drive is installed, the controller responds to all commands from the system. When two drives (and, therefore, two controllers) are installed, all commands from the system are received by both controllers. Each controller then must be set up to respond only to commands for itself. In this situation, one controller then must be designated as the master and the other as the slave. When the system sends a command for a specific drive, the controller on the other drive must remain silent while the selected controller and drive are functioning. Setting the jumper to master or slave allows discrimination between the two controllers by setting a special bit (the *DRV bit*) in the Drive/Head Register of a command block.

ATA I/O Connector The ATA interface connector is a 40-pin header-type connector that should be keyed to prevent the possibility of installing it upside down. A key is provided by the removal of pin 20, and the corresponding pin on the cable connector should be plugged in to prevent a backward installation. The use of keyed connectors and cables is highly recommended, because plugging an IDE cable in backward can damage both the drive and the bus adapter circuits (although I have done it myself many times with no smoked parts yet!).

Table 2.1 shows the ATA-IDE interface connector pinout.

Table 2.1 ATA Connector.

Signal Name	Pin	Pin	Signal Name
-RESET	1	2	GROUND
Data Bit 7	3	4	Data Bit 8
Data Bit 6	5	6	Data Bit 9
Data Bit 5	7	8	Data Bit 10
Data Bit 4	9	10	Data Bit 11
Data Bit 3	11	12	Data Bit 12
Data Bit 2	13	14	Data Bit 13
Data Bit 1	15	16	Data Bit 14
Data Bit 0	17	18	Data Bit 15
GROUND	19	20	KEY (pin missing)
DRQ 3	21	22	GROUND
-IOW	23	24	GROUND
-IOR	25	26	GROUND
I/O CH RDY	27	28	SPSYNC:CSEL

Table 2.1 ATA Connector (continued).

Signal Name	Pin	Pin	Signal Name
-DACK 3	29	30	GROUND
IRQ 14	31	32	-IOCS16
Address Bit 1	33	34	-PDIAG
Address Bit 0	35	36	Address Bit 2
-CS1FX	37	38	-CS3FX
-DA/SP	39	40	GROUND
+5 Vdc (Logic)	41	42	+5 Vdc (Motor)
GROUND	43	44	-TYPE (0=ATA)

ATA I/O Cable A 40-conductor ribbon cable is specified to carry signals between the bus adapter circuits and the drive (controller). To maximize signal integrity and to eliminate potential timing and noise problems, the cable should not be longer than 0.46 meters (18 inches).

ATA Signals This section describes some of the most important signals in more detail.

Pin 20 is used as a key pin for cable orientation and is not connected through in the interface. This pin should be missing from any ATA connectors, and the cable should have the pin-20 hole in the connector plugged off to prevent the cable from being plugged in backward.

Pin 39 carries the *Drive Active/Slave Present (DASP) signal*, which is a dual-purpose, time-multiplexed signal. During power-on initialization, this signal indicates whether a slave drive is present on the interface. After that, each drive asserts the signal to indicate that it is active. Early drives could not multiplex these functions and required special jumper settings to work with other drives. Standardizing this function to allow for compatible dual-drive installations is one of the features of the ATA standard.

Pin 28 carries the *Cable Select* or *Spindle Synchronization signal* (CSEL or SPSYNC), which is a dual-purpose conductor. A given installation, however, may use only one of the two functions. The CSEL function is the most widely used and is designed to control the designation of a drive as master (drive 0) or slave (drive 1) without requiring jumper settings on the drives. If a drive sees the CSEL as being grounded, the drive is a master. If CSEL is open, the drive is a slave.

You can install special cabling to ground CSEL selectively. This installation normally is accomplished through a Y-cable arrangement, with the IDE bus connector in the middle and each drive at opposite ends of the cable. One leg of the Y has the CSEL line connected through, indicating a master drive. The other leg has the CSEL line open (conductor interrupted or removed), making the drive at that end the slave.

ATA Commands One of the best features of the ATA IDE interface is the enhanced command set. The ATA IDE interface was modeled after the WD1003 controller that IBM used in the original AT system. All ATA IDE drives must support the original WD command set (eight commands), with no exceptions, which is why IDE drives are so easy to install in systems today. All IBM-compatible systems have built-in ROM BIOS support for the WD1003, which means that essentially, they support ATA IDE as well.

In addition to supporting all the WD1003 commands, the ATA specification added numerous other commands to enhance performance and capabilities. These commands are an optional part of the ATA interface, but several of them are used in most drives available today and are very important to the performance and use of ATA drives in general.

Perhaps the most important is the Identify Drive command. This command causes the drive to transmit a 512-byte block of data that provides all details about the drive. Through this command, any program (including the system BIOS) can find out exactly what type of drive is connected, including the drive manufacturer, model number, operating parameters, and even the serial number of the drive. Many modern BIOSs use this information to automatically receive and enter the drive's parameters into CMOS memory, eliminating the need for the user to enter these parameters manually during system configuration. This arrangement helps prevent mistakes that can later lead to data loss when the user no longer remembers what parameters he or she used during setup.

The Identify Drive data can tell you many things about your drive, including the following:

- Number of cylinders in the recommended (default) translation mode
- Number of heads in the recommended (default) translation mode
- Number of sectors per track in the recommended (default) translation mode
- Number of cylinders in the current translation mode
- Number of heads in the current translation mode
- Number of sectors per track in the current translation mode
- Manufacturer and model number
- Firmware revision
- Serial number
- Buffer type, indicating sector buffering or caching capabilities

Two other very important commands are the Read Multiple and Write Multiple commands. These commands permit multiple-sector data transfers and, when combined with block-mode Programmed I/O (PIO) capabilities in the system, can result in incredible data-transfer rates many times faster than single-sector PIO transfers.

Tip

If you want the ultimate in IDE performance and installation ease, make sure that your motherboard BIOS and IDE adapter supports ATA-2 or EIDE. This support allows your BIOS to execute data transfers to and from the IDE drive several times faster than normal, and also makes installation and configuration easier because the BIOS will be able to detect the drive-parameter information automatically. High-speed PIO and automatic detection of the drive type are included in the latest versions of most PC BIOSs.

There are many other enhanced commands, including room for a given drive manufacturer to implement what are called *vendor-unique commands*. These commands often are used by a particular vendor for features unique to that vendor. Often, features such as low-level formatting and defect management are controlled by vendor-unique commands. This is why low-level format programs can be so specific to a particular manufacturer's IDE drives and why many manufacturers make their own LLF programs available.

ATA IDE Drive Categories ATA-IDE drives can be divided into three main categories. These categories separate the drives by function (such as translation capabilities) and design (which can affect features such as low-level formatting):

- Non-Intelligent ATA-IDE drives
- Intelligent ATA-IDE drives
- Intelligent zoned recording ATA-IDE drives

Non-Intelligent IDE The ATA standard requires that the built-in controller respond exactly as though it were a Western Digital WD1003 controller. This controller responds to a command set of eight commands. Early IDE drives supported these commands and had few, if any, other options. These early drives actually were more like regular ST-506/412 or ESDI controllers bolted directly into the drive than the more intelligent drives that we consider today to be IDE. These drives were not considered to be intelligent IDE drives.

These drives could be low-level formatted in the same manner as any normal ST-506/412 or ESDI drive. They were universally low-level formatted at the factory, with factory-calculated optimum interleave (usually, 1:1) and head- and cylinder-skew factors. Also, factory defects were recorded in a special area on the drive; they no longer were written on a sticker pasted to the exterior. Unfortunately, this arrangement means that if you low-level format these drives in the field, you most likely will alter these settings (especially the skew factors) from what the factory set as optimum, as well as wipe out the factory-written defect table.

Some manufacturers released special low-level format routines that would reformat the drives while preserving these settings, but others did not make such programs available. Because they did not want you to overwrite the defect list or potentially slow the drive, most manufacturers stated that you should never low-level format their IDE drives.

This statement started a myth that the drives could somehow be damaged or rendered inoperable by such a format, which truly is not the case. One rumor was that the servo information could be overwritten, which would mean that you would have to send the drive back to the manufacturer for re-servoing. This also is not true; the servo information is protected and cannot be overwritten. The only consequence of an improper low-level format of these drives is the possible alteration of the skew factors and the potential loss of the factory defect maps.

Intelligent IDE Later IDE drives became known as *intelligent IDE drives*. These drives support enhanced ATA commands, such as the Identify Drive command, and sector-translation capabilities.

These drives can be configured in two ways: in raw physical mode or in translation mode. To configure the drive in raw physical mode, you simply enter the CMOS drive parameters during setup so that they match the true physical parameters of the drive. For example, if the drive physically has 800 cylinders, 6 heads, and 50 sectors per track, you enter these figures during setup. To configure the drive in translation mode, you simply enter any combination of cylinders, heads, and sectors that adds up to equal or less than the true number of sectors on the drive.

In the example I just used, the drive has a total of 240,000 sectors (800×6×50). All I have to do is figure out another set of parameters that adds up to equal or less than 240,000 sectors. The simplest way to do this is to cut the number of cylinders in half and double the number of heads. Thus, the new drive parameters become 400 cylinders, 12 heads, and 50 sectors per track. This method adds up to 240,000 sectors and enables the drive to work in translation mode.

When these drives are in translation mode, a low-level format cannot alter the interleave and skew factors, nor can it overwrite the factory defect-mapping information. A low-level format program can, however, perform additional defect mapping or sector sparing while in this mode.

If the drive is in true physical mode, a low-level format rewrites the sector headers and modifies the head and cylinder skewing. If performed incorrectly, the format can be repaired by a proper low-level format program that allows you to set the correct head and cylinder skew. This task can be accomplished automatically by the drive manufacturer's recommended low-level format program. To obtain the correct skew values, it is best to contact the drive manufacturer's technical support department. You can calculate the skew values if the manufacturer cannot provide them.

To protect the skew factors and defect information on intelligent IDE drives, all you have to do is run them in translation mode. In translation mode, this information cannot be overwritten.

Intelligent Zoned Recording IDE The last and most sophisticated IDE drives combine intelligence with *zoned recording*. With zoned recording, the drive has a variable number sectors per track in several zones across the surface of the drive. Because the PC BIOS can handle only a fixed number of sectors on all tracks, these drives always must run in translation mode. Because these drives are always in translation mode, you cannot alter the factory-set interleave and skew factors or wipe out the factory defect information.

You still can low-level format these drives, however, and use such a format to map or spare additional defective sectors that crop up during the life of the drive.

ATA-2 (Enhanced IDE) *ATA-2* is an extension of the original ATA (IDE) specification. The most important additions are performance enhancing features such as fast PIO and DMA modes. ATA-2 also features improvements in the Identify Drive command allowing a drive to tell the software exactly what its characteristics are; this is essential for both Plug and Play and compatibility with future revisions of the standard.

ATA-2 is often called Enhanced IDE (or EIDE). EIDE is technically a marketing program from Western Digital. Fast-ATA and Fast-ATA-2 are similar Seagate-inspired marketing programs, which are also endorsed by Quantum. As far as the hard disk and BIOS are concerned, these are all different terms for basically the same thing.

There are four main areas where ATA-2 and EIDE have improved the original ATA/IDE interface:

- Increased maximum drive capacity
- Faster data transfer
- Secondary two-device channel
- ATAPI (ATA Program Interface)

The following section describes these improvements.

Increased Drive Capacity ATA-2/EIDE allows for increased drive capacity over the original ATA/IDE specification. This is done through an Enhanced BIOS, which makes it possible to use hard disks exceeding the 504M barrier. The origin of this limit is the disk geometry (cylinders, heads, sectors) supported by the combination of an IDE drive and the BIOS' software interface. Both IDE and the BIOS are capable of supporting huge disks, but their combined limitations conspire to restrict the useful capacity to 504M.

An Enhanced BIOS circumvents this by using a different geometry when talking to the drive than when talking to the software. What happens in between is called *translation*. For example, if your drive has 2,000 cylinders and 16 heads, a translating BIOS will make programs think that the drive has 1,000 cylinders and 32 heads.

You can usually tell if your BIOS is enhanced by the ability to specify more than 1,024 cylinders in the BIOS setup, although this is not conclusive. If you see drive-related settings like LBA, ECHS, or even Large, these are tell-tale signs of a BIOS with translation support. Most BIOSs with a date of 1994 or later are enhanced. If your system currently does not have an Enhanced BIOS, you may be able to get an upgrade.

There are roughly three ways today's BIOSs can handle translation: standard CHS addressing, Extended CHS addressing, and LBA addressing. They are summarized in the following table:

BIOS Mode	Operating System to BIOS	BIOS to Drive Ports
Standard CHS	Logical CHS Parameters	Logical CHS Parameters
Extended CHS	Translated CHS Parameters	Logical CHS Parameters
LBA	Translated CHS Parameters	LBA Parameters

In Standard CHS, there is only one possible translation step internal to the drive. The drive's actual, physical geometry is completely invisible from the outside with all zoned recorded ATA drives today. The cylinders, heads, and sectors printed on the label for use in the BIOS setup are purely logical geometry, and do not represent the actual physical parameters. Standard CHS addressing is limited to 16 heads and 1,024 cylinders, which gives a limit of 504MB.

This is often called "Normal" in the BIOS setup, and causes the BIOS to behave like an old-fashioned one without translation. Use this setting if your drive has fewer than 1,024 cylinders or if you want to use the drive with a non-DOS operating system that doesn't understand translation.

In Extended CHS, a translated logical geometry is used to communicate between the drive and the BIOS, while a different translated geometry is used to communicate between the BIOS and everything else. In other words, there are normally two translation steps. The drive still translates internally, but has logical parameters that exceed the 1,024 cylinder limitation of the standard BIOS. In this case, the drive's cylinder count is usually divided by 2, and the head count is multiplied by 2 to get the translated values from those actually stored in the CMOS Setup. This type of setting breaks the 504/528MB barrier.

This is often called "Large" or "ECHS" in the BIOS setup, and tells the BIOS to use Extended CHS translation. It uses a different geometry (cylinders/heads/sectors) when talking to the drive than when talking to the BIOS. This type of translation should be used with drives that have more than 1,024 cylinders but that do not support LBA (Logical Block Addressing). Note that the geometry entered in your BIOS setup is the logical geometry, not the translated one.

LBA is a means of linearly addressing sector's addresses, beginning at Cylinder 0, Head 0, Sector 1 as LBA 0, and proceeding on to the last physical sector on the drive. This is new in ATA-2, but has always been the one and only addressing mode in SCSI.

With LBA, each sector on the drive is numbered starting from 0. The number is a 28-bit binary number internally, which translates to a sector number of from 0 to 268,435,456. Because each sector represents 512 bytes, this results in a maximum drive capacity of exactly 128GB, or 137 billion bytes. Unfortunately, the operating system still needs to see a translated CHS, so the BIOS determines how many sectors there are, and comes up with Translated CHS to match. The BIOS CHS limits are 1,024 cylinders, 256 heads, and 63 sectors per track, which limits total drive capacity to just under 8GB.

In other words, this breaks the 528MB barrier in essentially the same way as Extended CHS does. Because it is somewhat simpler to use a single linear number to address a sector on the hard disk compared to a CHS type address, this is the preferred method if the drive supports LBA.

Caution

A word of warning with these BIOS translation settings: If you switch between standard CHS, Extended CHS, or LBA, the BIOS may change the (translated) geometry. The same thing may happen if you transfer a disk that has been formatted on an old, non-LBA computer to a new one that uses LBA. This will cause the logical CHS geometry seen by the operating system to change, and will cause the date to appear in the wrong locations from where it actually is! This can cause you to lose access to your data if you are not careful. I always recommend recording the CMOS Setup screens associated with the hard disk configuration so that you can properly match the setup of a drive to the settings it was originally at.

Faster Data Transfer ATA-2/EIDE defines several high-performance modes for transferring data to and from the drive. These faster modes are the main part of the new specifications and were the main reason they were initially developed. Most of the faster drives on the market today will support either PI/O transfer Mode 3 or Mode 4, which results in a very fast transfer. The following section discusses these modes.

The PIO mode determines how fast data is transferred to and from the drive. In the slowest possible mode, PIO mode 0, the data cycle time cannot exceed 600 nanoseconds (ns). In a single cycle, 16 bits are transferred in or out of the drive making the theoretical transfer rate of PIO Mode 0 (600 ns cycle time) 3.3MBps. Most of the high-performance ATA-2 (EIDE) drives today support PIO Mode 4, which offers a 16.6MBps transfer rate.

The following table shows the PIO modes, with their respective transfer rates:

PIO Mode	Cycle Time (ns)	Transfer Rate (M/sec)	Specification
0	600	3.3	ATA
1	383	5.2	ATA
2	240	8.3	ATA

PIO Mode	Cycle Time (ns)	Transfer Rate (M/sec)	Specification (continued)
3	180	11.1	ATA-2
4	120	16.6	ATA-2

To run in Mode 3 or 4 requires that the IDE port on the system be a local bus port. This means that it must operate through either a VL-Bus or PCI bus connection. Some newer motherboards with ATA-2/EIDE support have dual IDE connectors on the motherboard, with only the primary connector running through the system's PCI local bus. The secondary connector usually runs through the ISA bus, and therefore supports up to Mode 2 operation only. Make sure you get clarification of this before you purchase that new motherboard!

When interrogated with an Identify Drive command, a hard disk returns, among other things, information about the PIO and DMA modes it is capable of using. Most enhanced BIOSs will automatically set the correct mode to match the capabilities of the drive. If you set a mode faster than the drive can handle, data corruption will result.

ATA-2 drives also perform *Block Mode PIO*, which means that they use the Read/Write Multiple commands that greatly reduce the number of interrupts sent to host processor. This lowers the overhead, and the resulting transfers are even faster.

DMA Transfer Modes Although it is not used by most operating system or BIOS software, ATA-2 drives also support Direct Memory Access transfers. DMA means that the data is transferred directly between drive and memory without using the CPU as an intermediary, as opposed to PIO.

There are two distinct types of direct memory access: *DMA* and *busmastering DMA*. Ordinary DMA relies on the DMA controller on the system's mainboard to perform the complex task of arbitration, grabbing the system bus, and transferring the data. In the case of busmastering DMA, all this is done by logic on the interface card itself. Of course, this adds considerably to the complexity and the price of a busmastering interface.

Unfortunately, the DMA controller on ISA systems is ancient and slow, and out of the question for use with a modern hard disk. Today, proper software support for DMA is still rare.

The newest development in DMA transfer modes is *Ultra-DMA* which can speed data transfer up to 33M/sec with an Ultra-DMA drive.

Many of these latest IDE versions require not only correct drive and BIOS support but chipset support as well. While there are many companies who make chipsets, Intel chipsets are found in most modern motherboards. The following table lists the recent Intel chipsets, which processors they are used with, and which recent IDE ATA extensions they support.

Chipset	Processor	ATA IDE Extensions Supported
430VX	Pentium	Bus Mastering
430TX	Pentium	Ultra DMA 33
430HX (Triton 2)	Pentium	Bus Mastering
430FX (Triton)	Pentium	Bus Mastering
430MX	Pentium	

Chipset	Processor	ATA IDE Extensions Supported (continued)
440FX	Pentium Pro or Pentium II	Bus Mastering
450 GX	Pentium Pro	*
450 KX	Pentium Pro	*
440LX	Pentium II	Bus Mastering and DMA 33

* Some motherboards with this chipset support bus mastering IDE

ATAPI (ATA Packet Interface) *ATAPI* is a standard designed for devices such as CD-ROMs and tape drives that plug into an ordinary ATA (IDE) connector. The principal advantage of ATAPI hardware is that it's cheap and works on your current adapter. For CD-ROMs, it has a somewhat lower CPU usage compared to proprietary adapters, but there's no performance gain otherwise. For tape drives, ATAPI has potential for superior performance and reliability compared to the popular "floppy" tape devices.

While ATAPI CD-ROMs use the hard disk interface, this does not mean that they look like an ordinary hard disk; to the contrary, from a software point of view they are a completely different kind of animal. They actually most closely resemble a SCSI device.

Caution

Intelligent caching controllers that are not ATAPI-aware will *not* work with these devices. This means that, at present, you cannot boot from an ATAPI CD-ROM and you still must load a driver to use it under DOS or Windows. Windows 95/NT has native ATAPI support, and the first ATAPI-aware BIOS that will even allow booting from an ATAPI CD-ROM are now available.

IDE Drive Configuration IDE drives can be both simple and troublesome to configure. Single-drive installations usually are very simple, with few if any special jumper settings to worry about. Multiple-drive configurations, however, can be a problem. Jumpers have to be set on both drives; the names, locations, and even functions of these jumpers can vary from drive to drive.

Because the CAM ATA IDE specification was ironed out only after many companies were already making and selling drives, many older IDE drives have problems in dual-drive installations, especially when the drives are from different manufacturers. In some cases, two particular drives may not function together at all. Fortunately, most of the newer drives follow the CAM ATA specification, which clears up this problem. Drives that follow the specification have no problems in dual-drive installations.

Cable Configuration The cable connection to IDE drives usually is very simple. There is a single 40-pin cable that normally has three pin-header style connectors on it. One of the connectors plugs into the IDE interface connector; the other two plug into the primary and secondary drives. The cable normally runs from the IDE connector to both drives in a daisy-chain arrangement. On one end, this cable plugs into the IDE interface connector, which is located on the motherboard in many systems but also can be located on an IDE interface adapter card. The cable then connects to the secondary (D) and primary (C) drives in succession, with the primary drive usually (but not always) being at the end of the cable opposite the IDE interface connector.

There are no terminating resistors to set with IDE drives; instead, a distributed termination circuit is built into all IDE drives. The last drive on the cable need not be the primary drive, so you actually may find the primary or secondary drive at either connector. Jumpers on the drives themselves normally control whether a drive responds as primary or secondary.

> **Caution**
>
> You may see a different arrangement of cable connections in some IDE installations. In some installations, the middle connector is plugged into the motherboard, and the primary and secondary drives are at opposite ends of the cable in a Y arrangement. If you see this arrangement, be careful; in some of these Y-cable installations, the cable, rather than jumpers on the drives, actually controls which drive is primary and which is secondary.

Controlling master/slave selection via the cable rather than jumpers on the drive is performed via a special signal on the IDE interface called CSEL, which is on pin 28 of the interface. If the CSEL line is connected through from the drive to the IDE interface connector, the drive automatically is designated as primary. If the CSEL line is open between a drive and the IDE interface connector, that drive automatically is designated as secondary.

In the Y-cable approach, the IDE interface connector is in the middle of the cable, and a separate length of cable goes to each drive. Study this type of cable closely. If one of the ends of the Y has line 28 open (usually a hole in the cable through that wire), only the secondary drive can be plugged into that connector. HP Vectra PC systems use exactly this type of IDE cable arrangement. This type of setup eliminates the need to set jumpers on the IDE drives to configure them for primary or secondary operation, but the setup can be troublesome if you do not know about it.

IDE Drive Jumper Settings Configuring IDE drives can be simple, as is the case with most single-drive installations, or troublesome, especially where it comes to mixing two drives from different manufacturers on a single cable.

Most IDE drives come in three configurations:

- Single-drive (master)
- Master (dual-drive)
- Slave (dual-drive)

Because each IDE drive has its own controller, you must specifically tell one drive to be the master and the other to be the slave. There's no functional difference between the two, except that the drive that's specified as the slave will assert the DASP signal after a system reset that informs the master that a slave drive is present in the system. The master drive then pays attention to the Drive Select line, which it otherwise ignores. Telling a drive that it's the slave also usually causes it to delay its spinup for several seconds to allow the master to get going and thus to lessen the load on the system's power supply.

Until the ATA IDE specification, no common implementation for drive configuration was in use. Some drive companies even used different master/slave methods for different models of drives. Because of these incompatibilities, some drives work together only in a specific master/slave or slave/master order. This situation affects mostly older IDE drives that were introduced before the ATA specification.

Most drives that fully follow the ATA specification now need only one jumper (Master/Slave) for configuration. A few also need a Slave Present jumper as well. Table 2.2 shows the jumper settings required by most ATA IDE drives.

Table 2.2 Jumper Settings for Most ATA IDE-Compatible Drives.

Jumper Name	Single-Drive	Dual-Drive Master	Dual-Drive Slave
Master (M/S)	On	On	Off
Slave Present (SP)	Off	On	Off

The Master jumper indicates that the drive is a master or a slave. Some drives also require a Slave Present jumper, which is used only in a dual-drive setup and then installed only on the master drive, which is somewhat confusing. This jumper tells the master that a slave drive is attached. With many ATA IDE drives, the Master jumper is optional and may be left off. Installing this jumper doesn't hurt in these cases and may eliminate confusion, so I recommend that you install the jumpers listed here.

Conner Peripherals Drives Because they were introduced before the ATA IDE specification was formalized, Conner Peripherals drives often are different in configuration from many other-brand drives. When you mix and match IDE hard drives from different manufacturers, the drives are not always fully compatible. Table 2.3 shows the jumper settings that are correct for most Conner IDE drive installations.

Table 2.3 Jumper Settings for Conner Peripherals IDE Drives.

Jumper Name	Single Drive	Dual-Drive Master	Dual-Drive Slave
Master or Slave (C/D)	On	On	Off
Drive Slave Present (DSP)	Off	On	Off
Host Slave Present (HSP)	Off	Off	On
Drive Active (ACT)	On	On	Off

The C/D jumper is used to determine whether the drive is a master (drive C) or a slave (drive D). The drive is configured as master when this jumper is on. The DSP jumper indicates that a slave drive is present. The HSP jumper causes the drive to send the Slave Present signal to the master drive. The ACT jumper enables the master drive to signal when it is active.

Some Conner drives are not set up to support the industry-standard CAM ATA interface by default. The problems show up when you attempt to connect another manufacturer's drive to some Conner drives in either a master or slave role. Fortunately, you can correct many of these situations by changing the configuration of the drive.

You can make this change in two ways. One way is to use a special program to semipermanently change the mode of the drive. A special file available on the Conner BBS, called FEATURE.COM, contains a program that displays the current ISA/ATACAM setting and allows the setting to be changed. The change actually is stored in a feature byte in the firmware of the drive, and after this byte is changed, most other manufacturers' drives will work with the Conner drives. The program also can be used to reset the feature byte to its original configuration, which is best when you are connecting to other Conner drives.

The second method for changing this configuration is available on some Conner drives. These drives also have a special jumper called ATA/ISA. This jumper almost always should be installed in the ATA position to provide compatibility with the ATA standard. If you are using only Conner drives, you can leave this jumper in ISA mode if you want. Some Conner drives have a separate jumper (E1) that can delay startup of the drive to minimize the load on the power supply. This jumper should be enabled on any drive that is configured as a slave. Most other drives automatically delay startup of the slave drive for a few seconds.

Most Conner drives also have a special 12-pin connector that is used to drive an optional LED (pin 1, LED +5v; and pin 2, ground), as well as to connect to special factory equipment for low-level formatting and configuration.

Introduction to SCSI

SCSI (pronounced "scuzzy") stands for *Small Computer System Interface*. This interface has its roots in *SASI*, the *Shugart Associates System Interface*. SCSI is not a disk interface, but a systems-level interface. SCSI is not a type of controller, but a bus that supports as many as eight devices. One of these devices, the host adapter, functions as the gateway between the SCSI bus and the PC system bus. The SCSI bus itself does not talk directly with devices such as hard disks; instead, it talks to the controller that is built into the drive.

A single SCSI bus can support as many as eight physical units, usually called *SCSI IDs*. One of these units is the adapter card in your PC; the other seven can be other peripherals. You can have hard disks, tape drives, CD-ROM drives, a graphics scanner, or other devices (up to seven total) attached to a single SCSI host adapter. Most systems support up to four host adapters, each with seven devices, for a total 28 devices! Some of the newer SCSI implementations allow for 15 devices on each bus.

When you purchase a SCSI hard disk, you usually are purchasing the drive, controller, and SCSI adapter in one circuit. This type of drive usually is called an *embedded SCSI drive*; the SCSI interface is built into the drive. Most SCSI hard drives actually are IDE drives with SCSI bus adapter circuits added. You do not need to know what type of controller is inside the SCSI drive, because your system cannot talk directly to the controller as though it were plugged into the system bus, like a standard controller. Instead, communications go through the SCSI host adapter installed in the system bus. You can access the drive only with the SCSI protocols.

SCSI is a standard, in much the same way that RS-232 is a standard. The SCSI standard (like the RS-232 standard), however, defines only the hardware connections, not the driver specifications required to communicate with the devices. Software ties the SCSI subsystem into your PC, but unfortunately, most of the driver programs work only for a specific device and a specific host adapter. For example, a graphics scanner comes with its own SCSI host adapter to connect to the system; a CD-ROM drive comes with another (different) SCSI host adapter and driver software that works only with that SCSI adapter. On a system with those two SCSI adapters, you would need a third SCSI host adapter to run SCSI hard disk drives, because the host adapters supplied by the scanner and CD-ROM companies do not include a built-in, self-booting BIOS that supports hard disk drives.

SCSI has become something of a mess in the IBM world because of the lack of a host-adapter standard, a software interface standard, and standard ROM BIOS support for hard disk drives attached to the SCSI bus. Fortunately, some simple recommendations can keep you from living this compatibility nightmare.

In the beginning, SCSI lacked the capability to run hard disks off the SCSI bus. To boot from these drives and use a variety of operating systems was a problem that resulted from the lack of an interface standard. The standard IBM XT and AT ROM BIOS software was designed to talk to ST-506/412 hard disk controllers. The software easily was modified to work with ESDI because ESDI controllers are similar to ST-506/412 controllers at the register level. (This similarity at the register level enabled manufacturers to easily design self-booting, ROM-BIOS-supported ESDI drives.) The same can be said of IDE, which completely emulates the WD1003 ST-506/412 controller interface and works perfectly with the existing BIOS as well. SCSI is so different from these other standard disk interfaces that a new set of ROM BIOS routines is necessary to support the system so it can self-boot. The newer IBM PS/2 systems that come with SCSI drives have this support built into the motherboard BIOS or as an extension BIOS on the SCSI host adapter.

Companies such as Adaptec and Future Domain have produced SCSI cards with built-in ROM BIOS support for several years, but these BIOS routines were limited to running the drives only under DOS. The BIOS would not run in the AT-protected mode, and other operating systems included drivers for only the standard ST-506/412 and ESDI controllers. Thus, running SCSI was impossible under many non-DOS operating systems. This situation has changed significantly, however; IBM now supports many third-party SCSI host adapters in OS/2, especially those from Adaptec and Future Domain. For compatibility reasons, I usually recommend using SCSI adapters from these two companies, or any other adapters that are fully hardware-compatible with the Adaptec and Future Domain adapters.

ANSI SCSI standards The original SCSI standard (ANSI X3.131-1986) was approved in 1986, SCSI-2 was approved in January of 1994, and a new revision called SCSI-3 is being developed. One problem with the original SCSI-1 document was that many of the commands and features were optional, and there was little or no guarantee that a particular peripheral would support the expected commands. This problem caused the industry as a whole to define a set of 18 basic SCSI commands called the *Common Command Set (CCS)*, which would become the minimum set of commands supported by all peripherals. CCS became the basis for what is now the SCSI-2 specification.

In addition to formal support for CCS, SCSI-2 provided additional definitions for commands to access CD-ROM drives (and their sound capabilities), tape drives, removable drives, optical drives, and several other peripherals. In addition, an optional higher speed called FAST SCSI-2 and a 16-bit version called WIDE SCSI-2 were defined. Another feature of SCSI-2 is *command queuing*, which enables a device to accept multiple commands and execute them in the order that the device deems to be most efficient. This feature is most beneficial when you are using a multitasking operating system that could be sending several requests on the SCSI bus at the same time.

Most companies indicate that their host adapters follow both the ANSI X3.131-1986 (SCSI-1) as well as the x3.131-1994 (SCSI-2) standards. Note that because virtually all parts of SCSI-1 are supported in SCSI-2, virtually any SCSI-1 device is also considered SCSI-2 by default. Many manufacturers advertise that their devices are SCSI-2, but this does not mean that they support any of the additional optional features that were incorporated in the SCSI-2 revision.

For example, an optional part of the SCSI-2 specification includes a fast synchronous mode that doubles the standard synchronous transfer rate from 5MBps to 10MBps. This Fast SCSI transfer mode can be combined with 16-bit Wide SCSI for transfer rates of up to 20MBps. There was an optional 32-bit version defined in SCSI-2, but component manufacturers have shunned this as too expensive. In essence, 32-bit SCSI was a stillborn specification. Most SCSI implementations are 8-bit standard SCSI or Fast/Wide SCSI. Even devices which support none of the Fast or Wide modes can still be considered SCSI-2.

The SCSI-3 standard is still being defined and is still a long way off from being approved. However, portions of this specification, although not final, are being sold in products today. One of these developments is the new Fast-20 mode, which is also called *Ultra-SCSI*. This essentially is quad-speed SCSI, and will run 20MBps on an 8-bit standard SCSI bus, and 40MBps on Wide (16-bit) SCSI.

Table 2.4 shows the maximum transfer rates for the SCSI bus at various speeds and widths, as well as the cable type required for the specific transfer widths.

Table 2.4 SCSI Data-Transfer Rates.

Bus Width	Standard SCSI	Fast SCSI	Fast-20 (Ultra) SCSI	Cable Type
8-bit	5MBps	10MBps	20MBps	A (50-pin)
16-bit (Wide)	10MBps	20MBps	40MBps	P (68-pin)

Note

The A cable is the standard 50-pin SCSI cable, whereas the P cable is a 68-pin cable designed for 16-bit. Maximum cable length is 6m (about 20 feet) for standard speed SCSI, and only 3m (about 10 feet) for Fast or Fast-20 (Ultra) SCSI.

So-called SCSI-1 adapters have no problems with SCSI-2 peripherals. In fact, as was stated earlier, virtually any SCSI-1 device can also legitimately be called SCSI-2 (or even SCSI-3). You can't take advantage of Fast, Fast-20, or Wide transfer capabilities, but the extra commands defined in SCSI-2 can be sent by means of a SCSI-1 controller. In other words, nothing is different between SCSI-1 and SCSI-2 compliant hardware. For example, I am running a Seagate Barracuda 4GB Fast SCSI-2 drive with my standard IBM SCSI-1 host adapter, and it runs fine. Most adapters are similar, in that they actually are SCSI-2 compatible, even if they advertise only SCSI-1 support.

Because the SCSI-2 standard was not actually approved before January 1994, any devices that claimed to be SCSI-2 before that time were not officially in compliance with the standard. This is really not a problem, however, because the SCSI-2 document had not changed appreciably since it was nearly approved in 1990. The same thing is currently happening with advertisers listing devices as "SCSI-3." The SCSI-3 specification is not yet approved, although certain areas are being worked out.

SCSI Hard Disk Evolution and Construction SCSI is not a disk interface, but a bus that supports SCSI bus interface adapters connected to disk and other device controllers. The first SCSI drives for PCs simply were standard ST-506/412 or ESDI drives with a separate SCSI bus interface adapter (sometimes called a *bridge controller*) that converted the ST-506/412 or ESDI interfaces to SCSI. This interface originally was in the form of a secondary logic board, and the entire assembly often was mounted in an external case.

The next step was to build the SCSI bus interface "converter" board directly into the drive's own logic board. Today, we call these drives *embedded SCSI drives*, because the SCSI interface is built in.

At that point, there was no need to conform to the absolute specifications of ST-506/412 or ESDI on the internal disk interface, because the only other device that the interface ever would have to talk to was built in as well. Thus, the disk-interface and controller-chipset manufacturers began to develop more customized chipsets that were based on the ST-506/412 or ESDI chipsets already available but offered more features and higher performance. Today, if you look at a typical SCSI drive, you often can identify the chip or chipset that serves as the disk controller on the drive as being exactly the same kind that would be used on an ST-506/412 or ESDI controller or as some evolutionary customized variation thereof.

If you follow the hard drive market, you usually will see that drive manufacturers offer most of their newer drives in both ATA-IDE and SCSI versions. In other words, if a manufacturer makes a particular 500MB IDE drive, you invariably will see that the company also make a SCSI model with the same capacity and specifications, which uses the same HDA (Head Disk Assembly) and even looks the same as the IDE version. If you study these virtually identical drives, the only major difference you will find is the additional chip on the logic board of the SCSI version, called a *SCSI Bus Adapter Chip (SBIC)*.

Most drive manufacturers design their ATA-IDE and SCSI drives the same way, often using the very same Western Digital chips as well as disk controller and SCSI bus interface chips from other manufacturers. You now should be able to understand that most SCSI drives simply are "regular" ATA-IDE drives with SCSI bus logic added. This fact will come up again later in this chapter in the section "SCSI versus IDE," which discusses performance and other issues differentiating these interfaces.

Now think about this from a performance standpoint. If virtually all SCSI drives really are ATA-IDE drives with a SCSI Bus Interface Controller chip added, what conclusions can you draw?

First, no drive can perform sustained data transfers faster than the data can actually be read from the disk platters. In other words, the HDA limits performance to whatever it is capable of achieving. Drives can transmit data in short bursts at very high speeds, because they often have built-in cache or read-ahead buffers that store data. Many of the newer high-performance SCSI and ATA-IDE drives have 1MB or more of cache memory on-board! No matter how big or intelligent the cache is, however, sustained data transfer still will be limited by the HDA.

Data from the HDA must pass through the disk controller circuits, which, as you have seen, are virtually identical between similar SCSI and ATA-IDE drives. In the ATA-IDE drive, this data then is presented directly to the system bus. In the SCSI drive, however, the data must pass through a SCSI Bus Interface adapter on the drive, travel through the SCSI bus itself, and then pass through another SCSI Bus Interface controller in the SCSI host adapter card in your system. The longer route that a

SCSI transfer must take makes this type of transfer slower than the much more direct ATA-IDE transfer.

The conventional wisdom has been that SCSI always is much faster than IDE; unfortunately, this wisdom usually is wrong! This incorrect conclusion was derived by looking at the raw SCSI and ISA bus performance capabilities. An 8-bit Fast SCSI-2 bus can transfer data at 10MBps , whereas the 16-bit ISA bus used directly by IDE drives can transfer data at rates ranging from 2MB to 8MBps. Based on these raw transfer rates, SCSI seems to be faster, but the raw transfer rate of the bus is not the limiting factor. Instead, the actual HDA and disk-controller circuitry place the limits on performance. Another point to remember is that unless you are using a PCI, VL-Bus, EISA, or 32-bit MCA SCSI adapter, the SCSI data-transfer speeds will be limited by the host bus performance as well as by the drive performance.

However, modern operating systems are multitasking, and SCSI devices (with all their additional controller circuitry) function independent of each other, unlike IDE. Therefore, data can be read and written to any of the SCSI devices *simultaneously*. This allows for smoother multitasking and increased overall data throughput. The most advanced operating systems, like Windows NT, even allow drive striping. A *strip drive set* is two or more drives that appear to the user as one drive. Data is split between the drives equally, again increasing overall throughput.

Single-Ended or Differential SCSI "Normal" SCSI also is called *single-ended SCSI*. For each signal that needs to be sent across the bus, a wire exists to carry it. With differential SCSI, for each signal that needs to be sent across the bus, a pair of wires exists to carry it. The first in this pair carries the same type of signal that the single-ended SCSI carries. The second in this pair, however, carries the logical inversion of the signal. The receiving device takes the difference of the pair (hence the name *differential*), which makes it less susceptible to noise and allows for greater cable length. Because of this, differential SCSI can be used with cable lengths up to 25m, whereas single-ended SCSI is good only for 6m with standard asynchronous or synchronous transfers or for only 3m for Fast SCSI.

You cannot mix single-ended and differential devices on a single SCSI bus; the result would be catastrophic. (That is to say, you probably will see smoke!) Notice that the cables and connectors are the same, so it's entirely possible to make this mistake. This usually is not a problem, however, because very few differential SCSI implementations exist. Especially with SCSI in the PC environment, single-ended is about all you will ever see. If, however, you to come upon a peripheral that you believe might be differential, there are a few ways to tell. One way is to look for a special symbol on the unit; the industry has adopted different universal symbols for single-ended and differential SCSI. Figure 2-1 shows these symbols.

If you do not see such symbols, you can tell whether you have a differential device by using an ohmmeter to check the resistance between pins 21 and 22 on the device connector. On a single-ended system, the pins should be tied together and also tied to the ground. On a differential device, the pins should be open or have significant resistance between them. Again, this generally should not be a problem, because virtually all devices used in the PC environment are single-ended.

SCSI-1 and SCSI-2 The SCSI-2 specification essentially is an improved version of SCSI-1 with some parts of the specification tightened and with several new features and options added. Normally, SCSI-1 and SCSI-2 devices are compatible, but SCSI-1 devices ignore the additional features in SCSI-2.

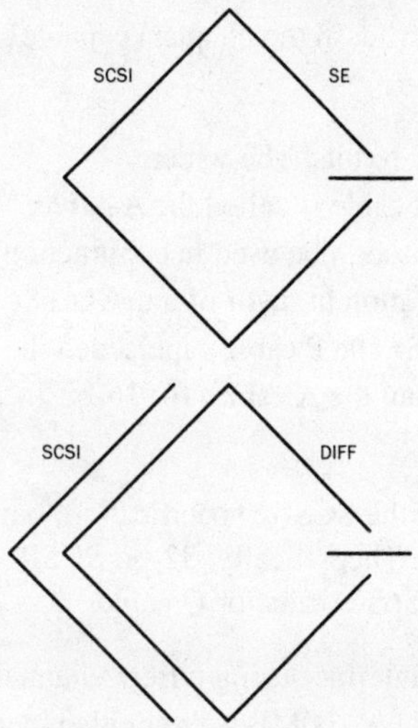

FIG. 2-1 Single-ended and differential SCSI universal symbols.

Some of the changes in SCSI-2 are very minor. For example, SCSI-1 allowed SCSI Bus parity to be optional, whereas parity must be implemented in SCSI-2. Another requirement is that initiator devices, such as host adapters, provide terminator power to the interface; most devices already did so.

SCSI-2 also has several optional features:

- Fast SCSI
- Wide SCSI
- Command queuing
- High-density cable connectors
- Improved Active (Alternative 2) termination

These features are not required; they are optional under the SCSI-2 specification. If you connect a standard SCSI host adapter to a Fast SCSI drive, for example, the interface will work, but only at standard SCSI speeds.

SCSI-3 *SCSI-3* is a term used to describe a set of standards currently being developed. Simply put, it is the next generation of documents a product conforms to. (See the "New Commands" section later in this chapter.)

Fast and Fast-Wide SCSI *Fast SCSI* refers to high-speed synchronous transfer capability. Fast SCSI achieves a 10MBps transfer rate on the standard 8-bit SCSI cabling. When combined with a 16-bit Wide SCSI interface, this configuration results in data-transfer rates of 20MBps (called *Fast/Wide*).

Fast-20 (Ultra) SCSI *Fast-20* or *Ultra SCSI* refers to high-speed synchronous transfer capability that is twice as fast as Fast-SCSI. This has been introduced in the Draft (unfinished) SCSI-3 specification and has already been adopted by the marketplace, especially for high-speed hard disks. Ultra SCSI achieves a 20MBps transfer rate on the standard 8-bit SCSI cabling. When combined with a 16-bit Wide SCSI interface, this configuration results in data-transfer rates of 40MBps (called *Ultra/Wide*).

Fast-40 SCSI Fast-40 SCSI is a future revision of SCSI-3 (mentioned earlier in the chapter) capable of achieving a 40MBps transfer rate.

Wide SCSI *Wide SCSI* allows for parallel data transfer at a bus width of 16 bits. The wider connection requires a new cable design. The standard 50-conductor 8-bit cable is called the *A cable.* SCSI-2 originally defined a special 68-conductor B cable that was supposed to be used in conjunction with the A cable for wide transfers, but the industry ignored this specification in favor of a newer 68-conductor P cable that was introduced as part of the SCSI-3 specification. The P cable superseded the A and B cable combination because the P cable can be used alone (without the A cable) for 16-bit Wide SCSI.

A 32-bit Wide SCSI version was originally defined on paper as a part of the SCSI-2 specification, but has not found popularity and probably never will in the PC environment. Theoretically, 32-bit SCSI implementations would require two cables: a 68-conductor P cable and a 68-conductor Q cable.

Fiber Channel SCSI Fiber Channel SCSI is a specification for a serial interface using a fiber channel physical and protocol characteristic, with SCSI command set. It can achieve 100MBps over either fiber or coaxial cable.

Termination The single-ended SCSI bus depends on very tight termination tolerances to function reliably. Unfortunately, the original 132-ohm passive termination defined in the SCSI-1 document was not designed for use at the higher synchronous speeds now possible. These passive terminators can cause signal reflections to cause errors when transfer rates increase or when more devices are added to the bus. SCSI-2 defines an active (voltage-regulated) terminator that lowers termination impedance to 110 ohms and improves system integrity.

Command Queuing In SCSI-1, an initiator device, such as a host adapter, was limited to sending one command per device. In SCSI-2, the host adapter can send as many as 256 commands to a single device, which will store and process those commands internally before responding on the SCSI bus. The target device even can resequence the commands to allow for the most efficient execution or performance possible. This feature is especially useful in multitasking environments, such as OS/2 and Windows NT, that can take advantage of this feature.

New Commands SCSI-2 took the Common Command Set that was being used throughout the industry and made it an official part of the standard. The CCS was designed mainly for disk drives and did not include specific commands designed for other types of devices. In SCSI-2, many of the old commands are reworked, and several new commands have been added. New command sets have been added for CD-ROMs, optical drives, scanners, communications devices, and media changers (jukeboxes).

SCSI-3 Even though the SCSI-2 specification has only recently been approved (although it has remained stable for some time), the SCSI-3 specification is already being developed. SCSI-3 will have everything that SCSI-2 has and definitely will add new commands, features, and implementations. For example, SCSI-3 will provide support for up to 32 devices on the bus instead of only eight.

One of the most exciting things about SCSI-3 is the proposed *Serial SCSI,* a scheme that may use only a six-conductor cable and that will be able to transfer data at up to 100MBps. The switch to serial instead of parallel is designed to control the delay, noise, and termination problems that have plagued SCSI-2, as well as to simplify the cable connection. Serial SCSI will be capable of transferring more data over six wires than 32-bit Fast Wide SCSI-2 can over 128 wires. The intention is that Serial SCSI be implemented on the motherboard of future systems, giving them incredible expansion and performance capabilities.

Although Serial SCSI may not make the older host adapters and cables obsolete overnight, it does make future cabling possibilities even more of a puzzle. Serial SCSI offers the possibility of longer cable lengths, less electromagnetic interference, and easier connections on laptops, notebooks, and docking stations. Expect SCSI-3 to offer almost pain-free installations with automatic Plug and Play SCSI ID setup and termination schemes.

In any practical sense, SCSI-3 is still some ways away from being approved. Because the standard exists in draft documents before being officially approved, if the portions of the standard become stable, we may very well see products claiming SCSI-3 compatibility well before the standard truly exists. Because SCSI-3 actually incorporates all of what is in SCSI-2, technically anybody can call any SCSI-1 or SCSI-2 device a SCSI-3 device as well. Beware of product hype along these lines. Some of the new SCSI-3 features will likely be incompatible with previous SCSI implementations, and may take a while to appear on the market.

SCSI Cables and Connectors The SCSI standards are very specific when it comes to cables and connectors. The most common connectors specified in this standard are the 50-position unshielded pin header connector for internal SCSI connections and the 50-position shielded Centronics latch-style connectors for external connections. The shielded Centronics style connector also is called *Alternative 2* in the official specification. Passive or Active termination (Active is preferred) is specified for both single-ended and differential buses. The 50-conductor bus configuration is defined in the SCSI-2 standard as the A-cabled.

The SCSI-2 revision added a high-density, 50-position, D-shell connector option for the A-cable connectors. This connector now is called *Alternative 1*. The Alternative 2 Centronics latch-style connector remains unchanged from SCSI-1. A 68-conductor B-cable specification was added to the SCSI-2 standard to provide for 16- and 32-bit data transfers; the connector, however, had to be used in parallel with an A cable. The industry did not widely accept the B cable option, which has been dropped from the SCSI-3 standard.

To replace the ill-fated B cable, a new 68-conductor P cable was developed as part of the SCSI-3 specification. Shielded and unshielded high-density D-shell connectors are specified for both the A cable and P cable. The shielded high-density connectors use a squeeze-to-release latch rather than the wire latch used on the Centronics-style connectors. Active termination for single-ended buses is specified, providing a high level of signal integrity.

SCSI Cable and Connector Pinouts There are two electrically different version of SCSI, *Single Ended* and *Differential*. These two versions are electrically incompatible, and must not be interconnected or damage will result. Fortunately, there are very few Differential SCSI applications available in the PC industry, so you will rarely (if ever) encounter it. Within each electrical type (Single Ended or Differential), there are basically three SCSI cable types:

- A Cable (Standard SCSI)
- P Cable (16- and 32-bit Wide SCSI)

- Q Cable (32-bit Wide SCSI)

The A cable is used in most SCSI-1 and SCSI-2 installations, and is the most common cable you will encounter. SCSI-2 Wide (16-bit) applications use a P cable instead, which completely replaces the A cable. You can intermix standard and Wide SCSI devices on a single SCSI bus by interconnecting A and P cables with special adapters. 32-bit wide SCSI-3 applications use both the P and Q cables in parallel to each 32-bit device. Today there are virtually no PC applications for 32-bit Wide SCSI-3, and because of the two cable requirement, it is not likely to catch on.

The A cables can have Pin Header (Internal) type connectors or External Shielded connectors, each with a different pinout. The P and Q cables feature the same connector pinout on either Internal or External cable connections.

Single-Ended SCSI Cables and Connectors The single-ended electrical interface is the most popular type for PC systems. The A cable is available in both internal unshielded as well as external shielded configurations. The pinouts for these are shown in Tables 2.5 and 2.6.

Table 2.5 A Cable (Single-Ended) Internal Unshielded Header Connector.

Signal Name	Pin	Pin	Signal Name
GROUND	1	2	-DB(0)
GROUND	3	4	-DB(1)
GROUND	5	6	-DB(2)
GROUND	7	8	-DB(3)
GROUND	9	10	-DB(4)
GROUND	11	12	-DB(5)
GROUND	13	14	-DB(6)
GROUND	15	16	-DB(7)
GROUND	17	18	-DB(Parity)
GROUND	19	20	GROUND
GROUND	21	22	GROUND
RESERVED	23	24	RESERVED
Open	25	26	TERMPWR
RESERVED	27	28	RESERVED
GROUND	29	30	GROUND
GROUND	31	32	-ATN
GROUND	33	34	GROUND
GROUND	35	36	-BSY
GROUND	37	38	-ACK
GROUND	39	40	-RST
GROUND	41	42	-MSG

Table 2.5 A Cable (Single-Ended) Internal Unshielded Header Connector (continued).

Signal Name	Pin	Pin	Signal Name
GROUND	43	44	-SEL
GROUND	45	46	-C/D
GROUND	47	48	-REQ
GROUND	49	50	-I/O

Table 2.6 A-Cable (Single-Ended) External Shielded Connector.

Signal Name	Pin	Pin	Signal Name
GROUND	1	26	-DB(0)
GROUND	2	27	-DB(1)
GROUND	3	28	-DB(2)
GROUND	4	29	-DB(3)
GROUND	5	30	-DB(4)
GROUND	6	31	-DB(5)
GROUND	7	32	-DB(6)
GROUND	8	33	-DB(7)
GROUND	9	34	-DB(Parity)
GROUND	10	35	GROUND
GROUND	11	36	GROUND
RESERVED	12	37	RESERVED
Open	13	38	TERMPWR
RESERVED	14	39	RESERVED
GROUND	15	40	GROUND
GROUND	16	41	-ATN
GROUND	17	42	GROUND
GROUND	18	43	-BSY
GROUND	19	44	-ACK
GROUND	20	45	-RST
GROUND	21	46	-MSG
GROUND	22	47	-SEL
GROUND	23	48	-C/D
GROUND	24	49	-REQ
GROUND	25	50	-I/O

IBM has standardized on the SCSI interface for virtually all PS/2 systems introduced since 1990. These systems use a Micro-Channel SCSI adapter or have the SCSI Host Adapter built into the motherboard. In either case, IBM's SCSI interface uses a special 60-pin mini-Centronics type external shielded connector that is unique in the industry. A special IBM cable is required to adapt this connector to the standard 50-pin Centronics style connector used on most external SCSI devices. The pinout of the IBM 60-pin mini-Centronics style External Shielded connector is shown Table 2.7. Notice that although the pin arrangement is unique, the pin number to signal designations correspond with the standard unshielded internal pin header type of SCSI connector.

Table 2.7 IBM PS/2 SCSI External Shielded 60-Pin Connector.

Signal Name	Pin	Pin	Signal Name
GROUND	1	60	Not Connected
-DB(0)	2	59	Not Connected
GROUND	3	58	Not Connected
-DB(1)	4	57	Not Connected
GROUND	5	56	Not Connected
-DB(2)	6	55	Not Connected
GROUND	7	54	Not Connected
-DB(3)	8	53	Not Connected
GROUND	9	52	Not Connected
-DB(4)	10	51	GROUND
GROUND	11	50	-I/O
-DB(5)	12	49	GROUND
GROUND	13	48	-REQ
-DB(6)	14	47	GROUND
GROUND	15	46	-C/D
-DB(7)	16	45	GROUND
GROUND	17	44	-SEL
-DB(Parity)	18	43	GROUND
GROUND	19	42	-MSG
GROUND	20	41	GROUND
GROUND	21	40	-RST
GROUND	22	39	GROUND
RESERVED	23	38	-ACK
RESERVED	24	37	GROUND
Open	25	36	-BSY
TERMPWR	26	35	GROUND
RESERVED	27	34	GROUND
RESERVED	28	33	GROUND
GROUND	29	32	-ATN
GROUND	30	31	GROUND

The P cable (Single-Ended) and connectors are used in 16-bit wide SCSI-2 applications (see table 2.8 for the pinout).

Table 2.8 P Cable (Single-Ended) Internal or External Shielded Connector.

Signal Name	Pin	Pin	Signal Name
GROUND	1	35	-DB(12)
GROUND	2	36	-DB(13)
GROUND	3	37	-DB(14)
GROUND	4	38	-DB(15)
GROUND	5	39	-DB(Parity 1)
GROUND	6	40	-DB(0)
GROUND	7	41	-DB(1)
GROUND	8	42	-DB(2)
GROUND	9	43	-DB(3)
GROUND	10	44	-DB(4)
GROUND	11	45	-DB(5)
GROUND	12	46	-DB(6)
GROUND	13	47	-DB(7)
GROUND	14	48	-DB(Parity 0)
GROUND	15	49	GROUND
GROUND	16	50	GROUND
TERMPWR	17	51	TERMPWR
TERMPWR	18	52	TERMPWR
RESERVED	19	53	RESERVED
GROUND	20	54	GROUND
GROUND	21	55	-ATN
GROUND	22	56	GROUND
GROUND	23	57	-BSY
GROUND	24	58	-ACK
GROUND	25	59	-RST
GROUND	26	60	-MSG
GROUND	27	61	-SEL
GROUND	28	62	-C/D
GROUND	29	63	-REQ
GROUND	30	64	-I/O
GROUND	31	65	-DB(8)
GROUND	32	66	-DB(9)
GROUND	33	67	-DB(10)
GROUND	34	68	-DB(11)

The Q Cable (Single-Ended) and connector is defined only for 32-bit SCSI implementations, which also require a P able as well. 32-bit SCSI applications are rare to virtually nonexistent.

Differential SCSI Differential SCSI is not normally used in a PC environment, but is very popular with minicomputer installations due to the very long bus lengths that are allowed.

The A cable (Differential) connector is available in both internal unshielded form as well as an external shielded form.

The P able (Differential) and Connector is used for 16-bit wide SCSI connections.

The Q cable (Differential) and connector is used only with the proposed 32-bit wide SCSI implementations (which have not been implemented by the marketplace as of yet), and in that case would also require a 16-bit wide P-cable.

Termination All buses need to be electrically terminated at each end; the SCSI bus is no exception. Improper termination still is one of the most common problems in SCSI installations. Three types of terminators typically are available for the SCSI bus:

- Passive
- Active (also called Alternative 2)
- Forced Perfect Termination (FPT): FPT-3, FPT-18, and FPT-27

Typical passive terminators (a network of resistors) allow signal fluctuations in relation to the terminator power signal on the bus. Usually, passive terminating resistors suffice over short distances, such as 2 or 3 feet, but for longer distances, active termination is a real advantage. Active termination is required with Fast SCSI.

An active terminator actually has one or more voltage regulators to produce the termination voltage, rather than resistor voltage dividers. This arrangement helps ensure that the SCSI signals always are terminated to the correct voltage level. Active terminators will usually have some sort of LED indicating the termination activity. The SCSI-2 specification recommends active termination on both ends of the bus and requires active termination whenever Fast or Wide SCSI devices are used. Most high-performance host adapters have an " auto-termination" feature so if it is the end of a chain, it will terminate itself.

A variation on active termination is available: *Forced Perfect Termination.* Forced Perfect Termination is an even better form of active termination, in which diode clamps are added to eliminate signal overshoot and undershoot. The trick is that instead of clamping to +5 and Ground, these terminators clamp to the output of two regulated voltages. This arrangement enables the clamping diodes to eliminate signal overshoot and undershoot, especially at higher signaling speeds and over longer distances.

FPT terminators are available in several versions. FPT-3 and FPT-18 versions are available for 8-bit standard SCSI, while the FPT-27 is available for 16-bit (Wide) SCSI. The FPT-3 version forces perfect the three most highly active SCSI signals on the 8-bit SCSI bus, while the FPT-18 forces perfect all the SCSI signals on the 8-bit bus except grounds. FPT-27 also forces perfect all of the 16-bit Wide SCSI signals except grounds.

> **Note**
>
> Several companies make high-quality terminators for the SCSI bus, including Aeronics and the Data Mate division of Methode. Both of these companies make a variety of terminators, but Aeronics is well-noted for some unique FPT versions that are especially suited to problem configurations that require longer cable runs or higher signal integrity. One of the best investments that you can make in any SCSI installation is in high-quality cables and terminators.

SCSI Drive Configuration SCSI drives are not too difficult to configure, especially compared with IDE drives. The SCSI standard controls the way that the drives must be set up. You need to set two or three items when you configure an SCSI drive:

- SCSI ID setting (0–7)
- Terminating resistors

The SCSI ID setting is very simple. Up to eight SCSI devices can be used on a single SCSI bus, and each device must have a unique SCSI ID address. The host adapter takes one address, so the rest are free for up to seven SCSI peripherals. Most SCSI host adapters are factory-set to ID 7, which is the highest-priority ID. All other devices must have unique IDs that do not conflict with one another. Some host adapters boot only from a hard disk set to a specific ID. In my system, for example, the IBM SCSI host adapter requires the boot drive to be set to ID 6. Newer IBM host adapters and systems enable you to boot from a hard disk at any SCSI ID. Older Adaptec host adapters required the boot hard disk to be ID 0; newer ones can boot from any ID.

Setting the ID usually involves changing jumpers on the drive itself. If the drive is installed in an external chassis, the chassis may have an ID selector switch that is accessible at the rear. This selector makes ID selection a simple matter of pressing a button or rotating a wheel until the desired ID number appears. If no external selector is present, you must open the external device chassis and set the ID via the jumpers on the drive.

Three jumpers are required to set the SCSI ID; the particular ID selected actually is derived from the binary representation of the jumpers themselves. For example, setting all three ID jumpers off results in a binary number of 000b, which translates to an ID of 0. A binary setting of 001b equals ID 1, 010b equals 2, 011b equals 3, and so on. (Notice that as I list these values, I append a lowercase b to indicate binary numbers.)

Unfortunately, the jumpers can appear either forward or backward on the drive, depending on how the manufacturer set them up. To keep things simple, I have recorded all the different ID jumper settings in the following tables. Table 2.9 shows the settings for drives that order the jumpers with the Most Significant Bit (MSB) to the left; Table 2.10 shows the settings for drives that have the jumpers ordered so that the MSB is to the right.

Table 2.9 SCSI ID Jumper Settings with the Most Significant Bit to the Left.

SCSI	ID	Jumper	Settings
0	0	0	0

Table 2.9 SCSI ID Jumper Settings with the Most Significant Bit to the Left (continued).

SCSI	ID	Jumper	Settings
1	0	0	1
2	0	1	0
3	0	1	1
4	1	0	0
5	1	0	1
6	1	1	0
7	1	1	1

1 = Jumper On, 0 = Jumper Off

Table 2.10 SCSI ID Jumper Settings with the Most Significant Bit to the Right.

SCSI	ID	Jumper	Settings
0	0	0	0
1	1	0	0
2	0	1	0
3	1	1	0
4	0	0	1
5	1	0	1
6	0	1	1
7	1	1	1

1 = Jumper On, 0 = Jumper Off

SCSI termination is very simple. Termination is required at both ends of the bus; there are no exceptions. If the host adapter is at one end of the bus, it must have termination enabled. If the host adapter is in the middle of the bus, and if both internal and external bus links are present, the host adapter must have its termination disabled, and the devices at each end of the bus must have terminators installed. Several types of terminators are available, differing both in quality and in appearance. Active terminators are the minimum recommended, and Forced Perfect Terminators (FPT) are considered the best available. For more information on the different types, see the previous section on terminators.

The rules are simple: Use the best terminators possible, and make sure that only the ends of the SCSI bus are terminated. The majority of problems that I see with SCSI installations are the result of improper termination. Some devices have built-in termination resistors that are enabled or disabled through a jumper or by being physically removed. Other devices do not have built-in terminating resistors; these devices instead rely on external terminator modules for termination.

When installing an external SCSI device, you will usually find the device in a storage enclosure with both input and output SCSI connectors, so that you can use the device in a daisy chain. If the enclosure is at the end of the SCSI bus, an external terminator module most likely will have to be plugged into the second (outgoing) SCSI port to provide proper termination at that end of the bus (see Figure 2-2).

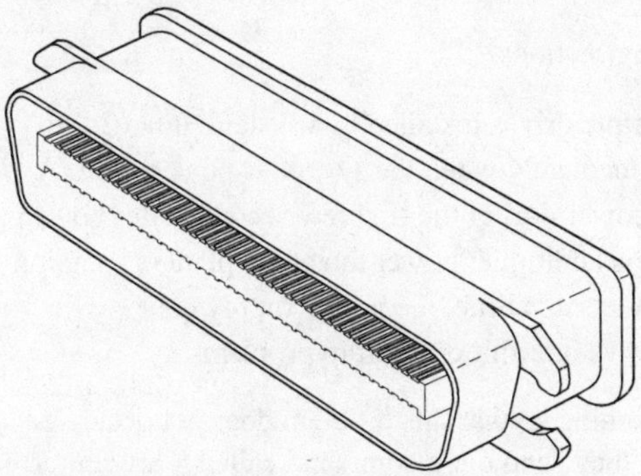

FIG. 2-2 External SCSI device terminator.

External terminator modules are available in a variety of connector configurations, including pass-through designs, which are needed if only one port is available. Pass-through terminators also are commonly used in internal installations in which the device does not have built-in terminating resistors. Many hard drives use pass-through terminators for internal installations to save space on the logic-board assembly (see Figure 2-3).

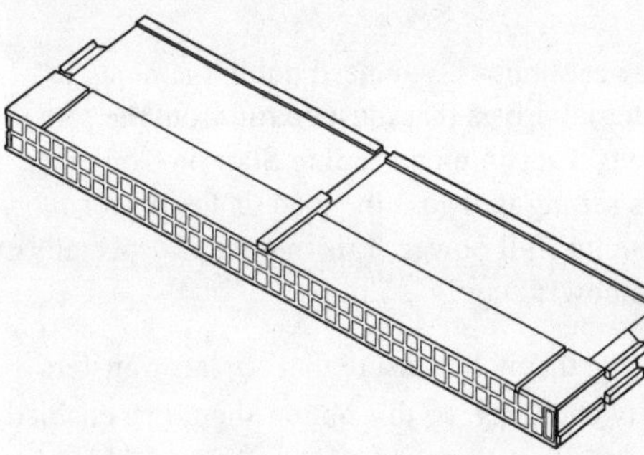

FIG. 2-3 Internal pin-header connector pass-through SCSI terminator.

The pass-through models are required when a device is at the end of the bus and only one SCSI connector is available.

> **Tip**
>
> Remember to stick with high-quality active or Forced Perfect terminators at each end of the bus, and you will eliminate most common termination problems.

Other configuration items on a SCSI drive can be set via jumpers. Following are several of the most common additional settings that you will find:

- Start on Command (delayed start)
- SCSI Parity

- Terminator Power
- Synchronous Negotiation

These configuration items are described in the following sections.

Start On Command (Delayed Start) If you have multiple drives installed in a system, it is wise to set them up so that all the drives do not start to spin immediately when the system is powered on. A hard disk drive can consume three or four times more power during the first few seconds after power-on than during normal operation. The motor requires this additional power to get the platters spinning quickly. If several drives are drawing all this power at the same time, the power supply may be overloaded, which can cause the system to hang or to have intermittent startup problems.

Nearly all SCSI drives provide a way to delay drive spinning so that this problem does not occur. When most SCSI host adapters initialize the SCSI bus, they send out a command called Start Unit to each of the ID addresses in succession. By setting a jumper on the hard disk, you can prevent the disk from spinning until it receives the Start Unit command from the host adapter. Because the host adapter sends this command to all the ID addresses in succession, from the highest-priority address (ID 7) to the lowest (ID 0), the higher-priority drives can be made to start first, with each lower-priority drive spinning up sequentially. Because some host adapters do not send the Start Unit command, some drives may simply delay spinup for a fixed number of seconds rather than wait for a command that never will arrive.

If drives are installed in external chassis with separate power supplies, you need not implement the delayed-start function. This function is best applied to internal drives that must be run from the same power supply that runs the system. For internal installations, I recommend setting Start on Command (delayed start) even if you have only one SCSI drive; this setting will ease the load on the power supply by spinning the drive up after the rest of the system has full power. This method is especially good for portable systems and other systems in which the power supply is limited.

SCSI Parity SCSI Parity is a limited form of error checking that helps ensure that all data transfers are reliable. Virtually all host adapters support SCSI parity checking, so this option should be enabled on every device. The only reason why it exists as an option is that some older host adapters do not work with SCSI parity, so the parity must be turned off.

Terminator Power The terminators at each end of the SCSI bus require power from at least one device on the bus. In most cases, the host adapter supplies this terminator power; in some cases, however, it does not. For example, parallel port SCSI host adapters typically do not supply terminator power. It is not a problem if more than one device supplies terminator power because each source is diode-protected. For simplicity's sake, many will configure all devices to supply terminator power. If no device supplies terminator power, the bus will not be terminated correctly and will not function properly.

SCSI Synchronous Negotiation The SCSI bus can run in two modes: *asynchronous* (the default) and *synchronous*. The bus actually switches modes during transfers through a protocol called *synchronous negotiation*. Before data is transferred across the SCSI bus, the sending device (called the *initiator*) and the receiving device (called the *target*) negotiate how the transfer will take place. If both devices support synchronous transfers, they will discover this fact through the negotiation, and the transfer will take place at the faster synchronous rate.

Unfortunately, some older devices do not respond to a request for synchronous transfer and can actually be disabled when such a request is made. For this reason, both host adapters and devices that support synchronous negotiation often have a jumper that can be used to disable this negotiation so that it can work with older devices. By default, all devices today should support synchronous negotiation, and this function should be enabled.

Plug and Play SCSI Plug and Play SCSI was originally released in April 1994. This specification allows SCSI device manufacturers to build Plug and Play peripherals that will automatically configure when used with a Plug and Play operating system. This will allow you to easily connect or reconfigure external peripherals, such as hard disk drives, backup tapes, and CD-ROMs.

To connect SCSI peripherals to the host PC, the specification requires a Plug and Play SCSI host adapter such as Plug and Play ISA or PCI. Plug and Play add-in cards enable a Plug and Play operating system to automatically configure software device drivers and system resources for the host bus interface.

The Plug and Play SCSI specification version 1.0 includes these technical highlights:

- A single cable-connector configuration
- Automatic termination of the SCSI bus
- SCAM (SCSI Configured Automagically) automatic ID assignment
- Full backward compatibility of Plug and Play SCSI devices with the installed base of SCSI systems.

This should go a long way in making SCSI easier to use for the normal user.

Each SCSI peripheral that you add to your SCSI bus (other than hard disk drives) requires an external driver to make the device work. Hard disks are the exception. Driver support for them normally is provided as part of the SCSI host adapter BIOS. These external drivers are specific not only to a particular device, but also to the host adapter.

Recently, two types of standard host adapter interface drivers have become popular, greatly reducing this problem. By having a standard host adapter driver to write to, peripheral makers can more quickly create new drivers that support their devices and then talk to the universal host adapter driver. This arrangement eliminates dependence on one particular type of host adapter. These primary or universal drivers link the host adapter and the operating system.

The *Advanced SCSI Programming Interface* (ASPI) currently is the most popular universal driver, with most peripheral makers writing their drivers to talk to ASPI. The "A" in ASPI used to stand for Adaptec, the company that introduced it, but other SCSI device vendors have licensed the right to use ASPI with their products. DOS does not support ASPI directly, but it does when the ASPI driver is loaded. Windows 95, Windows NT, and OS/2 2.1 and later versions provide automatic ASPI support for several SCSI host adapters.

Future Domain and NCR have created another interface driver called the *Common Access Method* (CAM). CAM is an ANSI-approved protocol that enables a single driver to control several host adapters. In addition to ASPI, OS/2 2.1 and later versions currently offer support for CAM. Future Domain also provides a CAM-to-ASPI converter in the utilities that go with its host adapters.

SCSI Configuration Tips When you are installing a chain of devices on a single SCSI bus, the installation can get complicated very quickly. Here are some tips for getting your setup to function quickly and efficiently:

- *Start by adding one device at a time.* Rather than plug numerous peripherals into a single SCSI card and then try to configure them at the same time, start by installing the host adapter and a single hard disk. Then you can continue installing devices one at a time, checking to make sure that everything works before moving on.
- *Keep good documentation.* When you add a SCSI peripheral, write down the SCSI ID address as well as any other switch and jumper settings, such as SCSI Parity, Terminator Power, and Delayed and/or Remote Start. For the host adapter, record the BIOS addresses, Interrupt, DMA channel, and I/O Port addresses used by the adapter, as well as any other jumper or configuration settings (such as termination) that might be important to know later.
- *Use proper termination.* Each end of the bus must be terminated, preferably with active or Forced Perfect (FPT) terminators. If you are using any Fast SCSI-2 device, you must use active terminators rather than the cheaper passive types. Even with standard (slow) SCSI devices, active termination is highly recommended. If you have only internal or external devices on the bus, the host adapter and last device on the chain should be terminated. If you have external and internal devices on the chain, you generally will terminate the first and last of these devices but not the SCSI host adapter itself (which is in the middle of the bus).
- *Use high-quality shielded SCSI cables.* Make sure that your cable connectors match your devices. Use high-quality shielded cables, and observe the SCSI bus-length limitations. Use cables designed for SCSI use, and if possible, stick to the same brand of cable throughout a single SCSI bus. Different brands of cables have different impedance values; this situation sometimes causes problems, especially in long or high-speed SCSI implementations.

Following these simple tips will help minimize problems and leave you with a trouble-free SCSI installation.

IDE Versus SCSI

When you compare the performance and capabilities of IDE and SCSI interfaced drives, you need to consider several factors. These two types of drives are the most popular drives used in PC systems today, and a single manufacturer may make identical drives in both interfaces. Deciding which drive type is best for your system is a difficult decision that depends on many factors.

In most cases, you will find that an IDE drive outperforms an equivalent SCSI drive at a given task or benchmark, and that IDE drives usually cost less than SCSI drives, thus offering better value. In some cases, however, SCSI drives have significant performance and value advantages over IDE drives.

Performance

ATA IDE drives currently are used in most PC configurations on the market today, because the cost of an IDE-drive implementation is low and the performance capabilities are high. In comparing any given IDE and SCSI drive for performance, you have to look at the capabilities of the HDAs that are involved.

To minimize the variables in this type of comparison, it is easiest to compare IDE and SCSI drives from the same manufacturer that also use the identical HDA. Because the SCSI version always has the additional overhead of the SCSI bus to go through, in almost all cases the directly attached IDE version performs faster.

SCSI Versus IDE: Advantages and Limitations

IDE drives have much less command overhead for a given sector transfer than do SCSI drives. In addition to the drive-to-controller command overhead that both IDE and SCSI must perform, a SCSI transfer involves negotiating for the SCSI bus; selecting the target drive; requesting data; terminating the transfer over the bus; and finally converting the logical data addresses to the required cylinder, head, and sector addresses. This arrangement gives IDE an advantage in sequential transfers handled by a single-tasking operating system. In a multitasking system that can take advantage of the extra intelligence of the SCSI bus, SCSI can have the performance advantage.

SCSI drives offer significant architectural advantages over IDE and other drives. Because each SCSI drive has its own embedded disk controller that can function independently from the system CPU, the computer can issue simultaneous commands to every drive in the system. Each drive can store these commands in a queue and then perform the commands simultaneously with other drives in the system. The data could be fully buffered on the drive and transferred at high speed over the shared SCSI bus when a time slot was available.

Although IDE drives also have their own controllers, they do not operate simultaneously, and command queuing is not supported. In effect, the dual controllers in a dual-drive IDE installation work one at a time so as not to step on each other.

Although SCSI drives require an additional-cost host adapter card, more and more PCs require tape-backup, CD-ROM, or optical-drive support and thus must still be configured with a SCSI host bus adapter. This means that the incremental cost of supporting SCSI drives is virtually nil, because the SCSI host bus adapter is shared with other devices, such as tape and optical drives. In addition, all major operating systems today include software support for a wide range of SCSI devices.

What are the limitations of IDE?

- IDE does not support overlapped, multitasked I/O.
- IDE does not support command queuing.

As you can see, SCSI has some advantages over IDE, especially where expansion is concerned, and also with regard to support for multitasking operating systems. Unfortunately, it also costs more to implement.

Disk Hardware and Software Limitations

By studying the capabilities of the different disk interfaces as well as the ROM BIOS and operating systems, it is possible to determine the limits on disk storage. The following section details the limits under the different interfaces and operating systems.

Disk Interface Capacity Limitations

Different disk interfaces have different limitations on the theoretical maximum drive capacities that they may support. These limitations are due to variations in the way that each interface operates at the hardware level. It is important to note that even though a particular interface may permit access to a given amount of disk real estate, the BIOS and DOS usually are much more limiting and end up being the true limits for system disk capacity.

ST-506/412, ESDI, and IDE To determine the capacity limits for the ST-506/412, ESDI, or IDE interface, you first need to determine the limits on the maximum number of cylinders, heads, and sectors per track. To do so, look at the size of the registers that hold this data in the controller. All these interfaces have the same controller register specifications, so the capacity limits calculated here apply to all of them. As you will see, the interface capacity limits are quite high. The drive parameter limits are as follows:

Cylinders (16 bits) = 65,536

Heads (4 bits) = 16

Sectors (8 bits) = 256

This calculates to a maximum theoretical drive size of:

65,536 cylinders×16 heads×256 sectors×512 bytes = 137,438,953,472 bytes (128GB)

Unfortunately, the maximum capacity—128GB—is limited by the BIOS. There are two different BIOS types with regards to disk size limitations. The standard BIOS built into most systems is limited to 1,024 cylinders, 16 heads, and 63 sectors per track. If the BIOS is an enhanced version, it will be limited to 1,024 cylinders, 256 heads, and 63 sectors per track. Combining the BIOS and interface limits results in the following maximum capacities (assuming 512-byte sectors):

Limit with Standard BIOS: 1,024 cylinders×16 heads×63 sectors = 528,482,304 bytes (504MB)

Limit with Enhanced BIOS: 1,024 cylinders×256 Heads×63 sectors = 8,455,716,864 bytes (7.88GB)

If you do not have enhanced BIOS support on your motherboard, you can add an IDE bus adapter that has an on-board enhanced BIOS. To get around such BIOS problems, some IDE drive implementations greater than 528 million bytes split the drive to act as two physical units. In this case, the drive would appear on the IDE bus connector as being both master and slave, and could be used only as two 504MB maximum-size drives.

ATA-2 has defined LBA support for EIDE. This is a Logical Block Address mode where each sector on the drive is numbered from 0 to *x*. The limitations are that *x* is a 28-bit number that has a maximum value of 268,435,456. Using 512-byte sectors, this brings the maximum drive capacity to 137,438,953,472. That is coincidentally the same as the IDE internal limit has always been. The LBA is translated by the ATA-2/EIDE enhanced BIOS to the extended Cylinder, Head, and Sector (CHS) parameters which allow a maximum of 1,024 cylinders, 256 heads, and 63 sectors, respectively.

SCSI According to the SCSI specification, drives are not addressed by cylinders, heads, and sectors but instead by what is called a Logical Block Address (LBA). This is a sector number in which all the sectors on the drive are numbered sequentially from start to finish. The LBA is specified by a 32-bit number and with 512-byte sectors, results in the following limitation:

4,294,967,296 LBAs (sectors)×512 bytes = 2,199,023,255,552 bytes (2,048GB or 2T)

As you can see, SCSI drive capacity limits are extremely high. However, because the SCSI drive must appear to the BIOS as being a given number of cylinders, heads, and sectors per track, the BIOS limits SCSI capacity. Virtually all SCSI adapters have an enhanced BIOS that supports a maximum drive capacity as follows (assuming 512-byte sectors):

> *SCSI with Enhanced BIOS*: 1,024 cylinders×256 heads×63 sectors = 8,455,716,864 bytes (7.88GB)

> **Tip**
>
> If you do not have enhanced BIOS support in your SCSI adapter or motherboard, in some cases you can load an external driver for your adapter to provide this support.

Most systems support up to four SCSI host adapters, each with up to seven hard disk drives, for a total 28 physically installed drives.

ROM BIOS Capacity Limitations

In addition to the capacity limit of 504MB, the standard ROM BIOS is limited to supporting only two hard disk drives. The enhanced BIOS is limited to 128 drives maximum. Most SCSI and IDE adapters get around the two-drive standard BIOS limits by incorporating an enhanced BIOS on board that takes over the disk interface. Some of the newer adapter on-board BIOS versions support booting from CD-ROM drives as well.

Operating System Capacity Limitations

IBM and Microsoft officially say that DOS 5 and later versions will support up to eight physical hard disks. IBM says that OS/2 1.30.1 and later versions (including 2.x) support up to 24 physical hard disks, and because OS/2 includes DOS, that implies that DOS under OS/2 would support 24 physical drives as well. OS/2 HPFS (High Performance File System) also supports a maximum partition size of 8GB and a maximum single-file size of 2GB, whereas DOS and OS/2 FAT partitions have a maximum size of 2GB and a maximum single-file size of 2GB. FAT32 in Windows 95 OSR2 and Windows 98 allows a single volume of 2T or 2048GB. As you have seen, BIOS limitations currently limit the maximum physical hard disk size to about 7.88GB (or about 8.46 million bytes).

Chapter 3

Hard Disk Drive Installation

This chapter thoroughly describes hard disk installation. In particular, the chapter examines the configuration, physical installation, and formatting of a hard disk drive. This chapter also covers the basic procedures necessary to install a hard disk drive into a PC system.

Hard Disk Installation Procedures

To install a hard drive in an IBM-compatible system, you must perform several procedures:

- Configure the drive
- Configure the controller or interface
- Physically install the drive
- Configure the system
- Low-level format the drive (not required with IDE and SCSI)
- Partition the drive
- High-level format the drive

Drive configuration was discussed in Chapter 2, "Hard Disk Interfaces." For complete configuration information, consult the section that covers the type of drive that you are installing.

To begin the setup procedure, you need to know several details about the hard disk drive, controller or host adapter, and system ROM BIOS, as well as most of the other devices in the system.

Controller Configuration

All hard disk controllers and SCSI host adapters require one or more of the following system resources:

- ROM addresses

- Interrupt Request Channel (IRQ)
- DMA channel (DRQ)
- I/O port addresses

Not all adapters use every one of these resources, but some will use them all. In most cases, these resources must be configured so that they are unique and cannot be shared among several adapters.

For most systems, you need the documentation for every adapter in the system to ensure that no conflicts exist and to find out how to reconfigure a card to eliminate a conflict. Software included with your system, such as MSD (Microsoft Diagnostics, which comes with Windows 3.x and DOS 6.x) or the Device Manager in Windows 95 and 98, can help when documentation is not available or is limited. Unless your system conforms to the Plug and Play (PnP) specification, software will normally not be able to identify direct conflicts, but if you install one board at a time, they can identify the addresses or resources that a given board is using.

ROM Addresses Many disk controllers and SCSI host adapters require an on-board BIOS to function. An on-board BIOS can provide many functions, including

- Low-level formatting
- Drive-type (parameter) control
- Adapter configuration
- Support for nonstandard I/O port addresses and interrupts

If the motherboard BIOS supports a hard disk controller, an on-board BIOS is not needed and, in fact, is undesirable because it uses memory in the Upper Memory Area (UMA). Fortunately, the on-board BIOS usually can be disabled if it is not required.

Only controllers that meet certain standards can run off the motherboard BIOS, including ST-506/412 controllers, ESDI controllers, and IDE bus adapters. These standards include the use of I/O port addresses 170–17Fh and interrupt 14. If you are installing a controller that uses other I/O port addresses or interrupt settings (such as when adding a second controller to a system), the motherboard BIOS will not be able to support it, and an on-board BIOS will be required. XT controllers universally need an on-board BIOS because the motherboard BIOS has no hard disk support whatsoever.

SCSI adapters normally do not emulate the WD1003-type disk interface and almost always require an on-board BIOS to provide disk driver functions. This on-board BIOS supports any of the adapter's settings. In most cases, multiple SCSI host adapters can use the BIOS of the first adapter, in which case the BIOSes on all but the first adapter can be disabled.

If an on-board BIOS is required and enabled, it will use specific memory address space in the UMA. The UMA is the top 384KB in the first megabyte of system memory. The UMA is divided into three areas of two 64KB segments each, with the first and last areas being used by the video-adapter circuits and the motherboard BIOS, respectively. Segments C000h and D000h are reserved for use by adapter ROMs such as those found on disk controllers or SCSI host adapters.

> **Note**
>
> You need to ensure that any adapters using space in these segments do not overlap with another adapter that uses this space. No two adapters can share this memory space. Most adapters have jumpers, switches, or even software that can adjust the configuration of the board and change the addresses that are used to prevent conflict.

Interrupt Request Channel (IRQ) All disk controllers and SCSI host adapters require an interrupt line to gain the system's attention. These devices invoke a hardware interrupt to gain timely access to the system for data transfers and control. The original 8-bit ISA systems have only eight interrupt levels, with interrupts 2–7 available to any adapter. AT bus (16-bit ISA), EISA, and MCA systems have 16 interrupt levels, with interrupts 3–7, 9–12, and 14 and 15 available to any adapter cards. IRQs 10–12 and 14 and 15 are 16-bit interrupts available only to 16- or 32-bit adapters.

Tables 3.1 and 3.2 show the normally used and normally available interrupts in ISA, EISA, and MCA systems and in 8-bit ISA systems. The tables list the default use for each interrupt and indicate whether the interrupt is available in a bus slot.

Table 3.1 ISA, EISA, and MCA Default Interrupt Assignments.

IRQ	Function	Bus Slot
0	System Timer	No
1	Keyboard Controller	No
2	Second IRQ Controller	No
8	Real-Time Clock	No
9	Network/Available (Redirected IRQ 2)	Yes (8-bit)
10	Available	Yes (16-bit)
11	SCSI/Available	Yes (16-bit)
12	Motherboard Mouse Port	Yes (16-bit)
13	Math Coprocessor	No
14	Hard Disk Controller	Yes (16-bit)
15	Secondary IDE	Yes (16-bit)
3	Serial Port 2 (COM2:)	Yes (8-bit)
4	Serial Port 1 (COM1:)	Yes (8-bit)
5	Sound/Parallel Port 2 (LPT2:)	Yes (8-bit)
6	Floppy Disk Controller	Yes (8-bit)
7	Parallel Port 1 (LPT1:)	Yes (8-bit)

Table 3.2 XT-Bus (8-Bit ISA) Default Interrupt Assignments.

IRQ	Function	Bus Slot
0	System Timer	No
1	Keyboard Controller	No
2	Network/Available	Yes (8-bit)
3	Serial Port 2 (COM2:)	Yes (8-bit)
4	Serial Port 1 (COM1:)	Yes (8-bit)
5	Hard Disk Controller	Yes (8-bit)
6	Floppy Disk Controller	Yes (8-bit)
7	Parallel Port 1 (LPT1:)	Yes (8-bit)

Notice that some of the interrupts are simply not available in slots; they are reserved for use only by the indicated system function. Any interrupt that is listed as being in use by an item that is not installed in your system would be available. For example, if your system does not have a motherboard mouse port, IRQ 12 would be available, and if your system does not have a second serial port, IRQ 3 would be available.

In a standard configuration, the hard disk controller uses IRQ 14. Any secondary controllers would have to use other interrupts. The standard interrupt for a secondary controller is IRQ 15. If the system does not support EIDE (Enhanced IDE) in the motherboard BIOS, any controllers that do not use IRQ 14 must have an on-board BIOS to function. The older motherboard BIOS supports disk controllers only at IRQ 14, whereas a BIOS with EIDE support will run IDE ports at both IRQ 14 and 15. Most newer systems have integral EIDE support and automatically include a secondary IDE port, which is at IRQ 15.

Standard IDE adapters come preconfigured for IRQ 14. Many SCSI host adapters, come configured to one of the other available 16-bit interrupts, such as IRQ 11. Old XT (8-bit) hard disk controllers normally use IRQ 5.

DMA Channel *Direct Memory Access* (DMA) is a technique for transferring blocks of data directly into system memory without the complete attention of the main processor. The motherboard has DMA control circuits that orchestrate and govern DMA transfers. In the original 8-bit XT bus, DMA was the highest-performance transfer method, and XT hard disk controllers universally used DMA channel 3 for high-speed transfers.

In AT-Bus (16-bit ISA) systems, most 16-bit disk controllers and SCSI host adapters do not use a DMA channel, partly because the performance of the AT Bus DMA circuitry turned out to be very poor. Therefore, most adapters use a technique called *Programmed I/O* (PIO), which simply sends bytes of data through the I/O ports. PIO transfers are faster than DMA transfers in most cases, especially if the motherboard BIOS and device support block-mode PIO, such as with the new IDE drives. If an adapter does not use DMA, you can assume that PIO is used as the data-transfer method and that no DMA channel is required.

Some adapters have found a way around the poor performance of the ISA bus by becoming what is known as a bus master. A *bus master* actually takes control of the bus and can override the DMA controller circuitry of the motherboard to perform fast DMA transfers. These transfers can exceed the performance of a PIO transfer (even block-mode PIO), so you will find that many of the highest-performing controllers have bus-master capabilities.

You have to select a DMA channel for bus-master adapters to use. In an 8-bit ISA bus, normally only DMA channel 1 is available; in a 16-bit ISA bus, however, DMA channels 0–1, 3, and 5–7 are available. DMA channels 5–7 are 16-bit channels that most high-performance bus-master adapters would want to use. XT disk controllers always use DMA channel 3, whereas most 16-bit AT or IDE controllers do not use DMA at all. This is not a concern in the newer PCI bus systems.

Tables 3.3 and 3.4 show the normally used and normally available DMA channels. The tables list the default use for each DMA channel and indicate whether the DMA channel is available in a bus slot.

Table 3.3 16-bit ISA Default DMA Channel Assignments.

DMA	Function	Transfer	Bus Slot
0	Available	8-bit	Yes(16-bit)
1	Sound/Available	8-bit	Yes (8-bit)
2	Floppy Disk Controller	8-bit	Yes (8-bit)
3	ECP Parallel/Available	8-bit	Yes (8-bit)
4	First DMA Controller	N/A	No
5	Sound/Available	16-bit	Yes(16-bit)
6	SCSI/Available	16-bit	Yes(16-bit)
7	Available	16-bit	Yes(16-bit)

Table 3.4 8-Bit ISA Default DMA Channel Assignments.

DMA	Function	Transfer	Bus Slot
0	Dynamic RAM Refresh	N/A	No
1	Sound/Available	8-bit	Yes (8-bit)
2	Floppy Disk Controller	8-bit	Yes (8-bit)
3	Hard Disk Controller	8-bit	Yes (8-bit)

Notice that some of the DMA channels simply are not available in slots. They are reserved for use only by the indicated system function.

PCI and EISA bus systems have additional DMA capabilities that support even faster transfers without the performance problems associated with non-bus-master cards. The PCI and EISA buses also provide even better support for bus-master devices that offer even higher performance.

To configure an adapter that requires a DMA channel, you first must find out which DMA channels currently are in use and which channels are available in your system. Unless you have an PCI or EISA system, software techniques for determining this are very limited. Most programs that claim to be capable of discovering which DMA channels are being used are only reporting what any standard configuration would be. In standard ISA systems, the only way to know for sure is to check the documentation for each adapter or to use a special hardware device that monitors DMA transfers.

PCI and EISA systems automatically set up the boards so that no DMA conflicts exist. This method works fully only in EISA systems with all EISA 32-bit adapters installed. PCI buses usually have at least two or more ISA slots that you still need to worry about.

I/O Port Addresses *I/O port addresses* are like mailboxes through which data and commands are sent to and from an adapter. These addresses are different from memory addresses. I/O ports must be used exclusively and cannot be shared among different adapters. Each adapter usually uses a group of sequential port addresses for communication with the bus.

The standard I/O port addresses used by disk controllers are 1F0–1F7h. These are the only addresses that the motherboard BIOS supports, so if you have a disk controller at any other port address, it must have an on-board BIOS. Obviously, if you are adding a secondary controller to a system, that controller must use different I/O addresses and also must have an on-board BIOS. Most controllers use 170–177h as secondary I/O addresses, which would be used if another disk controller were in the system; however, you can use any I/O addresses that are free.

I/O port conflicts are rare unless you are installing multiple disk controllers in a system. In that case, each controller needs different I/O port address settings so as not to conflict with the others. To determine what I/O ports are currently in use, you normally have to refer to the documentation that comes with each device in your system. Software normally cannot identify all used I/O port addresses unless you have a PCI or EISA system. When port conflicts exist, the devices in conflict do not function, or they function improperly.

Sidebar

Multi-function I/O cards are a combination floppy disk, hard disk, and serial and parallel port adapters all on one expansion card. Most newer motherboards have all these functions built right into the motherboard, but you may see a few of these multi-function boards in older PCI or VLB systems, and it is important to note that you will have to disable the hard disk controller on this board before installing a new controller.

Physical Installation

You must have the correct screws, brackets, and faceplates for the specific drive and system before you can install the drive. Some AT or Baby-AT type computer cases require plastic rails that are secured to the sides of the drives so that they can slide into the proper place in the system (see Figure 3.1). Compaq uses a different type of rail, as does Hewlett-Packard, Packard Bell, and so on. When you purchase a drive, the vendor usually includes the "Standard-type" rails, so be sure to specify whether you need the special manufacturer type. IBM PC-type and XT-type systems do not need rails, but they may need a bracket to enable double-stacking of half-height drives. Also, many newer after-market computer cases have eliminated the need for drive rails all together by making the expansion slot itself to the 3 1/2- or 5 1/4-inch drive specification, allowing you to bolt the new drive directly to the case itself.

Note

You should also note the length of the cable itself. In some cases, the drive cable is not long enough to reach the new drive location. You can try to reposition the original drive if you have the available expansion slots or just get a longer cable.

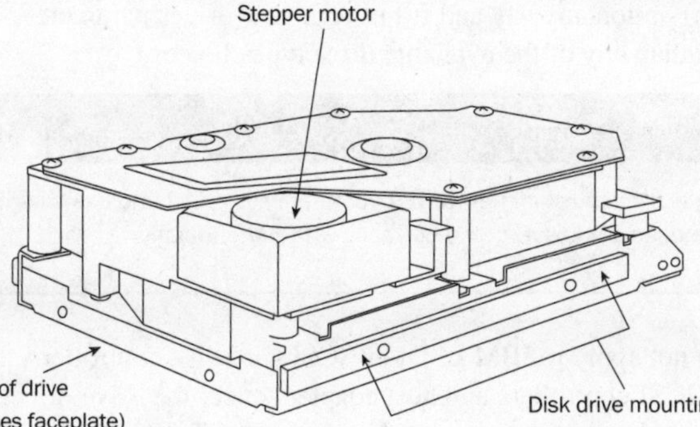

FIG. 3-1 A full-height hard disk with AT mounting rails.

Caution

Make sure you use only the screws that come with your new drive. Many drives come with a special short-length screw that can have the same thread as other screws you might use in your system, but the shorter screws will not drive too far into the drive, which otherwise could cause problems.

System Configuration

When the drive is physically installed, you can begin configuring the system to the drive. You have to tell the system about the drive so that the system can boot from it when it is powered on. How you set and store this information depends on the type of drive and system you have. Standard (IDE) setup procedures are used for most hard disks except SCSI drives. SCSI drives normally follow a custom setup procedure that varies depending on the host adapter that you are using. If you have SCSI drives, follow the instructions included with the host adapter to configure the drives.

Automatic Drive Typing If the system is an AT type and you are using the motherboard BIOS to support the hard disks, you need to know some information about the BIOS, such as what drives are supported in the hard drive table. Many BIOS versions now have user-definable drive types that enable you to enter any set of parameters required to match your drive. For IDE drives, all new BIOS versions have automatic typing, which interrogates the drive and automatically enters the parameter information returned by the drive. This procedure eliminates errors or confusion in parameter selection.

If you are dealing with an older motherboard that does not support automatic typing, the information about the drive table appears in the extensive tables of drive information that compose the bulk of the later portion of this book.

Manual Drive Typing After you collect the necessary information, the next step is to tell the system what kind of drive is attached so that the system can boot from the drive (eventually). With knowledge of drive interfacing, you can install just about any drive in any system.

After you find the information about the drive and controller, you need to match the drive's parameters to one of the drive-table entries in the motherboard ROM. However, most drive tables are now embedded in the CMOS of the motherboard, and usually you have to choose User Defined (if you don't have an Auto-Configure option) option anyway and fill in your new drives parameters because most new drives are much larger than any of the available drive type choices.

Tip

The landing-cylinder designation is superfluous because all new drives automatically parks and locks its heads at power-down, although it would be used if you ever ran a correctly written head-parking program.

This type of drive-table information does not apply to IBM ESDI or SCSI hard disk controllers, host adapters, and drives. Because ESDI and SCSI controllers and host adapters query the drive directly for the required parameters, no table-entry selection is necessary. The table for ST-506/412 drives, however, still appears in the ROM BIOS of most PS/2 systems, even if the model came standard with the ESDI or SCSI disk subsystem.

The manufacturers of most compatibles have enhanced the motherboard ROM BIOS tables in three ways:

- *Additional types.* The first thing that the manufacturers did was add more drive types to the table. Because the table had room for 47 or more entries, many compatible BIOS versions simply filled out all the entries with values that matched the most popular drives on the market, generally making drive installations easier. IBM tables often were short of the maximum number of possible entries.
- *User-definable drive types.* Most makers of compatibles then added a user-definable type, which used unused areas of the CMOS memory to store all the drive-parameter information. This was an excellent solution because during setup you can type a parameter that matches any drive on the market. The only drawback is that if the CMOS battery dies or the saved values are corrupted in some way, you would have to re-enter the information exactly as it was before to regain access to the drive. Many people did not write down the parameters that they used, or they used improper parameters that caused problems.
- *Automatic detection.* Most of the newer BIOS versions include a feature that is specific to IDE drives. Because most IDE drives are intelligent and will respond to a command called Identify Drive, the BIOS sends this command to the drive, which then responds with the correct parameters. This feature eliminates the need to type the parameters because the BIOS will accept what the drive tells it.

As mentioned earlier, most of the newer compatible BIOS versions have both the user-definable type feature and automatic detection for IDE drives.

ROM Replacement One way around the drive-table limits is to purchase and install a new ROM BIOS. A Phoenix ROM BIOS set, for example, costs about $50. These ROMs include a user-definable drive-type setting, which is the most elegant solution to this problem. A new set of ROMs probably will give you additional features, such as a built-in setup program, support for HD or ED 3 1/2-inch floppy drives, and Enhanced Keyboard support.

RLL/ESDI System Configuration RLL and ESDI drives usually are not represented in the internal drive tables of older BIOS versions. Consequently, the controllers for these drives often have an on-board ROM BIOS that either contains an internal list of choices for the interface or enables you to dynamically configure (define) the controller to the specific geometry of the drive.

If you have a motherboard BIOS with a user-defined drive type (recommended), you can simply enter the correct parameters and the drive will be supported. (Remember to write down the parameters that you use; if you lose them, you can lose access to the drive if you don't re-enter the parameters properly.) When using a user-definable type, you can disable the controller BIOS.

IDE System Configuration Intelligent IDE drives can use the geometry that represents their true physical parameters, or they can translate to other drive geometries that have the same number of sectors or fewer. Simply select a type, or enter a user-definable type that is less than or equal to the total capacity of the drive.

SCSI System Configuration Almost all SCSI drives use DRIVE TYPE 0 or NONE because the host adapter BIOS and the drive communicate to establish the drive geometry. The low-level formatting routines usually are accessed on the host adapter through a configuration, setup, and format program. All SCSI drives are low-level formatted at the factory.

Formatting and Software Installation

Proper setup and formatting are critical to a drive's performance and reliability. This section describes the procedures used to format a hard disk drive correctly. Use these procedures when you install a new drive in a system or immediately after you recover data from a hard disk that has been exhibiting problems.

Three major steps complete the formatting process for a hard disk drive subsystem:

- Low-level formatting
- Partitioning
- High-level formatting

Considerations Before Low-Level Formatting In a *low-level format* (LLF), which is a "real" format, the tracks and sectors of the disk are outlined and written. During the LLF, data is written across the entire disk. An improper LLF results in lost data and in many read and write failures. You need to consider several things before initiating an LLF.

Data Backup Low-level formatting is the primary standard repair procedure for hard disk drives that are having problems. Because data values are copied to the drive at every possible location during an LLF, before performing and LLF you should back up any data on the drive if you will ever need access to that data again.

Because an LLF overwrites all the data on a drive, it is a good way to erase an entire drive if you are trying to ensure that nobody will be able to get data from it.

System Temperature Sector header and trailer information is written or updated only during the LLF operation. During normal read and write activity, only the 512 bytes plus the Cyclic Redundancy Check (CRC) bytes in the trailer are written in a sector. Temperature-induced dimensional changes in the drive platters during read and write operations are not a problem with modern voice-coil actuators used today. However, it is still doesn't hurt to leave the system's power on for at least 30 minutes before performing an LLF on its hard disk. This step ensures that the platters are at a normal operating temperature.

Drive Operating Position Another consideration before formatting a drive is ensuring that the drive is formatted in the operating position it will have when it is installed in the system. Gravity can place on the head actuator different loads that can cause mistracking if the drive changes between a vertical and a horizontal position. This effect is minimized or even eliminated in most voice-coil drives, but this procedure cannot hurt.

Additionally, drives that are not properly shock-mounted should be formatted only when they are installed in the system because the installation screws exert twisting forces on the drive's Head Disk Assembly (HDA), which can cause mistracking. If you format the drive with the mounting screws installed tightly, it may not read with the screws out, and vice versa. Be careful not to overtighten the mounting screws because doing so can stress the HDA. This usually is not a problem if the drive's HDA is isolated from the frame by rubber bushings.

In summary, for a proper LLF, the drive should be

- At a normal operating temperature
- In a normal operating position
- Mounted in the host system (if the drive HDA is not shock-mounted or isolated from the drive frame by rubber bushings)

Low-Level Format Of these procedures, the LLF is most important to ensure trouble-free operation of the drive. This format is the most critical of the operations and must be done correctly for the drive to work properly. The LLF includes several subprocedures:

- Scanning for existing defect mapping
- Selecting the interleave
- Formatting and marking (or remarking) manufacturer defects
- Running a surface analysis

On all new systems, these subprocedures are performed automatically by the system's LLF program and require no user intervention, and you need not continue in this section. On older systems, you must take the initiative and should read the section on low level formatting included in the complete text of *Upgrading and Repairing PCs*, Eighth Edition.

All new drives are low-level formatted by the manufacturer. If you bought a system with a drive already installed by the manufacturer or dealer, an LLF probably was done for you. Most manufacturers no longer recommend you LLF any IDE type drive.

Why Low-Level Format? Even though it generally is not necessary (or even recommended) to LLF IDE or SCSI drives, there are a few good reasons to consider an LLF. One reason is that an LLF will wipe out all the data on a drive, ensuring that other people will not be able to read or recover that data. This procedure is useful if you are selling a system and do not want your data to be readable by the purchaser. Another reason for wiping all the data from a drive is to remove corrupted or non-DOS operating-system partitions and even virus infections. The best reason is for defect management. As you may have noticed, most ATA-IDE drives appear to have no "bytes in bad sectors" under CHKDSK or any other software.

Any defects that were present on the drive after manufacturing were reallocated by the factory LLF. Essentially, any known bad sectors are replaced by spare sectors stored in different parts of the drive. If any new defects occur, such as from a minor head/platter contact or drive mishandling, a proper IDE-aware LLF program can map the new bad sectors to other spares, hiding them and restoring the drive to what appears to be defect-free status.

Because the IDE (ATA) specification is an extension of the IBM/WD ST-506/412 controller interface, the specification includes several new CCB commands that were not part of the original INT 13h/CCB support. Some of these new CCB commands are vendor-specific and are unique to each IDE drive manufacturer. Some manufacturers use these special CCB commands for tasks such as rewriting the sector headers to flag bad sectors, which in essence means LLF. When using these commands, the drive controller can rewrite the sector headers and data areas and then carefully step over any servo information (if the drive uses an embedded servo).

IDE drives can be low-level formatted, although some drives require special vendor-specific commands to activate certain low-level formatting features and defect-management options. Seagate, Western Digital, Maxtor, IBM, and others make specific LLF and spare-sector defect-management software specific to their respective IDE drives. Conner drives are unique in that to actually LLF them, you need a special hardware device that attaches to a diagnostic port connector on the Conner IDE drive.

Intelligent IDE drives must be in *nontranslating*, or native, mode to LLF them. Zoned recording drives can perform only a partial LLF, in which the defect map is updated and new defective sectors can be marked or spared, but the sector headers usually are rewritten only partially, and only for the purpose of defect mapping. In any case, you are writing to some of the sector headers in one form, and physical (sector-level) defect mapping and sector sparing can be performed. This procedure is, by any standard definition, an LLF.

On an embedded servo drive, all the servo data for a track is recorded at the same time by a specialized (usually laser-guided) servowriter. This servo information is used to update the head position continuously during drive function so that the drive automatically compensates for thermal effects. As a result, all the individual servo bursts are in line on the track. Because the servo controls head position, there is no appreciable head-to-sector drift, as there could be on a nonservo drive.

This is why even though it is possible to LLF embedded servo drives, it rarely is necessary. The only purpose for performing an LLF on an embedded servo drive is to perform additional physical- (sector-) level defect mapping or sector sparing for the purpose of managing defects that occur after manufacture. Because no drift occurs, when a sector is found to contain a flaw, it should remain permanently marked bad. A physical flaw cannot be repaired by reformatting.

Most IDE drives have three to four spare sectors for each physical cylinder of the drive. These hundreds of spare sectors are more than enough to accommodate the original defects and any subsequent defects. If more sectors are required, the drive likely has serious physical problems that cannot be fixed by software.

SCSI Low-Level Format Software If you are using a SCSI drive, you must use the LLF program provided by the manufacturer of the SCSI host adapter. The design of these devices varies enough that a register-level program can work only if it is tailored to the individual controller. Fortunately, all SCSI host adapters include such format software, either in the host adapter's BIOS or in a separate disk-based program.

The interface to the SCSI drive is through the host adapter. SCSI is a standard, but there are no true standards for what a host adapter is supposed to look like. This means that any formatting or configuration software will be specific to a particular host adapter.

Some SCSI host adapters often include the complete setup, configuration, and formatting software in the host adapter's on-board ROM BIOS. Most of these adapters also include an INT 13h interface in the BIOS.

> **Note**
>
> Notice that SCSI format and configuration software is keyed to the host adapter and is not specific in any way to the particular SCSI hard disk drive that you are using.

IDE Low-Level Format Software IDE drive manufacturers have defined extensions to the standard WD1002/1003 AT interface, which was further standardized for IDE drives as the ATA (AT Attachment) interface. The ATA specification provides for vendor-unique commands, which are manufacturer proprietary extensions to the standard. To prevent improper low-level formatting, many of these IDE drives have special codes that must be sent to the drive to unlock the format routines. These codes vary among manufacturers. If possible, you should obtain LLF and defect-management software from the drive manufacturer; this software usually is specific to that manufacturer's products.

The custom nature of the ATA interface drives is the source of some myths about IDE. Many people say, for example, that you cannot perform an LLF on an IDE drive, and that if you do, you will wreck the drive. This statement is untrue! What can happen is that in some drives, you may be able to set new head and sector skew factors that are not as optimal for the drive as the ones that the manufacturer set, and you also may be able to overwrite the defect-map information. This situation is not good, but you still can use the drive with no problems provided that you perform a proper surface analysis.

Most ATA IDE drives are protected from any alteration to the skew factors or defect map erasure because they are in a translated mode. Zoned recording drives always are in translation mode and are fully protected. Most ATA drives have a custom command set that must be used in the format process. The standard format commands defined by the ATA specification usually do not work, especially with intelligent or zoned recording IDE drives. Without the proper manufacturer-specific format commands, you will not be able to perform the defect management by the manufacturer-specified method, in which bad sectors often can be spared.

Currently, the following manufacturers offer specific LLF and defect-management software for their own IDE drives:

- Seagate
- Maxtor
- Western Digital
- IBM

Nondestructive Formatters General-purpose, BIOS-level, nondestructive formatters such as Calibrate and SpinRite are not recommended in most situations for which a real LLF is required. These programs have several limitations and problems that limit their effectiveness. In some cases, they can even cause problems with the way defects are handled on a drive. These programs attempt to perform a track-by-track LLF by using BIOS functions, while backing up and restoring the track data as they go. These programs do not actually perform a complete LLF, because they do not even try to LLF the first track (Cylinder 0, Head 0) due to problems with some controller types that store hidden information on the first track.

These programs also do not perform defect mapping in the way that a standard LLF programs do, and they can even remove the carefully applied sector header defect marks during a proper LLF. This situation potentially allows data to be stored in sectors that originally were marked defective and may actually void the manufacturer's warranty on some drives. Another problem is that these programs work only on drives that are already formatted and can format only drives that are formattable through BIOS functions.

A true LLF program bypasses the system BIOS and send commands directly to the disk controller hardware. For this reason, many LLF programs are specific to the disk controller hardware for which they are designed. It is virtually impossible to have a single format program that will run on all different types of controllers. Many hard drives have been incorrectly diagnosed as being defective because the wrong format program was used and the program did not operate properly.

Drive Partitioning *Partitioning* a hard disk is the act of defining areas of the disk for an operating system to use as a volume. To DOS, a volume is an area of a disk denoted as a drive letter; for example, drive C is volume C, drive D is volume D, and so on. Some people think that you have to partition a disk only if you are going to divide it into more than one volume. This is a misunderstanding; a disk must be partitioned even if it will be the single volume C.

When a disk is partitioned, a master partition boot sector is written at Cylinder 0, Head 0, Sector 1—the first sector on the hard disk. This sector contains data that describes the partitions by their starting and ending cylinder, head, and sector locations. The partition table also indicates to the ROM BIOS which of the partitions is bootable and, therefore, where to look for an operating system to load. A single hard disk can have 1 to 24 partitions. This number includes all the hard drives installed in the system, which means that you can have as many as 24 separate hard disks with one partition each, a single hard disk with 24 partitions, or a combination of disks and partitions such that the total number of partitions is no more than 24. If you have more than 24 drives or partitions, DOS does not recognize them, although other operating systems may. What limits DOS is that a letter is used to name a volume, and the Roman alphabet ends with Z—the 24th volume, when you begin with C.

Today, PC operating systems use three common file systems:

- *FAT (File Allocation Table).* The standard file system used by DOS, Windows 95 (Non-OSR2 release), OS/2, and Windows NT. FAT partitions support file names of 11 characters maximum

(8+3 character extension) under DOS, and 255 characters under Windows 95 or NT 4.0 or later versions. Under the standard FAT system, 12-bit or 16-bit numbers are used to identify allocation units, resulting in a maximum volume size of 2GB.

- *FAT32 (File Allocation Table, 32-bit)*. An optional file system used by Windows 95 OSR2 (also called OEM Service Release 2 or Windows 95B) or later versions. Under FAT32, file allocation units are stored as 32-bit numbers, allowing for a single volume of 2TB or 2,048GB in size. FAT32 support will likely be added to Windows NT in the future.
- *HPFS (High Performance File System)*. A file system that's accessible only under OS/2 and Windows NT 3.51 or earlier. DOS applications running under OS/2 or Windows NT, or via a network, can access files in HPFS partitions, but straight DOS cannot. File names can be 256 characters long, and volume size is limited to 8GB.
- *NTFS (Windows NT File System)*. A UNIX-like file system that's accessible only under Windows NT. DOS cannot access these partitions, but DOS applications running under Windows NT or accessing a Windows NT volume from the network can. File names can be 256 characters long, and volume size is limited to 8GB.

Of these three file systems, the FAT file system still is by far the most popular (and recommended). The main problem with the original 16-bit FAT file system is that disk space is used in groups of sectors called *allocation units* or *clusters*. Because the total number of clusters is limited to 65,536 (the most which can be represented with a 16-bit number), larger drives required that the disk be broken into larger clusters. The larger cluster sizes required cause disk space to be used inefficiently. FAT32 solves this problem by allowing the disk to be broken up into over 4 billion clusters, so the cluster sizes can be kept smaller. Most FAT32 and NTFS volumes use 4KB clusters.

The term *cluster* was changed to *allocation unit* in DOS 4 and later versions. The newer term is appropriate because a single cluster is the smallest unit of the disk that DOS can allocate when it writes a file. A cluster is equal to one or more sectors, and although a cluster can be a single sector in some cases (specifically 1.2MB and 1.44MB floppies), it is usually more than one. Having more than one sector per cluster reduces the size and processing overhead of the FAT and enables DOS to run faster because it has fewer individual units of the disk to manage. The tradeoff is in wasted disk space. Because DOS and Windows can manage space only in full cluster units, every file consumes space on the disk in increments of one cluster.

Smaller clusters generate less slack (space wasted between the actual end of each file and the end of the cluster). With larger clusters, the wasted space grows larger. For hard disks, the cluster size varies with the size of the partition. Table 3.5 shows the default cluster sizes FDISK selects for a particular partition volume size.

Table 3.5 Default Cluster Sizes.

Hard Disk Partition Size	Cluster (Allocation Unit) Size	FAT Type
0 15MB	8 sectors or 4,096 (4KB) bytes	12-bit
16 128MB	4 sectors or 2,048 (2KB) bytes	16-bit
129 256MB	8 sectors or 4,096 (4KB) bytes	16-bit

Table 3.5 Default Cluster Sizes (continued).

Hard Disk Partition Size	Cluster (Allocation Unit) Size	FAT Type
257 512MB	16 sectors or 8,192 (8KB) bytes	16-bit
513 1,024MB	32 sectors or 16,384 (16KB) bytes	16-bit
1,025 2,048MB	64 sectors or 32,768 (32KB) bytes	16-bit
0 260MB	1 sector or 512 bytes	32-bit
260M 8GB	8 sectors or 4,096 (4KB) bytes	32-bit
8 16GB	16 sectors or 8,192 (8KB) bytes	32-bit
16 32GB	32 sectors or 16,384 (16KB) bytes	32-bit
32 2,048GB	64 sectors or 32,768 (32KB) bytes	32-bit

In most cases, these cluster sizes, which are selected by the FORMAT command, are the minimum possible for a given partition size. Therefore, 8KB clusters are the smallest possible for a partition size greater than 256MB. Note that FDISK creates a FAT using 12-bit numbers if the partition is 16MB or less, while all other FATs are created using 16-bit numbers, unless Large Disk Support is specifically enabled in Windows 95 OSR2 or later.

The effect of the larger cluster sizes on larger disk partitions can be substantial. There can be a significant amount of slack space in the leftover portions of larger clusters. On average the amount of slack space for a file is one half of the space of the last cluster the file uses. To calculate an estimate of slack space on an entire drive use the following formula:

slack space = #_files * cluster_size / 2

A drive partition of over 1GB and up to 2GB using FAT16 (thus 32KB clusters) containing 10,000 files wastes about 16KB per file or 160,000KB (160MB) total [10000*32KB/2]. If you were to repartition the drive into two separate partitions of less than or equal to 1GB each, then the cluster size would be cut in half, as would the total wasted slack space. You would therefore gain 80MB of disk space. The tradeoff is that managing multiple partitions is not as convenient as a single large partition. The only way you can control cluster or allocation unit sizing using FAT16 is by changing the sizes of the partitions.

If you were to reformat the drive using FAT32, the wasted space would drop to only 2KB per file or about 20MB total! In other words, by converting to FAT32, you would end up with approximately 140MB more disk space free in this example.

NTFS, HPFS, and FAT32 all dramatically reduce the slack space but also increase file management overhead because many more allocation units must be managed.

Despite the problem with slack space, the basic FAT file system is still often the most recommended for compatibility reasons. All the operating systems can access FAT volumes, and the file structures and data-recovery procedures are well known. Also note that data recovery can be difficult to impossible under the HPFS and NTFS systems; for those systems, good backups are imperative.

During partitioning, no matter what file system is specified, the partitioning software writes a special boot program and partition table to the first sector, called the Master Boot Sector (MBS). Because the term *record* sometimes is used to mean *sector*, this sector can also be called the Master Boot Record (MBR).

FDISK The DOS FDISK program is the accepted standard for partitioning hard disks. Partitioning prepares the boot sector of the disk in such a way that the DOS FORMAT program can operate correctly. It also enables different operating systems to coexist on a single hard disk.

If a disk is set up with two or more partitions, FDISK shows only two total DOS partitions: the primary partition and the extended partition. The extended partition then is divided into logical DOS volumes, which are partitions themselves. FDISK gives a false impression of how the partitioning is done. FDISK reports that a disk divided as C, D, E, and F is set up as two partitions, with a primary partition having a volume designator of C and a single extended partition containing logical DOS volumes D, E, and F. But in the real structure of the disk, each logical DOS volume is a separate partition with an extended partition boot sector describing it. Each drive volume constitutes a separate partition on the disk, and the partitions point to one another in a daisy-chain arrangement.

The minimum size for a partition in any version of DOS is one cylinder; however, FDISK in DOS 4 and later versions allocates partitions in megabytes, meaning that the minimum-size partition is 1MB.

FDISK Undocumented Functions FDISK is a very powerful program, and in DOS 5 and later versions, it gained some additional capabilities. Unfortunately, these capabilities were never documented in the DOS manual and remain undocumented even in DOS 7. The most important undocumented parameter in FDISK is the /MBR (Master Boot Record) parameter, which causes FDISK to rewrite the Master Boot Sector code area, leaving the partition tables intact.

Caution

/MBR will overwrite the partition tables if the two signature bytes at the end of the sector (55AAh) are damaged. This situation is highly unlikely, however. In fact, if these signature bytes were damaged, you would know; the system would not boot and would act as though there were no partitions at all.

The /MBR parameter seems to be tailor-made for eliminating boot-sector virus programs that infect the Master Partition Boot Sector (Cylinder 0, Head 0, Sector 1) of a hard disk. To use this feature, you simply enter **FDISK /MBR**

FDISK then rewrites the boot sector code, leaving the partition tables intact. This should not cause any problems on a normally functioning system, but just in case, I recommend backing up the partition table information to floppy disk before trying it. You can do this with the following command:

```
MIRROR /PARTN
```

This procedure uses the MIRROR command to store partition-table information in a file called PARTNSAV.FIL, which should be stored on a floppy disk for safekeeping. To restore the complete partition-table information, including all the master and extended partition boot sectors, you would use the UNFORMAT command as follows:

```
UNFORMAT /PARTN
```

This procedure causes the UNFORMAT command to ask for the floppy disk containing the PARTNSAV.FIL file and then to restore that file to the hard disk. Note that if you are using Windows 95, the MIRROR and UNFORMAT programs have been eliminated, and you will have to purchase Norton Utilities instead.

FDISK also has three other undocumented parameters: /PRI:, /EXT:, and /LOG:. These parameters can be used to have FDISK create master and extended partitions, as well as logical DOS volumes in the extended partition, directly from the command line rather than through the FDISK menus. This feature was designed so that you can run FDISK in a batch file to partition drives automatically. Some system vendors probably use these parameters (if they know about them, that is!) when setting up systems on the production line. Other than that, these parameters have little use for a normal user and, in fact, may be dangerous!

High-Level (Operating-System) Format The final step in the software preparation of a hard disk is the DOS high-level format. The primary function of the high-level format is to create a FAT and a directory system on the disk so that DOS can manage files.

Usually, you perform the high-level format with the standard DOS FORMAT program, using the following syntax:

```
FORMAT C: /S /V
```

This step high-level formats drive C (or volume C, in a multivolume drive), places the hidden operating-system files in the first part of this partition, and prompts for the entry of a volume label to be stored on the disk at completion.

The high-level format program performs the following functions and procedures:

1. Scans the disk (read-only) for tracks and sectors marked as bad during the LLF, and notes these tracks as being unreadable.
2. Returns the drive heads to the first cylinder of the partition and, at that Cylinder 1, Head 1, Sector 1, writes a DOS volume boot sector.
3. Writes a FAT at Head 1, Sector 2. Immediately after this FAT, it writes a second copy of the FAT. These FATs essentially are blank except for bad-cluster marks noting areas of the disk that were found to be unreadable during the marked-defect scan.
4. Writes a blank root directory.
5. If the /S parameter is specified, copies the system files (IBMBIO.COM and IBMDOS.COM or IO.SYS and MSDOS.SYS, depending on which DOS you run) and COMMAND.COM files to the disk (in that order).
6. If the /V parameter is specified, prompts the user for a volume label, which is written as the fourth file entry in the root directory.

Now DOS can use the disk for storing and retrieving files, and the disk is a bootable disk.

> **Note**
>
> The FORMAT command can be run through the Windows Explorer within Windows 95 even on hard disks, as long as no files are open. You cannot format the drive where Windows 95 resides.

During the first phase of the high-level format, a marked defect scan is performed. Defects marked by the LLF operation show up during this scan as being unreadable tracks or sectors. When the high-level format encounters one of these areas, it automatically performs up to five retries to read these tracks or sectors. If the unreadable area was marked by the LLF, the read fails on all attempts.

After five retries, the DOS FORMAT program gives up on this track or sector and moves to the next. Areas that remain unreadable after the initial read and the five retries are noted in the FAT as being bad clusters. DOS 3.3 and earlier versions can mark only entire tracks bad in the FAT, even if only one sector was marked in the LLF. DOS 4 and later versions individually check each cluster on the track and recover clusters that do not involve the LLF marked-bad sectors. Because most LLF programs mark all the sectors on a track as bad, rather than the individual sector that contains the defect, the result of using DOS 3.3 or 4 is the same: all clusters involving sectors on that track are marked in the FAT as bad.

> **Note**
>
> Some LLF programs mark only the individual sector that is bad on a track, rather than the entire track. This is true of the IBM PS/2 low-level formatters on the IBM PS/2 Advanced Diagnostics or Reference disk. In this case, high-level formatting with DOS 4 or later versions results in fewer lost bytes in bad sectors because only the clusters that contain the marked bad sectors are marked bad in the FAT. DOS 4 and later versions display the Attempting to recover allocation unit *x* message (in which *x* is the number of the cluster) in an attempt to determine whether a single cluster or all the clusters on the track should be marked bad in the FAT.

If the controller and LLF program together support sector and track sparing, the high-level format finds the entire disk defect-free because all the defective sectors have been exchanged for spare good ones.

If a disk has been low-level formatted correctly, the number of bytes in bad sectors is the same before and after the high-level format. If the number does change after you repeat a high-level format (reporting fewer bytes or none), the LLF was not done correctly. The manufacturer's defects were not marked correctly, or Norton, Mace, PC Tools, or a similar utility was used to mark defective clusters on the disk. The utilities cannot mark the sectors or tracks at the LLF level. The bad-cluster marks that they make are stored only in the FAT and erased during the next high-level format operation. Defect marks made in the LLF consistently show up as bad bytes in the high-level format, no matter how many times you run the format.

Only an LLF or a surface-analysis tool can correctly mark defects on a disk; anything else makes only temporary bad-cluster marks in the FAT. This kind of marking may be acceptable temporarily, but when additional bad areas are found on a disk, you should run a new LLF of the disk and either mark the area manually or run a surface analysis to place a more permanent mark on the disk.

Hard Disk Drive Troubleshooting and Repair

If a hard disk drive has a problem inside its sealed HDA, repairing the drive usually is not feasible. If the failure is in the logic board, you can replace that assembly with a new or rebuilt assembly easily. However, unless you have data you need to recover, it makes more sense to just buy a new drive, considering today's cost.

Most hard disk problems really are not hardware problems; instead, they are soft problems that can be solved by a new LLF and defect-mapping session. *Soft problems* are characterized by a drive that sounds normal but produces various read and write errors.

Hard problems are mechanical, such as when the drive sounds as though it contains loose marbles. Constant scraping and grinding noises from the drive, with no reading or writing capability, also qualify as hard errors. In these cases, it is unlikely that an LLF will put the drive back into service. If a hardware problem is indicated, first replace the logic-board assembly. You can make this repair yourself and, if successful, you can recover the data from the drive. If replacing the logic assembly does not solve the problem, contact the manufacturer or a specialized repair shop that has clean-room facilities for hard disk repair.

Most of the time, a *seek failure* indicates that the drive is not responding to the controller. This failure usually is caused by one of the following problems:

- Incorrect drive-select jumper setting
- Loose, damaged, or backward control cable
- Loose or bad power cable
- Friction between drive heads and platters
- Bad power supply

If a diagnostics cylinder read error occurs, the most likely problems are these:

- Incorrect drive-type setting
- Loose, damaged, or backward data cable

The methods for correcting most of these problems are obvious. If the drive-select jumper setting is incorrect, for example, correct it. If a cable is loose, tighten it. If the power supply is bad, replace it. You get the idea.

Chapter 4

Controller Card Settings

Interface Connections

Please consult Chapter Two, "Hard Disk Interfaces" for specific details on connectors for the IDE and SCSI interfaces.

Controller to Drive Power Connections

Figure 4-1 shows the connections that are common to all controllers with an on-board power source.

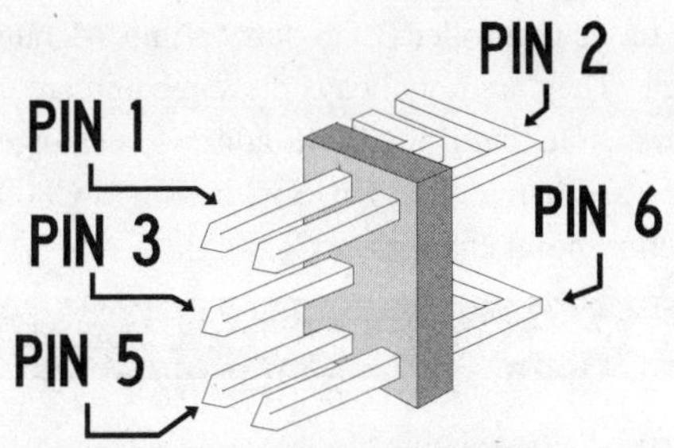

PIN	DESCRIPTION
1	+12 Volts DC
2	+12 Volts DC
3	+5 Volts DC
4	Not used
5	Ground
6	Ground

Figure 4-1: Controller Onboard Power Connector

Drive Activity Connector

Figure 4-2 shows connections that are common to all controllers with a 4-pin drive activity connector. A LED may be connected to either ground pin (pin 1 or pin 4) and either signal pin (pin 2 or pin 3).

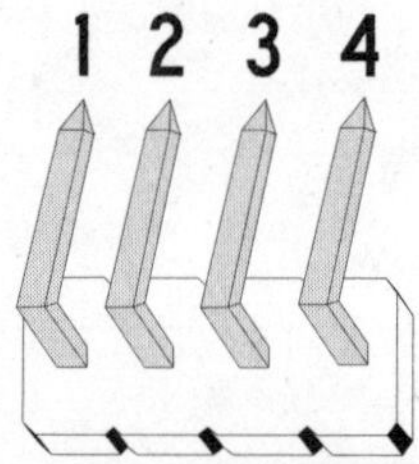

PIN	DESCRIPTION
1	Ground
2	+5 Volts DC
3	+5 Volts DC
4	Ground

(Lines 2 and 3 only go high during disk activity)

Figure 4-2: Controller Drive Activity LED Connector

Common BIOS Debug Format Codes

Table 4-1 is a general guide to the controller BIOS entry points (debug codes) on many of the listed manufacturers' products. They are not always the same and are usually only available on 8-bit RLL and MFM controllers. Note that if the BIOS address is changed on the controller, then the entry point listed below will also change (i.e.: BIOS address was C800, entry point was C800:5. Address changed to CC00, entry point changes to CC00:5).

Table 4-1: Common BIOS Debug Format Codes

Adaptec	G=C800:ccc or G=C800:6
Scientific Micro Systems-OMTI	G=C800:6
Seagate	G=C800:5
Western Digital	G=C800:5

To use one of the above, enter the DOS Debug utility and type the corresponding code at the dash (-) prompt.

Common Hardware Settings

Default Jumper Settings

The default or most common jumpers positions are denoted by the ⇨ symbol. Pin-1 and jumper 1 are indicated by an arrowhead symbol (▲).

Base I/O Address

This option allows the selection of the base I/O address from among the choices given. Available addresses are usually standardized to avoid conflict with other peripherals.

IRQ

IRQ is short for Interrupt Request, which is the standard method used by peripherals to access the CPU. Available IRQs are usually standardized to avoid conflict with other peripherals

BIOS Address

Controller cards with an onboard BIOS usually have an option to set the BIOS address. BIOS address is sometimes called "memory address."

Data Transfer Modes

Some controllers allow the selection of the data transfer mode between PIO and DMA. Be sure that the transfer mode selected is supported by all hardware elements in the system.

Non-Hard Drive I/O Options

Many controllers bundle other I/O options, such as floppy, parallel, and serial ports. Be sure to de-conflict any IRQ or base I/O address settings between these ports and any ports that may already exist in the system. COM refers to serial ports, and LPT refers to parallel ports.

Undocumented Jumpers

Some jumper settings are specified as "unidentified," "factory configured – do not alter," or may have a dash. Information was not available on these jumpers or settings and their configuration should not be altered.

Selected Controller Settings

The remainder of this chapter contains information on specific hardware settings for selected controllers.

ACCULOGIC, INC.

ISAPPORT

Data bus: 16-bit ISA
Size: Half-length, full-height card
Hard drive supported: Up to seven Fast SCSI-2 devices
Floppy drives supported: Two 360KB, 720KB, 1.2MB, 1.44MB, or 2.88MB drives

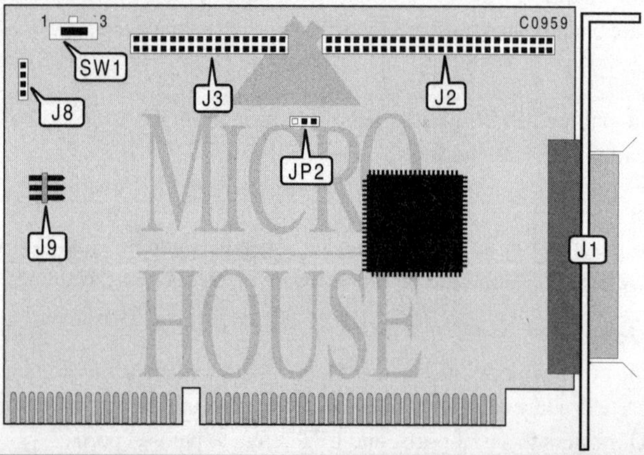

CONNECTIONS	
Function	**Location**
50-pin SCSI connector - external	J1
50-pin SCSI connector - internal	J2
34-pin cable connector - floppy drive	J3
4-pin power connector	J8
4-pin connector - drive active LED	J9

USER CONFIGURABLE SETTINGS		
Function	**Location**	**Setting**
⇨ Floppy drive interface enabled as primary	JP2	Pins 2 & 3 closed
Floppy drive interface disabled	JP2	Pins 1 & 2 closed
⇨ Normally configure board during installation	SW1	Position 2
Lock configuration during installation	SW1	Position 1
Reset configuration	SW1	Position 3

Note: SW1 is set to normal unless two ISApport boards are installed on the same system. In this situation, the boards are installed sequentially. First, board 1 is installed and configured. Then, its SW1 is set to lock, and board 2 is installed and configured with its SW1 set to normal. The reset setting for SW1 is not identified.

ACCULOGIC, INC.

SIDE-2/PCI

Data bus:	32-bit PCI
Size:	Half-length, half-height card
Hard drive supported:	Four IDE(AT) drives
Floppy drives supported:	None

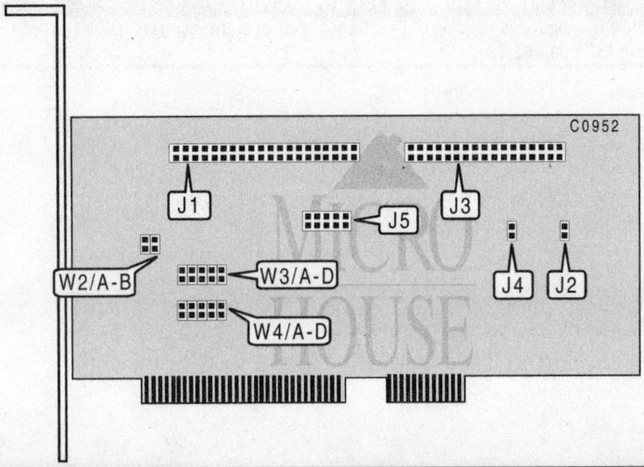

CONNECTIONS

Function	Location
40-pin primary IDE(AT) connector	J1
2-pin connector - drive active LED	J2
40-pin secondary IDE(AT) connector	J3
10-pin ISA IRQ board connector	J5

USER CONFIGURABLE SETTINGS

Function	Location	Setting
➪ Factory configured - do not alter	J4	N/A

ADDRESS MODE SELECTION

Mode	W2/A	W2/B
➪ ISA IDE	Off	On
Native mode	On	Off

PCI PRIMARY INTERRUPT SELECTION

INT	W3/A	W3/B	W3/C	W3/D
➪ ISA IRQ 14	Off	Off	Off	Off
A	On	Off	Off	Off
B	Off	On	Off	Off
C	Off	Off	On	Off
D	Off	Off	Off	On

PCI SECONDARY INTERRUPT SELECTION

INT	W4/A	W4/B	W4/C	W4/D
➪ ISA IRQ 15	Off	Off	Off	Off
A	On	Off	Off	Off
B	Off	On	Off	Off
C	Off	Off	On	Off
D	Off	Off	Off	On

Continued on next page. . .

ACCULOGIC, INC.

SIDE-2/PCI

. . .continued from previous page

MISCELLANEOUS TECHNICAL NOTES
If IRQ 14 or 15 is selected, the ISA IRQ board is required. This board connects the primary IDE interrupt to the ISA bus IRQ 14 and secondary IDE interrupt to ISA bus IRQ 15.

ACCULOGIC, INC.

SIDE-3/*PLUS*

Data bus: 16-bit, ISA
Size: Half-length, half-height card
Hard drives supported: Two IDE(AT) drives
Floppy drives supported: Two 360KB, 720KB, 1.2MB, 1.44MB or 2.88MB

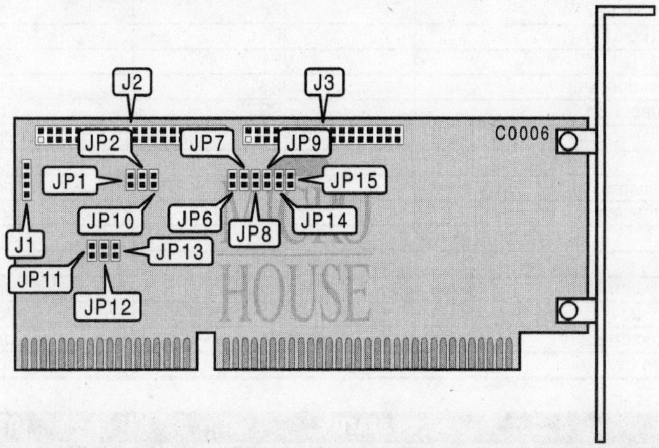

CONNECTIONS	
Function	**Location**
4-pin connector-Drive Active LED	J1
40-pin AT Interface (IDE) connector	J2
34-pin control cable connector-floppy drive	J3

BIOS ADDRESS		
Address	**Jumper JP12**	**Jumper JP13**
⇨ C800h	Open	open
CC00h	Open	closed
D800h	Closed	open
DC00h	Closed	closed

Continued on next page. . .

ACCULOGIC, INC.

SIDE-3/*PLUS*

. . .continued from previous page

USER CONFIGURABLE SETTINGS		
Function	**Location**	**Setting**
➪ Floppy drive port address is 3F0-3F7h	JP6	open
Floppy drive port address is 370-377h	JP6	closed
➪ Hard drive port address is 1F0-1F7h	JP7	open
Hard drive port address is 170-177h	JP7	closed
➪ Single speed floppy drives supported	JP8	open
Dual speed floppy drives supported	JP8	closed
➪ Floppy drive precompensation is 125ns	JP9	open
Floppy drive precompensation is 187ns	JP9	closed
➪ I/O Channel Ready signal disabled	JP10	open
I/O Channel Ready signal enabled	JP10	closed
➪ BIOS enabled	JP11	open
BIOS disabled	JP11	closed
➪ 2.88MB drives not supported on A:	JP14	open
2.88MB drives supported on A:	JP14	closed
➪ 2.88MB drives not supported on B:	JP15	open
2.88MB drives supported on B:	JP15	closed

IRQ14 STATUS		
Function	**Jumper JP1**	**Jumper JP2**
➪ IRQ14 normal	closed	open
IRQ14 delayed until ACTIVE- signal is removed	open	closed

MISCELLANEOUS TECHNICAL NOTES
This Controller has no on-board software.

ACCULOGIC, INC.

SIDE-4

Data bus: 16-bit, ISA
Size: Half-length, full-height card
Hard drives supported: Two IDE(AT) drives
Floppy drives supported: Two 360KB, 720KB, 1.2MB or 1.44MB drives

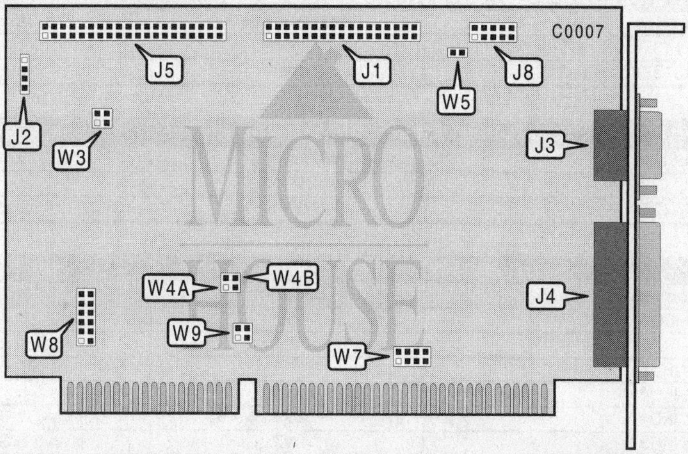

CONNECTIONS	
Function	**Location**
34-pin control cable connector-floppy drive	J1
4-pin connector-Drive Active LED	J2
9-pin serial port 1-external	J3
25-pin parallel port-external	J4
40-pin AT Interface (IDE) connector	J5
9-pin serial port 2-internal	J8

IRQ14 STATUS		
Function	**W3/Pins 1 & 2**	**W3/Pins 3 & 4**
➪ IRQ14 normal	open	closed
IRQ14 delayed until ACTIVE- signal is removed	closed	open

USER CONFIGURABLE SETTINGS		
Function	**Location**	**Setting**
➪ Floppy drive enabled	W4A	closed
Floppy drive disabled	W4A	open
➪ Floppy drive port address is 3F0-3F7h	W4B	open
Floppy drive port address is 370-377h	W4B	closed
➪ Single speed floppy drives supported	W5	open
Dual speed floppy drives supported	W5	closed
➪ Hard drive enabled	W9	closed
Hard drive disabled	W9	open

PARALLEL PORT (CN4) CONFIGURATION		
CN4	**W8/jumper 1**	**W8/jumper 2**
LPT1 enabled	closed	open
LPT2 enabled	closed	closed
Disabled	open	open

Continued on next page . . .

ACCULOGIC, INC.

SIDE-4

. . . continued from previous page

SERIAL PORT 1 (CN3) CONFIGURATION		
Function	**W8/Jumper 3**	**W8/Jumper 4**
➪ Serial port 1 set to COM1 (3F8h)	closed	open
Serial port 1 set to COM3 (3E8h)	open	closed

SERIAL PORT 2 (CN8) CONFIGURATION		
Function	**W8/Jumper 5**	**W8/Jumper 6**
➪ Serial port 2 set to COM2 (2F8h)	closed	open
Serial port 2 set to COM4 (2E8h)	open	closed

I/O PORT INTERRUPT REQUEST		
Function	**Location**	**Setting**
➪ Parallel port (CN4) uses IRQ7	W7	pins 3 & 4 closed
Parallel port (CN4) uses IRQ5	W7	pins 1 & 2 closed
➪ Serial port 2 (CN8) uses IRQ3	W7	pins 5 & 6 closed
Serial port 2 (CN8) uses IRQ4	W7	pins 5 & 7 closed
➪ Serial port 1 (CN3) uses IRQ4	W7	pins 7 & 8 closed
Serial port 1 (CN3) uses IRQ3	W7	pins 6 & 8 closed

MISCELLANEOUS TECHNICAL NOTES
This controller has no on-board software.

ACCULOGIC, INC.

SIDE-4/*PLUS*

Data bus: 16-bit, ISA
Size: Half-length, full-height card
Hard drives supported: Two IDE(AT) drives
Floppy drives supported: Two 360KB, 720KB, 1.2MB or 1.44MB drives

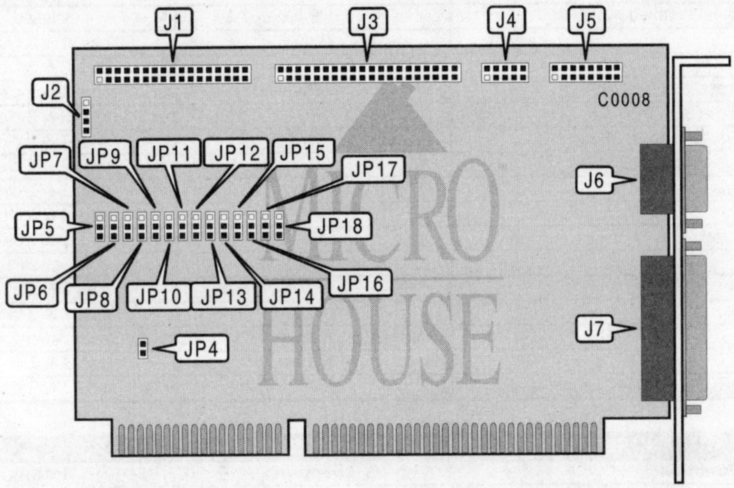

CONNECTIONS	
Function	**Location**
34-pin control cable connector-floppy drive	J1
4-pin connector-drive active LED	J2
40-pin AT Interface (IDE) connector	J3
9-pin serial port 2-internal	J4
15-pin game port-internal	J5
9-pin serial port 1-external	J6
25-pin parallel port-external	J7

Continued on next page . . .

ACCULOGIC, INC.

SIDE-4/*PLUS*

. . . continued from previous page

USER CONFIGURABLE SETTINGS		
Function	**Location**	**Setting**
➪ I/O Channel Ready signal disabled	JP4	open
I/O Channel Ready signal enabled	JP4	closed
➪ Floppy drive enabled	JP5	pins 1 & 2 closed
Floppy drive disabled	JP5	pins 2 & 3 closed
➪ Floppy drive port address is 3F0-3F7h	JP6	pins 1 & 2 closed
Floppy drive port address is 370-377h	JP6	pins 2 & 3 closed
➪ Single speed floppy drives supported	JP7	pins 1 & 2 closed
Dual speed floppy drives supported	JP7	pins 2 & 3 closed
➪ Hard drive enabled	JP8	pins 1 & 2 closed
Hard drive disabled	JP8	pins 2 & 3 closed
➪ IDE select mode enabled	JP9	pins 1 & 2 closed
Not used	JP9	open
➪ Hard drive port address is 1F0-1F7h	JP10	pins 1 & 2 closed
Hard drive port address is 170-177h	JP10	pins 2 & 3 closed
➪ Not used	JP11	open
➪ IRQ14 normal	JP12	pins 1 & 2 closed
IRQ14 delayed until ACTIVE- signal is removed	JP12	pins 2 & 3 closed

SERIAL & PARALLEL PORT CONFIGURATIONS		
Function	**Location**	**Setting**
➪ Serial port 1 (CN6) enabled	JP13	pins 1 & 2 closed
Serial port 1 (CN6) disabled	JP13	pins 2 & 3 closed
➪ Serial port 1 (CN6) set to COM1	JP14	pins 1 & 2 closed
Serial port 1 (CN6) set to COM3	JP14	pins 2 & 3 closed
➪ Serial port 2 (CN4) enabled	JP15	pins 1 & 2 closed
Serial port 2 (CN4) disabled	JP15	pins 2 & 3 closed
➪ Serial port 2 (CN4) set to COM2	JP16	pins 1 & 2 closed
Serial port 2 (CN4) set to COM4	JP16	pins 2 & 3 closed
➪ Parallel port (CN7) enabled	JP17	pins 1 & 2 closed
Parallel port (CN7) disabled	JP17	pins 2 & 3 closed
➪ Parallel port (CN7) address is LPT1	JP18	pins 1 & 2 closed
Parallel port (CN7) address is LPT2	JP18	pins 2 & 3 closed

ADAPTEC, INC.

AHA-1510

Data bus: 16-bit, ISA
Size: Half-length, full-height card
Hard drives supported: Up to seven SCSI devices
Floppy drives supported: None

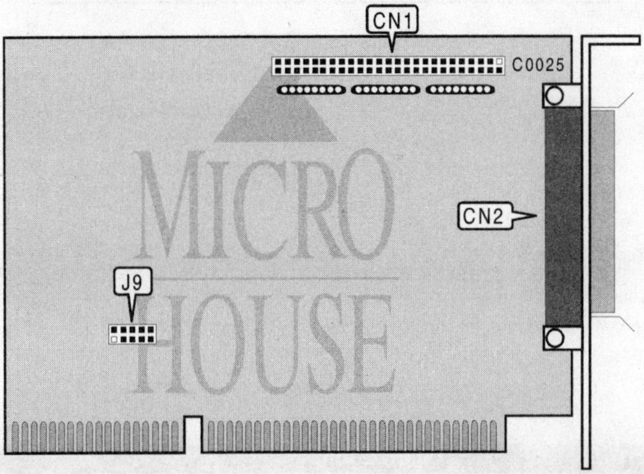

CONNECTIONS	
Function	**Location**
50-pin SCSI cable connector-internal	CN1
50-pin SCSI cable connector-external	CN2

USER CONFIGURABLE SETTINGS		
Function	**Location**	**Setting**
⇨ I/O port address 340h	J9/ umper 5	open
I/O port address 140h	J9/ umper 5	closed
Note: If the secondary address, 140h, is selected, the option must be set accordingly in ASPI.S S.		

INTERRUPT REQUEST				
Select	**J9/jumper 1**	**J9/jumper 2**	**J9/jumper 3**	**J9/jumper 4**
⇨ IRQ11	open	closed	open	open
IRQ12	closed	open	open	open
IRQ10	open	open	closed	open
IRQ9	open	open	open	closed

MISCELLANEOUS TECHNICAL NOTES
The SCSI Disconnection, SCSI Address, SCSI Parity, Sync. Negotiation, I/O Port Address, and Data Transfer Mode options are set by the software driver, ASPI.S S.

ADAPTEC, INC.

AHA-1510A

Data bus: 16-bit ISA
Size: Half-length, full-height card
Hard drive supported: Up to seven SCSI devices
Floppy drives supported: None

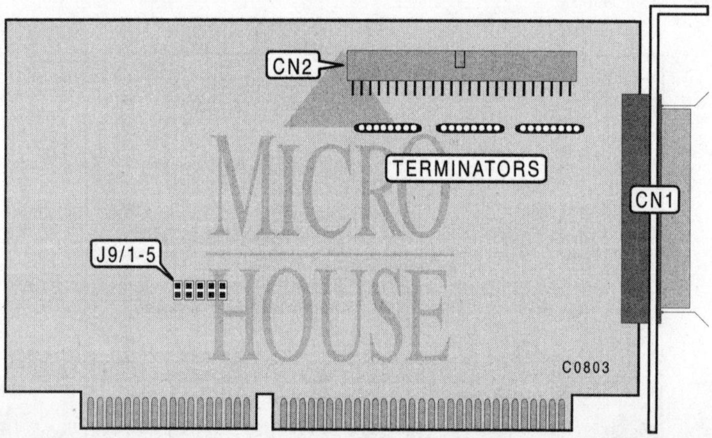

CONNECTIONS	
Function	**Location**
50-pin SCSI connector - external	CN1
50-pin SCSI connector - internal	CN2

USER CONFIGURABLE SETTINGS		
Function	**Location**	**Setting**
⇨ Adapter I/O address is 340h	J9/5	Open
Adapter I/O address is 140h	J9/5	Closed
Note: When connecting both external & internal devices to 1510A - remove terminators		

INTERRUPT SELECTION				
IRQ	**J9/1**	**J9/2**	**J9/3**	**J9/4**
⇨ 11	open	closed	open	Open
9	open	open	open	Closed
10	open	open	closed	Open
12	closed	open	open	Open

ADAPTEC, INC.

AHA-1520, AHA-1522

Data bus: 16-bit, ISA
Size: Half-length, full-height card
Hard drives supported: Up to seven SCSI devices
Floppy drives supported: Two 360KB, 720KB, 1.2MB or 1.44MB

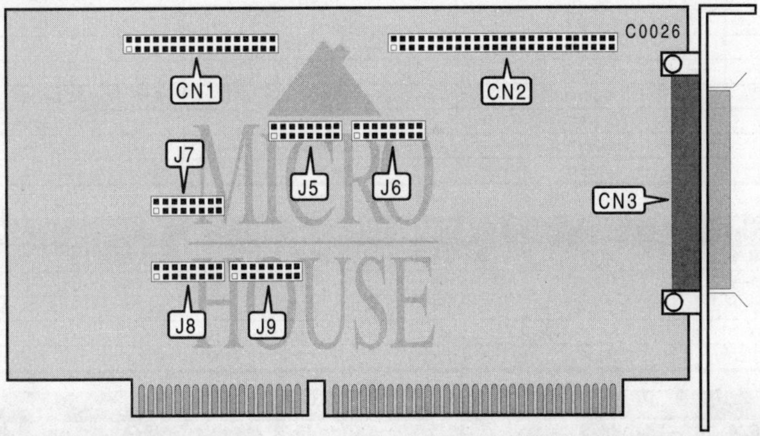

CONNECTIONS	
Function	**Location**
34-pin data cable connector-floppy drive	CN1
50-pin SCSI cable connector-internal	CN2
50-pin SCSI cable connector-external	CN3

USER CONFIGURABLE SETTINGS		
Function	**Location**	**Setting**
➪ Data transfer in DMA disabled	J5/ umper 1	open
Data transfer in DMA enabled	J5/ umper 1	closed
➪ Boot enabled (Intercept INT19 call)	J5/ umper 2	closed
Boot disabled (INT19 normal)	J5/ umper 2	open
➪ Synchronous negotiation enabled	J5/ umper 5	closed
Synchronous negotiation disabled	J5/ umper 5	open
➪ Disconnect by target / INT15 enabled	J5/ umper 6	closed
Disconnect by target / INT15 disabled	J5/ umper 6	open
➪ Factory configured - do not alter	J5/ umpers 7 & 8	open

USER CONFIGURABLE SETTINGS		
Function	**Location**	**Setting**
➪ SCSI parity check enabled	J6/ umper 8	open
SCSI parity check disabled	J6/ umper 8	closed
➪ Floppy drive enabled	J7/ umper 1	closed
Floppy drive disabled	J7/ umper 1	open
➪ Floppy drive Interrupt is IRQ6	J7/ umper 6	closed
Floppy drive Interrupt is IRQ10	J7/ umper 7	closed
➪ Floppy drive-dual speed disabled	J7/ umper 8	open
Floppy drive-dual speed enabled	J7/ umper 8	closed
➪ I/O port address 340h	J9/ umper 5	open
I/O port address 140h	J9/ umper 5	closed

Continued on next page . . .

ADAPTEC, INC.

AHA-1520, AHA-1522

. . . continued from previous page

SCSI ADDRESS			
SCSI ID	**J6/jumper 1**	**J6/jumper 2**	**J6/jumper 3**
0	open	open	open
1	closed	open	open
2	open	closed	open
3	closed	closed	open
4	open	open	closed
5	closed	open	closed
6	open	closed	closed
⇨ 7	closed	closed	closed

DMA CHANNEL						
DMA	**J6/6**	**J6/7**	**J8/1 & 5**	**J8/2 & 6**	**J8/3 & 7**	**J8/4 & 8**
7	closed	closed	closed	open	open	open
6	open	closed	open	closed	open	open
5	closed	open	open	open	closed	open
⇨ 0	open	open	open	open	open	closed

INTERRUPT REQUEST						
IRQ	**J6/4**	**J6/5**	**J9/1**	**J9/2**	**J9/3**	**J9/4**
IRQ9	open	open	open	open	open	closed
IRQ10	closed	open	open	open	closed	open
⇨ IRQ11	open	closed	open	closed	open	open
IRQ12	closed	closed	closed	open	open	open

BIOS ADDRESS			
Address	**J9/jumper 6**	**J9/jumper 7**	**J9/jumper 8**
⇨ DC000h	closed	closed	closed
C8000h	open	open	closed
CC000h	open	closed	closed
D8000h	closed	open	closed
Disabled	N/A	N/A	open

FLOPPY DRIVE DMA - J7				
	DRQ		**DACK**	
DMA	**Jumper 2**	**Jumper 3**	**Jumper 4**	**Jumper 5**
⇨ DMA Channel 2	closed	open	closed	open
DMA Channel 3	open	closed	open	closed

MESSAGE SELECT Msg #0 - Adaptec Header Msg #1 - Jumper config. info. Msg #2 - SCSI device info. Msg #3 - Boot Progress info. Msg #4 - Error messages		
Message	**JP5/jumper 3**	**JP5/jumper 4**
⇨ Msg 0 and 4	open	open
Msg 0, 3 and 4	open	closed
All messages displayed	closed	open
Msg 4 only	closed	closed

ADAPTEC, INC.

AHA-1520A, AHA1522A

Data bus: 16-bit ISA
Size: Half-length, full-height card
Hard drive supported: Up to seven SCSI devices
Floppy drives supported: None

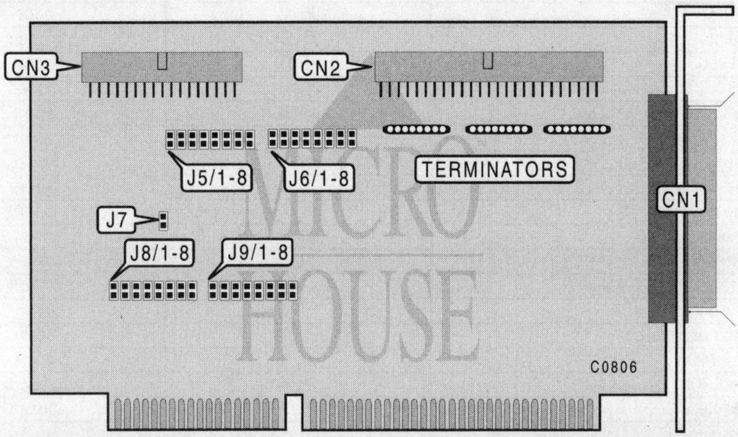

CONNECTIONS	
Function	**Location**
50-pin SCSI connector - external	CN1
50-pin SCSI connector - internal	CN2
34-pin floppy controller - 1522A only	CN3

SCSI ID SELECTION			
SCSI ID	**J6/1**	**J6/2**	**J6/3**
➪ 7	closed	closed	closed
0	open	open	open
1	closed	open	open
2	open	closed	open
3	closed	closed	open
4	open	open	closed
5	closed	pen	closed
6	open	closed	closed

Continued on next page. . .

ADAPTEC, INC.

AHA-1520A, AHA1522A

. . .continued from previous page

USER CONFIGURABLE SETTINGS		
Function	**Location**	**Setting**
➪ Factory configured - do not alter	J5/1	open
➪ INT 19 is re-routed	J5/2	closed
INT 19 is not re-routed	J5/2	open
➪ Sync. negotiation enabled	J5/5	closed
Sync. negotiation disabled	J5/5	open
➪ Disconnect by target SCSI device enabled	J5/6	closed
Disconnect by target SCSI device disabled	J5/6	open
➪ Factory configured - do not alter	J5/7	open
➪ Factory configured - do not alter	J5/8	open
➪ Factory configured - do not alter	J6/6	open
➪ Factory configured - do not alter	J6/7	open
➪ SCSI parity check enabled	J6/8	open
SCSI parity check disabled	J6/8	closed
➪ Floppy drive enabled (only on 1522A)	J7	closed
Floppy drive disabled (only on 1522A)	J7	open
➪ Factory configured - do not alter	J8/1-8	open
➪ Adapter I/O address is 340h	J9/5	open
Adapter I/O address is 140h	J9/5	closed
➪ Host adapter BIOS enabled	J9/8	closed
Host adapter BIOS disabled	J9/8	open
Note: When connecting both external & internal devices to the host adapter - remove terminators		

MESSAGES AT BOOT-UP		
Message	**J5/3**	**J5/4**
➪ Adapter header	open	open
Error messages	closed	closed
Boot progress report	open	closed
SCSI device information	closed	open

BIOS ADDRESS SELECTION		
Address	**J9/6**	**J9/7**
➪ DFFFFh	closed	closed
CBFFFh	open	open
CFFFFh	open	closed
DBFFFh	closed	open

INTERRUPT SELECTION						
IRQ	**J6/4**	**J6/5**	**J9/1**	**J9/2**	**J9/3**	**J9/4**
➪ 11	open	closed	open	closed	open	open
9	open	open	open	open	open	closed
10	closed	open	open	open	closed	open
12	closed	closed	closed	open	open	open

ADAPTEC, INC.

AHA-1540

Data bus:	16-bit ISA
Size:	Full-length, full-height card
Hard drive supported:	Up to seven SCSI drives
Floppy drives supported:	None

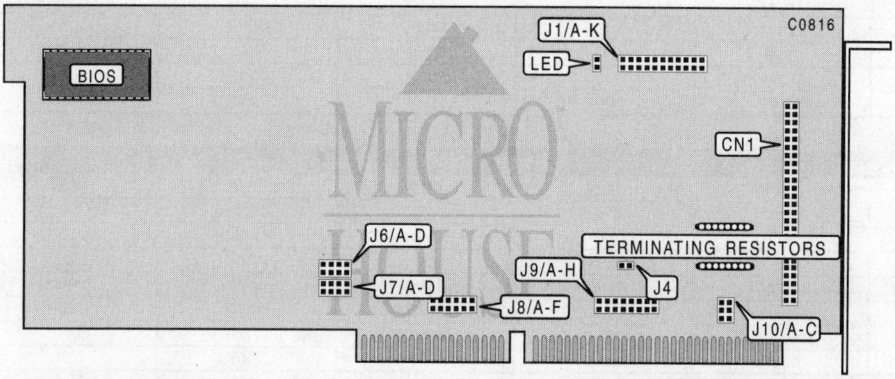

CONNECTIONS	
Function	**Location**
50-pin SCSI connector - internal	CN1
2-pin connector - drive active LED	LED

USER CONFIGURABLE SETTINGS		
Function	**Location**	**Setting**
➪ Synchronous negotiation disabled	J1/A	open
Synchronous negotiation enabled	J1/A	closed
➪ Factory configured - do not alter	J1/B	open
➪ Parity enabled	J1/C	open
Parity disabled	J1/C	closed
➪ Wait state disabled	J4	N/A
Wait state enabled	J4	N/A

scsi address configuration			
Address	**J1/D**	**J1/E**	**J1/F**
➪ 7	open	open	open
0	closed	closed	closed
1	open	closed	closed
2	closed	open	closed
3	open	open	closed
4	closed	closed	Open
5	open	closed	Open
6	closed	open	Open

DMA channel selection					
Channel	**J1/G**	**J1/H**	**J6/A**	**J6/B**	**J6/C**
➪ 5	open	closed	open	closed	Open
0	closed	closed	closed	open	Open
6	closed	open	open	open	Closed
7	open	open	open	open	Open

Continued on next page. . .

ADAPTEC, INC.

AHA-1540

. . .continued from previous page

DMA channel selection (con t)					
Channel	**J6/D**	**J7/A**	**J7/B**	**J7/C**	**J7/D**
➪ 5	open	open	closed	open	Open
0	open	closed	open	open	Open
6	open	open	open	closed	Open
7	closed	open	open	open	Closed

At interrupt selection					
IRQ	**J1/I**	**J1/J**	**J1/K**	**J8/A**	**J8/B**
➪ 11	open	closed	open	open	Open
9	open	open	open	closed	Open
10	closed	open	open	open	Closed
12	closed	closed	open	open	Open
14	open	open	closed	open	Open
15	closed	open	closed	open	Open
Not defined	open	closed	closed	N/A	N/A
Not defined	closed	closed	closed	N/A	N/A

At interrupt selection (con t)				
IRQ	**J8/C**	**J8/D**	**J8/E**	**J8/F**
➪ 11	closed	open	open	Open
9	open	open	open	Open
10	open	open	open	Open
12	open	closed	open	Open
14	open	open	closed	Open
15	open	open	open	Closed

port address selection								
Address	**J9/A**	**J9/B**	**J9/C**	**J9/D**	**J9/E**	**J9/F**	**J9/G**	**J9/H**
➪ 330h	open	open	closed	closed	open	open	closed	closed
004h	closed	open	open	open	open	open	open	open
008h	open	closed	open	open	open	open	open	open
010h	open	open	closed	open	open	open	open	open
020h	open	open	open	closed	open	open	open	open
040h	open	open	open	open	closed	open	open	open
080h	open	open	open	open	open	closed	open	open
100h	open	open	open	open	open	open	closed	open
200h	open	open	open	open	open	open	open	closed

AT bios address selection			
Address	**J10/A**	**J10/B**	**J10/C**
➪ DC000h	open	open	Open
C8000h	closed	open	Closed
D8000h	open	open	Closed
CC000h	closed	open	Open

ADAPTEC, INC.

AHA-1540A, AHA-1542A

Data bus: 16-bit, ISA
Size: Full-length, full-height card
Hard drives supported: Up to seven SCSI devices
Floppy drives supported: Two 360KB, 720KB, 1.2MB or 1.44MB

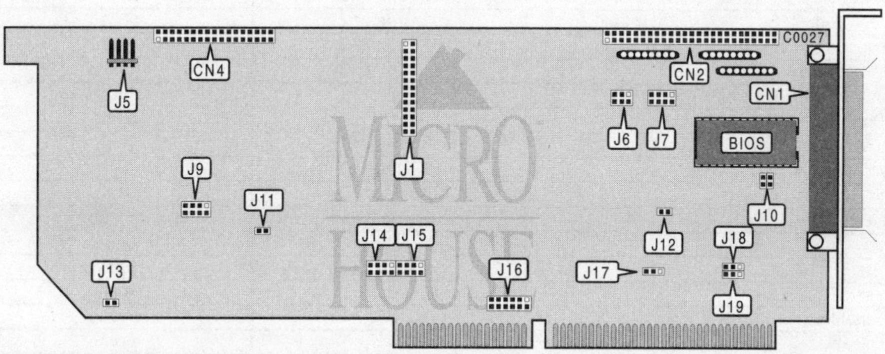

CONNECTIONS	
Function	**Location**
50-pin SCSI connector-external	CN1
50-pin SCSI connector-internal	CN2
34-pin data cable connector-floppy drive	CN4
4-pin connector-drive active LED	J5

SCSI ADDRESS			
SCSI ID	**J1/jumper 4**	**J1/jumper 5**	**J1/jumper 6**
➪ 7	open	open	open
6	closed	open	open
5	open	closed	open
4	closed	closed	open
3	open	open	closed
2	closed	open	closed
1	open	closed	closed
0	closed	closed	closed

Continued on next page . . .

ADAPTEC
AHA-1540A, AHA-1542A

. . . continued from previous page

USER CONFIGURABLE SETTINGS		
Function	**Location**	**Setting**
➪ Synchronous negotiation disabled	J1/ umper 1	open
Synchronous negotiation enabled	J1/ umper 1	closed
➪ Factory test disabled	J1/ umper 2	open
Factory test enabled	J1/ umper 2	closed
➪ SCSI parity checking enabled	J1/ umper 3	open
SCSI parity checking disabled	J1/ umper 3	closed
➪ LA enable (Only on early W1542A)	J8	closed
LA disable (Only on early W1542A)	J8	open
➪ Auto request sense enabled	J9/ umper 1	open
Auto request sense disabled	J9/ umper 1	closed
➪ Factory configured - do not alter	J9/ umpers 2-4	open
➪ BOIS enabled	J11	closed
BIOS disabled	J11	open
➪ Floppy address 3F0-3F7h	J12	open
Floppy address 370-377h	J12	closed
➪ Floppy drive enabled	J13	closed
Floppy drive disabled	J13	open
➪ Floppy drive Interrupt is IRQ6	J17	pins 1 & 2 closed
Floppy drive Interrupt is IRQ10	J17	pins 2 & 3 closed
➪ Floppy drive DMA Request is DREQ2	J18	pins 1 & 2 closed
Floppy drive DMA Request is DREQ3	J18	pins 2 & 3 closed
➪ Floppy drive DMA Acknowledge Is DACK2	J19	pins 1 & 2 closed
Floppy drive DMA Acknowledge is DACK3	J19	pins 2 & 3 closed

DMA CHANNEL				
DRQ	**J1/Jumper 7**	**J1/Jumper 8**	**J14**	**J15**
0	closed	closed	Jumper 1-closed	Jumper 1-closed
➪ 5	open	closed	Jumper 2-closed	Jumper 2-closed
6	closed	open	Jumper 3-closed	Jumper 3-closed
7	open	open	Jumper 4-closed	Jumper 4-closed

INTERRUPT REQUEST				
IRQ	**J1/Jumper 9**	**J1/Jumper 10**	**J1/Jumper 11**	**J16**
9	open	open	open	Jumper 1-closed
10	closed	open	open	Jumper 2-closed
➪ 11	open	closed	open	Jumper 3-closed
12	closed	closed	open	Jumper 4-closed
14	open	open	closed	Jumper 5-closed
15	closed	open	closed	Jumper 6-closed

DMA TRANSFER RATE		
Speed	**J1/jumper 12**	**J1/jumper 13**
➪ 5.0	open	open
5.7	closed	open
6.7	open	closed
8.0	closed	closed

Continued on next page . . .

ADAPTEC

AHA-1540A, AHA-1542A

. . . continued from previous page

I/O PORT ADDRESS			
Address	**J6/Jumper 1**	**J6/Jumper 2**	**J6/Jumper 3**
334h	open	open	open
➪ 330h	open	open	closed
234h	open	closed	open
230h	open	closed	closed
134h	closed	open	open
130h	closed	open	closed

BUS WAIT STATE				
Delay	**J7/jumper 1**	**J7/jumper 2**	**J7/jumper 3**	**J7/jumper 4**
➪ Disable	closed	open	open	open
100nsec	open	closed	open	open
200nsec	open	open	closed	open
300nsec	open	open	open	closed

BIOS ADDRESS		
Address	**J10/jumper 1**	**J10/jumper 2**
➪ DC000h	open	open
CC000h	closed	open
D8000h	open	closed
C8000h	closed	closed

ADAPTEC, INC.

AHA-1540B, AHA-1542B
(Also Sold as the Maynard HA-101S)

Data bus: 16-bit, ISA
Size: Three-quarter length, full-height card
Hard drives supported: Up to seven SCSI devices
Floppy drives supported: Two 360KB, 720KB, 1.2MB or 1.44MB (AHA-1542B)
None (AHA-1540B, HA-101S)

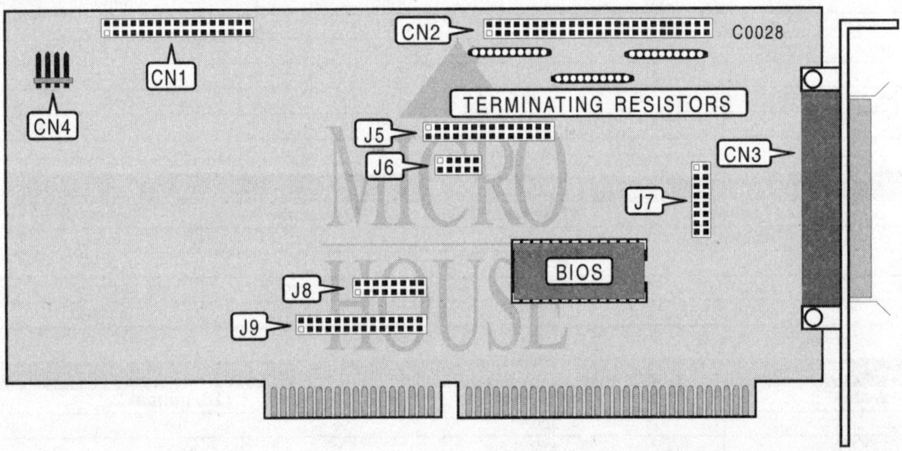

CONNECTIONS	
Function	**Location**
34-pin data cable connector-floppy drive	CN1
50-pin SCSI connector-internal	CN2
50-pin SCSI connector-external	CN3
4-pin connector-drive active LED	CN4

SCSI ADDRESS			
Select	**J5/jumper 4**	**J5/jumper 5**	**J5/jumper 6**
⇨ 7	open	open	open
6	closed	open	open
5	open	closed	open
4	closed	closed	open
3	open	open	closed
2	closed	open	closed
1	open	closed	closed
0	closed	closed	closed

Continued on next page . . .

ADAPTEC, INC.

AHA-1540B, AHA-1542B
(Also Sold as the Maynard HA-101S)

. . . continued from previous page

USER CONFIGURABLE SETTINGS		
Function	**Location**	**Setting**
➪ SCSI synchronous negotiation disabled	J5/ umper 1	open
SCSI synchronous negotiation enabled	J5/ umper 1	closed
➪ Normal operation	J5/ umper 2	open
Self diagnostics enabled	J5/ umper 2	closed
➪ SCSI parity check enabled	J5/ umper 3	open
SCSI parity check disabled	J5/ umper 3	closed
➪ BIOS enabled	J6/ umper 1	closed
BIOS disabled	J6/ umper 1	open
➪ Not used	J6/ umpers 2-4	open
➪ Auto sense enabled	J6/ umper 5	open
Auto sense disable	J6/ umper 5	closed
➪ Floppy drive port address 3F0-3F7h	J7/ umper 1	open
Floppy drive port address 370-377h	J7/ umper 1	closed
➪ Floppy drive enabled	J8/ umper 1	closed
Floppy drive disabled	J8/ umper 1	open
➪ Floppy drive DMA Request is DREQ2	J8/ umper 2	closed
Floppy drive DMA Request is DREQ3	J8/ umper 3	closed
➪ Floppy drive DMA Acknowledge Is DACK2	J8/ umper 4	closed
Floppy drive DMA Acknowledge is DACK3	J8/ umper 5	closed
➪ Floppy drive Interrupt is IRQ6	J8/ umper 6	closed
Floppy drive Interrupt is IRQ10	J8/ umper 7	closed
➪ Floppy drive dual speed disabled	J8/ umper 8	open
Floppy drive dual speed enabled	J8/ umper 8	closed
➪ AT Bus DMA Request is DREQ5	J9/ umper 2	closed
AT Bus DMA Request is DREQ0	J9/ umper 1	closed
AT Bus DMA Request is DREQ6	J9/ umper 3	closed
AT Bus DMA Request is DREQ7	J9/ umper 4	closed
AT Bus DMA Acknowledge is DACK0	J9/ umper 5	closed
➪ AT Bus DMA Acknowledge is DACK5	J9/ umper 6	closed
AT Bus DMA Acknowledge is DACK6	J9/ umper 7	closed
AT Bus DMA Acknowledge is DACK7	J9/ umper 8	closed
Hard drive Interrupt is IRQ9	J9/ umper 9	closed
Hard drive Interrupt is IRQ10	J9/ umper 10	closed
➪ Hard drive Interrupt is IRQ11	J9/ umper 11	closed
Hard drive Interrupt is IRQ12	J9/ umper 12	closed
Hard drive Interrupt is IRQ14	J9/ umper 13	closed
Hard drive Interrupt is IRQ15	J9/ umper 14	closed

DMA CHANNEL		
DRQ	**J5/jumper 7**	**J5/jumper 8**
➪ 5	open	closed
6	closed	open
7	open	open
0	closed	closed

Continued on next page . . .

ADAPTEC

AHA-1540B, AHA-1542B
(Also Sold as the Maynard HA-101S)

. . . continued from previous page

INTERRUPT REQUEST			
IRQ	**J5/jumper 9**	**J5/jumper 10**	**J5/jumper 11**
9	open	open	open
10	closed	open	open
⇨ 11	open	closed	open
12	closed	closed	open
14	open	open	closed
15	closed	open	closed

DMA TRANSFER RATE		
Speed	**J5/jumper 12**	**J5/jumper 13**
5.0Mb/s	open	open
⇨ 5.7Mb/s	closed	open
6.7Mb/s	open	closed
8.0Mb/s	closed	closed

I/O PORT ADDRESS			
Address	**J7/jumper 2**	**J7/jumper 3**	**J7/jumper 4**
⇨⇨ 334h	open	open	open
⇨ 330h	closed	open	open
234h	open	closed	open
200h	closed	closed	open
134h	open	open	closed
130h	closed	open	closed
Not used	closed	closed	closed
Note: ⇨⇨ denotes default for 1540BS11/1542BS11			

BUS WAIT STATE		
Wait Time	**J7/jumper 5**	**J7/jumper 6**
⇨ Disable	open	open
100nsec	closed	open
200nsec	open	closed
300nsec	closed	closed

MISCELLANEOUS TECHNICAL NOTE
Note: Manufacturer documentation does not specify possible BIOS memory addresses, but a selection of DC00:6 has been recommended by some card users.

ADAPTEC, INC.

AHA-1542CF, AHA-1540CP, AHA-1542CF, AHA-1542CP

Data bus:	16-bit, ISA
Size:	Half-length, full-height card
Hard drive supported:	Up to seven SCSI drives
Floppy drives supported:	Two 360KB, 720KB, 1.2MB, or 1.44MB drives

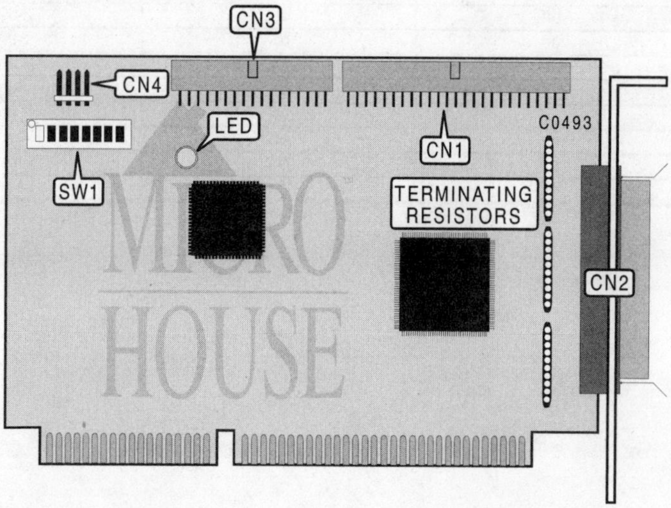

CONNECTIONS	
Function	**Location**
50-pin SCSI connector - internal	CN1
50-pin SCSI connector - external	CN2
34-pin control cable connector - floppy drive	CN3
4-pin connector - drive active LED	CN4

USER CONFIGURABLE SETTINGS		
Function	**Location**	**Setting**
➪ SCSI Termination set through software	SW1/1	off
SCSI Termination enabled	SW1/1	on
➪ Floppy drive enabled	SW1/5	off
Floppy drive disabled	SW1/5	on

I/O PORT ADDRESS SELECTION			
Address	**SW1/2**	**SW1/3**	**SW1/4**
➪ 330-333h	off	off	off
334-337h	on	off	off
230-233h	off	on	off
234-237h	on	on	off
130-133h	off	off	on
134-137h	on	off	on
Reserved	off	on	on
Reserved	on	on	on

Continued on next page . . .

ADAPTEC, INC.

AHA-1542CF, AHA-1540CP, AHA-1542CF, AHA-1542CP

. . . continued from previous page

BIOS ADDRESS SELECTION			
Address	**SW1/6**	**SW1/7**	**SW1/8**
⇨ DC000h	off	off	off
D8000h	on	off	off
D4000h	off	on	off
D0000h	on	on	off
CC000h	off	off	on
C8000h	on	off	on
Reserved	off	on	on
Disabled	on	on	on

ADAPTEC, INC.

AHA-1740/AHA-1742A (VER. 2)

Data bus: 32-bit EISA
Size: Three-quarter length, Full-height card
Hard drive supported: Up to fourteen SCSI devices
Floppy drives supported: 360KB, 720KB, 1.2MB, 1.44MB

CONNECTIONS	
Function	**Location**
50-pin SCSI interface - external	CN1
50-pin SCSI interface - internal	CN2
34-pin control cable connector - floppy drive	CN3
Terminating resistor 1	RN5
Terminating resistor 2	RN6
Terminating resistor 3	RN7

USER CONFIGURABLE SETTINGS		
Function	**Location**	**Setting**
➪ Factory configured - do not alter	JP1/A	N/A
➪ Factory configured - do not alter	JP1/B	N/A
➪ Factory configured - do not alter	JP1/C	N/A
➪ Factory configured - do not alter	JP1/D	N/A
➪ Floppy drive primary address select	JP1/E	open
Floppy drive alternate address select	JP1/E	closed
➪ Floppy drive interface enabled	JP1/F	closed
Floppy drive interface disabled	JP1/F	open
➪ TERMPWR supplied by host interface	JP2	closed
TERMPWR not supplied by host interface	JP2	open

ADAPTEC, INC.

AHA-1740A, AHA-1742A

Data bus:	32-bit, EISA
Size:	Full-length, full-height card
Hard drive supported:	Up to seven SCSI devices
Floppy drives supported:	Two 360KB, 720KB, 1.2MB, or 1.44MB drives

CONNECTIONS	
Function	**Location**
50-pinSCSI connector - external	CN1
50-pinSCSI connector - internal	CN2
34-pin control cable connector - floppy drive	CN3
4-pin connector - drive active LED	CN4

USER CONFIGURABLE SETTINGS		
Function	**Location**	**Setting**
⇨ Floppy drive enabled	J6/ umper 1	closed
Floppy drive disabled	J6/ umper 1	open
⇨ Floppy drive port address is 3F0-3F7h	J6/ umper 2	open
Floppy drive port address is 370-377h	J6/ umper 2	closed

ADAPTEC, INC.

AHA-2740T, AHA-2742, AHA-2742T, AHA-2740A, AHA-2742A, AHA-2740A-TWIN, AHA-2742A-TWIN

Data bus: 32-bit, EISA
Size: Half-length, full-height card
Hard drive supported: Up to seven SCSI devices
Floppy drives supported: Two 360KB, 720KB, 1.2MB, or 1.44MB drives

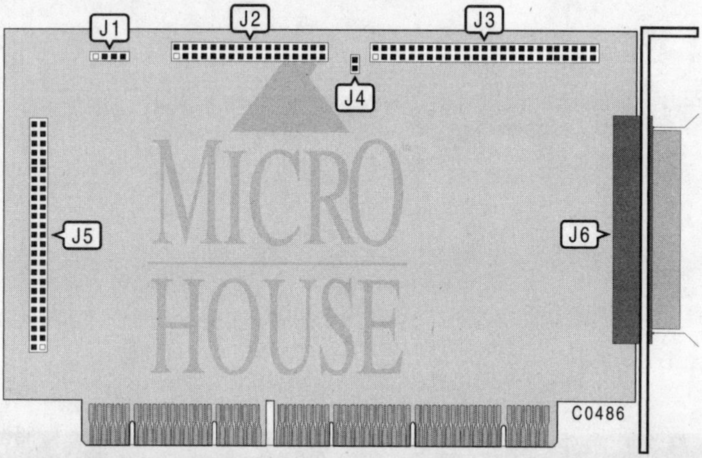

CONNECTIONS	
Function	**Location**
4-pin connector - drive active LED	J1
34-pin control cable connector - floppy drive (optional)	J2
50-pin SCSI Channel A connector internal	J3
50-pin SCSI Channel B connector - internal (optional)	J5
50-pin SCSI Channel A connector external	J6

USER CONFIGURABLE SETTINGS		
Function	**Location**	**Setting**
➪ Floppy drive enabled	J4	closed
Floppy drive disabled	J4	open
Note: Jumper J4 is only present on those boards with a floppy drive.		

ADAPTEC, INC.

AHA-2742W, AHA-2744W

Data bus:	32-bit EISA
Size:	Three/quarter-length, full-height card
Hard drive supported:	Up to fifteen SCSI devices
Floppy drives supported:	Two 360KB, 720KB, 1.2MB, or 1.44MB drives

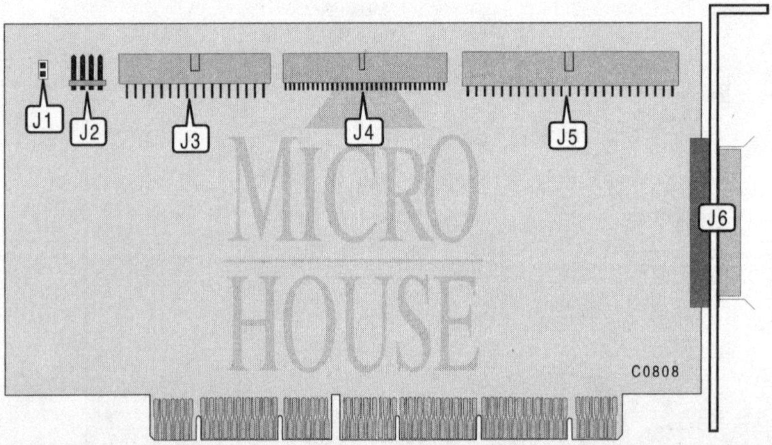

CONNECTIONS

Function	Location
4-pin connector - drive active LED	J2
34-pin floppy drive connector	J3
68-pin wide SCSI connector - internal	J4
50-pin SCSI connector - internal	J5
68-pin wide SCSI connector - external	J6

USER CONFIGURABLE SETTINGS

Function	Location	Setting
⇨ Floppy drive controller enabled	J1	Closed
Floppy drive controller disabled	J1	Open

MISCELLANEOUS TECHNICAL NOTES

Only single-ended SCSI devices can be connected to 2742W.
Only two of the three SCSI connector can be used at any given time.

ADAPTEC, INC.

AHA-2840VL, AHA-2842VL

Data bus:	32-bit VL-Bus
Size:	Full-lengh, full-height card
Hard drive supported:	Up to seven SCSI devices
Floppy drives supported:	Two 360KB, 720KB, 1.2MB, or 1.44MB drives

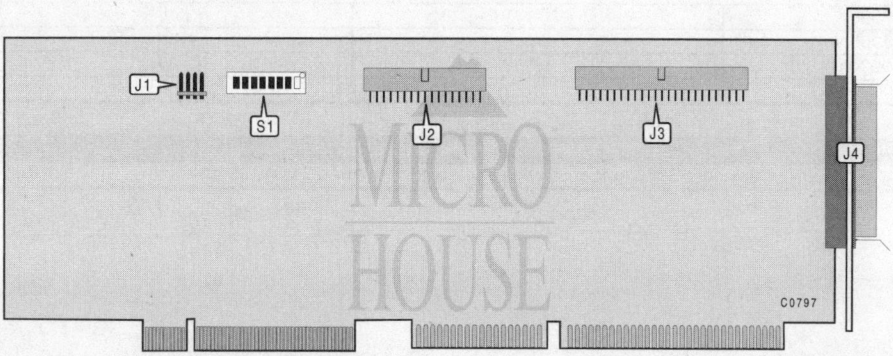

CONNECTIONS	
Function	**Location**
4-pin connector - drive active LED	J1
34-pin floppy drive connector	J2
50-pin SCSI connector internal	J3
50-pin SCSI connector external	J4

USER CONFIGURABLE SETTINGS		
Function	**Location**	**Setting**
➪ Floppy drive enabled	S1/switch 5	off
Floppy drive disabled	S1/switch 5	on
➪ Host adapter BIOS enabled	S1/switch 8	off
Host adapter BIOS disabled	S1/switch 8	on
Note: Disabling the host adapter BIOS for non-disk devices such as CD-ROM drives, tape drives, will reduce boot up time.		

I/O ADDRESS SELECTION				
Address	**S1/Switch 1**	**S1/Switch 2**	**S1/Switch 3**	**S1/Switch 4**
➪ 1C000h	off	off	off	off
2C000h	on	off	off	off
3C000h	off	on	off	off
4C000h	on	on	off	off
5C000h	off	off	on	off
6C000h	on	off	on	off
7C000h	off	on	on	off
8C000h	on	on	on	off
9C000h	off	off	off	on
AC000h	on	off	off	on
BC000h	off	on	off	on
CC000h	on	on	off	on
DC000h	off	off	on	on
EC000h	on	off	on	on
Reserved	off	on	on	on
Reserved	on	on	on	on

Continued on next page. . .

ADAPTEC, INC.

AHA-2840VL, AHA-2842VL

. . .continued from previous page

BASE ADDRESS SELECTION		
Address	**S1/Switch 6**	**S1/Switch 7**
D8000h	Off	off
C8000h	on	off
D0000h	off	on
E0000h	on	on

MISCELLANEOUS TECHNICAL NOTES
Only single-ended SCSI devices can be connected to the AHA-2840VL/2842VL.

ADAPTEC, INC.

AHA-2940W

Data bus: PCI local-bus
Size: Half-length, full-height card
Hard drive supported: Up to seven SCSI devices
Floppy drives supported: None

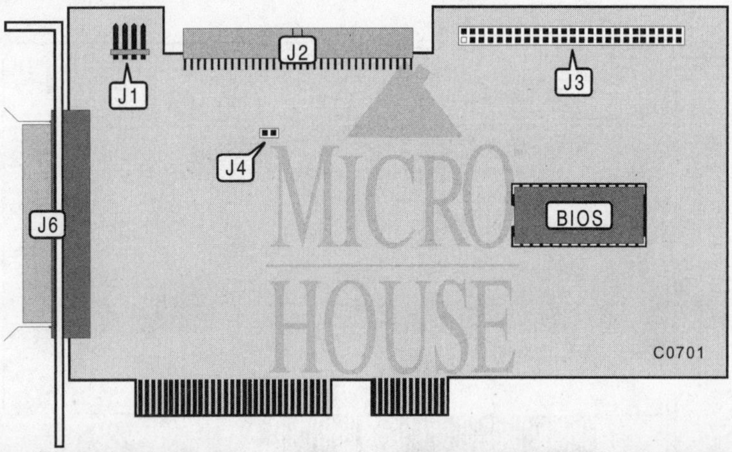

CONNECTIONS	
Function	**Location**
4-pin connector - drive active LED	J1
68-pin SCSI (Wide) connector internal	J2
50-pin SCSI connector internal	J3
68-pin SCSI (Wide) connector - external	J6

USER CONFIGURABLE SETTINGS		
Function	**Location**	**Setting**
⇨ TERMPWR disabled	J4	open
TERMPWR enabled	J4	closed

ADAPTEC, INC.

AHA-2944W

Data bus:	32-bit PCI
Size:	Half-length, full-height card
Hard drive supported:	Up to fifteen SCSI-2 devices
Floppy drives supported:	None

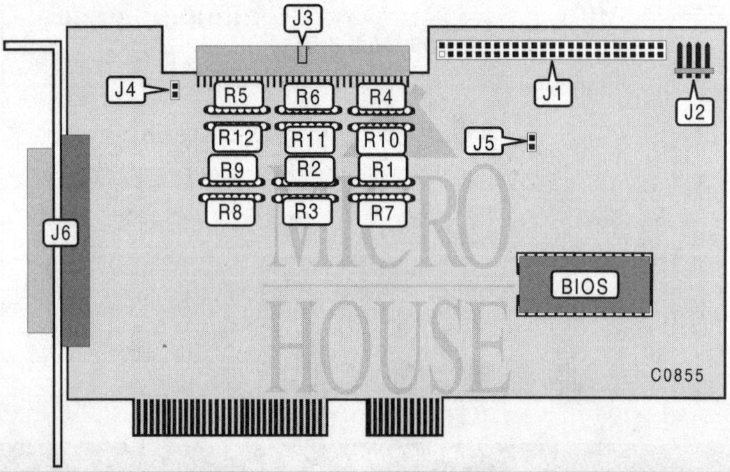

CONNECTIONS	
Function	**Location**
50-pin SCSI connector - internal	J1
4-pin connector - drive active LED	J2
68-pin SCSI connector - internal	J3
68-pin SCSI connector - external	J6

USER CONFIGURABLE SETTINGS		
Function	**Location**	**Setting**
Termination power for J3, when power off enabled	J4	closed
Termination power for J3, when power off disabled	J4	open
Termination power for J1, when power off enabled	J5	closed
Termination power for J1, when power off disabled	J5	open
Low byte & high byte set to off	R1-R12	removed
Low byte & high byte set to on	R1-R12	installed
Low byte set to off, high byte set to on	R9, R12, R3, R6	installed
	R1, R2, R4, R5, R7, R8, R10, R11	removed

MISCELLANEOUS TECHNICAL NOTES
R1, R2, R9, R10, R11, R12 are 330-ohm. R3, R4, R5, R6, R7, R8 are 150-ohm.
When using only the 68-pin external connector low & high bytes must be set to on.
When using the 68-pin internal & external connectors or when using the 50-pin & 68-pin internal connectors low byte must be set to off, high byte to on.
When using the 50-pin internal & 68-pin external low & high bytes must be set to off.

ADAPTEC, INC.
AHA-3980, AHA-3985

Data Bus	32-bit PCI
Card Size	Three-quarter length, full-height card
Card Drives supported	Up to 14 Fast SCSI devices (model AHA-3980) Up to 21 Fast SCSI devices (model AHA-3985)
Floppy drives supported	None

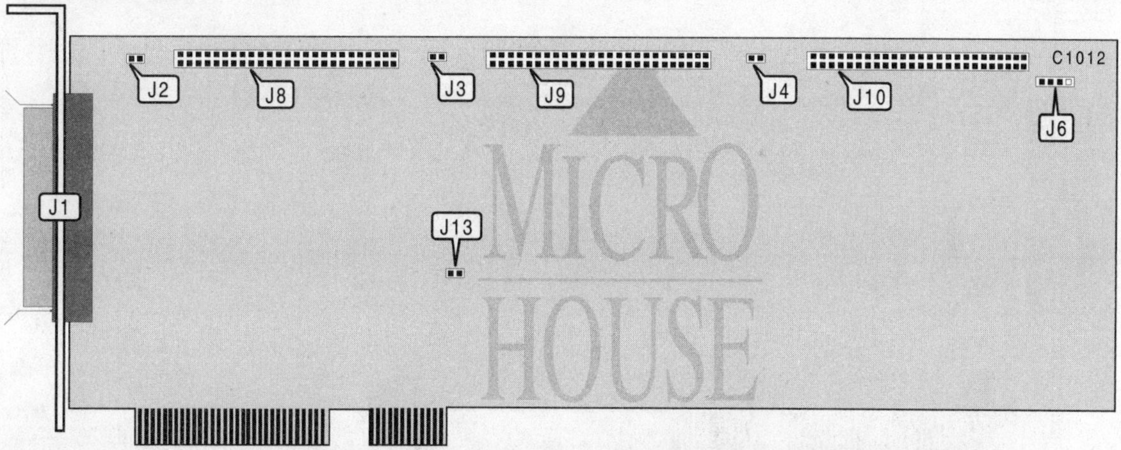

CONNECTIONS			
Function	**Location**	**Function**	**Location**
50-pin SCSI channel A connector - external	J1	50-pin SCSI channel B connector - internal	J9
4-pin connector - drive active LED	J6	50-pin SCSI channel C connector - internal (model AHA-3985 only)	J10
50-pin SCSI channel A connector - internal	J8		

USER CONFIGURABLE SETTINGS		
Function	**Label**	**Position**
➪ Channel A bus is connected to one computer only	J2	Open
Channel A bus is shared by two computers	J2	Closed
➪ Channel B bus is connected to one computer only	J3	Open
Channel B bus is shared by two computers	J3	Closed
➪ Channel C bus is connected to one computer (AHA-3985 only)	J4	Open
Channel C bus is shared by two computers (AHA-3985 only)	J4	Closed
➪ Host computer supports PCI interrupts A, B, C, & D	J13	Open
Host computer supports PCI interrupts A & B only	J13	Closed

MISCELLANEOUS TECHNICAL NOTES
Jumpers J2, J3 and J4 enable other adapter cards on different computers to share devices on channels A, B, or C (model AHA-3985) with this adapter card. Device access is sequential.

ADAPTEC, INC.

AHA-3980W, AHA-3985W

Data Bus	32-bit PCI
Card Size	Three-quarter length, full-height card
Card Drives supported	Up to 30 Fast and Wide SCSI devices (model AHA-3980W) Up to 45 Fast and Wide SCSI devices (model AHA-3985W)
Floppy drives supported	None

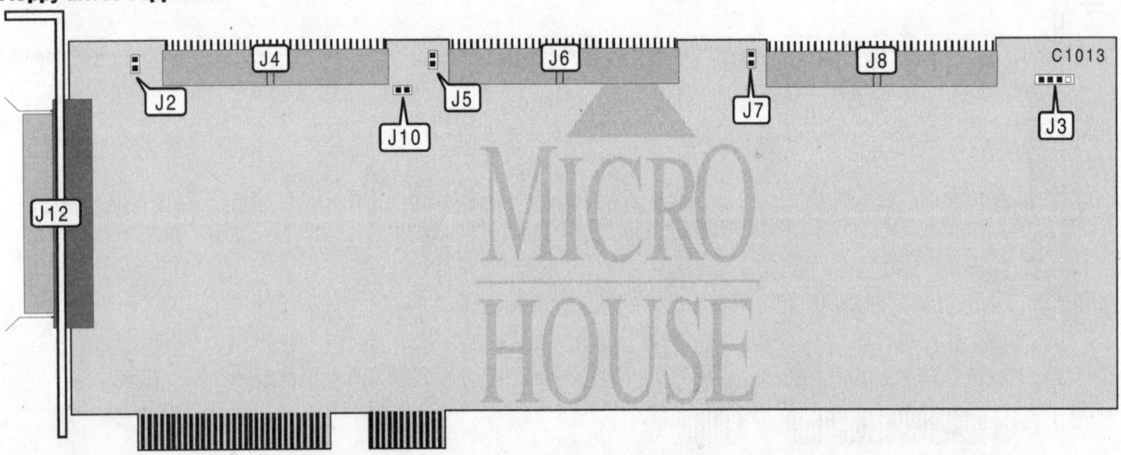

CONNECTIONS			
Function	**Location**	**Function**	**Location**
4-pin connector - drive active LED	J3	68-pin SCSI channel C connector - internal (model AHA-3985W only)	J8
68-pin SCSI channel A connector - internal	J4	68-pin SCSI channel A connector - external	J12
68-pin SCSI channel B connector - internal	J6		

USER CONFIGURABLE SETTINGS		
Function	**Label**	**Position**
⇨ Channel A bus is connected to one computer only	J2	Open
Channel A bus is shared by two computers	J2	Closed
⇨ Channel B bus is connected to one computer only	J5	Open
Channel B bus is shared by two computers	J5	Closed
⇨ Channel C bus is connected to one computer (AHA-3895W only)	J7	Open
Channel C bus is shared by two computers (AHA-3895W only)	J7	Closed
⇨ Host computer supports PCI interrupts A, B, C, & D	J10	Open
Host computer supports PCI interrupts A & B only	J10	Closed

MISCELLANEOUS TECHNICAL NOTES
Jumpers J2, J5 and J7 enable other adapter cards on different computers to share devices on channels A, B, and C (AHA-3895W only) with this adapter card. Device access is sequential.

ADAPTEC, INC.
AMM-1510M

Card Type	Hard drive controller/Sound card
Hard Drives supported	Up to 7 SCSI devices
Floppy drives supported	None
Data Bus	16-bit ISA
Card Size	Three-quarter length, full-height

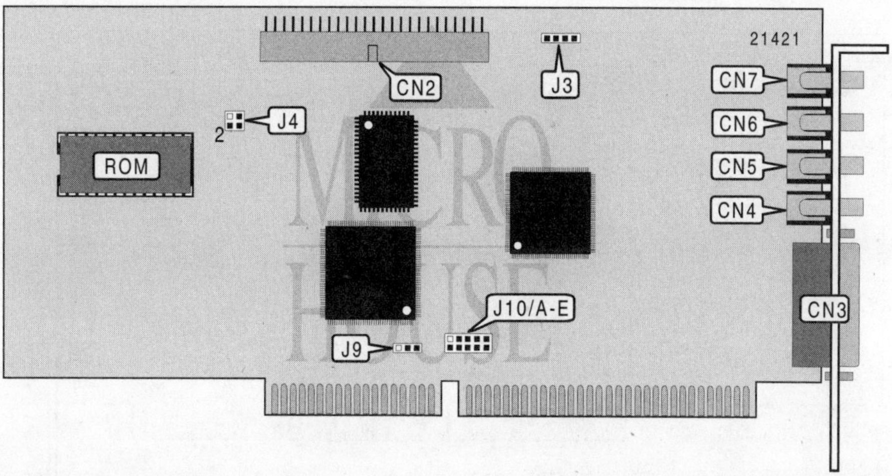

CONNECTIONS

Function	Location	Function	Location
50-pin SCSI connector - internal	CN2	Line out	CN6
16-pin game/MIDI port	CN3	Headphones/speakers	CN7
Microphone in	CN4	CD audio in	J3
Line in	CN5		

USER CONFIGURABLE SETTINGS

Function	Label	Position
➪ Factory configured - do not alter	J4/pins 1 - 4	Unidentified
➪ Sound chip base address is 220h	J9	Pins 1 & 2 closed
Sound chip base address is 240h	J9	Pins 2 & 3 closed
➪ I/O port address is 340h	J10/E	Open
I/O port address is 140h	J10/E	Closed

SCSI INTERRUPT SELECTION

IRQ	J10/A	J10/B	J10/C	J10/D
IRQ9	Open	Open	Open	Closed
IRQ10	Open	Open	Closed	Open
➪IRQ11	Open	Closed	Open	Open
IRQ12	Closed	Open	Open	Open

MISCELLANEOUS TECHNICAL NOTES

SCSI termination on the controller is always enabled and cannot be changed since the controller is always at one end of the SCSI bus.

ADAPTEC, INC.

AMM-1525M

Card Type	Hard drive controller/Sound card
Hard Drives supported	Two IDE/ATAPI drives Up to 7 SCSI devices
Floppy drives supported	None
Data Bus	16-bit ISA
Card Size	Three-quarter length, full-height

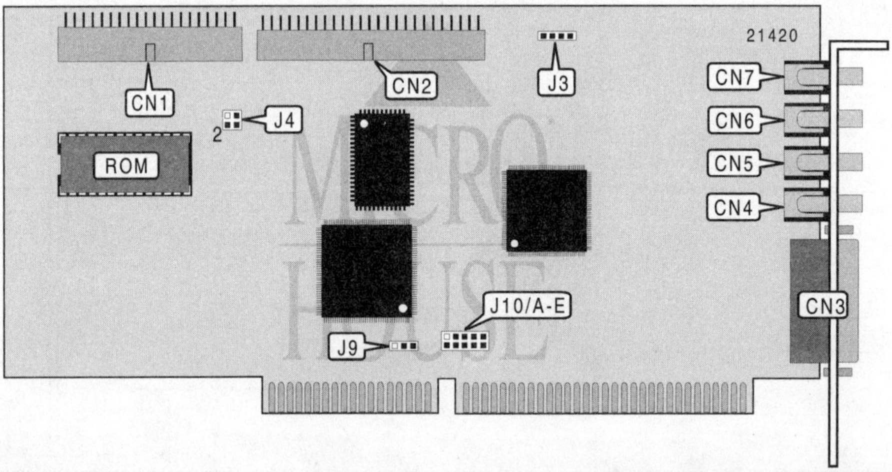

CONNECTIONS			
Function	**Location**	**Function**	**Location**
40-pin primary IDE/ATAPI connector	CN1	Line in	CN5
50-pin SCSI connector - internal	CN2	Line out	CN6
16-pin game/MIDI port	CN3	Headphones/speakers	CN7
Microphone in	CN4	CD audio in	J3

USER CONFIGURABLE SETTINGS		
Function	**Label**	**Position**
➩ IDE port address enabled	J4/pins 1 & 2	Open
IDE port address disabled	J4/pins 1 & 2	Closed
➩ IDE is secondary in system at 170h port address	J4/pins 3 & 4	Closed
IDE is primary in system at 1F0h port address	J4/pins 3 & 4	Open
➩ Sound chip base address is 220h	J9	Pins 1 & 2 closed
Sound chip base address is 240h	J9	Pins 2 & 3 closed
➩ I/O port address is 340h	J10/E	Open
I/O port address is 140h	J10/E	Closed

SCSI INTERRUPT SELECTION				
IRQ	**J10/A**	**J10/B**	**J10/C**	**J10/D**
IRQ9	Open	Open	Open	Closed
IRQ10	Open	Open	Closed	Open
➩ IRQ11	Open	Closed	Open	Open
IRQ12	Closed	Open	Open	Open

MISCELLANEOUS TECHNICAL NOTES
SCSI termination on the controller is always enabled and cannot be changed since the controller is always at one end of the SCSI bus.

ADAPTEC, INC.

AMM-1570

Data bus: 16-bit ISA
Size: Full-length, full-height card
Hard drive supported: Up to seven SCSI drives
Floppy drives supported: Two 360KB, 720KB, 1.2MB, 1.44MB, or 2.88MB drives

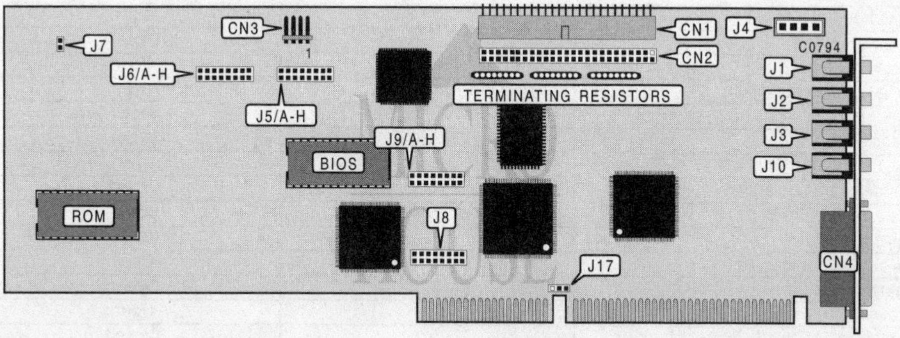

CONNECTIONS	
Function	**Location**
50-pin SCSI connector - internal	CN1
50-pin SCSI connector - external	CN2
4-pin connector - drive active LED	CN3
15-pin game port - external	CN4
Headphone/speaker jack	J1
Line out	J2
Line in	J3
CD audio in	J4
Mic in	J10

Continued on next page. . .

ADAPTEC,INC.

AMM-1570

. . .continued from previous page

USER CONFIGURABLE SETTINGS		
Function	**Location**	**Setting**
➪ Factory configured - do not alter	J5/A	Open
➪ INT 19h reroute enabled	J5/B	Closed
INT 19h reroute disabled	J5/B	Open
➪ Factory configured - do not alter	J5/C	Open
➪ Fast SCSI disabled	J5/D	Open
Fast SCSI enabled	J5/D	Closed
➪ Synchronous negotiation disabled	J5/E	Open
Synchronous negotiation enabled	J5/E	Closed
➪ SCSI disconnection enabled	J5/F	Closed
SCSI disconnection disabled	J5/F	Open
➪ Floptical support disabled	J5/G	Open
Floptical support enabled	J5/G	Closed
➪ Extended translation disabled	J5/H	Open
Extended translation enabled	J5/H	Closed
➪ Factory configured - do not alter	J6/F	Open
➪ Factory configured - do not alter	J6/G	Open
➪ SCSI parity checking enabled	J6/H	Open
SCSI parity checking disabled	J6/H	Closed
➪ Floppy drive enabled	J7	Closed
Floppy drive disabled	J7	Open
➪ Factory configured - do not alter	J8	Open
➪ I/O port address is 340h - 35Eh	J9/E	Open
I/O port address is 140h - 15Eh	J9/E	Closed
➪ Host adapter BIOS enabled	J9/H	Closed
Host adapter BIOS disabled	J9/H	Open

SCSI ID configuration			
ID	**J6/A**	**J6/B**	**J6/C**
➪ 7	closed	closed	Closed
0	open	open	Open
1	closed	open	Open
2	open	closed	Open
3	closed	closed	Open
4	open	open	Closed
5	closed	open	Closed
6	open	closed	Closed

interrupt selection						
IRQ	**J6/D**	**J6/E**	**J9/A**	**J9/B**	**J9/C**	**J9/D**
➪ 11	open	closed	open	closed	open	open
9	open	open	open	open	open	closed
10	closed	open	open	open	closed	open
12	closed	closed	closed	open	open	open

Continued on next page. . .

ADAPTEC,INC.
AMM-1570

. . .continued from previous page

bios base address selection		
Address	**J9/F**	**J9/G**
⇨ DC000h - DFFFFh	closed	Closed
C8000h - CBFFFh	open	Open
CC000h - CFFFFh	open	Closed
D8000h - DBFFFh	closed	Open

sound chip base address	
Address	**J17**
⇨ 220h	pins 1 & 2 closed
240h	pins 2 & 3 closed
Invalid	Open

ADAPTEC, INC.

AMM-1572

Data bus: 16-bit ISA
Size: Full-length, full-height card
Hard drive supported: Up to seven SCSI drives
Floppy drives supported: Two 360KB, 720KB, 1.2MB, 1.44MB, or 2.88MB drives

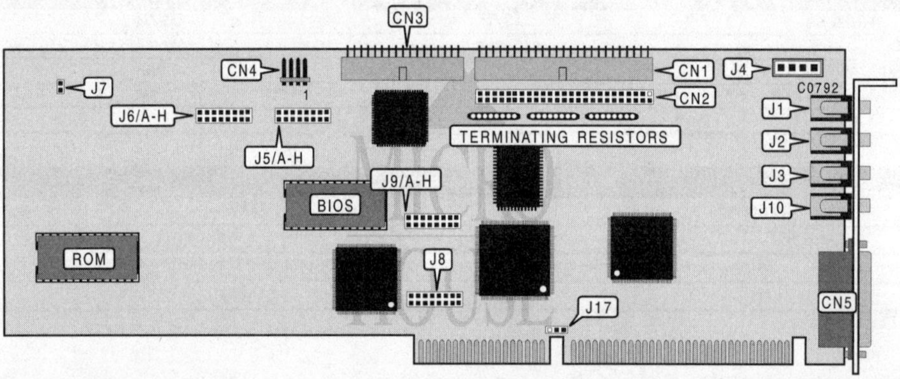

CONNECTIONS	
Function	**Location**
50-pin SCSI connector - internal	CN1
50-pin SCSI connector - external	CN2
34-pin control cable connector - floppy drive	CN3
4-pin connector - drive active LED	CN4
15-pin game port - external	CN5
Headphone/speaker jack	J1
Line out	J2
Line in	J3
CD audio in	J4
Mic in	J10

Continued on next page. . .

ADAPTEC,INC.
AMM-1572

. . .continued from previous page

USER CONFIGURABLE SETTINGS		
Function	**Location**	**Setting**
➪ Factory configured - do not alter	J5/A	Open
➪ INT 19h reroute enabled	J5/B	Closed
INT 19h reroute disabled	J5/B	Open
➪ Factory configured - do not alter	J5/C	Open
➪ Fast SCSI disabled	J5/D	Open
Fast SCSI enabled	J5/D	Closed
➪ Synchronous negotiation disabled	J5/E	Open
Synchronous negotiation enabled	J5/E	Closed
➪ SCSI disconection enabled	J5/F	Closed
SCSI disconection disabled	J5/F	Open
➪ Floptical support disabled	J5/G	Open
Floptical support enabled	J5/G	Closed
➪ Extended translation disabled	J5/H	Open
Extended translation enabled	J5/H	Closed
➪ Factory configured - do not alter	J6/F	Open
➪ Factory configured - do not alter	J6/G	Open
➪ SCSI parity checking enabled	J6/H	Open
SCSI parity checking disabled	J6/H	Closed
➪ Floppy drive enabled	J7	Closed
Floppy drive disabled	J7	Open
➪ Factory configured - do not alter	J8	Open
➪ I/O port address is340h - 35Eh	J9/E	Open
I/O port address is 140h - 15Eh	J9/E	Closed
➪ Host adapter BIOS enabled	J9/H	Closed
Host adapter BIOS disabled	J9/H	Open

SCSI ID configuration			
ID	**J6/A**	**J6/B**	**J6/C**
➪ 7	closed	closed	Closed
0	open	open	Open
1	closed	open	Open
2	open	closed	Open
3	closed	closed	Open
4	open	open	Closed
5	closed	open	Closed
6	open	closed	Closed

interrupt selection						
IRQ	**J6/D**	**J6/E**	**J9/A**	**J9/B**	**J9/C**	**J9/D**
➪ 11	open	closed	open	closed	open	open
9	open	open	open	open	open	closed
10	closed	open	open	open	closed	open
12	closed	closed	closed	open	open	open

bios base address selection		
Address	**J9/F**	**J9/G**
➪ DC000h - DFFFFh	closed	Closed
C8000h - CBFFFh	open	Open
CC000h - CFFFFh	open	Closed
D8000h - DBFFFh	closed	Open

Continued on next page. . .

ADAPTEC,INC.
AMM-1572

. . .continued from previous page

sound chip base address	
Address	**J17**
⇨ 220h	pins 1 & 2 closed
240h	pins 2 & 3 closed
Invalid	Open

ADAPTEC, INC.

AVA-1505

Data bus: 16-bit ISA
Size: Half-length, full-height card
Hard drive supported: Up to seven SCSI devices
Floppy drives supported: None

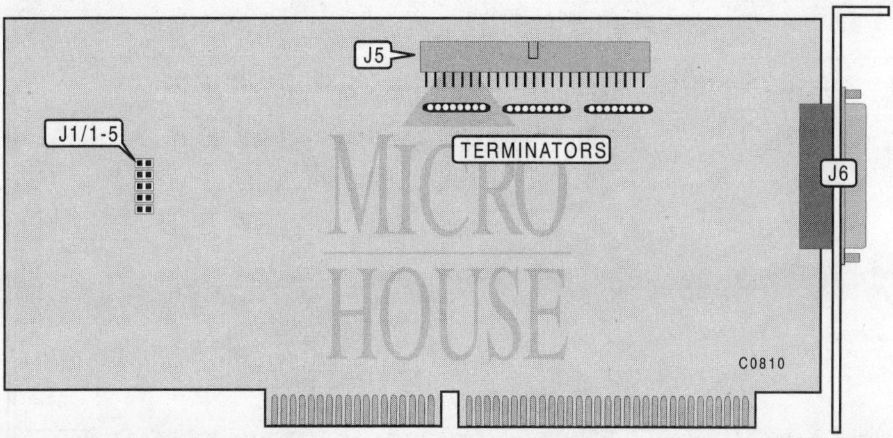

CONNECTIONS	
Function	**Location**
50-pin SCSI connector - external	J5
50-pin SCSI connector - internal	J6

USER CONFIGURABLE SETTINGS		
Function	**Location**	**Setting**
➩ Adapter I/O address is 340h	J1/1	Open
Adapter I/O address is 140h	J1/1	Closed

INTERRUPT SELECTION				
IRQ	**J1/2**	**J1/3**	**J1/4**	**J1/5**
➩ 11	open	closed	open	open
9	open	open	open	closed
10	open	open	closed	open
12	closed	open	open	open

MISCELLANEOUS TECHNICAL NOTES
Only single-ended devices are supported. When connecting both external & internal devices to AVA-1505 remove terminators.

ADAPTEC, INC.

AVA-2825

Hard Drives supported	Two IDE (AT) drives Up to 7 Fast SCSI-2 devices
Floppy drives supported	Two 360KB, 720KB, 1.2MB, or 1.44MB drives
Data Bus	32-bit VESA Local Bus
Card Size	Three-quarter length, full-height

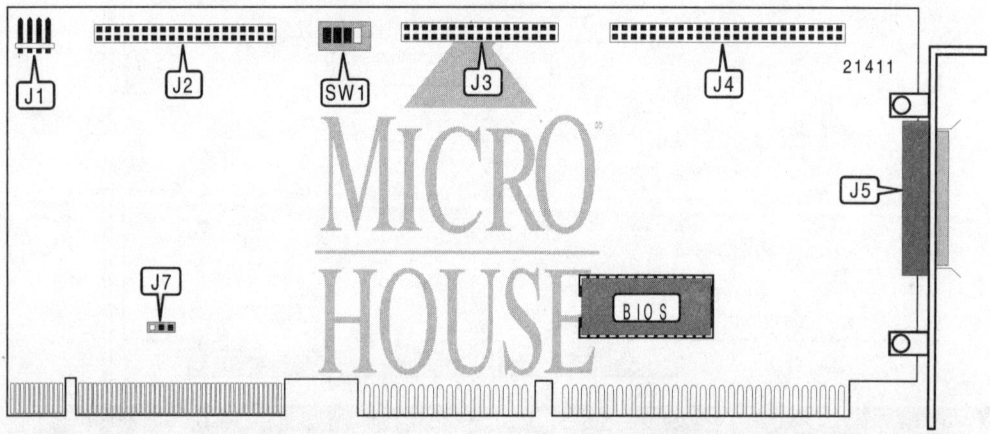

CONNECTIONS			
Function	**Location**	**Function**	**Location**
4-pin connector - drive active LED	J1	50-pin SCSI connector - internal	J4
40-pin IDE(AT) connector	J2	50-pin SCSI connector - external	J5
34-pin cable connector - floppy drive	J3		

USER CONFIGURABLE SETTINGS		
Function	**Label**	**Position**
⇨ SCSI I/O base address is 340h-35Fh	J7	Pins 1 & 2 closed
SCSI I/O base address is 140h-15Fh	J7	Pins 2 & 3 closed
⇨ Floppy drive interface enabled	SW1/1	On
Floppy drive interface disabled	SW1/1	Off
⇨ IDE(AT) interface enabled	SW1/2	Off
IDE(AT) interface disabled	SW1/2	On

BIOS ADDRESS SELECTION		
Address	**SW1/3**	**SW1/4**
C8000h	Off	Off
CC000h	Off	On
D8000h	On	Off
⇨ DC000h	On	On

BUSLOGIC, INC.

BT-545S

Data bus: 16-bit, ISA
Size: Three/quarter-length, full-height card
Hard drive supported: Up to seven SCSI drives
Floppy drives supported: Two 360KB, 720KB, 1.2MB or 1.44MB drives

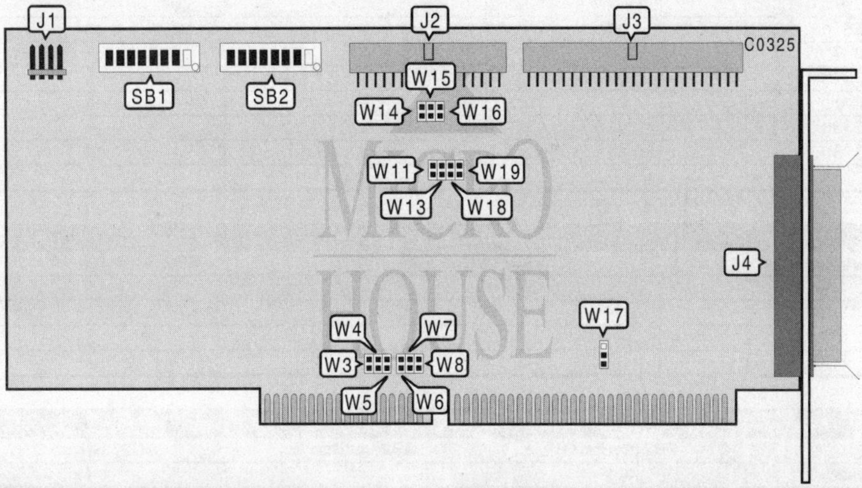

CONNECTIONS	
Function	**Location**
4-pin connector - Drive Active LED	J1
34-pin cable connector - floppy drive	J2
50-pin SCSI connector - internal	J3
50-pin SCSI connector - external	J4

SCSI ADDRESS			
Address	**SB1/switch 1**	**SB1/switch 2**	**SB1/switch 3**
0	off	off	off
1	off	off	on
2	off	on	off
3	off	on	on
4	on	off	off
5	on	off	on
6	on	on	off
⇨ 7	on	on	on

DMA CHANNEL		
DRQ	**SB1/switch 7**	**SB1/switch 8**
Disabled	off	off
DRQ7	on	off
DRQ6	off	on
⇨ DRQ5	on	on

Continued on next page . . .

BUSLOGIC, INC.
BT-545S

. . . continued from previous page

I/O PORT ADDRESS			
Address	SB2/switch 1	SB2/switch 2	SB2/switch 3
Reserved	off	off	off
134-137h	off	off	on
234-237h	off	on	off
334-337h	off	on	on
Reserved	on	off	off
130-133h	on	off	on
230-233h	on	on	off
➩ 330-333h	on	on	on

BIOS ADDRESS		
Address	SB2/switch 4	SB2/switch 5
C800h	off	off
Disabled	on	off
D800h	off	on
➩ DC00h	on	on

INTERRUPT REQUEST			
IRQ	SB2/switch 6	SB2/switch 7	SB2/switch 8
Reserved	off	off	off
Reserved	off	off	on
IRQ15	off	on	off
IRQ14	off	on	on
IRQ12	on	off	off
IRQ9	on	off	on
IRQ10	on	on	off
➩ IRQ11	on	on	on
Jumpers W3-W8 must be set in accordance with SB2/switches 6-8			

USER CONFIGURABLE SETTINGS		
Function	Location	Setting
➩ SCSI parity check enabled	SB1/switch 4	on
SCSI parity check disabled	SB1/switch 4	off
>1GB drive support disabled	SB1/switch 5	on
>1GB drive support enabled*	SB1/switch 5	off
➩ Sync Negotiation initiated by BT-545S disabled	SB1/switch 6	on
Sync Negotiation initiated by BT-545S enabled	SB1/switch 6	off
➩ Factory Configured - do not alter	W2	open
➩ Factory Configured - do not alter	W9	open
➩ Floppy drive port address is 3F0-3F7h	W14	open
Floppy drive port address is 370-377h	W14	closed
➩ Floppy drive enabled	W15 & W16	closed
Floppy drive disabled	W15 & W16	open
➩ IOCHRDY enabled	W17	Pins 1-2 closed
IOCHRDY disabled	W17	Pins 2-3 closed
➩ Factory Configured - do not alter	W18	closed
➩ Sync transfer rate is 5MB/sec	W19	closed
Sync transfer rate is 10MB/sec	W19	open
Note: The SCO UNIX device driver does not support drives greater than 1GB.		

Continued on next page . . .

BUSLOGIC, INC.
BT-545S

. . . continued from previous page

DMA TRANSFER RATE		
Speed	**W11**	**W13**
➪ 5MB/sec	open	open
5.7MB/sec	open	closed
6.7MB/sec	closed	open
8.0MB/sec	closed	closed

INTERRUPT REQUEST						
IRQ	**W3**	**W4**	**W5**	**W6**	**W7**	**W8**
IRQ15	closed	open	open	open	open	open
IRQ14	open	closed	open	open	open	open
IRQ12	open	open	closed	open	open	open
➪ IRQ11	open	open	open	closed	open	open
IRQ10	open	open	open	open	closed	open
IRQ9	open	open	open	open	open	closed
SB2/switches 6-8 must be set in accordance with Jumpers W3-W8						

BUSLOGIC, INC.
BT-757C, BT-757CD

Data bus: 32-bit EISA
Size: Three-quarter length, full-height card
Hard drive supported: Up to fifteen SCSI devices
Floppy drives supported: Two 360KB, 720KB, 1.2MB, or 1.44MB drives

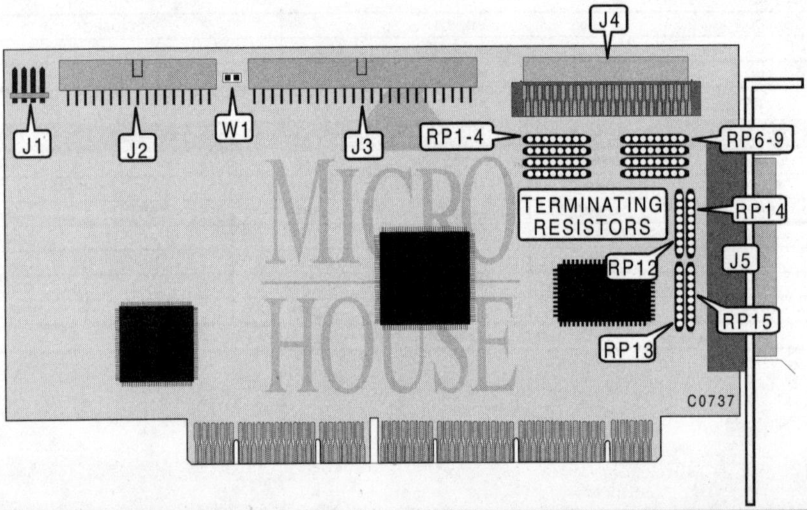

CONNECTIONS	
Function	**Location**
4-pin connector-drive active LED	J1
34-pin connector-floppy drive interface	J2
50-pin connector-SCSI interface (internal)	J3
68-pin connector-SCSI interface (internal)	J4
68-pin connector-SCSI interface (external)	J5

USER CONFIGURABLE SETTINGS		
Function	**Jumper**	**Setting**
⇨ Floppy drive interface enabled	W1	closed
Floppy drive interface disabled	W1	open
Note: The location of W1 is different in the C and CD models. For the C model, W1 is located between the floppy drive interface and the drive active LED connector.		

TERMINATION POWER CONFIGURATION	
Setting	**Terminators Installed**
Data bits 0-7 terminated	RP1, RP2, RP3, RP4, RP6, RP7, RP8, RP9
Data bits 8-15 terminated	RP12, RP13, RP14, RP15
All signals terminated	All installed
Note: These settings apply only to the BT-757CD.	

MISCELLANEOUS TECHNICAL NOTES
The BT-757C and BT-757CD differ in that the C model incorporates a single-ended interface and software controlled active termination, while the CD model utilizes a differential interface and standard (i.e. differential) termination.

BUSLOGIC, INC.

BT-946C

Data bus: 32-bit, PCI-bus
Size: Full-length, full-height card
Hard drive supported: Up to seven SCSI devices
Floppy drives supported: None

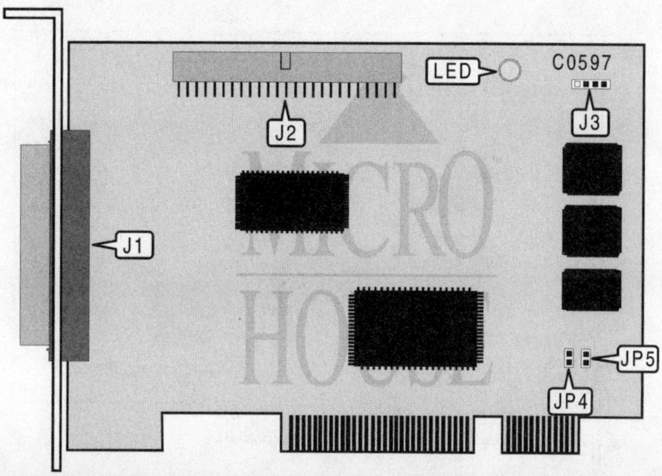

Function	Location
	J1
50-pin connector - Internal SCSI interface	
4-pin connector - Drive activity LED	J3

BIOS ADDRESS CONFIGURATION

BIOS Address	JP4	JP5
➪ DC00	Closed	Closed
C800	Closed	Open
D800	Open	Closed
System BIOS	Open	Open

MISCELLANEOUS TECHNICAL NOTE

Note: The BT-946C can be installed in a master slot only, not a slave slot.

BUSLOGIC, INC.
BT-956C

Data bus: 32-bit PCI
Size: Half-length, full-height card
Hard drive supported: Up to fifteen SCSI devices
Floppy drives supported: Two 360KB, 720KB, 1.2MB, or 1.44MB drives

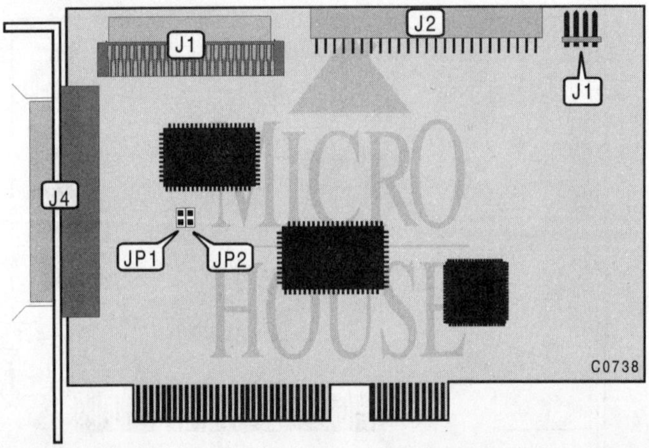

CONNECTIONS	
Function	**Location**
68-pin connector-SCSI interface (internal)	J1
50-pin connector-SCSI interface (internal)	J2
4-pin connector-drive active LED	J3
68-pin connector-SCSI interface (external)	J4

BASE MEMORY ADDRESS SELECTION		
Address	**JP1**	**JP2**
⇨ System BIOS	open	open
C800h	closed	open
D800h	open	closed
DC00h	closed	closed

DATA TECHNOLOGY CORPORATION

DTC-2130D

Data bus:	32-bit PCI
Size:	Half-length, half-height card
Hard drive supported:	Four IDE(AT) interface drives
Floppy drives supported:	None

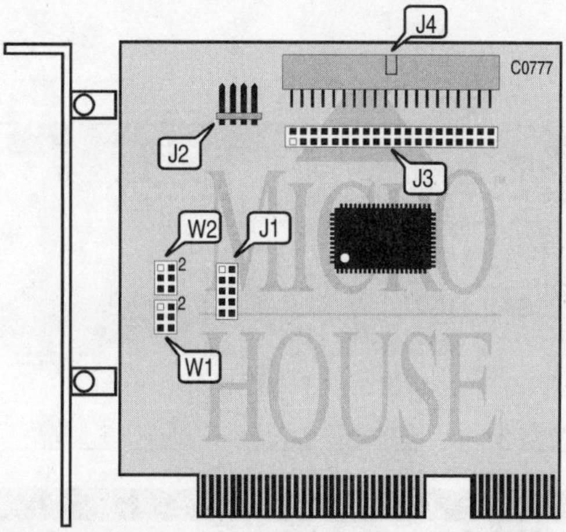

CONNECTIONS	
Function	**Location**
DTC-1415 ISA B Interrupt adapter connector	J1
4-pin connector - drive active LED	J2
40-pin connector - primary IDE(AT) interface	J3
40-pin connector - secondary IDE(AT) interface	J4
Note: Connector J1 is intended for use in systems that do not support redirected PCI interrupts.	

USER CONFIGURABLE SETTINGS		
Function	**Location**	**Setting**
Primary port (IRQ14) INTA select	W1	pins 1 & 3 closed
Primary port (IRQ14) INTB select	W1	pins 2 & 4 closed
Primary port (IRQ14) INTC select	W1	pins 3 & 5 closed
Primary port (IRQ14) INTD select	W1	pins 4 & 6 closed
Secondary port (IRQ15) INTA select	W2	pins 1 & 3 closed
Secondary port (IRQ15) INTB select	W2	pins 2 & 4 closed
Secondary port (IRQ15) INTC select	W2	pins 3 & 5 closed
Secondary port (IRQ15) INTD select	W2	pins 4 & 6 closed

DATA TECHNOLOGY CORPORATION

DTC 2278EB

Data bus: 32-bit VL-Bus
Size: Three/quarter-length, half-height card
Hard drive supported: Four IDE(AT) or EIDE interface drives
Floppy drives supported: Two 360KB, 720KB, 1.2MB, 1.44MB drives

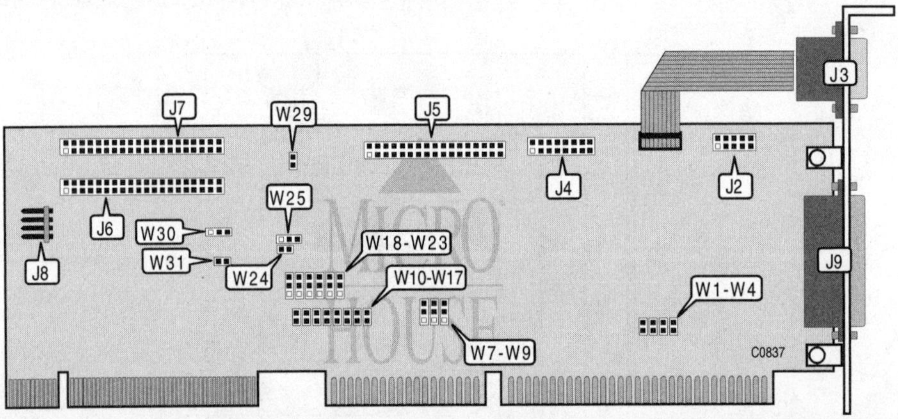

CONNECTIONS	
Function	**Location**
10-pin serial port 2 - internal	J2
9-pin serial port 1 - external	J3
16-pin game port	J4
34-pin cable connector - floppy drive	J5
40-pin connector - secondary IDE(AT) interface	J6
40-pin connector - primary IDE(AT) interface	J7
4-pin connector - drive active LED	J8
25-pin parallel port - external	J9

USER CONFIGURABLE SETTINGS		
Function	**Location**	**Setting**
➪ Game port enabled	W29	closed
Game port disabled	W29	open
➪ Floppy controller enabled	W30/pins 1 & 2	closed
Floppy controller disabled	W30/pins 2 & 3	closed

BIOS ADDRESS SELECTION			
Address	**W7**	**W8**	**W24**
➪ C8000-CBFFFh	pins 2 & 3 closed	pins 2 & 3 closed	closed
CC000-CFFFFh	pins 2 & 3 closed	pins 1 & 2 closed	closed
D8000-DBFFFh	pins 1 & 2 closed	pins 2 & 3 closed	closed
DC000-DFFFFh	pins 1 & 2 closed	pins 1 & 2 closed	closed
Disabled	N/A	N/A	open

PARALLEL PORT ADDRESS SELECTION		
LPT	**W18**	**W19**
➪ LPT1 (378h)	pins 2 & 3 closed	pins 1 & 2 closed
LPT2 (278h)	pins 1 & 2 closed	pins 1 & 2 closed
LPT3 (3BCh)	pins 1 & 2 closed	pins 2 & 3 closed
Disabled	pins 2 & 3 closed	pins 2 & 3 closed

Continued on next page. . .

DATA TECHNOLOGY CORPORATION

DTC 2278EB

...continued from previous page

SERIAL PORT 1 ADDRESS SELECTION		
COM	**W22**	**W23**
➪ COM1 (3F8h)	pins 1 & 2 closed	pins 1 & 2 closed
COM2 (2F8h)	pins 1 & 2 closed	pins 2 & 3 closed
COM3 (3E8h)	pins 2 & 3 closed	pins 1 & 2 closed
Disabled	pins 2 & 3 closed	pins 2 & 3 closed

SERIAL PORT 2 ADDRESS SELECTION		
COM	**W20**	**W21**
➪ COM2 (2F8h)	pins 1 & 2 closed	pins 1 & 2 closed
COM1 (3F8h)	pins 1 & 2 closed	pins 2 & 3 closed
COM4 (2E8h)	pins 2 & 3 closed	pins 1 & 2 closed
Disabled	pins 2 & 3 closed	pins 2 & 3 closed

SERIAL PORT 1 INTERRUPT SELECTION			
IRQ	**W15**	**W16**	**W17**
➪ IRQ4	open	open	closed
IRQ3	open	closed	open
IRQ5	closed	open	open

SERIAL PORT 2 INTERRUPT SELECTION			
IRQ	**W12**	**W13**	**W14**
➪ IRQ3	open	open	closed
IRQ4	open	closed	open
IRQ5	closed	open	open

PARALLEL PORT INTERRUPT SELECTION		
IRQ	**W10**	**W11**
➪ IRQ7	open	closed
IRQ5	closed	open

PARALLEL PORT MODE CONFIGURATION			
Mode	**W9**	**W25**	**W31**
Printer (uni-directional)	pins 2 & 3 closed	pins 2 & 3 closed	open
EPP & bi-directional	pins 1 & 2 closed	pins 2 & 3 closed	open
ECP & bi-directional	pins 2 & 3 closed	pins 1 & 2 closed	closed
EPP & ECP	pins 1 & 2 closed	pins 1 & 2 closed	closed
Note: You must select a DMA channel when using ECP mode.			

DMA SELECTION				
Channel	**W1**	**W2**	**W3**	**W4**
1	closed	open	closed	open
3	open	closed	open	closed

DATA TECHNOLOGY CORPORATION

DTC2280 PLUS

Data bus:	16-bit ISA
Size:	Half-length, half-height card
Hard drive supported:	Two IDE(AT) interface drives
Floppy drives supported:	Two 360KB, 720KB, 1.2MB, 1.44MB

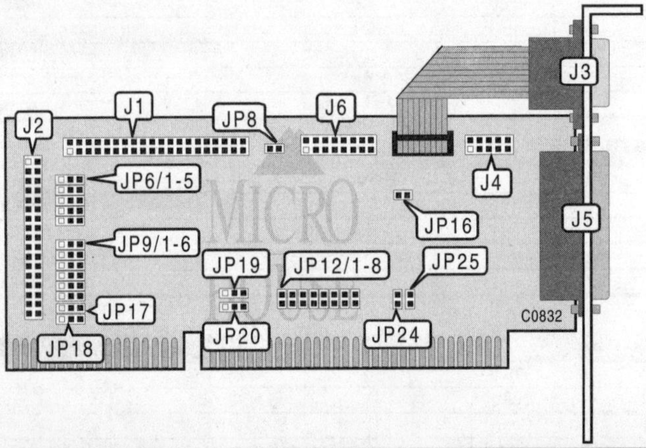

CONNECTIONS	
Function	**Location**
40-pin IDE(AT) interface connector	J1
34-pin cable connector - floppy drive	J2
9-pin serial port 1 - external	J3
10-pin serial port 2 - internal	J4
25-pin parallel port - external	J5
16-pin game port - internal	J6
2-pin connector - drive active LED	JP8

USER CONFIGURABLE SETTINGS		
Function	**Location**	**Setting**
➪ Floppy drives A & B normal	JP6/1	pins 1 & 2 closed
Floppy drives A & B reversed	JP6/1	pins 2 & 3 closed
➪ Floppy drive interface address is 3F0h	JP6/2	pins 2 & 3 closed
Floppy drive interface address is 370h	JP6/2	pins 1 & 2 closed
➪ IDE(AT) interface enabled	JP6/3	pins 1 & 2 closed
IDE(AT) interface disabled	JP6/3	pins 2 & 3 closed

Continued on next page. . .

DATA TECHNOLOGY CORPORATION
DTC2280 PLUS

. . .continued from previous page

USER CONFIGURABLE SETTINGS(CONTINUED)		
Function	**Location**	**Setting**
➪ IDE(AT) interface address is 1F0h	JP6/4	pins 2 & 3 closed
IDE(AT) interface address is 170h	JP6/4	pins 1 & 2 closed
➪ Floppy drive interface enabled	JP6/5	pins 1 & 2 closed
Floppy drive interface disabled	JP6/5	pins 2 & 3 closed
➪ Game port enabled	JP16	closed
Game port disabled	JP16	open
➪ Factory configured - do not alter	JP17	N/A
➪ Factory configured - do not alter	JP18	N/A
➪ Hard drive interrupt IRQ14 select	JP19	pins 1 & 2 closed
Hard drive interrupt IRQ15 select	JP19	pins 2 & 3 closed
➪ Factory configured - do not alter	JP20	N/A
➪ Factory configured - do not alter	JP24	N/A
➪ Factory configured - do not alter	JP25	N/A

SERIAL PORT 1 CONFIGURATION		
Port 1	**JP9/1**	**JP9/2**
➪ COM1	pins 1 & 2 closed	pins 2 & 3 closed
COM3	pins 2 & 3 closed	pins 1 & 2 closed
Disabled	pins 2 & 3 closed	pins 2 & 3 closed

SERIAL PORT 2 CONFIGURATION		
Port 2	**JP9/3**	**JP9/4**
➪ COM2	pins 1 & 2 closed	pins 2 & 3 closed
COM4	pins 2 & 3 closed	pins 1 & 2 closed
Disabled	pins 2 & 3 closed	pins 2 & 3 closed

PARALLEL PORT CONFIGURATION		
LPT	**JP9/5**	**JP9/6**
➪ 1 (378h)	pins 2 & 3 closed	pins 1 & 2 closed
2 (278h)	pins 1 & 2 closed	pins 2 & 3 closed
3 (BCh)	pins 1 & 2 closed	pins 1 & 2 closed

SERIAL PORT 1 INTERRUPT SELECTION			
IRQ	**JP12/1**	**JP12/2**	**JP12/3**
➪ IRQ4	closed	open	open
IRQ3	open	closed	open
IRQ5	open	open	closed

SERIAL PORT 2 INTERRUPT SELECTION			
IRQ	**JP12/4**	**JP12/5**	**JP12/6**
➪ IRQ3	closed	open	open
IRQ4	open	closed	open
IRQ5	open	open	closed

PARALLEL PORT INTERRUPT SELECTION		
IRQ	**JP12/7**	**JP12/8**
➪ 7	closed	open
5	open	closed

DISTRIBUTED PROCESSING TECHNOLOGY

PM2024, PM2024-MAC

Data bus: 32-bit PCI
Size: Three-quarter length, full-height card
Hard drive supported: Up to seven Fast SCSI-2 devices
Floppy drives supported: None

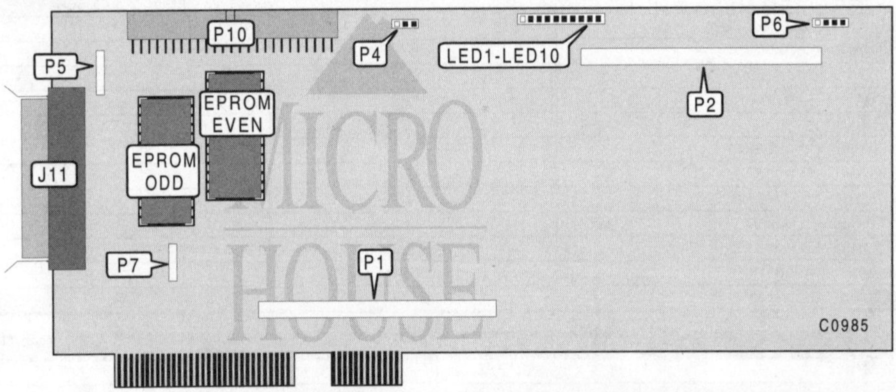

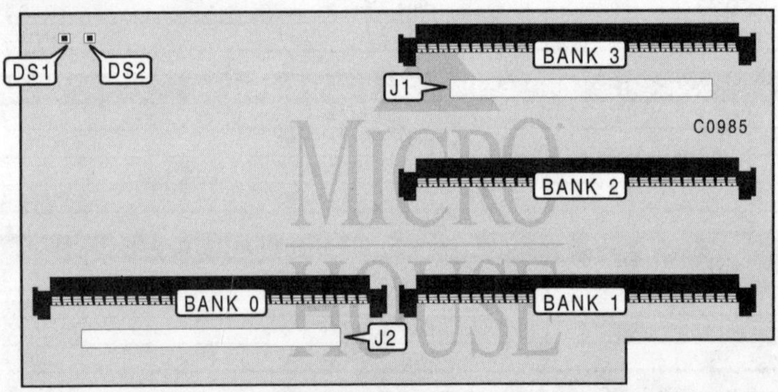

CM4000 Caching Module

CONNECTIONS (CM4000 CACHING MODULE)	
Function	**Location**
Fault LED (red)	DS1
ECC enabled LED (green)	DS2
Proprietary CM4000 connector - caching board to controller	J1
Proprietary CM4000 connector - caching board to controller	J2
50-pin SCSI connector - external	J11
Note: Proprietary connectors J1 and J2 are on the reverse side of the board.	

Continued on next page. . .

DISTRIBUTED PROCESSING TECHNOLOGY
PM2024, PM2024-MAC

. . .continued from previous page

CONNECTIONS (CONTROLLER)	
Function	**Location**
Proprietary CM4000 connector - controller to caching board	P1
Proprietary CM4000 connector - controller to caching board	P2
3-pin connector - factory test LED	P4
4-pin connector - drive active LED	P6
Proprietary DM4000 (disk array module) connector	P5
Proprietary DM4000 (disk array module) connector	P7
50-pin SCSI connector - internal	P10
LED - adapter busy	LED1
LED - computer bus transfer to adapter	LED2
LED - computer bus transfer from adapter	LED3
LED - cache hit	LED4
LED - disk read/ahead active	LED5
LED - disk read	LED6
LED - disk write	LED7
LED - adapter reset	LED8
LED - interrupt to computer pending	LED9
LED - DRQ to computer asserted	LED10

Continued on next page. . .

DISTRIBUTED PROCESSING TECHNOLOGY

PM2024, PM2024-MAC

..continued from previous page

CACHING MODULE DRAM CONFIGURATION				
Size	**Bank 0**	**Bank 1**	**Bank 2**	**Bank 3**
1MB	(1) 256K x 36	None	None	None
1MB	None	(1) 256K x 36	None	None
1MB	None	None	(1) 256K x 36	None
1MB	None	None	None	(1) 256K x 36
2MB	(1) 256K x 36	(1) 256K x 36	None	None
2MB	None	(1) 256K x 36	(1) 256K x 36	None
2MB	None	None	(1) 256K x 36	(1) 256K x 36
3MB	(1) 256K x 36	(1) 256K x 36	(1) 256K x 36	None
3MB	None	(1) 256K x 36	(1) 256K x 36	(1) 256K x 36
3MB	(1) 256K x 36	None	(1) 256K x 36	(1) 256K x 36
3MB	(1) 256K x 36	(1) 256K x 36	None	(1) 256K x 36
4MB	(1) 256K x 36	(1) 256K x 36	(1) 256K x 36	(1) 256K x 36
4MB	(1) 1M x 36	None	None	None
4MB	None	(1) 1M x 36	None	None
4MB	None	None	(1) 1M x 36	None
4MB	None	None	None	(1) 1M x 36
8MB	(1) 1M x 36	(1) 1M x 36	None	None
8MB	None	(1) 1M x 36	(1) 1M x 36	None
8MB	None	None	(1) 1M x 36	(1) 1M x 36
12MB	(1) 1M x 36	(1) 1M x 36	(1) 1M x 36	None
12MB	None	(1) 1M x 36	(1) 1M x 36	(1) 1M x 36
12MB	(1) 1M x 36	None	(1) 1M x 36	(1) 1M x 36
12MB	(1) 1M x 36	(1) 1M x 36	None	(1) 1M x 36
16MB	(1) 1M x 36	(1) 1M x 36	(1) 1M x 36	(1) 1M x 36
16MB	(1) 4M x 36	None	None	None
16MB	None	(1) 4M x 36	None	None
16MB	None	None	(1) 4M x 36	None
16MB	None	None	None	(1) 4M x 36
32MB	(1) 4M x 36	(1) 4M x 36	None	None
32MB	None	(1) 4M x 36	(1) 4M x 36	None
32MB	None	None	(1) 4M x 36	(1) 4M x 36
48MB	(1) 4M x 36	(1) 4M x 36	(1) 4M x 36	None
48MB	None	(1) 4M x 36	(1) 4M x 36	(1) 4M x 36
48MB	(1) 4M x 36	None	(1) 4M x 36	(1) 4M x 36
48MB	(1) 4M x 36	(1) 4M x 36	None	(1) 4M x 36
64MB	(1) 4M x 36	(1) 4M x 36	(1) 4M x 36	(1) 4M x 36

Notes: Although not shown above, 1MB, 4MB and 16MB SIMMs can be combined and DPT ECC SIMMs can replace parity SIMMs of 4MB or 16MB.
Single sided SIMMs are preferred, but low-profile double sided 2MB and 8MB increments are supported.

DISTRIBUTED PROCESSING TECHNOLOGY

PM2044UW/PM2044W/PM2144UW/PM2144W PM2044UW-MAC/PM2044W-MAC/PM2144UW-MAC/PM2144W-MAC

Data bus: 32-bit PCI
Size: Half length, full-height card
Hard drive supported: Fifteen SCSI-Wide, or eight SCSI-Wide with seven SCSI-2 Narrow SCSI-3 and Fast SCSI devices also supported
Floppy drives supported: Nones

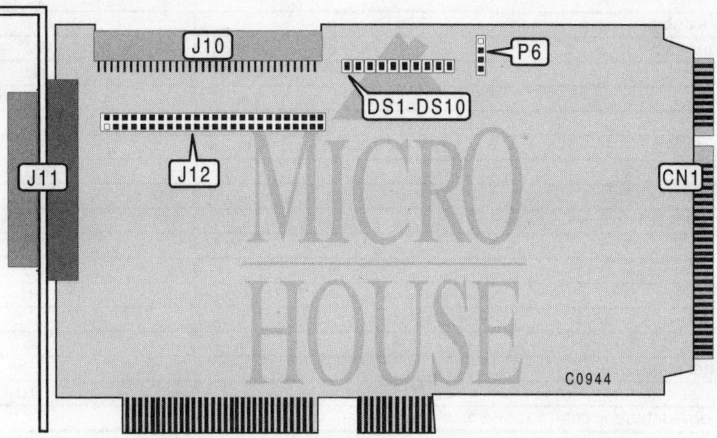

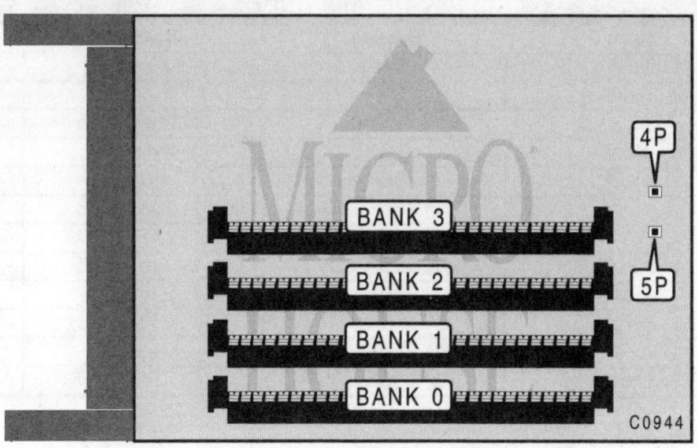

RC4040 - Optional RAID & Caching Module

Continued on next page. . .

DISTRIBUTED PROCESSING TECHNOLOGY

PM2044UW/PM2044W/PM2144UW/PM2144W PM2044UW-MAC/PM2044W-MAC/PM2144UW-MAC/PM2144W-MAC

. . .continued from previous page

CONNECTIONS	
Function	**Location**
68-pin SCSI connector - internal	J10
68-pin SCSI connector - external	J11
50-pin SCSI connector internal	J12
RC4040 connector	CN1
LED - ECC enabled (green)	P4
LED - memory fault (red)	P5
4-pin connector - drive active LED	P6
LED - adapter busy	DS1
LED - computer bus transfer to adapter	DS2
LED - computer bus transfer from adapter	DS3
LED - cache hit	DS4
LED - disk read/ahead active	DS5
LED - disk read	DS6
LED - disk write	DS7
LED - adapter reset	DS8
LED - interrupt to computer pending	DS9
LED - DRQ to computer asserted	DS10
Note: Location of floppy drive interface connector P8 is unidentified.	

DRAM CONFIGURATION				
Size	**Bank 0**	**Bank 1**	**Bank 2**	**Bank 3**
1MB	(1) 256K x 36	None	None	None
1MB	None	(1) 256K x 36	None	None
1MB	None	None	(1) 256K x 36	None
1MB	None	None	None	(1) 256K x 36
2MB	(1) 256K x 36	(1) 256K x 36	None	None
2MB	None	(1) 256K x 36	(1) 256K x 36	None
2MB	None	None	(1) 256K x 36	(1) 256K x 36
3MB	(1) 256K x 36	(1) 256K x 36	(1) 256K x 36	None
3MB	None	(1) 256K x 36	(1) 256K x 36	(1) 256K x 36
3MB	(1) 256K x 36	None	(1) 256K x 36	(1) 256K x 36
3MB	(1) 256K x 36	(1) 256K x 36	None	(1) 256K x 36
4MB	(1) 256K x 36	(1) 256K x 36	(1) 256K x 36	(1) 256K x 36

Continued on next page. . .

DISTRIBUTED PROCESSING TECHNOLOGY
PM2044UW/PM2044W/PM2144UW/PM2144W PM2044UW-MAC/PM2044W-MAC/PM2144UW-MAC/PM2144W-MAC

...continued from previous page

DRAM CONFIGURATION CONTINUED				
Size	**Bank 0**	**Bank 1**	**Bank 2**	**Bank 3**
4MB	(1) 1M x 36	None	None	None
4MB	None	(1) 1M x 36	None	None
4MB	None	None	(1) 1M x 36	None
4MB	None	None	None	(1) 1M x 36
8MB	(1) 1M x 36	(1) 1M x 36	None	None
8MB	None	(1) 1M x 36	(1) 1M x 36	None
8MB	None	None	(1) 1M x 36	(1) 1M x 36
12MB	(1) 1M x 36	(1) 1M x 36	(1) 1M x 36	None
12MB	None	(1) 1M x 36	(1) 1M x 36	(1) 1M x 36
12MB	(1) 1M x 36	None	(1) 1M x 36	(1) 1M x 36
12MB	(1) 1M x 36	(1) 1M x 36	None	(1) 1M x 36
16MB	(1) 1M x 36	(1) 1M x 36	(1) 1M x 36	(1) 1M x 36
16MB	(1) 4M x 36	None	None	None
16MB	None	(1) 4M x 36	None	None
16MB	None	None	(1) 4M x 36	None
16MB	None	None	None	(1) 4M x 36
32MB	(1) 4M x 36	(1) 4M x 36	None	None
32MB	None	(1) 4M x 36	(1) 4M x 36	None
32MB	None	None	(1) 4M x 36	(1) 4M x 36
48MB	(1) 4M x 36	(1) 4M x 36	(1) 4M x 36	None
48MB	None	(1) 4M x 36	(1) 4M x 36	(1) 4M x 36
48MB	(1) 4M x 36	None	(1) 4M x 36	(1) 4M x 36
48MB	(1) 4M x 36	(1) 4M x 36	None	(1) 4M x 36
64MB	(1) 4M x 36	(1) 4M x 36	(1) 4M x 36	(1) 4M x 36

Notes: Although not shown above, 1MB, 4MB and 16MB SIMMs can be combined and DPT ECC SIMMs can replace parity SIMMs of 4MB or 16MB.
Single sided SIMMs are preferred, but low-profile double sided 2M and 8MB increments are supported.

DISTRIBUTED PROCESSING TECHNOLOGY

PM2124, PM2124W, PM2124-MAC, PM2124W-MAC

Data bus:	32-bit PCI
Size:	Three-quarter length, full-height card
Hard drive supported:	Up to seven SCSI-2 or Fast SCSI-2 devices (PM2124/PM2124-MAC) Up to fifteen Wide SCSI-2 devices (PM2124W/PM2124W-MAC)
Floppy drives supported:	None

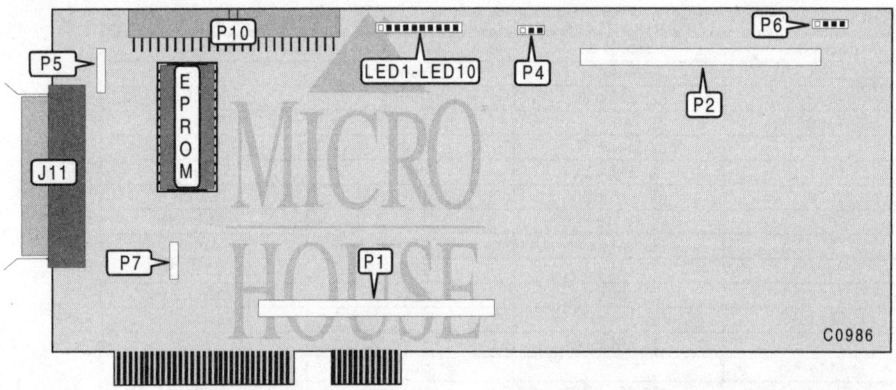

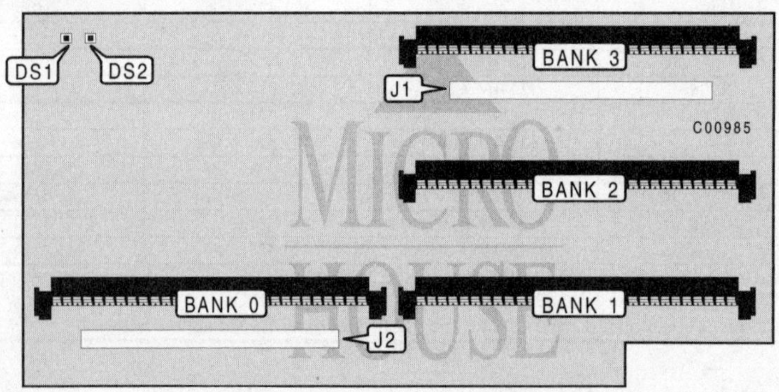

CM4000 Caching Module

CONNECTIONS (CM4000 CACHING MODULE)	
Function	**Location**
Fault LED (red)	DS1
ECC enabled LED (green)	DS2
Proprietary CM4000 connector - caching board to controller	J1
Proprietary CM4000 connector - caching board to controller	J2
Note: Proprietary connectors J1 and J2 are on the reverse side of the board.	

Continued on next page. . .

DISTRIBUTED PROCESSING TECHNOLOGY

PM2124, PM2124W, PM2124-MAC, PM2124W-MAC

...continued from previous page

CONNECTIONS (CONTROLLER)	
Function	**Location**
50/68-pin SCSI connector - external	J11
Proprietary CM4000 connector - controller to caching board	P1
Proprietary CM4000 connector - controller to caching board	P2
3-pin connector - factory test LED	P4
4-pin connector - drive active LED	P6
Proprietary DM4000 (disk array module) connector	P5
Proprietary DM4000 (disk array module) connector	P7
50/68-pin SCSI connector - internal	P10
LED - adapter busy	LED1
LED - computer bus transfer to adapter	LED2
LED - computer bus transfer from adapter	LED3
LED - cache hit	LED4
LED - disk read/ahead active	LED5
LED - disk read	LED6
LED - disk write	LED7
LED - adapter reset	LED8
LED - interrupt to computer pending	LED9
LED - DRQ to computer asserted	LED10
Note: 50-pin ports are on models PM2124 and PM2124-MAC. 68-pin ports are on the PM2124W and PM2124W-MAC.	

CACHING MODULE DRAM CONFIGURATION				
Size	**Bank 0**	**Bank 1**	**Bank 2**	**Bank 3**
1MB	(1) 256K x 36	None	None	None
1MB	None	(1) 256K x 36	None	None
1MB	None	None	(1) 256K x 36	None
1MB	None	None	None	(1) 256K x 36
2MB	(1) 256K x 36	(1) 256K x 36	None	None

Continued on next page...

DISTRIBUTED PROCESSING TECHNOLOGY

PM2124, PM2124W, PM2124-MAC, PM2124W-MAC

. . .continued from previous page

DRAM CONFIGURATION (CONTINUED)				
Size	**Bank 0**	**Bank 1**	**Bank 2**	**Bank 3**
2MB	None	(1) 256K x 36	(1) 256K x 36	None
2MB	None	None	(1) 256K x 36	(1) 256K x 36
3MB	(1) 256K x 36	(1) 256K x 36	(1) 256K x 36	None
3MB	None	(1) 256K x 36	(1) 256K x 36	(1) 256K x 36
3MB	(1) 256K x 36	None	(1) 256K x 36	(1) 256K x 36
3MB	(1) 256K x 36	(1) 256K x 36	None	(1) 256K x 36
4MB	(1) 256K x 36	(1) 256K x 36	(1) 256K x 36	(1) 256K x 36
4MB	(1) 1M x 36	None	None	None
4MB	None	(1) 1M x 36	None	None
4MB	None	None	(1) 1M x 36	None
4MB	None	None	None	(1) 1M x 36
8MB	(1) 1M x 36	(1) 1M x 36	None	None
8MB	None	(1) 1M x 36	(1) 1M x 36	None
8MB	None	None	(1) 1M x 36	(1) 1M x 36
12MB	(1) 1M x 36	(1) 1M x 36	(1) 1M x 36	None
12MB	None	(1) 1M x 36	(1) 1M x 36	(1) 1M x 36
12MB	(1) 1M x 36	None	(1) 1M x 36	(1) 1M x 36
12MB	(1) 1M x 36	(1) 1M x 36	None	(1) 1M x 36
16MB	(1) 1M x 36	(1) 1M x 36	(1) 1M x 36	(1) 1M x 36
16MB	(1) 4M x 36	None	None	None
16MB	None	(1) 4M x 36	None	None
16MB	None	None	(1) 4M x 36	None
16MB	None	None	None	(1) 4M x 36
32MB	(1) 4M x 36	(1) 4M x 36	None	None
32MB	None	(1) 4M x 36	(1) 4M x 36	None
32MB	None	None	(1) 4M x 36	(1) 4M x 36
48MB	(1) 4M x 36	(1) 4M x 36	(1) 4M x 36	None
48MB	None	(1) 4M x 36	(1) 4M x 36	(1) 4M x 36
48MB	(1) 4M x 36	None	(1) 4M x 36	(1) 4M x 36
48MB	(1) 4M x 36	(1) 4M x 36	None	(1) 4M x 36
64MB	(1) 4M x 36	(1) 4M x 36	(1) 4M x 36	(1) 4M x 36
Notes: Although not shown above, 1MB, 4MB and 16MB SIMMs can be combined and DPT ECC SIMMs can replace parity SIMMs of 4MB or 16MB. Single sided SIMMs are preferred, but low-profile double sided 2MB and 8MB increments are supported.				

DTK COMPUTER, INC.

PTI-227

Data bus: 16-bit, ISA
Size: Half-length, full-height card
Hard drives supported: Two IDE (AT) Interface drives
Floppy drives supported: Two 360KB, 720KB, 1.2MB or 1.44MB drives

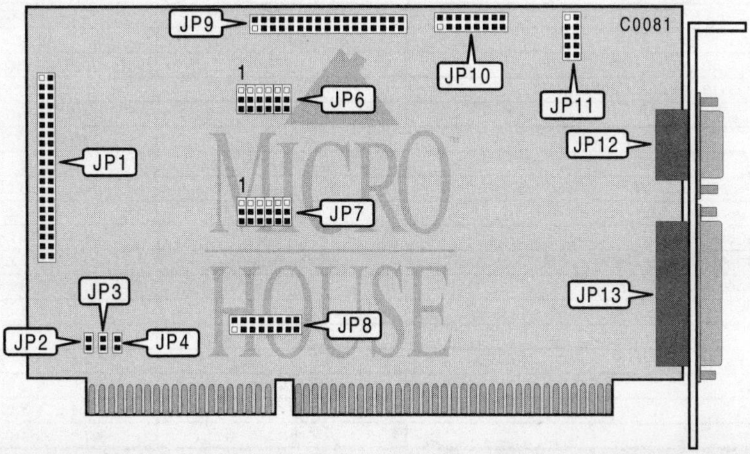

CONNECTIONS	
Function	**Location**
40-pin IDE (AT) Interface connector	JP1
2-pin connector-Drive Active LED	JP2
34-pin control cable connector-floppy drive	JP9
15-pin game port-internal	JP10
9-pin serial port 2 (COM2)-internal	JP11
9-pin serial port 1 (COM1)-external	JP12
25-pin parallel port-external	JP13

Continued on next page . . .

DTK COMPUTER, INC.

PTI-227

. . . continued from previous page

USER CONFIGURABLE SETTINGS		
Function	**Location**	**Setting**
➪ Single IDE drive on chain	JP3	closed
Two IDE drives on chain	JP3	open
➪ Parallel port (CN13) is an output port	JP4	closed
Parallel port (CN13) is an input or output port	JP4	open
➪ Supports IBM BIOS	JP5	pins 2 & 3 closed
Supports other BIOS	JP5	pins 1 & 2 closed
➪ Hard drive port address is 1F0-1F7h, 3F6-3F7h	JP7/jumper 1	pins 1 & 2 closed
Hard drive port address is 170-177h, 376-377h	JP7/jumper 1	pins 2 & 3 closed
➪ Floppy drive assignment changed (A: = B:; B: = A:)	JP7/jumper 2	pins 1 & 2 closed
Floppy drive assignment normal	JP7/jumper 2	pins 2 & 3 closed
➪ Floppy drive port address is 3F0-3F7h	JP7/jumper 3	pins 1 & 2 closed
Floppy drive port address is 370-377h	JP7/jumper 3	pins 2 & 3 closed
➪ Hard drive enabled	JP7/jumper 4	pins 2 & 3 closed
Hard drive disabled	JP7/jumper 4	pins 1 & 2 closed
➪ Floppy drive enabled	JP7/jumper 5	pins 2 & 3 closed
Floppy drive disabled	JP7/jumper 5	pins 1 & 2 closed
➪ AT compatible	JP7/jumper 6	pins 2 & 3 closed
XT compatible	JP7/jumper 6	pins 1 & 2 closed

SERIAL PORT 1 CONFIGURATION		
COM	**JP6/jumper 1**	**JP6/jumper 2**
➪ COM1 (3F8)	pins 2 & 3 closed	pins 1 & 2 closed
COM3 (3E8)	pins 1 & 2 closed	pins 2 & 3 closed
COM4 (2E8)	pins 2 & 3 closed	pins 2 & 3 closed
Disabled	pins 1 & 2 closed	pins 1 & 2 closed

SERIAL PORT 1 INTERRUPT REQUEST			
IRQ	**JP8/jumper 1**	**JP8/jumper 2**	**JP8/jumper 3**
➪ IRQ4	closed	open	open
IRQ3	open	cloqed	open
IRQ5	open	open	closed

SERIAL PORT 2 CONFIGURATION		
COM	**JP6/jumper 3**	**JP6/jumper 4**
➪ COM2 (2F8)	pins 2 & 3 closed	pins 1 & 2 closed
COM3 (3E8)	pins 2 & 3 closed	pins 2 & 3 closed
COM4 (2E8)	pins 1 & 2 closed	pins 2 & 3 closed
Disabled	pins 1 & 2 closed	pins 1 & 2 closed

SERIAL PORT 2 INTERRUPT REQUEST			
IRQ	**JP8/jumper 4**	**JP8/jumper 5**	**JP8/jumper 6**
➪ IRQ3	closed	open	open
IRQ4	open	closed	open
IRQ5	open	open	closed

Continued on next page . . .

DTK COMPUTER, INC.
PTI-227

... continued from previous page

PARALLEL PORT CONFIGURATION		
LPT	**JP6/jumper 5**	**JP6/jumper 6**
⇨ LPT1 (378)	pins 1 & 2 closed	pins 2 & 3 closed
LPT1 (3BC)	pins 2 & 3 closed	pins 2 & 3 closed
LPT2 (278)	pins 2 & 3 closed	pins 1 & 2 closed
Disabled	pins 1 & 2 closed	pins 1 & 2 closed

PARALLEL PORT INTERRUPT REQUEST		
IRQ	**JP8/jumper 5**	**JP8/jumper 6**
⇨ IRQ7	closed	open
IRQ5	open	closed

LAVA COMPUTER MFG, INC.

COMPLETE-550

Data Bus: 16-bit ISA
Size: Half-length, Full-height card
Hard drives supported: Two IDE (AT) Interface drives
Floppy drives supported: Two 360KB, 720KB, 1.2MB, or 1.44MB drives

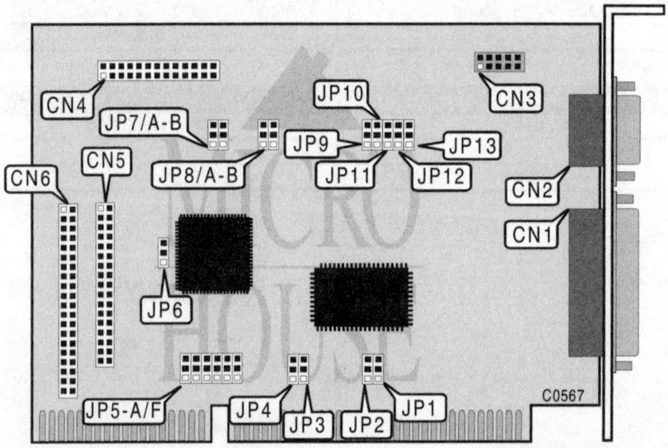

CONNECTIONS	
Function	**Location**
DB-25 female connector-parallel port 1	CN1
DB-9 male connector-serial port 1	CN2
10-pin connector-serial port 2 internal	CN3
26-pin connector-parallel port 2 internal	CN4
34-pin data cable connector-floppy drive internal	CN5
40-pin connector IDE (AT) Interface internal	CN6

USER CONFIGURABLE SETTINGS		
Function	**Location**	**Setting**
➪ Parallel port 1 enabled.	JP11	Pins 2 & 3 closed
Parallel port 1 disabled.	JP11	Pins 1 & 2 closed
➪ Floppy drive enabled.	JP12	Pins 2 & 3 closed
Floppy drive disabled.	JP12	Pins 1 & 2 closed
➪ Hard drive enabled.	JP13	Pins 2 & 3 closed
Hard drive disabled.	JP13	Pins 1 & 2 closed

PRIMARY PARALLEL PORT ADDRESS SELECT	
Function	**JP10**
LPT1	Pins 2 & 3 closed
LPT2	Pins 1 & 2 closed

ALTERNATE PARALLEL PORT ADDRESS SELECT	
Function	**JP9**
LPT1	Pins 2 & 3 closed
LPT2	Pins 1 & 2 closed

Continued on next page. . .

LAVA COMPUTER MFG, INC.
COMPLETE-550

. . .continued from previous page

SERIAL PORT 2 CONFIGURATION		
COM	**JP8/A**	**JP8/B**
⇨ COM1	Pins 1 & 2 closed	All open
COM2	Pins 2 & 3 closed	All open
COM3	All open	Pins 1 & 2 closed
COM4	All open	Pins 2 & 3 closed

SERIAL PORT 1 CONFIGURATION		
COM	**JP7/A**	**JP7/B**
⇨ COM2	Pins 2 & 3 closed	All open
COM1	Pins 1 & 2 closed	All open
COM3	All open	Pins 1 & 2 closed
COM4	All open	Pins 2 & 3 closed

PRIMARY SERIAL PORT INTERUPT SELECT								
IRQ	**JP5/A**	**JP5/B**	**JP5/C**	**JP5/D**	**JP5/E**	**JP5/F**	**JP4**	**JP3**
IRQ2	All open	All open	All open	All open	All open	2 & 3	All open	All open
IRQ3	All open	All open	All open	All open	All open	All open	All open	2 & 3
IRQ4	All open	All open	All open	All open	All open	All open	2 & 3	All open
IRQ5	All open	All open	All open	All open	2 & 3	All open	All open	All open
IRQ10	All open	All open	All open	2 & 3	All open	All open	All open	All open
IRQ11	All open	All open	2 & 3	All open	All open	All open	All open	All open
IRQ12	All open	2 & 3	All open	All open	All open	All open	All open	All open
IRQ13	2 & 3	All open	All open	All open	All open	All open	All open	All open
Note: Pins designated are in the closed position.								

ALTERNATE SERIAL PORT INTERUPT SELECT								
IRQ	**JP5/A**	**JP5/B**	**JP5/C**	**JP5/D**	**JP5/E**	**JP5/F**	**JP4**	**JP3**
IRQ2	All open	All open	All open	All open	All open	1 & 2	All open	All open
IRQ3	All open	All open	All open	All open	All open	All open	All open	1 & 2
IRQ4	All open	All open	All open	All open	All open	All open	1 & 2	All open
IRQ5	All open	All open	All open	All open	1 & 2	All open	All open	All open
IRQ10	All open	All open	All open	1 & 2	All open	All open	All open	All open
IRQ11	All open	All open	1 & 2	All open	All open	All open	All open	All open
IRQ12	All open	1 & 2	All open	All open	All open	All open	All open	All open
IRQ13	1 & 2	All open	All open	All open	All open	All open	All open	All open
Note: Pins designated are in the closed position.								

PRIMARY PARALLEL PORT IRQ SELECT	
IRQ	**JP1**
IRQ7	Pins 2 & 3 closed
IRQ5	Pins 1 & 2 closed

ALTERNATE PARALLEL PORT IRQ SELECT	
IRQ	**JP2**
IRQ7	Pins 2 & 3 closed
IRQ5	Pins 1 & 2 closed

Continued on next page. . .

LAVA COMPUTER MFG, INC.

COMPLETE-550

. . .continued from previous page

ALTERNATE PARALLEL PORT DIRECTION SETTING	
Function	**JP6**
➪ Unidirectional	Pins 2 & 3 closed
Bi-directional	Pins 1 & 2 closed
Note: The primary parallel port is always Uni-directional.	

MONOLITHIC SYSTEMS, INC. (COLORADO MSI)

MICROPAQ 452 REV. B

Data bus: 16-bit, ISA
Size: Full-length, full-height card
Hard drive supported: Two IDE (AT) Interface drives
Floppy drives supported: Two 360KB, 720KB, 1.2MB, or 1.44MB drives

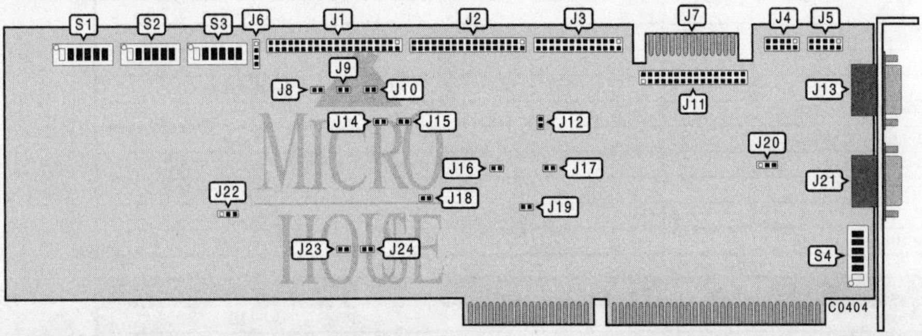

CONNECTIONS	
Function	**Location**
40-pin IDE(AT) Interface connector	J1
34-pin control cable connector - floppy drive	J2
26-pin parallel port (LPT1/2/3) - internal	J3
10-pin serial port 1 (COM1/2/3/4) - internal	J4
10-pin serial port 2 (COM1/2/3/4) - internal	J5
4-pin connector - Drive active LED	J6
28-pin VESA(VGA) feature connector - internal	J7
26-pin VESA(EGA) feature connector - internal	J11
15-pin analog VGA port - external	J13
9-pin digital EGA port- external	J21

USER CONFIGURABLE SETTINGS		
Function	**Location**	**Setting**
➪ Pin 34 on IDE interface grounded	J10	closed
Pin 34 on IDE interface not grounded	J10	open
➪ Floppy drive precompensation is 125ns	J12	open
Floppy drive precompensation is 187ns	J12	closed
➪ Hard drive enabled	J14	closed
Hard drive disabled	J14	open
➪ On-board video enabled	J15	closed
On-board video disabled	J15	open
➪ Floppy drive Interrupt IRQ6 enabled	J16	closed
Floppy drive Interrupt IRQ6 disabled	J16	open

Continued on next page . . .

MONOLITHIC SYSTEMS, INC. (COLORADO MSI)

MICROPAQ 452 REV. B

. . . continued from previous page

USER CONFIGURABLE SETTINGS		
Function	**Location**	**Setting**
➪ Single speed floppy drives supported	J17	open
Dual speed floppy drives supported	J17	closed
➪ 16-bit video bus enabled	J18	open
8-bit video bus enabled	J18	closed
➪ Floppy drive change signal sense enabled	J19	closed
Floppy drive change signal sense disabled	J19	open
➪ VGA clock enabled (normal operation)	J22	pins 1 & 2 closed
Feature connector clock enabled	J22	pins 2 & 3 closed
➪ Hard drive Interrupt IRQ14 enabled	J24	closed
Hard drive Interrupt IRQ14 disabled	J24	open

IDE IOCHRDY PIN SELECTION		
Function	**J8**	**J9**
➪ IDE IOCHRDY from IDE pin 21 enabled	off	on
IDE IOCHRDY from IDE pin 27 enabled	on	off

Pin 21 setting is for Seagate, CDC, Imprimis, and Miniscribe hard drives. Pin 27 setting is for Quantum hard drives. Conner and WD hard drives ignore the setting of these switches.

SERIAL PORT 1 CONFIGURATION				
COM	**S1/switch 1**	**S1/switch 2**	**S1/switch 3**	**S1/switch 4**
➪ 1	on	off	off	off
2	off	on	off	off
3	off	off	on	off
4	off	off	off	on
Disabled	off	off	off	off

SERIAL PORT 1 (J4) INTERRUPT REQUEST		
IRQ	**S1/switch 5**	**S1/switch 6**
➪ 3	on	off
4	off	on
Disabled	off	off

PARALLEL PORT (J3) CONFIGURATION - S2			
LPT	**S2/switch 1**	**S2/switch 2**	**S2/switch 3**
➪ 1	on	off	off
2	off	on	off
3	off	off	on
Disabled	off	off	off

PARALLEL PORT (J3) INTERRUPT REQUEST		
IRQ	**S2/switch 5**	**S2/switch 6**
➪ 7	off	on
5	on	off
Disabled	off	off

Continued on next page . . .

MONOLITHIC SYSTEMS, INC. (COLORADO MSI)

MICROPAQ 452 REV. B

. . . continued from previous page

SERIAL PORT 2 CONFIGURATION				
COM	**S3/switch 1**	**S3/switch 2**	**S3/switch 3**	**S3/switch 4**
⇨ 1	on	off	off	off
2	off	on	off	off
3	off	off	on	off
4	off	off	off	on
Disabled	off	off	off	off

SERIAL PORT 2 (J5) INTERRUPT REQUEST		
IRQ	**S3/switch 5**	**S3/switch 6**
⇨ 4	off	on
3	on	off
Disabled	off	off

VIDEO DISPLAY CONFIGURATION						
Primary Adapter	**Secondary Adapter**	**JP20**	**S4/1**	**S4/2**	**S4/3**	**S4/4**
*Monochrome	EGA w/CGA 40 x 25	1 & 2	on	on	on	on
*Monochrome	EGA w/CGA 80 x 25	1 & 2	off	on	on	on
*Monochrome	EGA w/EGA 200 line	1 & 2	on	off	on	on
*Monochrome	EGA w/EGA 350 line	1 & 2	off	off	on	on
*CGA 40 x 25	EGA w/monochrome	2 & 3	on	on	off	on
*CGA 80 x 25	EGA w/monochrome	2 & 3	off	on	off	on
EGA w/CGA 40 x 25	*Monochrome or none	1 & 2	on	off	off	on
EGA w/CGA 80 x 25	*Monochrome or none	1 & 2	off	off	off	on
EGA w/EGA 200 line	*Monochrome or none	1 & 2	on	on	on	off
⇨ EGA w/EGA 350 line	*Monochrome or none	1 & 2	off	on	on	off
EGA w/monochrome	*CGA 40 x 25 or none	2 & 3	on	off	on	off
EGA w/monochrome	*CGA 80 x 25 or none	2 & 3	off	off	on	off
VGA enable (all other modes disabled)		S4/switch 5 on				
All other modes enable (VGA mode disabled)		S4/switch 5 off				

For JP20 pins designated should be in a closed position.
When VGA mode is enabled, the settings of JP20 and S4/switch 1 - 4 are ignored.
The options with a (*) in front of them require an add-on card to function.

MISCELLANEOUS TECHNICAL NOTES

The function of S2/switch 4 is undocumented.
All other jumpers are for factory tests and should remain factory-configured.

PROMISE TECHNOLOGY, INC.

DRIVECACHE DC2040

Data bus: 16-bit, ISA
Size: Full-length, full-height card
Hard drives supported: Seven SCSI hard drives
Floppy drives supported: Two 360KB, 720KB, 1.2MB, or 1.44MB drives
One floppy tape drive

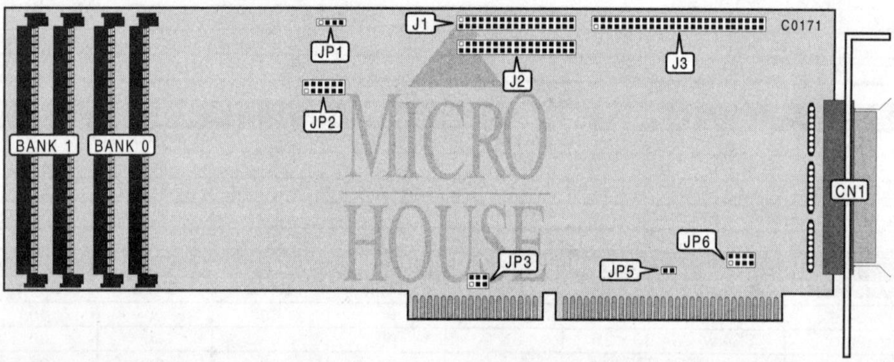

CONNECTIONS	
Function	**Location**
DB-25 SCSI control cable connector-external	CN1
34-pin control cable connector-floppy drive	J1
34-pin control cable connector-floppy tape drive	J2
50-pin SCSI control cable connector-internal	J3
4-pin connector-drive active LED	JP1

USER CONFIGURABLE SETTINGS		
Function	**Location**	**Setting**
⇨ Factory configured - do not alter	JP2/jumper 1	open
⇨ Floppy and tape drive enabled	JP5	closed
Floppy and tape drive disabled	JP5	open

INTERRUPT REQUEST					
IRQ	**JP2/2**	**JP2/3**	**JP3/1**	**JP3/2**	**JP3/3**
IRQ12	closed	open	closed	open	open
⇨ IRQ14	open	open	open	closed	open
IRQ15	open	closed	open	open	closed

SCSI ADDRESS			
Address	**JP2/jumper 4**	**JP2/jumper 5**	**JP2/jumper 6**
0	open	open	open
1	closed	open	open
2	open	closed	open
3	closed	closed	open
4	open	open	closed
5	closed	open	closed
6	open	closed	closed
⇨ 7	closed	closed	closed

Continued on next page . . .

PROMISE TECHNOLOGY, INC.

DRIVECACHE DC2040

. . . continued from previous page

BIOS ADDRESS				
Address	**JP6/jumper 1**	**JP6/jumper 2**	**JP6/jumper 3**	**JP6/jumper 4**
C800h	closed	open	closed	closed
CA00h	closed	open	closed	open
CC00h	closed	open	open	closed
CE00h	closed	open	open	open
D000h	open	closed	closed	closed
D200h	open	closed	closed	open
D400h	open	closed	open	closed
D600h	open	closed	open	open
⇨ D800h	open	open	closed	closed
DA00h	open	open	closed	open
DC00h	open	open	open	closed
DE00h	open	open	open	open

ON-BOARD CACHE CONFIGURATION		
Size	**Bank 0**	**Bank 1**
512KB	(2) 256KB x 9	NONE
1MB	(2) 256KB x 9	(2) 256KB x 9
2MB	(2) 1MB x 9	NONE
2.5MB	(2) 256KB x 9	(2) 1MB x 9
4MB	(2) 1MB x 9	(2) 1MB x 9
8MB	(2) 4MB x 9	NONE
8.5MB	(2) 256KB x 9	(2) 4MB x 9
10MB	(2) 1MB x 9	(2) 4MB x 9
16MB	(2) 4MB x 9	(2) 4MB x 9
Note: Uses only 80nsec or faster SIMM's; 256K, 1M, or 4M (either x8 or x9).		

PROMISE TECHNOLOGY, INC.

EIDEMAX

Data bus: 16-bit, ISA
Size: Half-length, half-height card
Hard drive supported: Two IDE (AT) Interface drives
Floppy drives supported: None

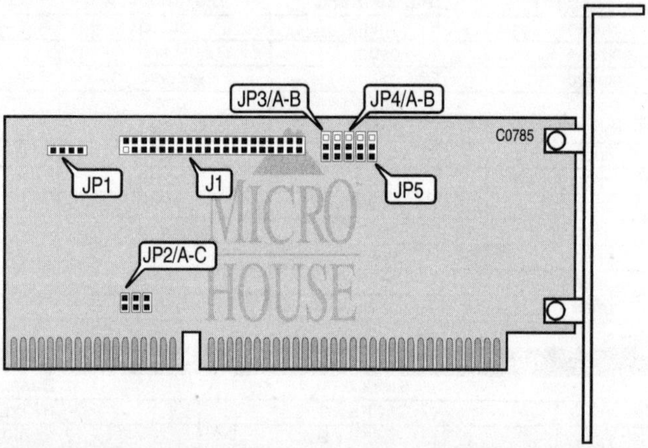

CONNECTIONS	
Function	**Location**
40-pin IDE(AT) interface connector	J1
4-pin connector - drive active LED	JP1

USER CONFIGURABLE SETTINGS		
Function	**Location**	**Setting**
⇨ IDE(AT) interface enabled	JP3/A	pins 1 & 2 closed
IDE(AT) interface disabled	JP3/A	pins 2 & 3 closed
⇨ EIDEMAX is primary IDE(AT) interface in system	JP3/B	pins 1 & 2 closed
EIDEMAX is secondary IDE(AT) interface in system	JP3/B	pins 2 & 3 closed
⇨ On-board BIOS enabled	JP5	pins 1 & 2 closed
On-board BIOS disabled	JP5	pins 2 & 3 closed

INTERRUPT SELECTION			
IRQ	**JP2/A**	**JP2/B**	**JP2/C**
⇨ IRQ15	closed	open	open
IRQ12	open	open	closed
IRQ14	open	closed	open

BASE MEMORY ADDRESS SELECTION		
Address	**JP4/A**	**JP4/B**
⇨ D0000h	pins 2 & 3 closed	pins 2 & 3 closed
D4000h	pins 2 & 3 closed	pins 1 & 2 closed
D8000h	pins 1 & 2 closed	pins 2 & 3 closed
DC000h	pins 1 & 2 closed	pins 1 & 2 closed

PROMISE TECHNOLOGY, INC.

SCSIMAX PREMIER SCSI
SCSI ULTRA PREMIER SCSI

Data Bus 32-bit PCI
Card Size Half-height, quarter-length card
Hard Drives supported Up to 15 Fast & Wide SCSI 2 drives (SCSIMAX)
Up to 15 Ultra SCSI-2/SCSI-3 drives (SCSIULTRA)
Floppy drives supported None

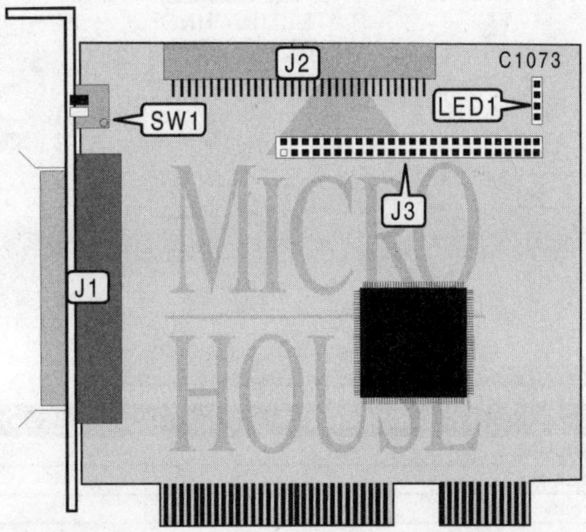

CONNECTIONS			
Function	**Location**	**Function**	**Location**
68-pin SCSI connector - external	J1	50-pin SCSI connector - internal	J3
68-pin SCSI connector - internal	J2	4-pin connector - drive active LED	LED1

USER CONFIGURABLE SETTINGS		
Function	**Label**	**Position**
➪ The termination is enabled	SW1	On
The termination is disabled	SW1	Off
Note: Facing the bracket, SW1 On position is to the left and Off position is to the right.		

MISCELLANEOUS TECHNICAL NOTES
Note: When combining internal and external devices, only one of the internal connectors can be used.

RANCHO TECHNOLOGY, INC.

RT1600

Data bus:	16-bit, ISA
Size:	Half-length, full-height card
Hard drive supported:	Up to seven SCSI devices
Floppy drives supported:	Four 360KB, 720KB, 1.2MB, or 1.44MB drives

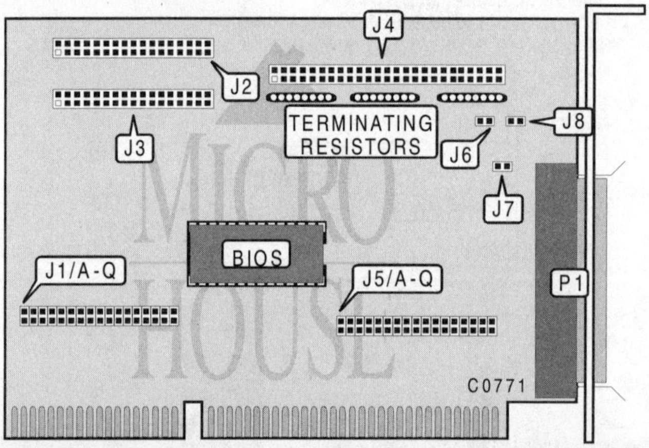

CONNECTIONS

Function	Location
34-pin control cable connector - floppy drive 0	J2
34-pin control cable connector - floppy drive 1	J3
50-pin SCSI connector-internal	J4
50-pin SCSI connector-external	P1
2-pin connector - chassis ground	J7
2-pin connector - drive active LED	J5/jumper O
2-pin connector - drive active LED	J5/jumper P

USER CONFIGURABLE SETTINGS

Function	Location	Setting
➪ Factory configured - do not alter	J1/jumper B	open
➪ Factory configured - do not alter	J1/jumper E	open
➪ Factory configured - do not alter	J1/jumper G	open
➪ Factory configured - do not alter	J1/jumper H	open
➪ Factory configured - do not alter	J1/jumper I	open
➪ Factory configured - do not alter	J1/jumper J	open
➪ Factory configured - do not alter	J1/jumper K	open
➪ Factory configured - do not alter	J1/jumper E	open

Continued on next page . . .

RANCHO TECHNOLOGY, INC.
RT1600

. . . continued from previous page

USER CONFIGURABLE SETTINGS		
Function	**Location**	**Setting**
➪ Factory configured - do not alter	J1/jumper O	open
➪ Factory configured - do not alter	J1/jumper Q	open
BIOS installed	J5/jumper A	closed
BIOS not installed	J5/jumper A	open
➪ Floppy drive enabled	J5/jumper B	closed
Floppy drive disabled	J5/jumper B	open
➪ ATA disabled	J5/jumper C	closed
ATA enabled	J5/jumper C	open
➪ I/O address = 330h	J5/jumper D	open
I/O address = 230h	J5/jumper D	closed
➪ Factory configured - do not alter	J5/jumper G	closed
➪ Factory configured - do not alter	J5/jumper H	closed
➪ SCSI active termination enabled	J5/jumper I	closed
SCSI active termination disabled	J5/jumper I	open
➪ Factory configured - do not alter	J5/jumper J	open
➪ PIO data transfer mode enabled	J5/jumper K	closed
DMA data transfer mode enabled	J5/jumper K	open
Zero wait state enabled	J5/jumper Q	closed
Zero wait state disabled	J5/jumper Q	open
➪ Remote terminator power disabled	J6	open
Remote terminator power enabled	J6	closed
➪ Current back-flow diode CR1 enabled	J8	open
Current back-flow diode CR1 disabled	J8	closed

DMA CHANNEL					
DMA	**J1/jumper A**	**J1/jumper C**	**J1/jumper D**	**J1/jumper F**	**J5/jumper L**
➪ DMA5	closed	open	closed	open	closed
DMA7	open	closed	open	closed	open
Disabled	open	open	open	open	n/a

INTERRUPT REQUEST						
IRQ	**J1/L**	**J1/M**	**J1/N**	**J1/P**	**J5/M**	**J5/N**
➪ IRQ10	closed	open	open	open	open	open
IRQ11	open	closed	open	open	closed	closed
IRQ12	open	open	closed	open	open	closed
IRQ15	open	open	open	closed	closed	open

MEMORY ADDRESS - J5		
Address	**Jumper E**	**Jumper F**
C800h	open	open
D800h	closed	open

SYMBIOS LOGIC

NCR8250S

Data bus: 32-bit, PCI
Size: Half-length, full-height card
Hard drive supported: Up to 15 SCSI interface drives
Floppy drives supported: None

CONNECTIONS	
Function	**Location**
50-pin low-density SCSI interface - internal	J2
50-pin high-density SCSI interface - external	J3

USER CONFIGURABLE SETTINGS		
Function	**Location**	**Setting**
SCSI single-ended active termination - enabled	E1	closed
SCSI single-ended active termination - disabled	E1	open
Note: You must disable SCSI termination when connecting both internal & external SCSI interfaces.		

TYAN COMPUTER

S1366

Data bus: 32-bit PCI
Size: Half-length, full-height card
Hard drive supported: Two IDE (AT) Interface drives
Floppy drives supported: Two 360KB, 720KB, 1.2MB, or 1.44MB drives

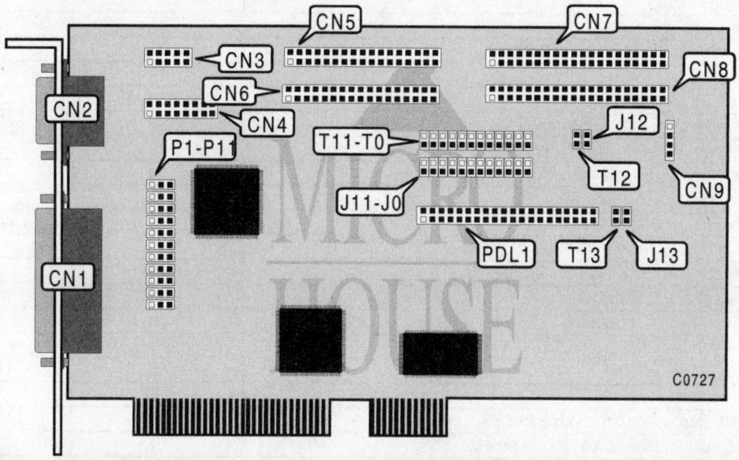

CONNECTIONS	
Function	**Location**
25 pin parallel port - external	CN1
9 pin serial port - external	CN2
10 pin serial port - internal	CN3
16 pin game port - internal	CN4
34 pin data cable connector - primary floppy drive interface	CN5
34 pin data cable connector - secondary floppy drive interface	CN6
40 pin connector - primary IDE(AT) interface	CN7
40 pin connector - secondary IDE(AT) interface	CN8
4 pin connector - drive active LED	CN9
S136x paddle card daughterboard connector	PDL1

Continued on next page . . .

TYAN COMPUTER

S1366

. . . continued from previous page

USER CONFIGURABLE SETTINGS		
Function	**Location**	**Setting**
➪ Factory configured - do not alter	J0	N/A
➪ Floppy drive interface DSKCHG signal not intercepted	J1	open
Floppy drive interface DSKCHG signal intercepted	J1	closed
➪ I/O address fixed	J2	closed
I/O address relocatable	J2	open
➪ Primary IDE(AT) channel legacy IRQ tied to IRQ14	J5	closed
No primary IDE(AT) channel legacy IRQ tied to IRQ14	J5	open
➪ Primary IDE(AT) legacy ISA IRQ14 not buffered	J6	open
Primary IDE(AT) legacy ISA IRQ14 buffered	J6	closed
➪ Secondary IDE(AT) channel legacy IRQ tied to IRQ15	J7	closed
No secondary IDE(AT) channel legacy IRQ tied to IRQ15	J7	open
➪ Secondary IDE(AT) legacy ISA IRQ15 not buffered	J8	open
Secondary IDE(AT) legacy ISA IRQ15 buffered	J8	closed
➪ IDE(AT) interrupt native mode not utilized	J9	open
IDE(AT) interrupt native mode utilized to PCI INTA	J9	closed
➪ Factory configured - do not alter	J10	N/A
➪ AT printer mode(ISA) enabled	J11	open
Normal mode enabled, floppy drive interface pin 29 grounded	J11	closed
➪ Floppy drive interface normal DRATE out mode select	P1	pins 2 & 3 closed
Floppy drive interface media ID input OS2 enhanced mode	P1	pins 1 & 2 closed
➪ Floppy drive interface I/O address 3F0h select	P2	pins 2 & 3 closed
Floppy drive interface I/O address 370h select	P2	pins 1 & 2 closed
➪ Floppy drive interface enabled	P3	pins 2 & 3 closed
Floppy drive interface disabled	P3	pins 1 & 2 closed
➪ Factory configured - do not alter	T0	N/A
➪ ISA AT printer mode enabled	T1	open
PS/2 bi-directional printer mode enabled	T1	closed
➪ Parallel port interrupt IRQ7 select	T2	closed
Parallel port interrupt disabled	T2	open
➪ Serial port 1 interrupt IRQ4 select	T4	closed
Serial port 1 interrupt disabled	T4	open
➪ Serial port 2 interrupt IRQ3 select	T5	closed
Serial port 2 interrupt disabled	T5	open
➪ Game port enabled	T6	closed
Game port disabled	T6	open

PCI IDE CONFIGURATION		
Setting	**J3**	**J4**
➪ PCI IDE channels enabled	closed	open
PCI IDE channels disabled	open	closed

SECONDARY IDE PCI INTERRUPT CONFIGURATION		
Setting	**J12**	**J13**
➪ Secondary IDE interrupt not inverted(PCI bus INTB#)	open	N/A
Secondary IDE interrupt inverted(PCI bus INTB#)	N/A	closed

PRIMARY IDE PCI INTERRUPT CONFIGURATION		
Setting	**T12**	**T13**
➪ Primary IDE interrupt not inverted(PCI bus INTA#)	open	N/A
Primary IDE interrupt inverted(PCI bus INTA#)	N/A	closed

Continued on next page . . .

TYAN COMPUTER

S1366

. . . continued from previous page

PARALLEL PORT I/O ADRESS SELECT		
Address	**P8**	**P9**
➪ 378-37Fh	pins 2 & 3 closed	pins 2 & 3 closed
278-27Fh	pins 2 & 3 closed	pins 1 & 2 closed
3BC-3BEh	pins 1 & 2 closed	pins 1 & 2 closed
Disabled	pins 1 & 2 closed	pins 2 & 3 closed

PRIMARY SERIAL PORT I/O ADRESS SELECT		
Address	**P4**	**P5**
➪ 3F8h (COM1)	pins 2 & 3 closed	pins 2 & 3 closed
2F8h (COM2)	pins 2 & 3 closed	pins 1 & 2 closed
3E8h (COM3)	pins 1 & 2 closed	pins 2 & 3 closed
Disabled	pins 1 & 2 closed	pins 1 & 2 closed

SECONDARY SERIAL PORT I/O ADRESS SELECT		
Address	**P6**	**P7**
➪ 2F8h (COM2)	pins 2 & 3 closed	pins 2 & 3 closed
2E8h (COM4)	pins 1 & 2 closed	pins 2 & 3 closed
3F8h (COM1)	pins 2 & 3 closed	pins 1 & 2 closed
Disabled	pins 1 & 2 closed	pins 1 & 2 closed

PARALLEL PORT INTERRUPT SELECT		
IRQ	**T2**	**T3**
➪ IRQ7	closed	open
IRQ5	open	closed
Disabled	open	open

ENHANCED PARALLEL PORT MODE CONFIGURATION		
Modes Supported	**P10**	**P11**
➪ PRN	pins 2 & 3 closed	pins 2 & 3 closed
PRN, EPP	pins 1 & 2 closed	pins 2 & 3 closed
PRN, ECP	pins 2 & 3 closed	pins 1 & 2 closed
PRN, ECP, EPP	pins 1 & 2 closed	pins 1 & 2 closed

ECP MODE CONFIGURATION					
Setting	**T7**	**T8**	**T9**	**T10**	**T11**
➪ ECP mode disabled	open	open	open	open	open
ECP mode w/DMA1	closed	open	closed	open	closed
ECP mode w/DMA3	open	closed	open	closed	closed

Continued on next page . . .

TYAN COMPUTER

S1366

. . . continued from previous page

SERIAL PORT INTERRUPT SELECT

Port 1	Port 2	T3	T4	T5
IRQ4	IRQ3	open	closed	closed
IRQ4	Disabled	open	closed	open
Disabled	IRQ3	open	open	closed
IRQ3	IRQ4	open	T4 pin 1 to T5 pin 1 T4 pin 2 to T5 pin 2	
IRQ5	IRQ3	open	T4 pin 1 to T5 pin 1	
Disabled	IRQ4	open	T4 pin 2 to T5 pin 2	
IRQ4	IRQ5	T3 pin 2 to T4 pin 2		closed
Disabled	IRQ5	T3 pin 2 to T4 pin 2		open

MISCELLANEOUS TECHNICAL NOTES

Note: In order to utilize the S1366 s IDE and I/O functions, the S136x paddle card must be installed on connector PDL1

ULTRASTOR CORPORATION

ULTRA15C

Data bus: 16-bit, PC/AT
Size: Half-length, full-height card
Hard drives supported: Two IDE (AT) Interface drives
Floppy drives supported: Three 360KB, 720KB, 1.2MB, or 1.44MB

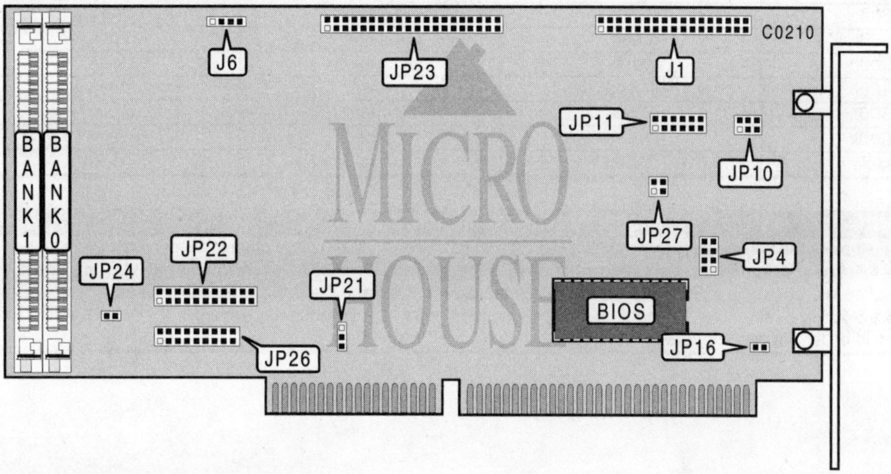

CONNECTIONS	
Function	**Location**
34-pin control cable connector-floppy drive	J1
4-pin connector-drive active LED	J6
40-pin IDE (AT) Interface control cable connector	JP23

USER CONFIGURABLE SETTINGS		
Function	**Location**	**Setting**
Supports a third floppy on single twisted cable (DS4)	JP11/jumper 1	closed
Supports a third floppy on double twisted cable (DS2)	JP11/jumper 1	open
➪ Hard drive port address is 1F0-1F7h	JP11/jumper 2	open
Hard drive port address is 170-177h	JP11/jumper 2	closed
➪ Floppy drive enabled	JP11/jumper 3	closed
Floppy drive disabled	JP11/jumper 3	open
➪ Floppy drive port address is 3F0-3F7h	JP11/jumper 4	open
Floppy drive port address is 370-377h	JP11/jumper 4	closed
➪ Single speed floppy drive supported	JP11/jumper 5	open
Dual speed floppy drive supported	JP11/jumper 5	closed
➪ Floppy drive precomp. is scaled by data rate: 500KHz=125ns, 300KHz=208ns, 250KHz=250ns	JP11/jumper 6	open
Floppy drive precomp. is 125ns	JP11/jumper 6	closed
➪ Chassis ground connection	JP16	open
➪ Hard drive interrupt is IRQ14	JP21	pins 2 & 3 closed
Hard drive interrupt is IRQ15	JP21	pins 1 & 2 closed

Continued on next page. . .

ULTRASTOR CORPORATION

ULTRA15C

. . .continued from previous page

BIOS ADDRESS			
Address	**JP10/jumper 1**	**JP10/jumper 2**	**JP10/jumper 3**
Disabled	open	open	open
➪ C800h	open	open	closed
CC00h	open	closed	open
D000h	open	closed	closed
D400h	closed	open	open
D800h	closed	open	closed
DC00h	closed	closed	open
Disabled	closed	closed	closed

MISCELLANEOUS TECHNICAL NOTES
Both SIMM's must be of the same size (256K, 1MB, or 4MB) and must be rated at 100ns or less. Reserved jumper settings are as follows: JP4/jumpers 1-4 are reserved and must be left "open". JP24 is reserved and must be left "open". JP27 is reserved and must remain as is. JP22 and JP26 are factory reserved and the defaults are as follows: JP22/jumpers 1 & 2, 5 & 6, and 10 & 11 should be "open", jumpers 3 & 4, 7 & 9 should be "closed." JP26/jumpers 2 through `6 should be left "open", jumpers 1, 7 through 9 should be left "closed"

XIRLINK, INC.

SCSI CONTROLLER XL-221

Data bus:	16-bit ISA
Size:	Half-length, half-height card
Hard drive supported:	Up to fourteen SCSI devices
Floppy drives supported:	None

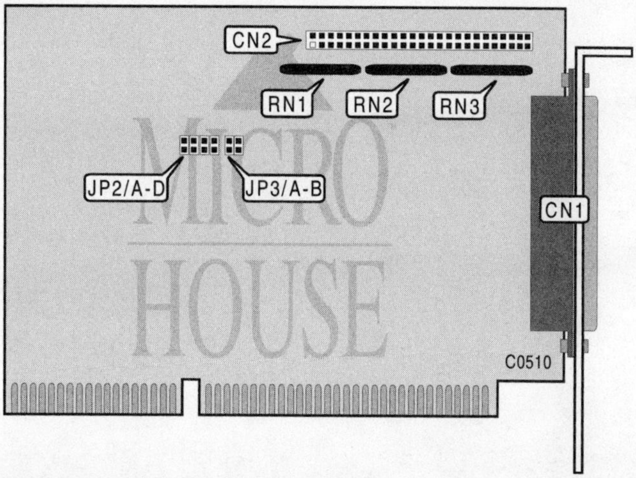

CONNECTIONS	
Function	**Location**
50-pin SCSI connector - external	CN1
50-pin SCSI connector - internal	CN2

USER CONFIGURABLE SETTINGS		
Function	**Location**	**Setting**
On-card TERMPWR enabled	RN1 - RN3	installed
On-card TERMPWR disabled	RN1 - RN3	Not installed

BASE MEMORY ADDRESS SELECT		
Address	**JP3/A**	**JP3/B**
➪ C800h	open	open
CC00h	open	closed
D800h	closed	open
DC00h	closed	closed

INTERRUPT SELECT				
IRQ	**JP2/A**	**JP2/B**	**JP2/C**	**JP2/D**
➪ IRQ11	open	open	closed	open
IRQ10	open	open	open	closed
IRQ12	open	closed	open	open
IRQ15	closed	open	open	open

Chapter 5

Hard Drive Settings

Interface Connections

Please consult Chapter Two "Hard Disk Interfaces" for specific details on connectors for IDE and SCSI interfaces.

Default Jumper Settings

The default or most common jumper positions are indicated by the ⇨ symbol. Pin-1 and jumper 1 are indicated by the ⚠ symbol.

Jumper Types

See Figure 5-1 for examples of common devices used to change the hardware configuration of hard drives.

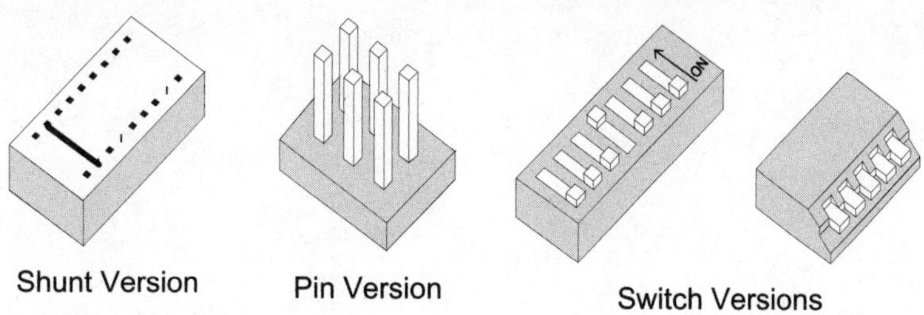

Figure Chapter 5-1: Common Jumper Types

Non-Documented Jumpers

Some jumper settings are specified as "unidentified," "factory configured – do not alter," or may have a dash. Information was not available on these jumpers or settings. These jumpers may be used by the factory for diagnostic or low-level purposes, and should be left as received out of the box. Also, some connection points for test equipment look like jumpers. In most cases, the diagrams do not show these test points.

Typical Jumpered Options

Most of the jumper settings in this chapter are self-explanatory. The most important hardware option to set will be Drive Select (SCSI, ESDI, ST-506/412), or Master/Slave (IDE).

Selected Drive Settings

Jumper settings and diagrams for the industry's most popular hard drives are provided in this chapter. The drives are listed in alphabetical order by manufacturer.

The diagrams only reflect the important access points and are not shown in fine detail. Some terminator resistor packs and jumper block locations change from one Printed Circuit Board (PCB) revision to another, but the functions of each remain the same.

CONNER PERIPHERALS, INC.

CFA-1080A, CFA-170A, CFA-270A, CFA-340A, CFA-540A CFA-810A, CFS-210A, CFS-420A, CP30174, CP-30344

Interface: IDE(AT)
Size: 3.5 in. third height
View: Bottom

DRIVE CONFIGURATION	
Selection	**C/D**
➪ Single drive only	closed
Master drive in a two drive system	closed
Slave drive in a two drive system	open

USER CONFIGURABLE SETTINGS		
Function	**Location**	**Setting**
Conner master/slave mode is used	ATA/ISA	open
ATA/CAM master/slave mode is used	ATA/ISA	closed
➪ Factory configured - do not alter	all other jumpers	N/A

CONNER PERIPHERALS, INC.

CFA850S, CFA1275S, CFP1080S

Interface: SCSI-2
Size: 3.5 in. Third Height
View: Bottom

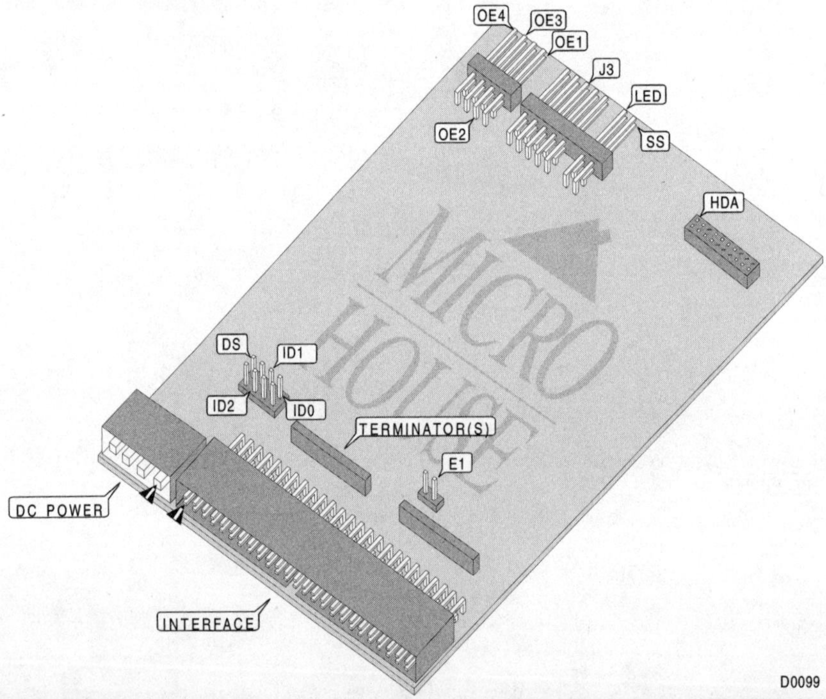

D0099

PRIMARY DRIVE SELECTION			
Drive Number	**ID0**	**ID1**	**ID2**
0	open	open	open
1	closed	open	open
2	open	closed	open
3	closed	closed	open
4	open	open	closed
5	closed	open	closed
6	open	closed	closed
7	closed	closed	closed
Note: Primary and secondary ID jumpers are mutually exclusive. Do not set both jumper blocks.			

USER CONFIGURABLE SETTINGS		
Function	**Location**	**Setting**
➪ TERMPWR enabled	E1	open
TERMPWR disabled	E1	closed

CONNECTIONS	
Function	**Location**
External LED	LED
Spindle Sync	SS

Continued on next page. . .

CONNER PERIPHERALS, INC.
CFA850S, CFA1275S, CFP1080S

. . .continued from previous page

PRIMARY SPINDLE STARTUP	
Setting	**DS**
Spindle starts on power-up	Open
Spindle starts on command only	closed
Note: Primary and secondary spindle startup jumpers are mutually exclusive. Do not set both jumpers.	

SECONDARY DRIVE SELECTION			
Drive Number	**OE1**	**OE2**	**OE3**
0	open	open	open
1	closed	open	open
2	open	closed	open
3	closed	closed	open
4	open	open	closed
5	closed	open	closed
6	open	closed	closed
7	closed	closed	closed
Note: Primary and secondary ID jumpers are mutually exclusive. Do not set both jumper blocks.			

SECONDARY SPINDLE STARTUP	
Setting	**OE4**
Spindle starts on power-up	open
Spindle starts on command only	closed
Note: Primary and secondary spindle startup jumpers are mutually exclusive. Do not set both jumpers.	

MISCELLANIOUS TECHNICAL NOTE
Spindle startup can be configured through the DSPN bit in the MODE SELECT page 0.

FACTORY CONFIGURED SETTING	
Jumper	**Setting**
J3/jumper 1	open
J3/jumper 2	open
J3/jumper 3	open
J3/jumper 4	Open
J3/jumper 5	Open

CONNER PERIPHERALS, INC.

CFP2107S, CFP4207S

Interface: SCSI 2 Fast
Size: 3.5 in. third height
View: Bottom

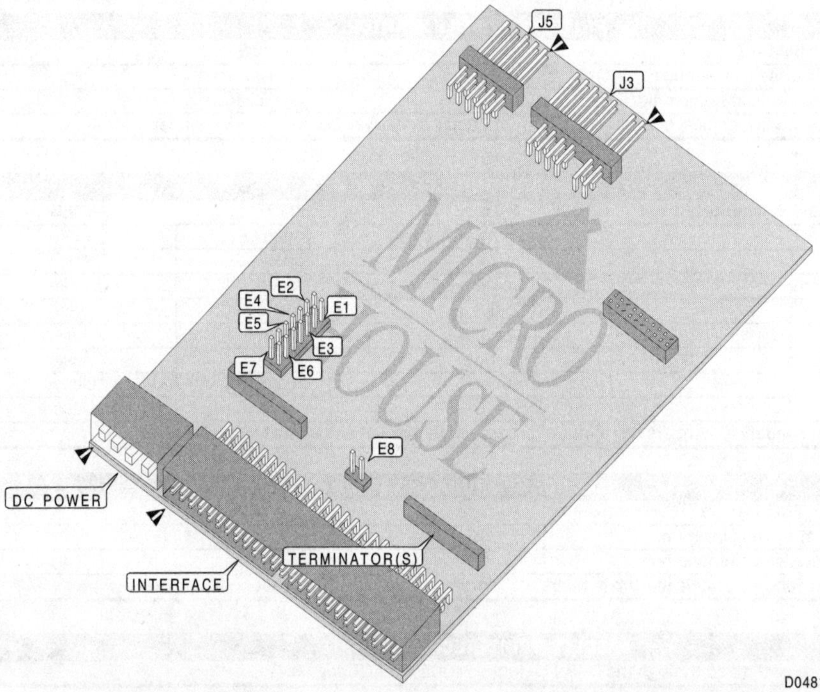

DRIVE SELECTION			
Drive Number	**E1**	**E2**	**E3**
0	open	open	open
1	closed	open	open
2	open	closed	open
3	closed	closed	open
4	open	open	closed
5	closed	open	closed
6	open	closed	closed
7	closed	closed	closed

Continued on next page...

CONNER PERIPHERALS, INC.

CFP2107S, CFP4207S

. . .continued from previous page

USER CONFIGURABLE SETTINGS		
Function	**Location**	**Setting**
➪ Factory oonfigured - do not alter	E4	open
➪ Spindle motor starts on power-up	E5	open
Spindle motor starts on command only	E5	closed
➪ Spindle motor delayed (Drive ID x 4 sec.)	E6	closed
Spindle motor not delayed	E6	open
➪ SCSI parity check enabled	E7	open
SCSI parity check disabled	E7	closed
➪ TERMPWR enabled	E8	closed
TERMPWR disabled	E8	open

CONNECTIONS	
Purpose	**Location**
Spindle Synch connector	J3/jumper 1
External LED	J3/jumper 2

MISCELLANEOUS TECHNICAL NOTE
Note: Jumpers E1-E3 and the first three jumpers of J5 are mutually exclusive. Do not set both. All other jumpers and settings are undocumented.

CONNER PERIPHERALS, INC.

CFS1081A, CFS1621A

Interface: EIDE
Size: 3.5 in. third height
View: Bottom

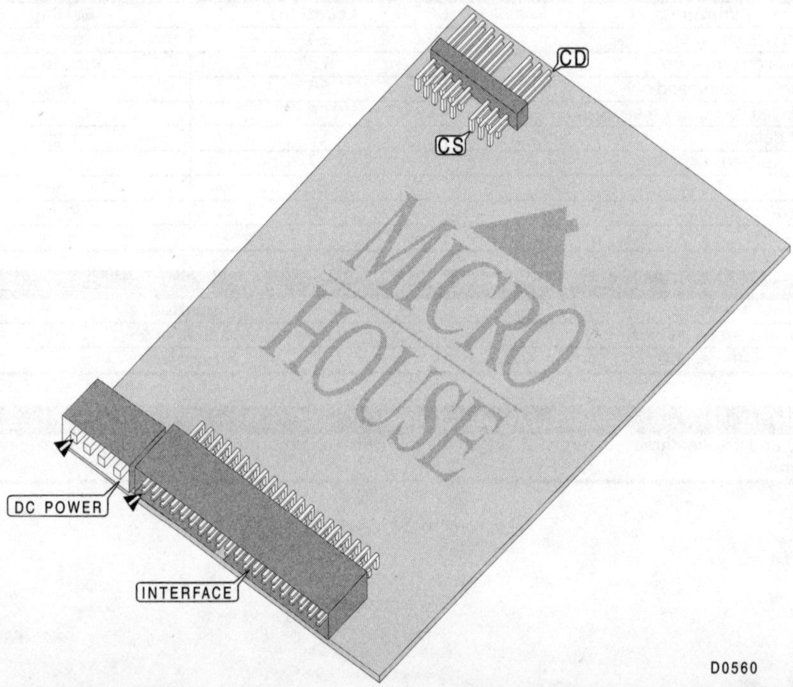

DRIVE SELECT		
Setting	**CD**	**CS**
⇨ Single drive in a system	closed	open
Master drive in a two drive system	closed	open
Slave drive in a two drive system	open	open
Cable Select is used to determine master/slave	open	closed
Note: All other jumper are factory configured - do not alter.		

CONNER PERIPHERALS, INC.

CP30540, CP31370

Interface:	SCSI
Size:	3.5 in. third height
View:	Bottom

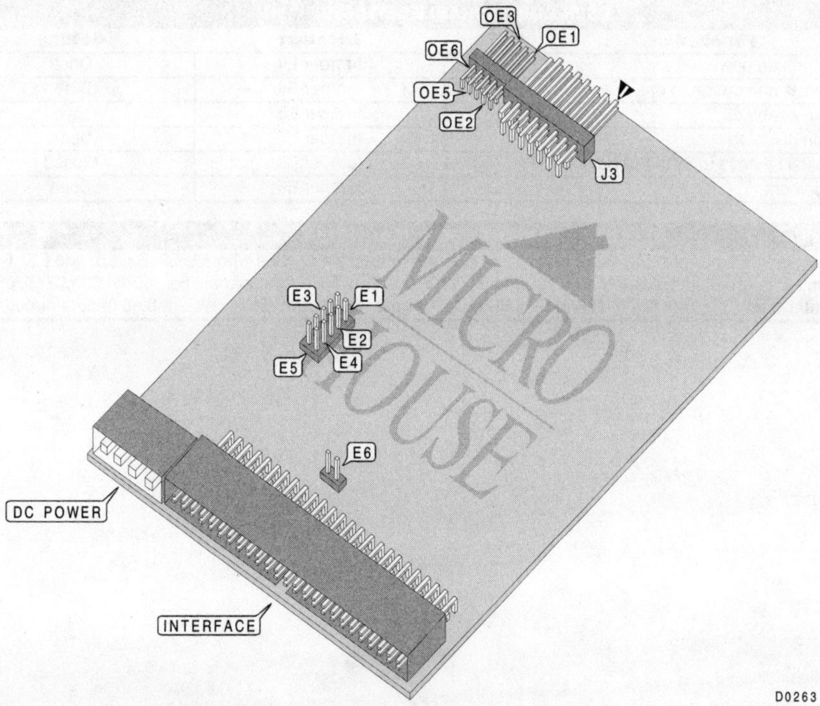

DRIVE SELECTION			
Drive Number	**Jumper E1**	**Jumper E2**	**Jumper E3**
0	open	open	open
1	closed	open	open
2	open	closed	open
3	closed	closed	open
4	open	open	closed
5	closed	open	closed
6	open	closed	closed
7	closed	closed	closed

CONNECTIONS	
Function	**Location**
External Spindle Sync	J3/jumper 1
Hard drive active LED	J3/jumper 2
Serial I/O connector (for factory use)	J3/jumper 3 - 7

Continued on next page . . .

CONNER PERIPHERALS, INC.

CP30540, CP31370

. . . continued from previous page

USER CONFIGURABLE SETTINGS		
Function	**Location**	**Setting**
➪ Spindle motor starts on power-up	Jumper E4	Open
Spindle motor starts on command only	Jumper E4	Closed
➪ Drive Termination enabled	Jumper E5	Closed
Drive Termination disabled	Jumper E5	Open
➪ TERMPWR supplied by drive to SCSI Interface pin 26	Jumper E6	Closed
TERMPWR not present	Jumper E6	Open

MISCELLANEOUS TECHNICAL NOTES
OE1, OE2, OE3, OE5, and OE6 perform the same function as and are mutually exclusive with E1, E2, E3, E5, and E6. For model CP30540, on PCBA revisions greater than 09230-003 Y1, OE6 is the LED output signal. For model CP31370, on PCBA revisions greater than 09370-002 Y1, OE6 is the LED output signal. SCSI parity is always enabled in both directions.

FUJITSU CORPORATION

M2691EQ/ER, M2692EQ/ER, M2693EQ/ER, M2694EQ/ER

Interface: SCSI
Size: 3.5 in. half height
View: Bottom

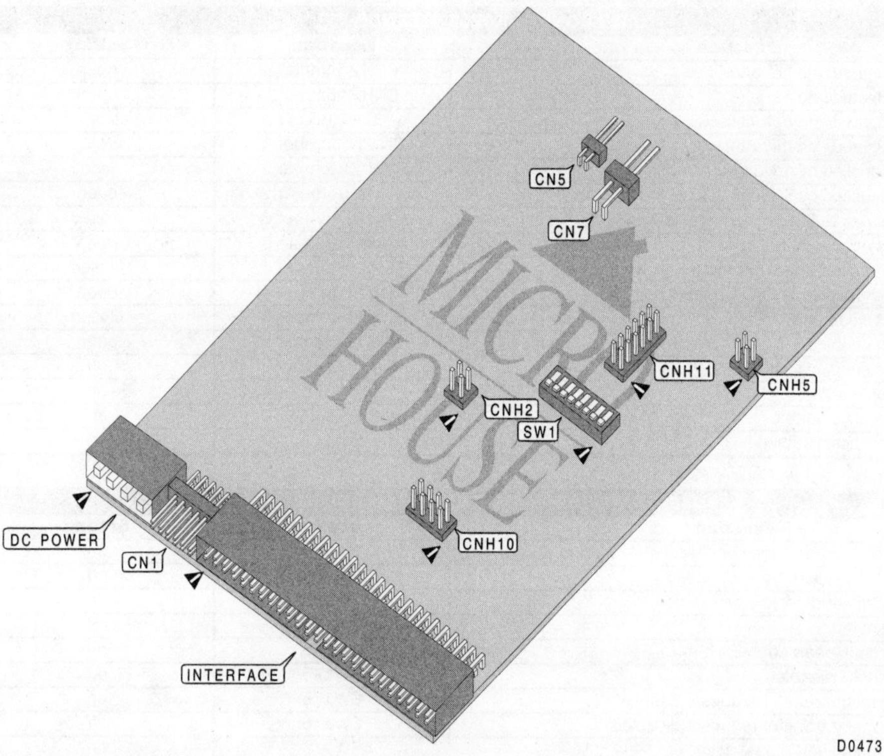

DRIVE SELECTION - CNH11				
SCSI ID Number	**Jumper 1**	**Jumper 2**	**Jumper 3**	**Jumper 4**
⇨ 0	open	open	open	open
1	closed	open	open	open
2	open	closed	open	open
3	closed	closed	open	open
4	open	open	closed	open
5	closed	open	closed	open
6	open	closed	closed	open
7	closed	closed	closed	open
8	open	open	open	closed
9	closed	open	open	closed
10	open	closed	open	closed
11	closed	closed	open	closed
12	open	open	closed	closed
13	closed	open	closed	closed
14	open	closed	closed	closed
15	closed	closed	closed	closed

Continued on next page. . .

FUJITSU CORPORATION

M2691EQ/ER, M2692EQ/ER, M2693EQ/ER, M2694EQ/ER

. . .continued from previous page

USER CONFIGURABLE SETTINGS		
Function	**Location**	**Setting**
➪ Factory configured - do not alter	CNH2/jumper 1	open
➪ Factory configured - do not alter	CNH2/jumper 2	open
➪ Factory configured - do not alter	CNH5	all open
➪ Factory configured - do not alter	CNH10/jumper 1	N/A
➪ TERMPWR supplied by drive (ER models only)	CNH10/jumper 2	closed
TERMPWR supplied by SCSI pin 26 (ER models only)	CNH10/jumper 2	open
➪ TERMPWR is tied to SCSI pin 26 and terminators	CNH10/jumper 3	closed
TERMPWR is not tied to SCSI pin 26 and terminators	CNH10/jumper 3	open
➪ Drive terminators enabled (ER models only)	CNH10/jumper 4	closed
Drive terminators disabled (ER models only)	CNH10/jumper 4	open
➪ Write protect disabled	CNH11/jumper 5	closed
Write protect enabled	CNH11/jumper 5	open
➪ SCSI-2 mode enabled	SW1/1	off
SCSI mode enabled	SW1/1	on
➪ Offline diagnostics enabled	SW1/2	off
Offline diagnostics disabled	SW1/2	on

USER CONFIGURABLE SETTINGS (CONTINUED)		
Function	**Location**	**Setting**
➪ UNIT ATTENTION report mode enabled	SW1/3	on
UNIT ATTENTION report mode disabled	SW1/3	off
➪ Retries are unlimited	SW1/4	on
Retries are limited to 10	SW1/4	off
➪ SCSI parity check enabled	SW1/5	on
SCSI parity check disabled	SW1/5	off
➪ Synch/Wide mode transfer requests enabled	SW1/6	on
Synch/Wide mode transfer requests disabled	SW1/6	off
➪ LED configured to light during activity	SW1/7	on
LED configured to light when ready	SW1/7	off

TERMINATING POWER CONFIGURATION(M269xEQ MODELS ONLY) - CNH10		
Function	**Jumper 2**	**Jumper 3**
➪ TERMPWR connected to SCSI pin 26 and drive terminators	closed	closed
TERMPWR connected to drive terminators	closed	open
TERMPWR connected to SCSI pin 26 and terminators	open	closed

SPINDLE MOTOR CONFIGURATION		
Setting	**CNH11/Jumper 6**	**SW1/8**
➪ Spindle motor starts on power-up	closed	on
Spindle motor starts on command only	open	off
Note: Setting CNH11/jumper 6 to open will inhibit user settings.		

CONNECTIONS	
Purpose	**Location**
External operator panel connector	CN1
External operator panel connector	CN5
Spindle synch connector	CN7

FUJITSU CORPORATION
M2952Q, M2954Q

Device Type	Hard Drive
Interface	Ultra SCSI Wide
Size	3.5 in. third height
View	Bottom

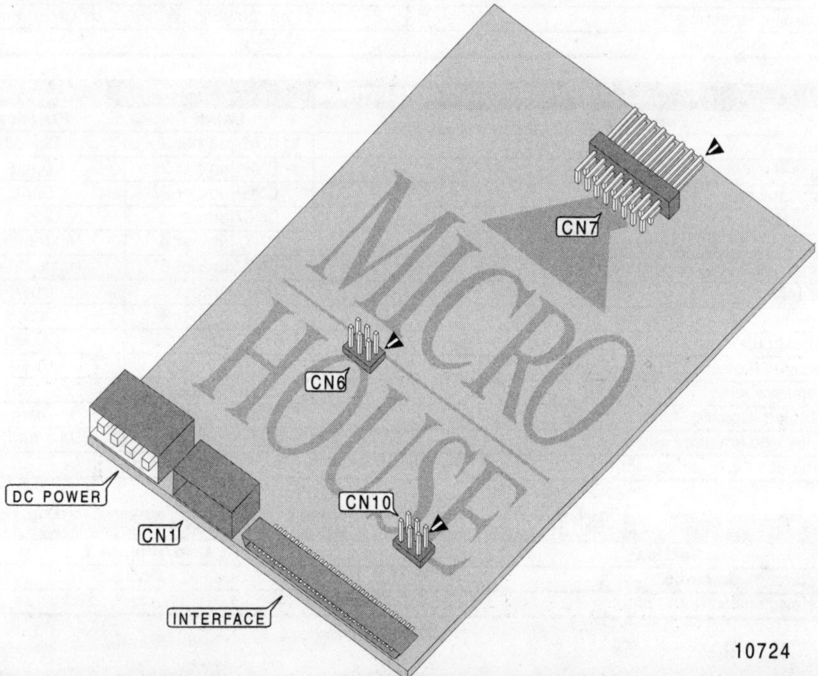

DRIVE SELECTION				
SCSI ID	**CN7/jumper 1**	**CN7/jumper 2**	**CN7/jumper 3**	**CN7/jumper 4**
0	Open	Open	Open	Open
1	Closed	Open	Open	Open
2	Open	Closed	Open	Open
3	Closed	Closed	Open	Open
4	Open	Open	Closed	Open
5	Closed	Open	Closed	Open
6	Open	Closed	Closed	Open
7	Closed	Closed	Closed	Open
8	Open	Open	Open	Closed
9	Closed	Open	Open	Closed
10	Open	Closed	Open	Closed
11	Closed	Closed	Open	Closed
12	Open	Open	Closed	Closed
13	Closed	Open	Closed	Closed
14	Open	Closed	Closed	Closed
15	Closed	Closed	Closed	Closed

Continued on next page . . .

FUJITSU CORPORATION
M2952Q, M2954Q

... continued from previous page

CONNECTIONS			
Function	**Label**	**Function**	**Label**
External operator panel connector	CN1	Remote LED	CN7/jumper 8
Note: Size of CN1 is unidentified.			

USER CONFIGURABLE SETTINGS		
Function	**Label**	**Position**
➪ SCSI parity enabled	CN6/jumper 2	Closed
SCSI parity disabled	CN6/jumper 2	Open
➪ Self-diagnostic disabled	CN6/jumper 3	Open
Self-diagnostic enabled	CN6/jumper 3	Closed
➪ Factory configured - do not alter (spindle sync pulse signal)	CN7/jumper 5	Unidentified
➪ Factory configured - do not alter (IDD reset signal)	CN7/jumper 6	Unidentified
➪ Write protect disabled	CN7/jumper 7	Open
Write protect enabled	CN7/jumper 7	Closed
➪ Termination enabled (Lower)	CN10/jumper 1	Closed
Termination disabled (Lower)	CN10/jumper 1	Open
➪ Termination enabled (Upper)	CN10/jumper 2	Closed
Termination disabled (Upper)	CN10/jumper 2	Open
➪ Factory configured - do not alter	CN10/jumper 3	Unidentified
Note: When termination is enabled - terminator needs to be installed.		

SPINDLE MOTOR SELECTION	
Setting	**CN6/jumper 1**
➪ Spindle motor starts on power-up	Closed
Spindle motor starts on command only	Open

HEWLETT-PACKARD COMPANY

HP C3324A 044, HP C3325A 044, HP C3724WC, HP C3725WC

Device Type	Hard Drive
Interface	SCSI-2 Wide/Differential with SCA
Size	3.5 in. third height
View	Bottom

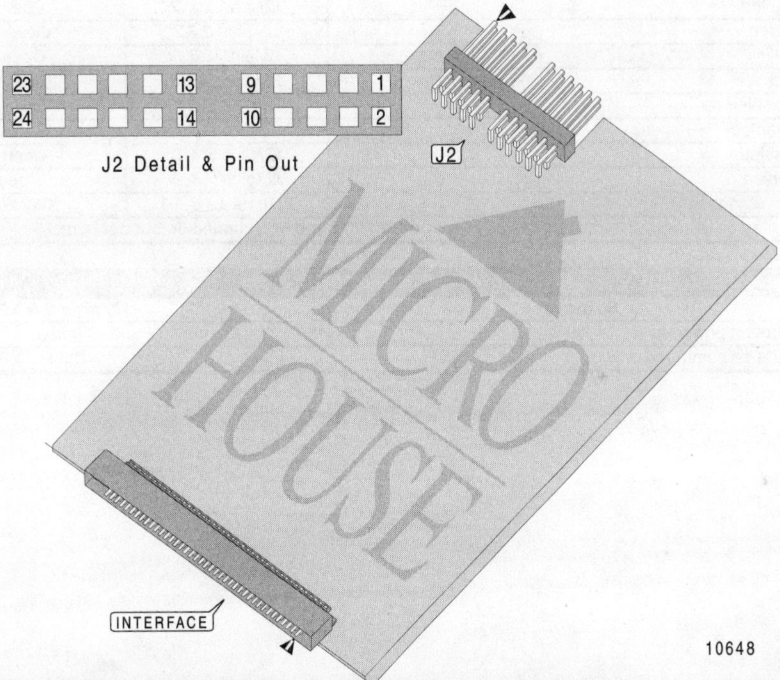

DRIVE SELECTION				
SCSI ID	**J2/pins 17 & 18**	**J2/pins 19 & 20**	**J2/pins 21 & 22**	**J2/pins 23 & 24**
0	Open	Open	Open	Open
1	Open	Open	Open	Closed
2	Open	Open	Closed	Open
3	Open	Open	Closed	Closed
4	Open	Closed	Open	Open
5	Open	Closed	Open	Closed
6	Open	Closed	Closed	Open
7	Open	Closed	Closed	Closed
8	Closed	Open	Open	Open
9	Closed	Open	Open	Closed
10	Closed	Open	Closed	Open
11	Closed	Open	Closed	Closed
12	Closed	Closed	Open	Open
13	Closed	Closed	Open	Closed
14	Closed	Closed	Closed	Open
15	Closed	Closed	Closed	Closed

Continued on next page . . .

HEWLETT-PACKARD COMPANY

HP C3324A 044, HP C3325A 044,
HP C3724WC, HP C3725WC

. . . continued from previous page

USER CONFIGURABLE SETTINGS		
Function	**Label**	**Position**
➪ Write protect - disabled	J2/pins 1 & 2	Open
Write protect - enabled	J2/pins 1 &2	Closed
➪ SCSI mode is SCSI-2	J2/pins 1 & 3	Open
SCSI mode is SCSI-1	J2/pins 1 & 3	Closed
SDTR message - disabled	J2/pins 5 & 6	Open
SDTR message - enabled	J2/pins 5 & 6	Closed
Unit attention - disabled	J2/pins 5 & 6	Open
Unit attention - enabled	J2/pins 5 & 6	Closed
➪ Factory configured - do not alter	J2/pins 13 & 14	Closed
Note: The LED driver may be connected across J2/pins 7 & 8, or driven from the SCA connector, but not both.		

SPINDLE MOTOR SELECTION	
Setting	**J2/pins 9 & 10**
➪ Spindle motor starts on power-up	Closed
Spindle motor starts on command only	Open

HITACHI CORPORATION

DK-315C-10M, DK-315C-11, DK-315C-14, DK-325C-57

Interface: SCSI
Size: 3.5 in. Half Height - DK-315C
3.5 in. Third Height - DK-325C

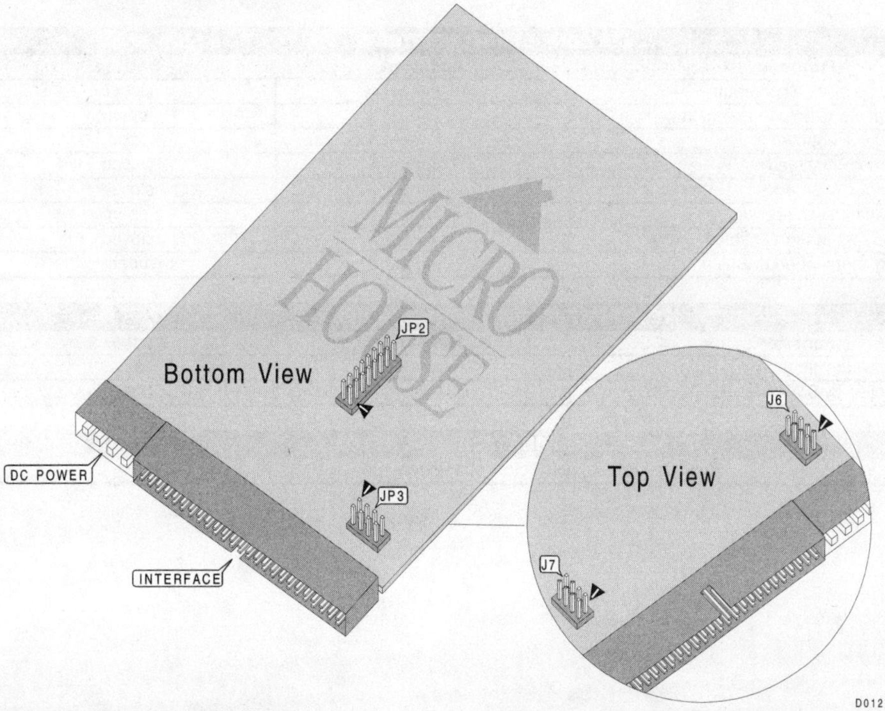

DRIVE CONFIGURATION			
Drive Number	**Jumper 1**	**Jumper 2**	**Jumper 3**
0	open	open	open
1	closed	open	open
2	open	closed	open
3	closed	closed	open
4	open	open	closed
5	closed	open	closed
6	open	closed	closed
7	closed	closed	closed

SPINDLE SYNC CONTROL - JP2		
Function	**Jumper 1**	**Jumper 2**
➪ Spindle Sync disabled	closed	closed
Spindle Sync enabled	closed	open
Slave mode enabled	open	closed

Continued on next page . . .

HITACHI CORPORATION

DK-315C-10M, DK-315C-11, DK-315C-14, DK-325C-57

. . . continued from previous page

USER CONFIGURABLE SETTINGS		
Function	**Location**	**Setting**
➪ Spindle motor starts on power-up	JP2/jumper 3	closed
Spindle motor starts on command only	JP2/jumper 3	open
➪ SCSI parity check enabled	JP2/jumper 4	open
SCSI parity check disabled	JP2/jumper 4	closed
➪ SCSI termination enabled	JP3/jumper 3	closed
SCSI termination disabled	JP3/jumper 3	open
➪ Spindle Sync. termination enabled	JP3/jumper 4	closed
Spindle Sync. termination disabled	JP3/jumper 4	open

TERMINATION POWER SETTINGS - JP3		
Function	**Jumper 1**	**Jumper 2**
TERMPWR supplied by drive	closed	open
TERMPWR supplied by SCSI interface pin 26	open	closed

MISCELLANEOUS TECHNICAL NOTES
All other Jumpers are factory configured - do not alter.

HITACHI CORPORATION

DK511-3, DK511-5

Interface: ST506/412
Size: 5.25 in. full height
View: Bottom

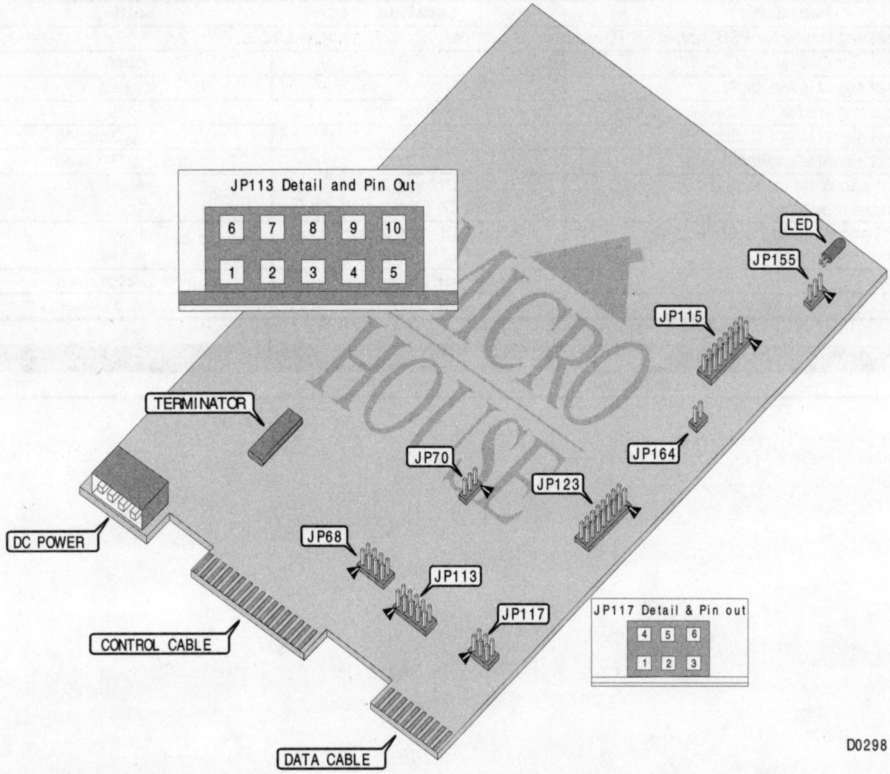

DRIVE CONFIGURATION				
Drive Number	**JP68/jumper 1**	**JP68/jumper 2**	**JP68/jumper 3**	**JP68/jumper 4**
➪ 0	closed	open	open	open
1	open	closed	open	open
2	open	open	closed	open
3	open	Open	open	closed

Continued on next page . . .

HITACHI CORPORATION

DK511-3, DK511-5

. . . continued from previous page

USER CONFIGURABLE SETTINGS		
Function	**Location**	**Setting**
The following setting is only for PCB Rev. 3 - JP70 is a two-pin jumper, JP182A is not available		
➪ General reset signal is not available	JP70	open
General reset signal is available	JP70	closed
The following setting are for PCB Rev. 4 and later - JP70 is a three-pin jumper		
➪ General reset signal is not available	JP70	pins 1 & 2 closed
General reset signal is available	JP70	pins 2 & 3 closed
➪ Write protect and Write protected signals not available on the interface	JP113/pins 2 & 9 JP117/pins 2 & 3	closed closed
Write protect and Write protected signals available on the interface	JP113/pins 1 & 2 JP117/pins 2 & 5	closed closed
➪ LED enabled	JP182A	pins 1 & 2 closed
LED disabled	JP182A	pins 2 & 3 closed

MISCELLANEOUS TECHNICAL NOTES
All other jumpers are factory configured - do not alter.

MAXTOR CORPORATION

250837, 251005, 251340

Interface: IDE 44-PIN
Size: 2.5 in. 0.5 in. height
View: Bottom

DRIVE CONFIGURATION - J2				
Selection	**Pins 1 & 2**	**Pins 3 & 4**	**Pins 1 & 3**	**Pins 2 & 4**
Single drive only	closed	open	N/A	N/A
Single drive only	open	open	open	open
Master drive in a two drive system	closed	open	N/A	N/A
Master drive in a two drive system	open	open	open	open
Slave drive in a two drive system	open	closed	N/A	N/A
Drive configuration determined by Cable Select	N/A	N/A	open	closed

MAXTOR CORPORATION

71000A, 71050A, 71050AT, 71084A, 71084AP, 71260A, 71260AP, 71626A, 71626AP, 7541A, 7541AP

Interface: IDE(AT)
Size: 3.5 in. third height
View: Bottom

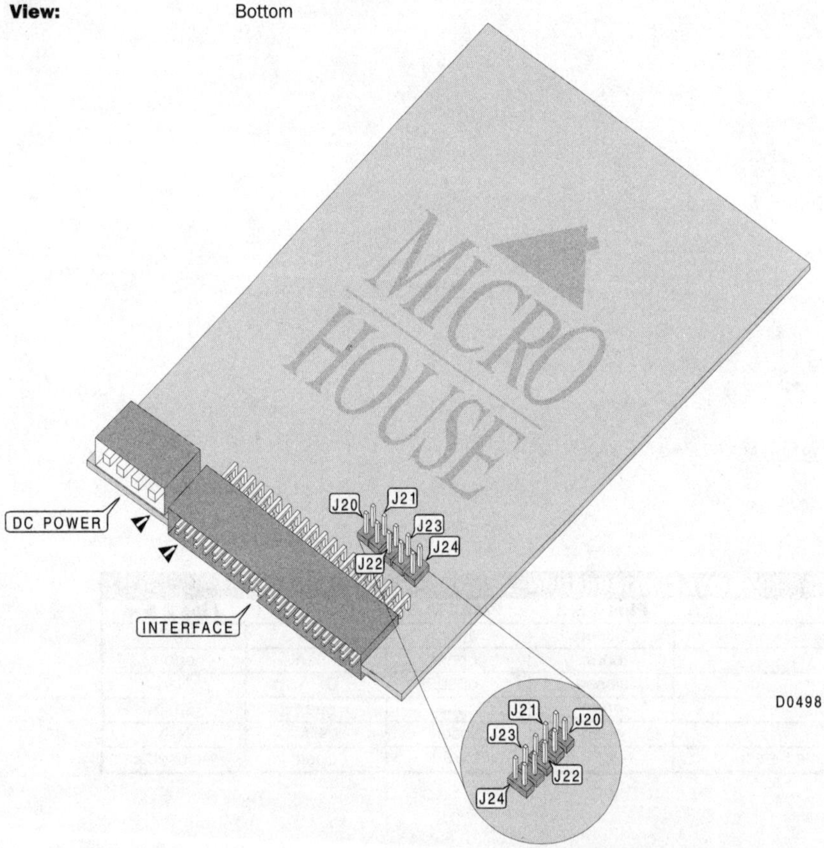

In some PCB versions the orientation of the jumper block is as shown above

DRIVE CONFIGURATION		
Setting	**J20**	**J24**
⇨ Single drive in a system	closed	open
Master drive in a two drive system	closed	open
Slave drive in a two drive system	open	open
Cable select	open	closed

Continued on next page...

MAXTOR CORPORATION

71000A, 71050A, 71050AT, 71084A, 71084AP, 71260A, 71260AP, 71626A, 71626AP, 7541A, 7541AP

...continued from previous page

USER CONFIGURABLE SETTINGS		
Function	**Location**	**Setting**
⇨ Factory configured-do not alter	J21	open
⇨ Factory configured-do not alter	J22	open
⇨ Write Cache enabled	J23	open
Write Cache disabled	J23	closed

MISCELLANEOUS TECHNICAL NOTE
A spare shunt, which is placed horizontally over J23 and J24 is shipped with the drives.

MAXTOR CORPORATION

71350AP/A, 71687AP/A, 72025AP, 72577AP, 72700AP, 80875A, 81312A, 81750A, 82187A, 82625A, 83062A, 83500A

Interface IDE(AT)
Size 3.5 in. third height
View Top

DRIVE CONFIGURATION		
Setting	**J48**	**J50**
➪ Single drive only	Open	Closed
Drive is master in two drive system	Open	Closed
Drive is slave in two drive system	Open	Open
Master/Slave determined by Cable Select	Closed	Open

USER CONFIGURABLE SETTINGS		
Function	**Label**	**Position**
➪ Write Cache enabled	J46	Open
Write Cache disabled	J46	Closed
➪ Factory configured - do not alter	J44	Open
➪ 4092 Cylinder Option disabled	J42	Open
4092 Cylinder Option enabled	J42	Closed

MAXTOR CORPORATION

81280A2, 81620A3, 81081A2, 82560A4, 82580A5, 82100A4, 83840A6, 83201A6, 84000A6, 84320A8, 85120A8

Device Type	Hard Drive
Interface	EIDE
Size	3.5 in. third height
View	Bottom

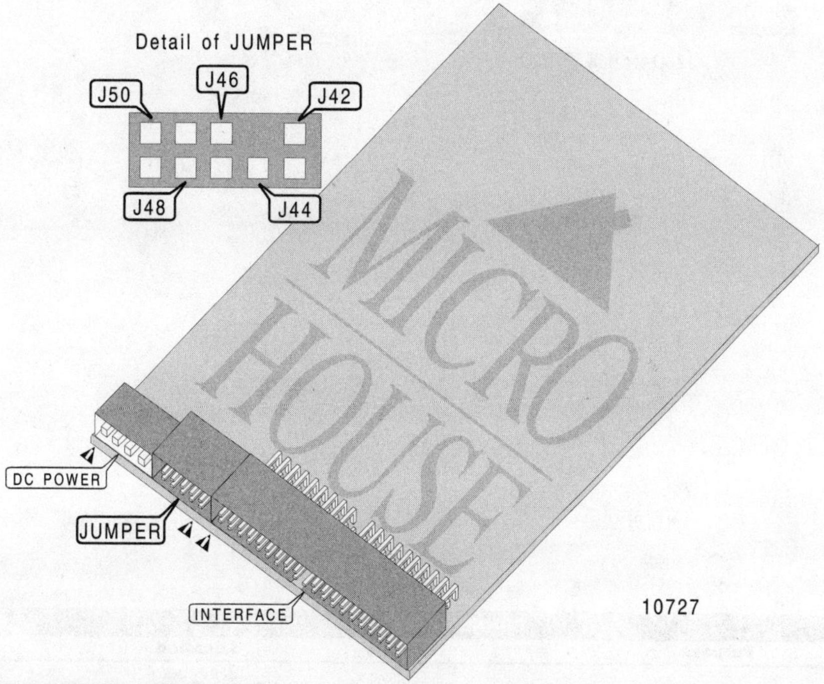

DRIVE CONFIGURATION		
Setting	**J48**	**J50**
➪ Drive is master	Open	Closed
Drive is slave	Open	Open
Cable Select	Closed	Open

USER CONFIGURABLE SETTINGS		
Function	**Label**	**Position**
➪ Factory configured - do not alter.	J42	Unidentified
➪ Factory configured - do not alter.	J44	Unidentified
4092 Cylinder limitation disabled	J46	Open
4092 Cylinder limitation enabled	J46	Closed
Note: If system hangs during boot, before accessing BIOS, or during boot after BIOS SETUP has been run, close jumper J46 and run Max-Blast software to complete hard drive installation.		

MAXTOR CORPORATION

MX9217SSW

Interface: SCA SCSI-2 wide
Size: 3.5 in. third height
View: Bottom

CONNECTIONS	
Purpose	**Location**
LED	J2/jumper 4
Spindle sync	J2/jumper 7

DRIVE CONFIGURATION				
SCSI Address	**J2/Jumper 9**	**J2/Jumper 10**	**J2/Jumper 11**	**J2/Jumper 12**
0	open	open	open	open
1	open	open	open	closed
2	open	open	closed	open
3	open	open	closed	closed
4	open	closed	open	open
5	open	closed	open	closed
6	open	closed	closed	open
7	open	Closed	closed	closed
8	closed	Open	open	open
9	closed	Open	open	closed
10	closed	Open	closed	open
11	closed	Open	closed	closed
12	closed	Closed	open	open
13	closed	Closed	open	closed
14	closed	Closed	closed	open
15	closed	Closed	closed	closed

Continued on next page . . .

MAXTOR CORPORATION
MX9217SSW

. . . continued from previous page

USER CONFIGURABLE SETTINGS		
Function	**Location**	**Setting**
➪ Write protect disabled	J2/jumper 1	open
Write protect enabled	J2/jumper 1	closed
➪ SCSI mode is SCSI-2 (if SCSI1/2 pin is active)	J2/pins 1 & 3	open
SCSI mode is SCSI-1 (if SCSI1/2 pin is active)	J2/pins 1 & 3	closed
SDTR message disabled (if SDTR function is active)	J2/jumper 3	open
SDTR message enabled (if SDTR function is active)	J2/jumper 3	closed
Unit attention enabled (if U/A function is active)	J2/jumper 3	open
Unit attention disabled (if U/A function is active)	J2/jumper 3	closed
Spindle motor starts on power-up	J2/jumper 5	open
Spindle motor starts on command only	J2/jumper 5	closed
Termination enabled	J4/jumper 1	closed
Termination disabled	J4/jumper 1	open
➪ Factory configured - do not alter	J4/jumper 2	unidentified
➪ Factory configured - do not alter	J4/jumper 3	unidentified

MISCELLANEOUS TECHNICAL NOTES
On models with SCA connector only passive termination is available.

MICROPOLIS CORPORATION

4221, 4221AV

Interface: SCSI-2 Fast
Size: 3.5 in. third height
View: Bottom

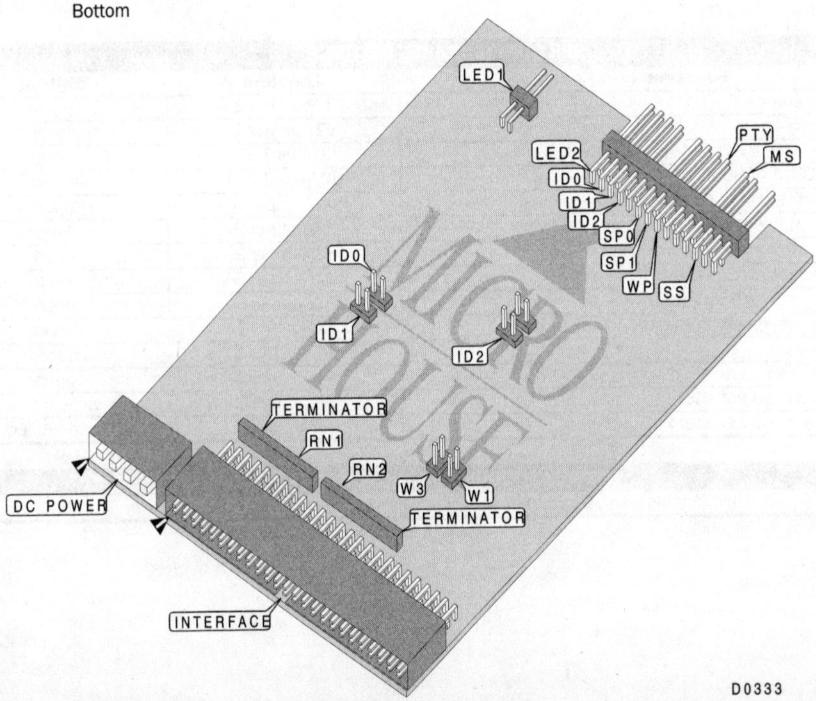

D0333

DRIVE SELECTION			
Drive Number	**ID0**	**ID1**	**ID2**
⇨ 0	open	open	open
1	closed	open	open
2	open	closed	open
3	closed	closed	open
4	open	open	closed
5	closed	open	closed
6	open	closed	closed
7	closed	closed	closed
Note: Only one set of ID jumpers should be used at any time.			

CONNECTIONS	
Purpose	**Location**
External LED connector	LED1

Continued on next page . . .

MICROPOLIS CORPORATION
4221, 4221AV

. . . continued from previous page

USER CONFIGURABLE SETTINGS		
Function	Location	Setting
➪ Front LED disabled	LED2	open
Front LED enabled	LED2	closed
➪ Spindle sync termination disabled	MS & SS	closed
Spindle sync termination enabled	MS & SS	open
➪ SCSI parity check enabled	PTY	open
SCSI parity check disabled	PTY	closed
➪ Drive is sole drive or at end of chain	RN1 & RN2	installed
Drive is not sole drive or at end of chain	RN1 & RN2	not installed
➪ Write protect disabled	WP	open
Write protect enabled	WP	closed

SPINDLE MOTOR CONFIGURATION		
Function	**SP0**	**SP1**
➪ Spindle motor starts on power-up	open	open
Spindle motor starts on command only	closed	open
Spindle motor delay by 12sec. x SCSI ID	open	closed

TERMINATION POWER SETTINGS

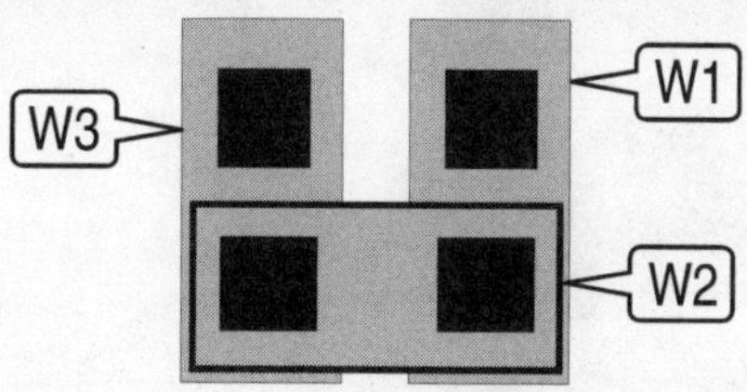

Setting	**W1**	**W2**	**W3**
➪ TERMPWR supplied by drive	closed	open	N/A
TERMPWR supplied by SCSI interface pin 26	open	closed	N/A
TRMPWR supplied by drive to bus pin 26	N/A	N/A	closed

MICROPOLIS CORPORATION

4525A (VER. 1), 4540A (VER. 1), 4550A (VER. 1)

Device Type Hard Drive
Interface EIDE
Size 3.5 in. third height
View Bottom

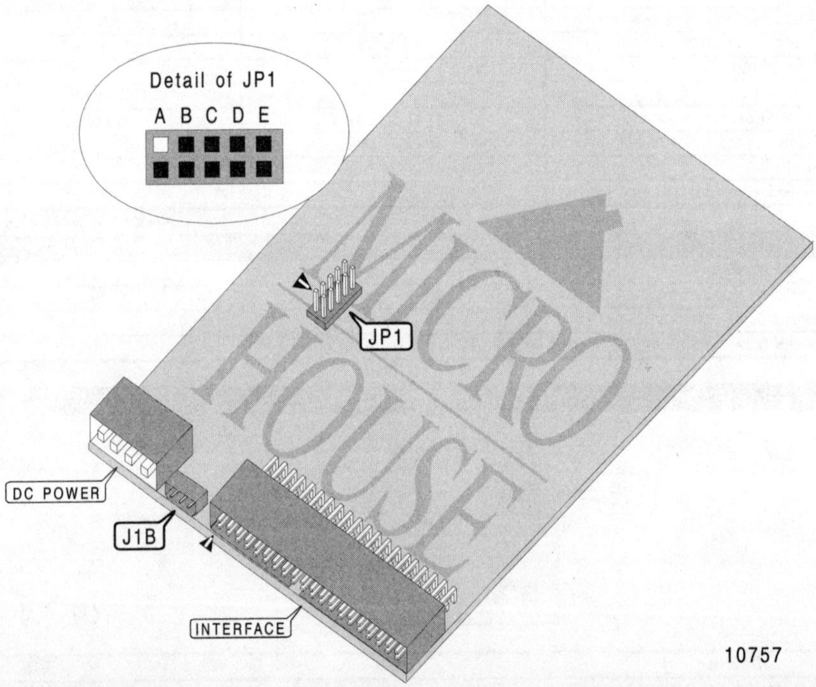

DRIVE CONFIGURATION			
Setting	**JP1/C**	**JP1/D**	**JP1/E**
⇨ Single drive only	Open	Closed	Open
Drive is master in two drive system	Closed	Closed	Open
Drive is slave in two drive system	Open	Open	Open
Master/Slave determined by Cable Select	Open	Open	Closed

MISCELLANEOUS TECHNICAL NOTES
Function of J1B, JP1/A, and, JP1/B is unidentified.

MICROPOLIS CORPORATION
4743SS

Device Type Hard Drive
Interface Ultra SCSI
Size 3.5 in. third height
View Bottom

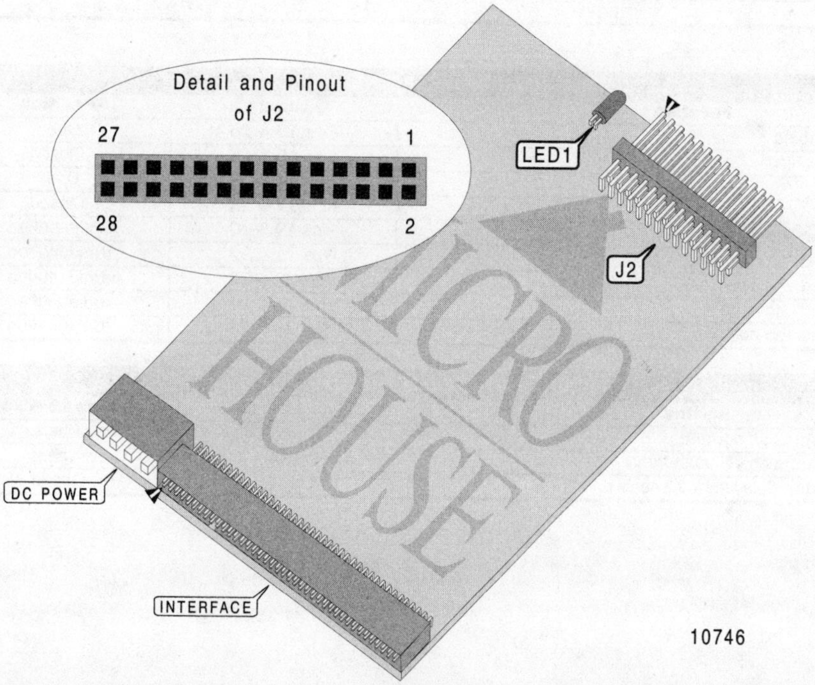

DRIVE SELECTION (J2)				
SCSI ID	Pins 3 & 4	Pins 5 & 6	Pins 7 & 8	Pins 9 & 10
0	Open	Open	Open	Open
⇨ 1	Closed	Open	Open	Open
2	Open	Closed	Open	Open
3	Closed	Closed	Open	Open
4	Open	Open	Closed	Open
5	Closed	Open	Closed	Open
6	Open	Closed	Closed	Open
7	Closed	Closed	Closed	Open
8	Open	Open	Open	Closed
9	Closed	Open	Open	Closed
10	Open	Closed	Open	Closed
11	Closed	Closed	Open	Closed
12	Open	Open	Closed	Closed
13	Closed	Open	Closed	Closed
14	Open	Closed	Closed	Closed
15	Closed	Closed	Closed	Closed

Continued on next page . . .

MICROPOLIS CORPORATION

4743SS

. . . continued from previous page

CONNECTIONS	
Function	Label
LED connector	J2/Pins 1 & 2

USER CONFIGURABLE SETTINGS (J2)		
Function	Label	Position
➪ Write protect disabled	Pins 15 & 16	Open
Write protect enabled	Pins 15 & 16	Closed
➪ Parity enabled	Pins 17 & 18	Open
Parity disabled	Pins 17 & 18	Closed
➪ Factory configured - do not alter.	Pins 19 & 20	Unidentified
➪ Factory configured - do not alter.	Pins 21 & 22	Unidentified
➪ Factory configured - do not alter.	Pins 23 & 24	Unidentified
➪ Factory configured - do not alter.	Pins 25 & 26	Unidentified
➪ Factory configured - do not alter.	Pins 27 & 28	Unidentified

SPINDLE MOTOR SELECTION (J2)		
Setting	Pins 11 & 12	Pins 13 & 14
➪ Spindle motor starts on power-up	Open	Open
Spindle motor starts on command only	Closed	Open
Spindle motor start delayed by 12 sec x drive ID	Open	Closed

QUANTUM CORPORATION

420AT, 635AT, 850AT (TRAILBLAZER)
540AT, 640AT, 1080AT, 1280AT (FIREBALL)

Interface:	IDE(AT)
Size:	3.5 in. third height
View:	Bottom

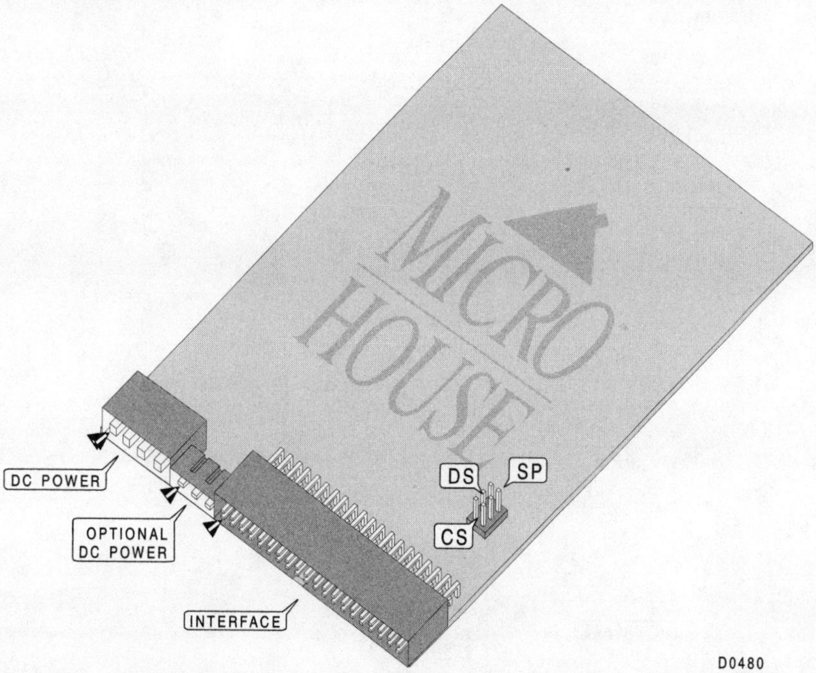

DRIVE CONFIGURATION				
Selection		**DS**	**SP**	
➪ Drive is master		closed	open	
Slave drive in a two drive system		open	open	
Master drive when non-CAM drive is slave		closed	closed	
Drive is master (cable select enabled)		N/A	open	
Slave in a two drive system (cable select enabled)		N/A	open	
Master drive when non-CAM drive is slave (cable select enabled)		N/A	closed	

Note: When configuring master/slave with the Cable Select jumper (CS), a cable that provides the correct signal to pin 28 needs to be used.

QUANTUM CORPORATION
420S, 840S, 850S (TRAILBLAZER)

Interface: SCSI-2 (Trail Blazer) SCSI-3 (Fireball)
3.5 in. third height

View:

D0486

DRIVE SELECTION

Drive ID		A1	A2
6	open		closed
	open	open	
1		open	open
2	open		open
	closed	closed	
4		open	closed
5	closed		closed
	closed	closed	

USER CONFIGURABLE SETTINGS

Function	Location	Setting
➪ Active termination enabled	TE	closed
Active termination disabled	TE	open
Factory configured	PK	N/A
Note: PK jumper can be used to store an extra shunt.		

QUANTUM CORPORATION

ATLAS WIDE SCSI SERIES: XP31070W/WD, XP32150W/WD, XP34300W/WD
ATLAS II SERIES: XP32181W/WD, XP34361W/WD, XP39100W/WD

Interface: SCSI-2 Fast/Wide (ATLAS W models)
SCSI-2 Fast/Wide/Differential (ATLAS WD models)
SCSI-3 Wide (ATLAS II W models)
SCSI-3 Wide/Differential (ATLAS II WD models)
Size: 3.5 in. third height (XP34300W/WD is half height)
View: Bottom

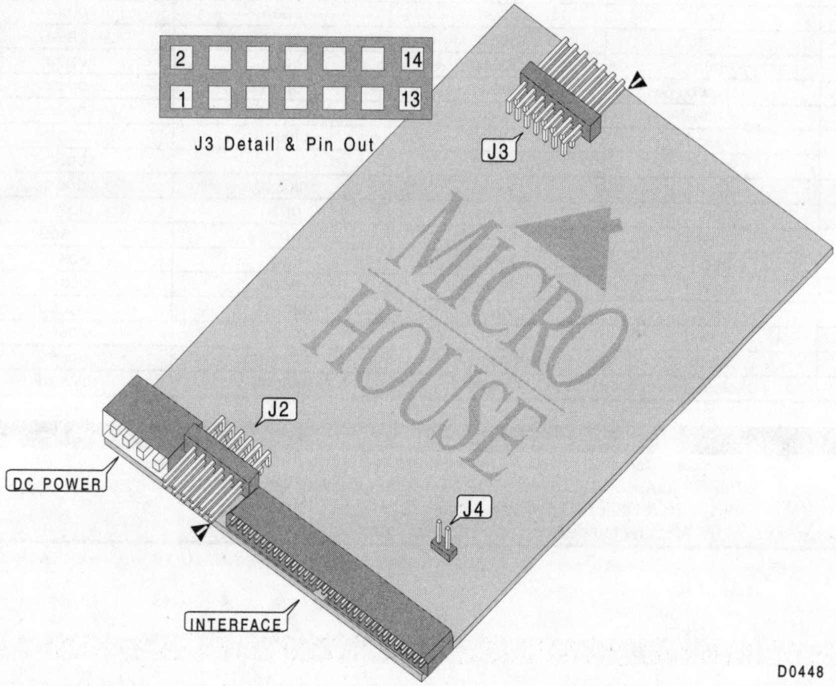

USER CONFIGURABLE SETTINGS	
Function	**Location**
Remote Busy LED	J3/pins 11 & 13
Remote Fault LED	J3/pins 13 & 14
Spindle sync connection	J3/pins 5 & 6

USER CONFIGURABLE SETTINGS		
Function	**Location**	**Setting**
Active termination enabled	J2/jumper 5	closed
Active termination disabled	J2/jumper 5	open
Spindle motor starts on power-up	J3/jumper 1	closed
Spindle motor starts on command only	J3/jumper 1	open
Write protect enabled	J3/jumper 4	closed
Write protect disabled	J3/jumper 4	open
TRMPWR supplied to SCSI bus	J4	closed
TRMPWR is not supplied to SCSI bus	J4	open

Continued on next page . . .

QUANTUM CORPORATION

ATLAS WIDE SCSI SERIES: XP31070W/WD, XP32150W/WD, XP34300W/WD
ATLAS II SERIES: XP32181W/WD, XP34361W/WD, XP39100

. . . continued from previous page

DRIVE SELECTION - J2				
Drive ID	**Jumper 1**	**Jumper 2**	**Jumper 3**	**Jumper 4**
0	open	open	open	open
1	closed	open	open	open
2	open	closed	open	open
3	closed	closed	open	open
4	open	open	closed	open
5	closed	open	closed	open
6	open	closed	closed	open
7	closed	closed	closed	open
8	open	open	open	closed
9	closed	open	open	closed
10	open	closed	open	closed
11	closed	closed	open	closed
12	open	open	closed	closed
13	closed	open	closed	closed
14	open	closed	closed	closed
15	closed	closed	closed	closed

MISCELLANEOUS TECHNICAL NOTES
The wide differential drives (WD) do not provide active termination. TERMPWR jumper (J4) is oriented differently on wide differential drives. Busy LED & Fault LED can also be connected in J2 jumper block. All other jumpers and their function are unidentified.

QUANTUM CORPORATION

BIGFOOT: 1275/1275AT, 2550/2550AT

Interface Fast-ATA-2
Size 5.25 in. quarter height
View Bottom

DRIVE CONFIGURATION			
Setting	**CS**	**DS**	**SP**
⇨ Single drive only	Open	Closed	Open
Drive is master in two drive system	Open	Closed	Open
Drive is slave in two drive system	Open	Open	Open
Master with Slave that does not use DSAP	Open	Closed	Closed

QUANTUM CORPORATION
DSP3053L/LD, 3107L/LD, 3133L/LD, DSP3160, DSP3210

Interface: SCSI-2
Size: 3.5 in. Half/Third Height
View: Bottom

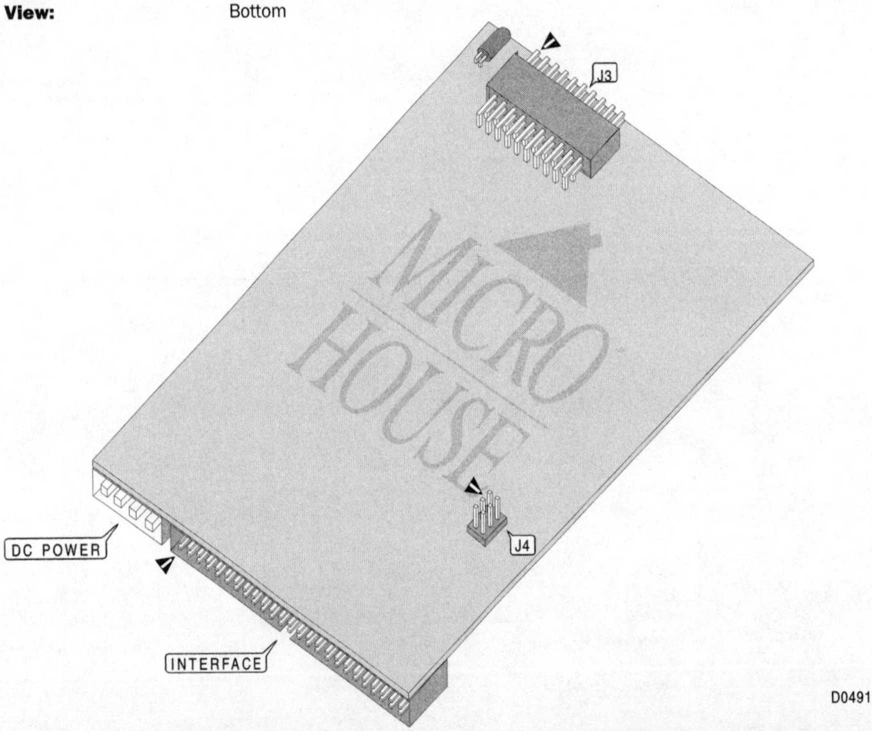

DRIVE SELECTION - J3			
Drive Number	**Jumper 1**	**Jumper 2**	**Jumper 3**
0	open	open	open
1	closed	open	open
2	open	closed	open
3	closed	closed	open
4	open	open	closed
5	closed	open	closed
6	open	closed	closed
7	closed	closed	closed

USER CONFIGURABLE SETTINGS		
Function	**Location**	**Setting**
Spindle motor starts on power-up	J3/Jumper 7	closed
Spindle motor starts on command only	J3/Jumper 7	open
Write protection enabled	J3/Jumper 7	closed
Write protection disabled	J3/Jumper 7	open
Note: Write protection and spindle motor options are not supported simultaneously.		

Continued on next page . . .

QUANTUM CORPORATION

DSP3053L/LD, 3107L/LD, 3133L/LD, DSP3160, DSP3210

. . . continued from previous page

TERMINATION POWER - J4			
Setting	**Jumper 1**	**Jumper 2**	**Jumper 3**
TERMPWR supplied by drive to SCSI bus, active termination is disabled.	closed	Open	open
TERMPWR supplied SCSI bus, active termination is enabled	open	Open	closed
TERMPWR supplied by drive to SCSI bus, active termination is enabled.	closed	Open	closed
TERMPWR disabled	open	Open	open
Note: LD models do not support active termination.			

MISCELLANEOUS TECHNICAL NOTES
(1) Models DSP3160 and DSP3210 are half-height, all other models are third-height.
(2) J4 may be installed as 2, 4, or 6 pin jumper.
(3) All other jumpers and their significance are unidentified.

QUANTUM CORPORATION

EMPIRE 540S, 1080S, 1400S, 1440S, 2100S, 2160S
PRODRIVE 1080S

Interface: SCSI-3 Fast
Size: 3.5 in. Half Height
View: Bottom

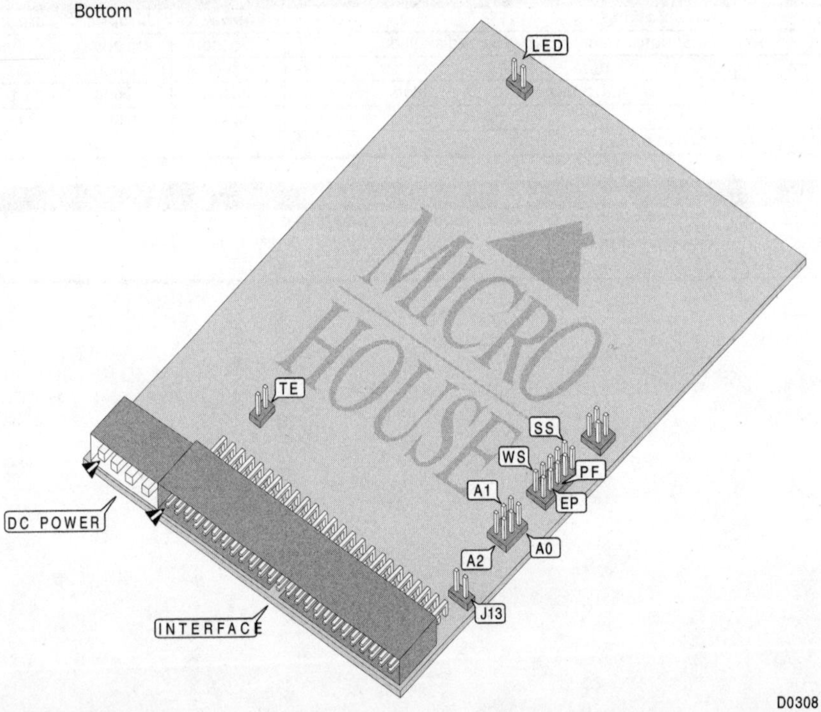

DRIVE CONFIGURATION			
Drive Number	**Jumper A0**	**Jumper A1**	**Jumper A2**
⇨ 6	open	closed	closed
0	open	open	open
1	closed	open	open
2	open	closed	open
3	closed	closed	open
4	open	open	closed
5	closed	open	closed
7	closed	closed	closed

USER CONFIGURABLE SETTINGS		
Function	**Location**	**Setting**
⇨ SCSI parity check enabled	EP	closed
SCSI parity check disabled	EP	open
⇨ TERMPWR supplied by the SCSI Interface pin 26	J13	closed
TERMPWR supplied by drive	J13	open
⇨ SCSI termination enabled	TE	closed
SCSI termination disabled	TE	open
⇨ Spindle motor starts on power-up	WS	open
Spindle motor starts on command only	WS	closed

Continued on next page . . .

QUANTUM CORPORATION
EMPIRE 540S, 1080S, 1400S, 1440S, 2100S, 2160S
PRODRIVE 1080S

. . . continued from previous page

MISCELLANEOUS TECHNICAL NOTES
All unlabeled jumpers are factory-configured - do not alter. The locations of jumper TE and J13 differ somewhat in older versions of the Empire series. These versions also have a factory configured jumper, A3, that is located next to A2, between the jumper block and the interface connector.

QUANTUM CORPORATION

EUROPA 540AT, 810AT, 1080AT

Interface:	Fast ATA-2
Size:	2.5 in. forth height (810AT, 1080AT) 2.5 in. fifth height (540AT)
View:	Bottom

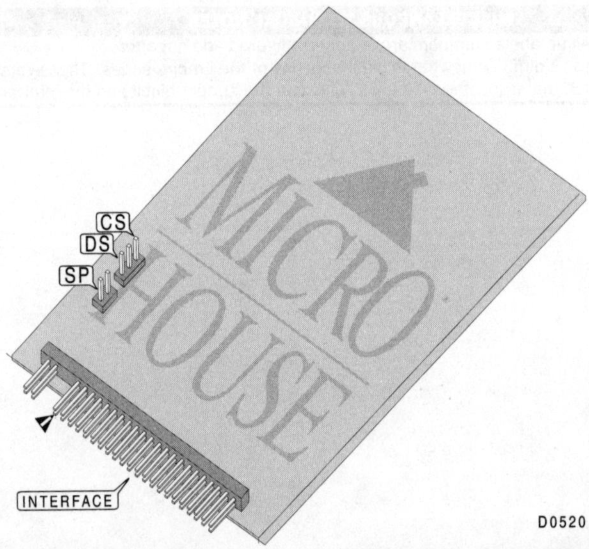

DRIVE CONFIGURATION			
Selection	**CS**	**DS**	**SP**
Single drive in a system	open	closed	open
Master drive when slave does not support PDIAG/DASP	open	closed	closed
Slave drive in a system	open	open	open
Master/slave is configured by cable select	closed	open	open

QUANTUM CORPORATION

FIREBALL TM 1.0AT, FIREBALL TM 1.2AT, FIREBALL TM 2.1AT
FIREBALL TM 2.5AT, FIREBALL TM 3.2AT, FIREBALL TM 3.8AT

Interface: Fast-ATA-2
Size: 3.5 in. third height
View: Bottom

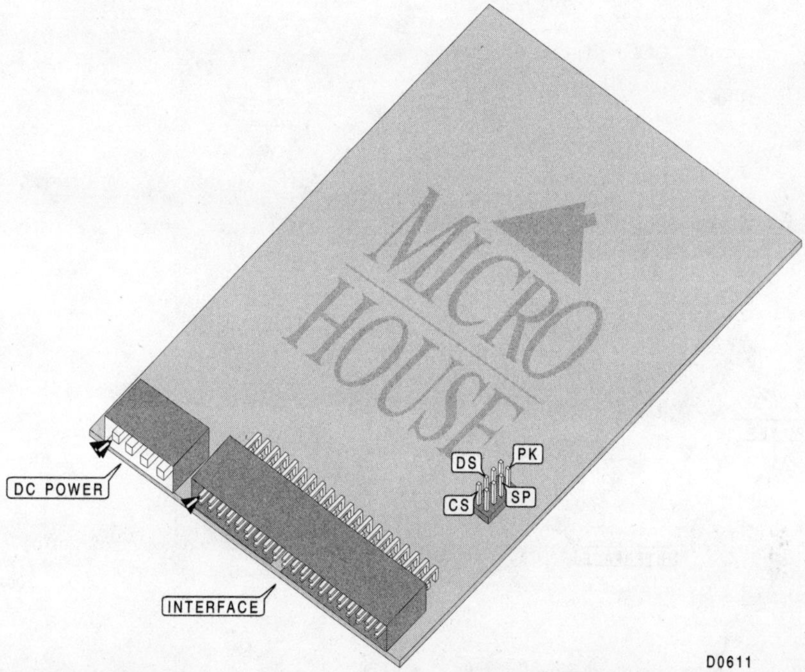

DRIVE CONFIGURATION			
Selection	**CS**	**DS**	**SP**
⇨ Drive is master	Open	Closed	Open
Slave drive in a two drive system	Open	Open	Open
Master drive when slave drive does not use DASP	Open	Closed	Closed
Cable Select	Closed	Open	Open
Note: Cable Select requires special cable and for both drives on the interface to support this feature.			

MISCELLANEOUS TECHNICAL NOTE
Note: The TM series has an extra jumper (PK) that is used to store spare jumpers.

QUANTUM CORPORATION

GRAND PRIX XP32151S, XP34301S

Interface: SCSI
Size: 3.5 in. half height
View: Bottom

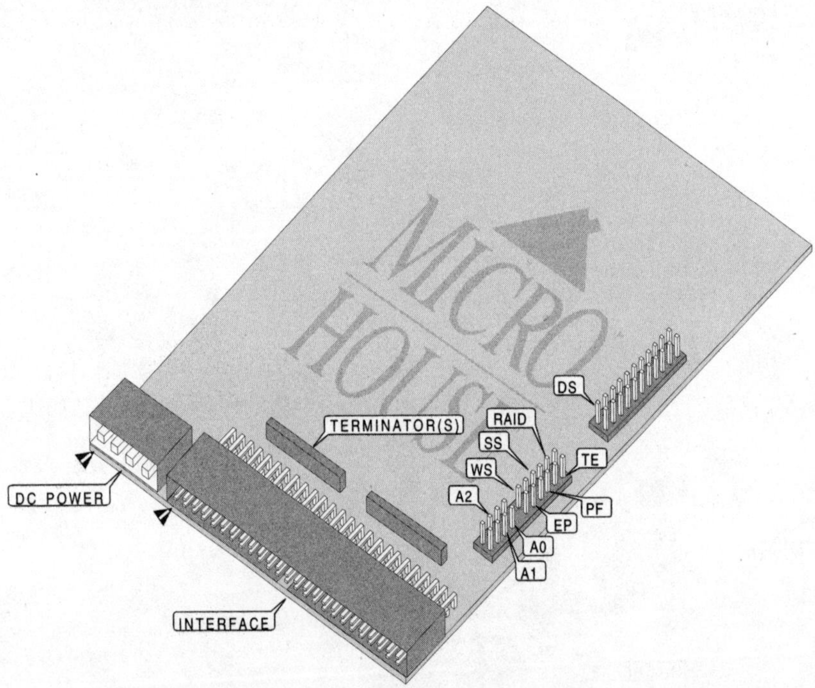

D0485

DRIVE SELECTION			
Drive number	**A0**	**A1**	**A2**
0	open	open	open
1	closed	open	open
2	open	closed	open
3	closed	closed	open
4	open	open	closed
5	closed	open	closed
6	open	closed	closed
7	closed	closed	closed

Continued on next page. . .

QUANTUM CORPORATION

GRAND PRIX XP32151S, XP34301S

. . .continued from previous page

USER CONFIGURABLE SETTINGS		
Function	**Location**	**Setting**
⇨ Spindle motor starts on power-up	WS	open
Spindle motor starts on command only	WS	closed
⇨ SCSI parity check enabled	EP	closed
SCSI parity check disabled	EP	open
⇨ Self seek test disabled	SS	open
Self seek test enabled	SS	Closed
⇨ Spindle motor delay disabled	DS	open
Spindle motor delay enabled(8 sec. x SCSI ID #)	DS	Closed
⇨ On-board TERMPWR enabled	TE	Closed
On-board TERMPWR disabled	TE	open
⇨ Reserved–to be used later	RAID	N/A
⇨ Factory configured- do not alter	PF	open
Note: A blinking LED during the self-seek test is an indication of drive error.		

MISCELLANEOUS TECHNICAL NOTES
Note: All other jumpers and their significance are unidentified.

QUANTUM CORPORATION

LIGHTNING 365AT, 540AT, 730AT
MAVERICK 270AT, 540AT
PRODRIVE ELS 42AT, 85AT, 127AT, 170AT
PRODRIVE LPS 120AT, 127AT, 170AT, 210AT, 240AT, 270AT, 340AT, 420AT, 540AT

Interface:	IDE(AT)
Size:	3.5 in. half height 3.5 in. third height (LPS)
View:	Bottom

DRIVE CONFIGURATION		
Selection	**DS**	**SP**
➩ Single drive only	closed	open
Master drive in a two drive system	closed	closed
Slave drive in a two drive system	open	open
Note: When Jumper SP is "closed" in Slave mode the drive will perform a selftest in seek mode.		

USER CONFIGURABLE SETTINGS		
Function	**Location**	**Setting**
➩ Drive order is set by jumpers DS & SP.	CS	open
Drive order is set by pin 28 of the AT interface.	CS	closed

Continued on next page . . .

QUANTUM CORPORATION
LIGHTNING 365AT, 540AT, 730AT
MAVERICK 270AT, 540AT
PRODRIVE ELS 42AT, 85AT, 127AT, 170AT
PRODRIVE LPS 120AT, 127AT, 170AT, 210AT, 240AT, 270AT, 340AT, 420AT, 540AT

. . . continued from previous page

MISCELLANEOUS TECHNICAL NOTES
Note: P0 is a product options set of jumpers and are not used. ELS drives do not have this jumper.

QUANTUM CORPORATION
MAVERICK 270S, 540S
PRODRIVE LPS 127S, 170S, 340S

Interface: SCSI-2
Size: 3.5 third height
View: Bottom

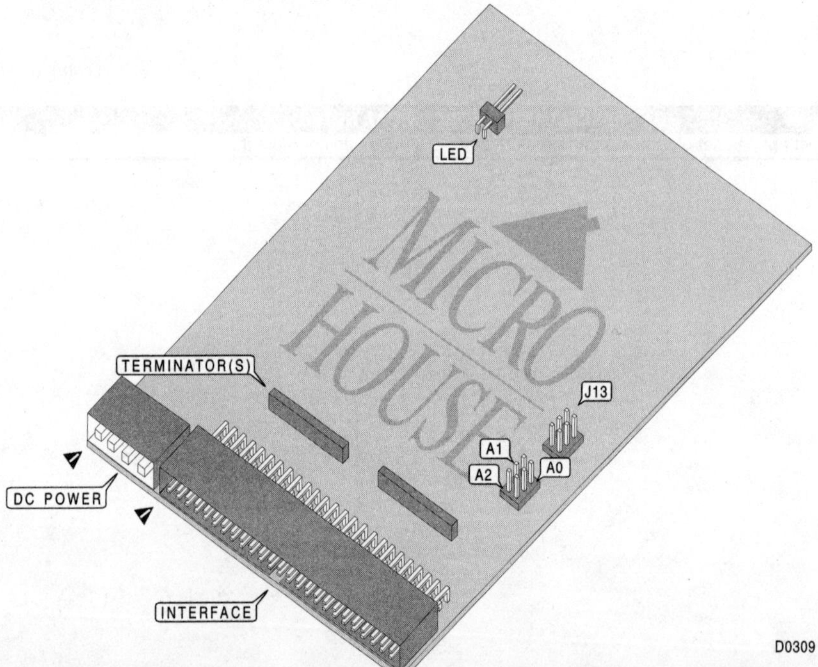

DRIVE SELECTION			
Drive Number	**A0**	**A1**	**A2**
0	open	open	open
1	closed	open	open
2	open	closed	open
3	closed	closed	open
4	open	open	closed
5	closed	open	closed
6	open	closed	closed
7	closed	closed	closed

MISCELLANEOUS TECHNICAL NOTES
Jumper J13 is factory configured - do not alter

QUANTUM CORPORATION

PIONEER SG 1080AT,PIONEER SG 2110AT

Device Type Hard Drive
Interface Fast-ATA
Size 3.5 in. third height
View Bottom

10748

DRIVE CONFIGURATION		
Setting	**JP1/A**	**JP1/B**
➪ Drive is master	Closed	Open
Drive is slave	Open	Open
Master/Slave determined by Cable Select	Open	Closed
Note: JP1/C is declared as parking position and should be used to store spare jumpers.		
Note: JP1/D is declared as reserved and should be left open.		

QUANTUM CORPORATION

ProDrive 525S Rev. 1, ProDrive 1800S
PRODRIVE LPS 525S

Interface:	SCSI-2
Size:	1800s: 3.5 in. Half Height 525S: 3.5 in. Third Height
View:	Bottom

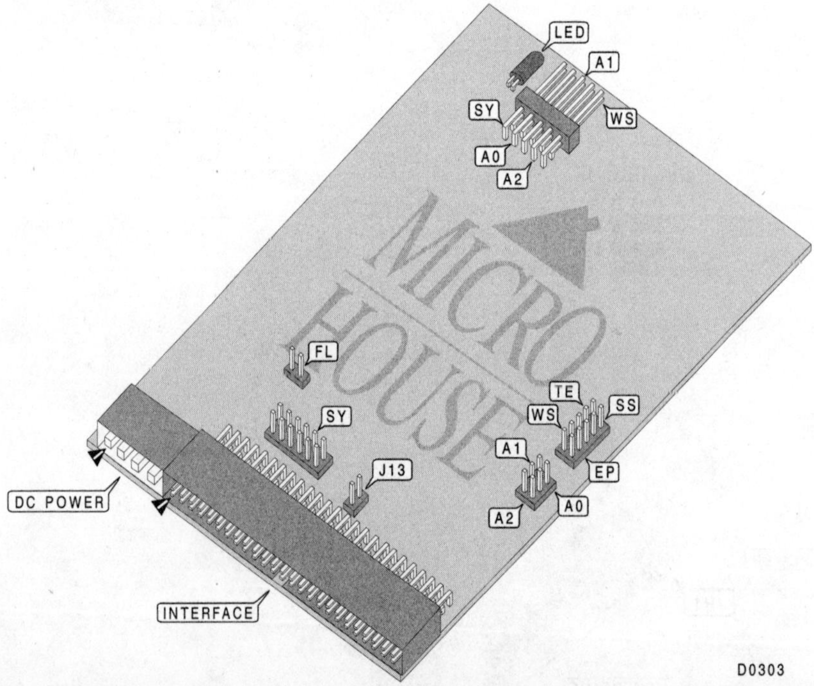

DRIVE CONFIGURATION			
Drive Number	**Jumper A0**	**Jumper A1**	**Jumper A2**
0	open	open	open
1	closed	open	open
2	open	closed	open
3	closed	closed	open
4	open	open	closed
5	closed	open	Closed
⇨ 6	open	closed	Closed
7	closed	closed	Closed

Continued on next page . . .

QUANTUM CORPORATION

PRODRIVE 525S REV. 1, PRODRIVE 1800S
PRODRIVE LPS 525S

. . . continued from previous page

USER CONFIGURABLE SETTINGS		
Function	**Location**	**Setting**
➪ SCSI parity check enabled	EP	closed
SCSI parity check disabled	EP	open
➪ Flash memory programming disabled	FL	open
Flash memory programming enabled	FL	closed
➪ TERMPWR supplied by drive	J13	open
TERMPWR supplied by the SCSI Interface pin 26	J13	closed
➪ Spindle motor synchronization disabled	SY	open
Spindle motor synchronization enabled	SY	closed
➪ SCSI termination enabled	TE	closed
SCSI termination disabled	TE	open
➪ Spindle motor starts on power-up	WS	open
Spindle motor starts on command only	WS	closed

MISCELLANEOUS TECHNICAL NOTES
All unlabeled jumpers are factory-configured - do not alter. Jumpers WS, A0, A1, A2, and SY are available in more than one location on this board. Do not install a jumper for any of these at more than one location at a time. Jumper SS is factory configured - do not alter.

QUANTUM CORPORATION

SATURN VP31080, SATURN VP32170

Interface SCSI-2 Fast
Size 3.5 in. third height
View Bottom

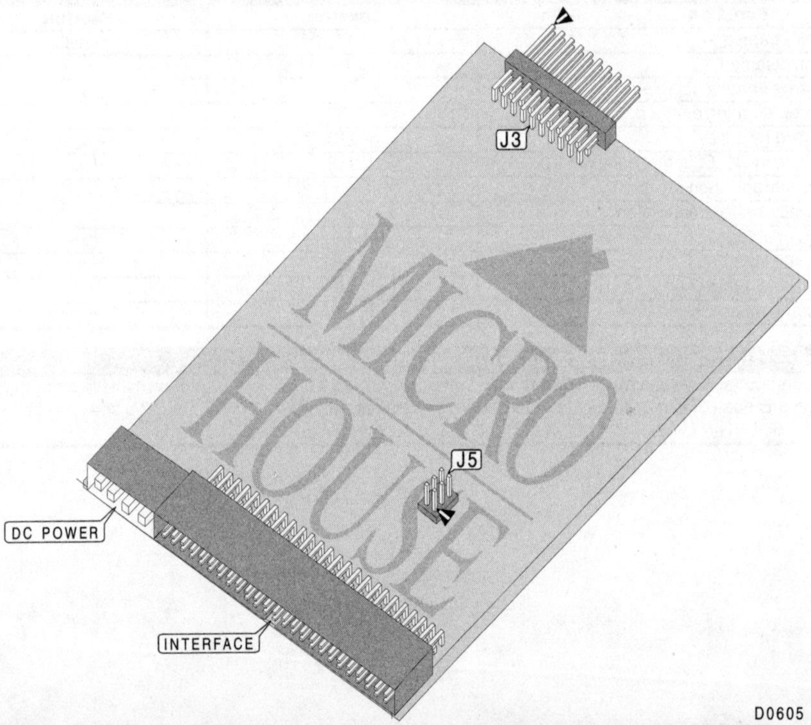

DRIVE SELECTION			
SCSI ID	**J3/Pins 1&2**	**J3/Pins 3&4**	**J3/Pins 5&6**
0	Open	Open	Open
⇨ 1	Closed	Open	Open
2	Open	Closed	Open
3	Closed	Closed	Open
4	Open	Open	Closed
5	Closed	Open	Closed
6	Open	Closed	Closed
7	Closed	Closed	Closed

USER CONFIGURABLE SETTINGS		
Function	**Label**	**Position**
Write Protect disabled	J3/Pins 17&18	Open
Write Protect enabled	J3/Pins 17&18	Closed
Active termination enabled	J5/pins 1 & 2	Closed
Active termination disabled	J5/pins 1 & 2	Open
TERMPWR supplied by drive to SCSI interface	J5/pins 5 & 6	Closed
TERMPWR not supplied by drive to SCSI interface	J5/ pins 5 & 6	Open

Continued on next page

QUANTUM CORPORATION
SATURN VP31080, SATURN VP32170

continued from previous page.

SPINDLE MOTOR SELECTION	
Setting	**J3/Pins 13&14**
➪ Spindle motor start delayed by 12 sec x drive ID	Closed
Spindle motor starts on command only	Open

QUANTUM CORPORATION

SATURN VP31080S, SATURN VP32170S

Interface SCSI-2
Size 3.5 in. third height
View Bottom

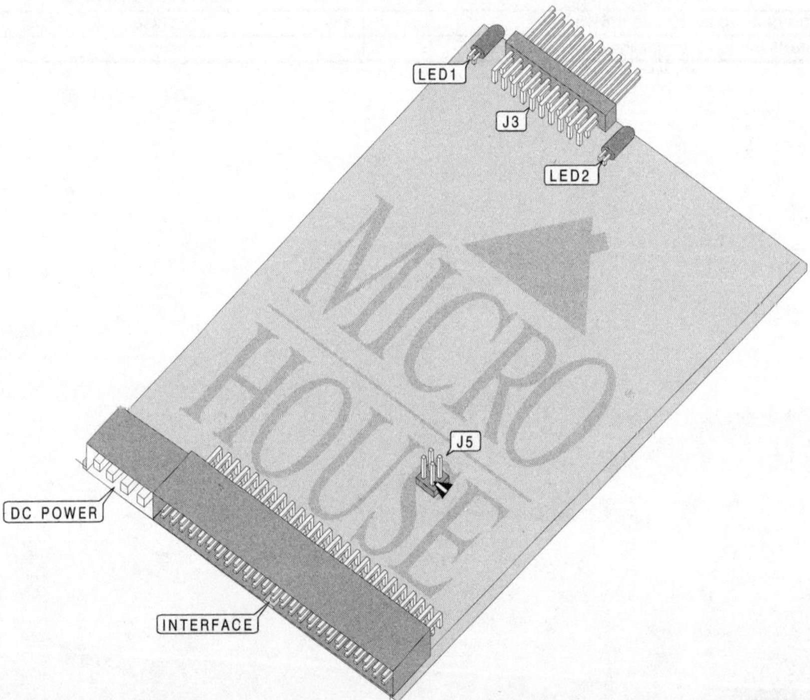

DRIVE SELECTION			
SCSI ID	**J3/Pins 1-2**	**J3/Pins 3-4**	**J3/Pins 5-6**
0	Open	Open	Open
➪ 1	Closed	Open	Open
2	Open	Closed	Open
3	Closed	Closed	Open
4	Open	Open	Closed
5	Closed	Open	Closed
6	Open	Closed	Closed
7	Closed	Closed	Closed

USER CONFIGURABLE SETTINGS		
Function	**Label**	**Position**
➪ Write Protect disabled	J3/Pins 17-18	Open
Write Protect enabled	J3/Pins 17-18	Closed
➪ Termination enabled	J5/Pins 1-2	Closed
Termination disabled	J5/Pins 1-2	Open
➪ Termination Power enabled	J5/Pins 3-4	Closed
Termination Power disable	J5/Pins 3-4	Open
Notes: Only the last drive in a series of SCSI devices must be terminated.		

Continued on next page

QUANTUM CORPORATION
SATURN VP31080S, SATURN VP32170S

Continued from previous page.

LEDs			
Function	**Label**	**Function**	**Label**
Busy LED	LED1	Fault LED	LED2

SPINDLE MOTOR SELECTION	
Setting	**J3/Pins 13-14**
➪ Spindle motor start on power up	Closed
Spindle motor starts on command only	Open

QUANTUM CORPORATION

SIROCCO 1700AT, 2550AT

Interface	ATA-2
Size	3.5 in. third height
View	Bottom

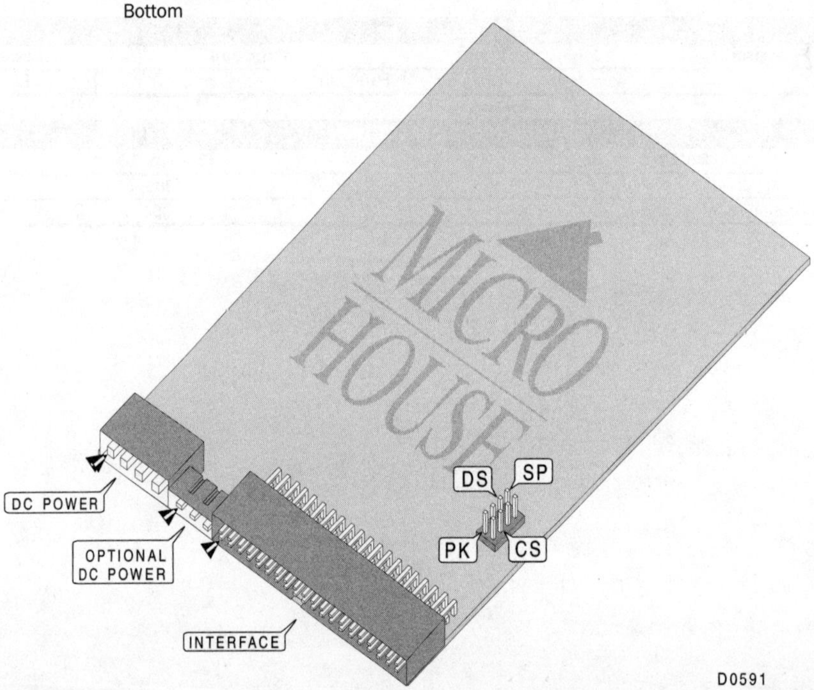

DRIVE CONFIGURATION				
Setting	**CS**	**DS**	**SP**	**Pin 28**
➪ Drive is master in a two drive system	Open	Closed	Open	N/A
Drive is master in a two drive system	Closed	Open	Open	Ground
Drive is master when slave does not use DASP signal	Closed	Open	Closed	Ground
Drive is master when slave does not use DASP signa	Open	Closed	Closed	N/A
Drive is slave in a two drive system	Closed	Open	Open	High
Drive is slave in a two drive system	Open	Open	Open	N/A
Note: When using the Cable Select option (CS closed) a special cable must be used.				

MISCELLANEOUS TECHNICAL NOTES
Note: Jumper PK is for parking an unused shunt.

QUANTUM CORPORATION

CAPELLA VP31110, VP31110S, VP32210, VP32210S

Interface: SCSI-2 Fast
Size: 3.5 in. half height
View: Bottom

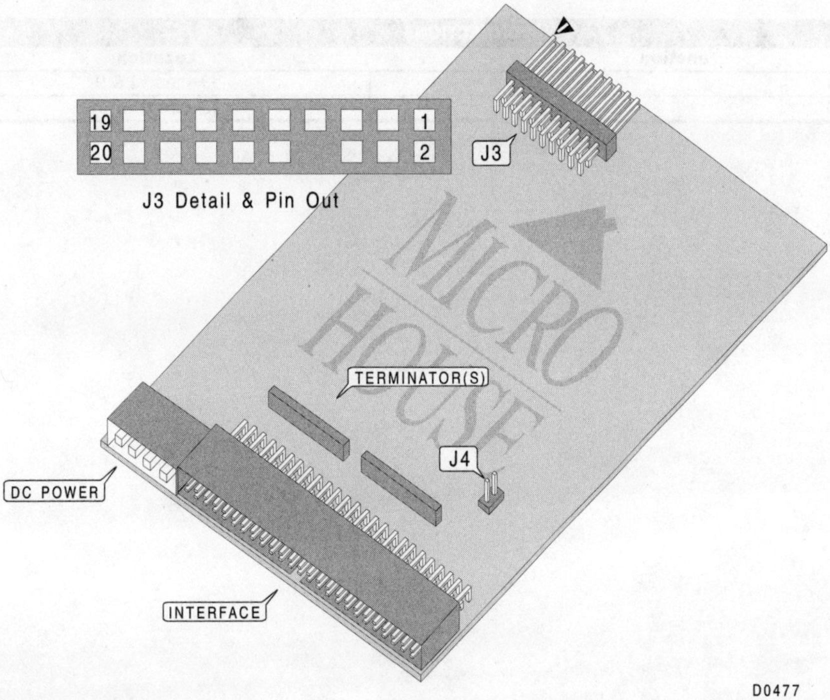

USER CONFIGURABLE SETTINGS		
Function	**Location**	**Setting**
TRMPWR supplied by drive to SCSI bus	J4	closed
TRMPWR is not supplied by drive to SCSI bus	J4	open
Spindle motor starts on power-up with delay of SCSI ID x 12 sec.	J3/jumper 7	closed
Spindle motor starts on command only	J3/jumper 7	open
Drive is first or last drive in chain	terminators	installed
Drive is not first or last drive in chain	terminators	removed
Note: Jumpers not mentioned in this document are unidentified.		

DRIVE SELECTION - J3			
Drive ID	**Jumper 1**	**Jumper 2**	**Jumper 3**
0	open	open	open
1	closed	open	open
2	open	closed	open
3	closed	closed	open
4	open	open	closed
5	closed	open	closed
6	open	closed	closed
7	closed	closed	closed

Continued on next page . . .

QUANTUM CORPORATION

CAPELLA VP31110, VP31110S, VP32210, VP32210S

. . . continued from previous page

CONNECTIONS	
Function	**Location**
Remote Busy LED	J3/pins 11 & 9
Remote Fault LED	J3/pins 11 & 7

QUANTUM CORPORATION

VP31110W/WD, VP32210W/WD (CAPELLA WIDE SCSI SERIES)

Interface: SCSI-2 F/W/D
Size: 3.5 in. third height
View: Bottom

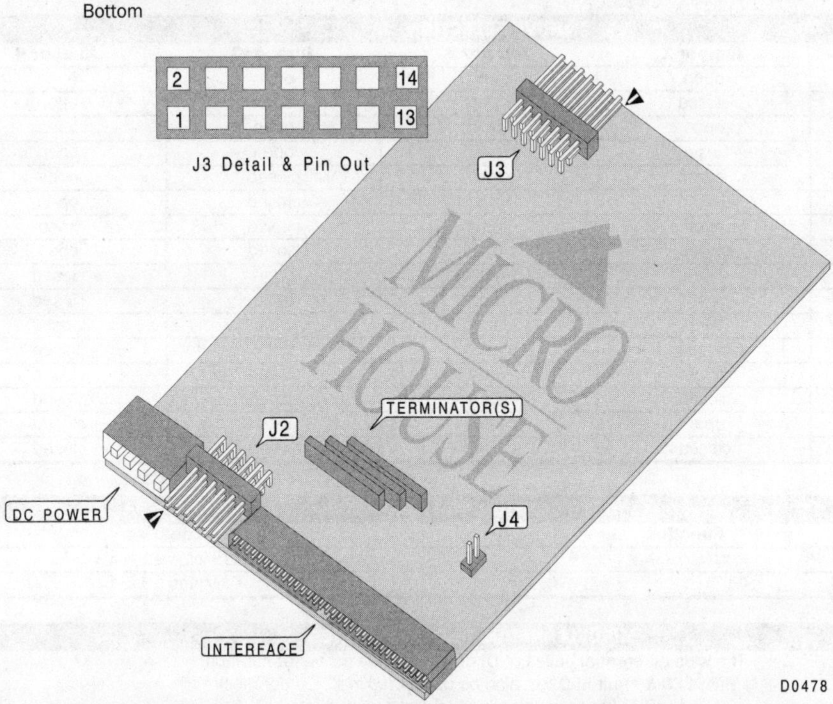

USER CONFIGURABLE SETTINGS		
Function	**Location**	**Setting**
Active termination enabled	J2/jumper 5	closed
Active termination disabled	J2/jumper 5	open
Spindle motor starts on power-up	J3/jumper 1	closed
Spindle motor starts on command only	J3/jumper 1	open
TRMPWR supplied to SCSI bus	J4	closed
TRMPWR is not supplied to SCSI bus	J4	open
Drive is first or last drive in chain	terminators	installed
Drive is not first or last drive in chain	terminators	removed

Continued on next page . . .

QUANTUM CORPORATION
VP31110W/WD, VP32210W/WD
(CAPELLA WIDE SCSI SERIES)

. . . continued from previous page

DRIVE SELECTION - J2				
Drive ID	**Jumper 1**	**Jumper 2**	**Jumper 3**	**Jumper 4**
0	open	open	open	open
1	closed	open	open	open
2	open	closed	open	open
3	closed	closed	open	open
4	open	open	closed	open
5	closed	open	closed	open
6	open	closed	closed	open
7	closed	closed	closed	open
8	open	open	open	closed
9	closed	open	open	closed
10	open	closed	open	closed
11	closed	closed	open	closed
12	open	open	closed	closed
13	closed	open	closed	closed
14	open	closed	closed	closed
15	closed	closed	closed	closed

USER CONFIGURABLE SETTINGS	
Function	**Location**
Remote Busy LED	J3/pins 11 & 13
Remote Fault LED	J3/pins 13 & 14

MISCELLANEOUS TECHNICAL NOTES
The wide differential drives (WD) do not provide active termination. Busy LED & Fault LED can also be connected in J2 jumper block. All other jumpers and their function are unidentified.

QUANTUM CORPORATION

ATLAS XP SERIES: XP31070S, XP32150S, XP34300S
ATLAS II SERIES: XP32181S, XP34361S, XP39100S

Interface: SCSI-2 Fast (ATLAS XP)
SCSI-3 (ATLAS II)
Size: 3.5 in. half height
View: Bottom

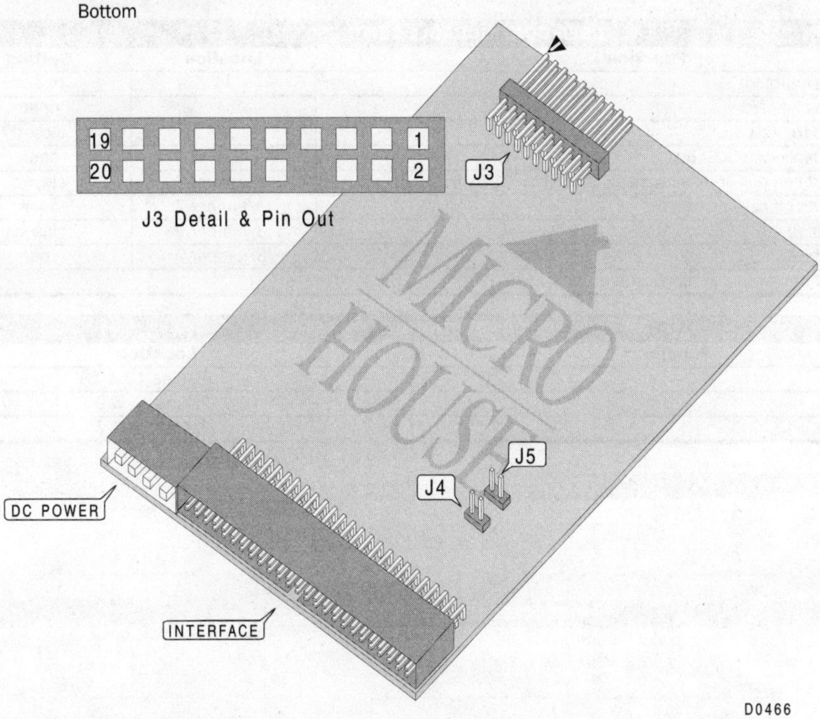

DRIVE SELECTION - J3			
Drive ID	**Jumper 1**	**Jumper 2**	**Jumper 3**
0	open	open	Open
1	closed	open	Open
2	open	closed	Open
3	closed	closed	Open
4	open	open	Closed
5	closed	open	Closed
6	open	closed	Closed
7	closed	closed	Closed

Continued on next page . . .

QUANTUM CORPORATION

ATLAS XP SERIES: XP31070S, XP32150S, XP34300S
ATLAS II SERIES: XP32181S, XP34361S, XP39100S

. . . continued from previous page

USER CONFIGURABLE SETTINGS		
Function	**Location**	**Setting**
Active termination enabled	J4	closed
Active termination disabled	J4	open
TRMPWR supplied to SCSI bus	J5	closed
TRMPWR is not supplied to SCSI bus	J5	open
Spindle motor starts on power-up with delay of SCSI ID x 12 sec.	J3/jumper 7	closed
Spindle motor starts on command only	J3/jumper 7	open
Write protect enabled	J3/jumper 9	closed
Write protect disabled	J3/jumper 9	open
Note: Jumpers not mentioned in this document are unidentified.		

CONNECTIONS	
Function	**Location**
Spindle sync	J3/pins 10 & 19
Remote Busy LED	J3/pins 11 & 9
Remote Fault LED	J3/pins 11 & 7

QUANTUM CORPORATION

ATLAS II SERIES: XP32181 SCA, XP34361 SCA, XP39100 SCA

Interface: SCSI-3 80-pin
Size: 3.5 in. third height (XP9100R is half height)
View: Bottom

SPINDLE MOTOR CONFIGURATION		
Settings	**Pin 38**	**Pin 78**
Spindle motor starts on power up	open	open
Spindle motor delay enabled (Y sec. x (SCSI ID +1))	open	ground
Spindle motor starts on command only	ground	open

Continued on next page . . .

QUANTUM CORPORATION

ATLAS II SERIES: XP32181 SCA, XP34361 SCA, XP39100 SCA

. . . continued from previous page

DRIVE SELECTION				
Drive ID	ID0/Pin 39	ID1/Pine 79	ID2/Pine 40	ID3/Pin 80
0	Open	open	open	open
1	Ground	open	open	open
2	Open	ground	open	open
3	Ground	ground	open	open
4	Open	open	ground	open
5	Ground	open	ground	open
6	Open	ground	ground	open
7	Ground	ground	ground	open
8	Open	open	open	ground
9	Ground	open	open	ground
10	Open	ground	open	ground
11	Ground	ground	open	ground
12	Open	open	ground	ground
13	Ground	open	ground	ground
14	Open	ground	ground	ground
15	Ground	ground	ground	ground

CONNECTIONS	
Function	Location
Drive active LED	pin 77
Spindle sync connection	pin 37

MISCELLANEOUS TECHNICAL NOTES
The 80-pin SCA connector is configured with a special cable that provides the appropriate signals to the desired pins. SCA drives do not provide TERMPWR, active termination or write protect. Pins not mentioned are unidentified.

QUANTUM CORPORATION

ATLAS II XP32275S, XP34550S, XP39100S
ATLAS II XP32275D, XP34550D, XP39100D

Interface	Ultra SCSI, Fiber Channel
Size	3.5 in. quarter height
View	Top

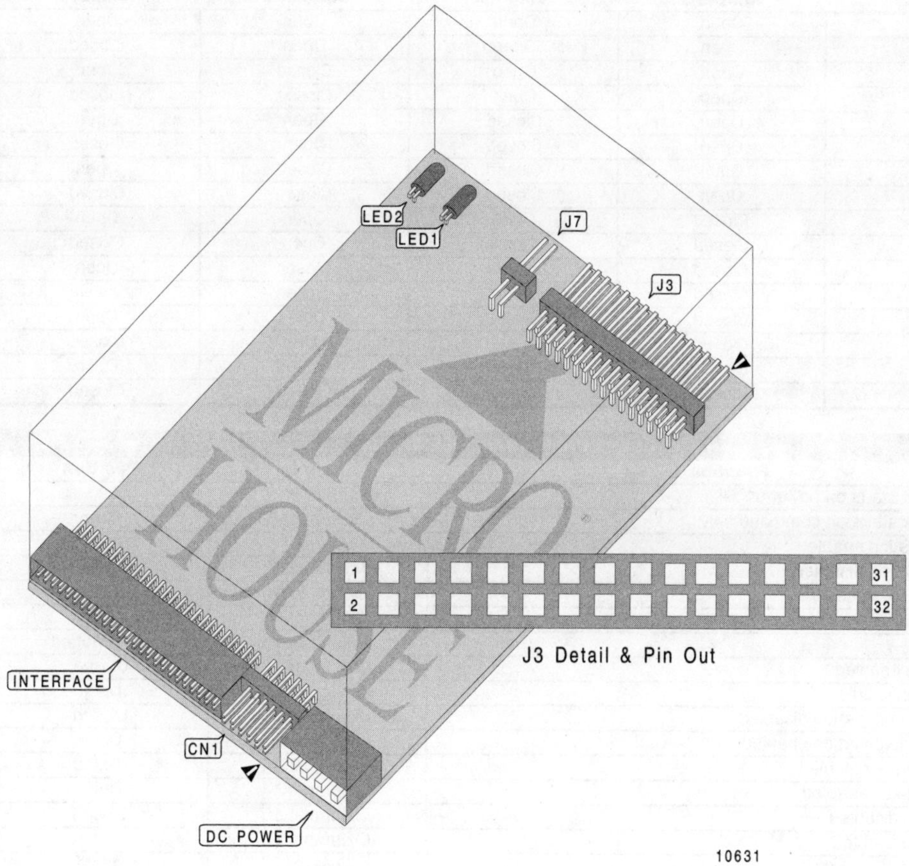

Continued on next page. . .

QUANTUM CORPORATION

ATLAS II XP32275S, XP34550S, XP39100S
ATLAS II XP322750, XP34550D, XP39100D

. . .continued from previous page

DRIVE SELECTION - J3				
SCSI ID	**Jumper 1**	**Jumper 2**	**Jumper 3**	**Jumper 4**
0	Open	Open	Open	Open
1	Open	Open	Open	Closed
2	Open	Open	Closed	Open
3	Open	Open	Closed	Closed
4	Open	Closed	Open	Open
5	Open	Closed	Open	Closed
6	Open	Closed	Closed	Open
7	Open	Closed	Closed	Closed
8	Closed	Open	Open	Open
9	Closed	Open	Open	Closed
10	Closed	Open	Closed	Open
11	Closed	Open	Closed	Closed
12	Closed	Closed	Open	Open
13	Closed	Closed	Open	Closed
14	Closed	Closed	Closed	Open
15	Closed	Closed	Closed	Closed

USER CONFIGURABLE SETTINGS		
Function	**Label**	**Position**
Spindle motor starts on power-up	J3/jumper 6	Closed
Spindle motor starts on command only	J3/jumper 6	Open
Active termination enabled	J3/jumper 7	Closed
Active termination disabled	J3/jumper 7	Open
Write protect disabled	J3/jumper 10	Closed
Write protect enabled	J3/jumper 10	Open
Stagger Spin	J3/jumper 11	Unidentified
Narrow mode enabled	J3/jumper 12	Closed
Wide mode enabled	J3/jumper 12	Open
Synchronous negotiation disabled	J3/jumper 13	Closed
Synchronous negotiation enabled	J3/jumper 13	Open
SCSI parity check disabled	J3/jumper 14	Closed
SCSI parity check enabled	J3/jumper 14	Open
Unit attention disabled	J3/jumper 15	Closed
Unit attention enabled	J3/jumper 15	Open
Customize	J3/jumper 16	Unidentified
Note: The exact purpose and settings of J3/jumper 11 and J3/jumper 16 are unidentified. D revisions of these drives to not support active termination. J3 pin 14 supplies logical ground for D versions of these drives.		

TERMINATION POWER SELECTION	
Setting	**J7**
TERMPWR supplied by drive	Closed
TERMPWR supplied by SCSI interface	Open

Continued on next page. . .

QUANTUM CORPORATION

ATLAS II XP32275S, XP34550S, XP39100S
ATLAS II XP32275D, XP34550D, XP39100D

. . .continued from previous page

CONNECTIONS			
Function	**Label**	**Function**	**Label**
Not Used	J3/pin 9	Busy LED	J3/pin 17
Fault LED	J3/pin 10	+5 VDC	J3/pin 18
Spindle Synch	J3/pins 15 & 16		

OPTION CONNECTOR CN1 PINOUT			
Function	**Label**	**Function**	**Label**
ID 0	CN1/pin 1	Fault LED out	CN1/pin 7
ID 1	CN1/pin 2	Logic ground	CN1/pin 8
ID 2	CN1/pin 3	Logic ground	CN1/pin 9
ID 3	CN1/pin 4	Busy LED out	CN1/pin 10
Enable Termination	CN1/pin 5	Logic ground	CN1/pin 11
+5 VDC out	CN1/pin 6	Reserved for future use	CN1/pin 12

DIAGNOSTIC LED(S)	
LED	**Significance**
LED1	Busy LED
LED2	Fault LED

SAMSUNG ELECTRONICS, INC.

PLS-30854S, PLS-31084S, PLS-31274S

Interface:	SCSI-2 , Fast SCSI-2 (PLS-30854S)
Size:	3.5 in. third height
View:	Bottom

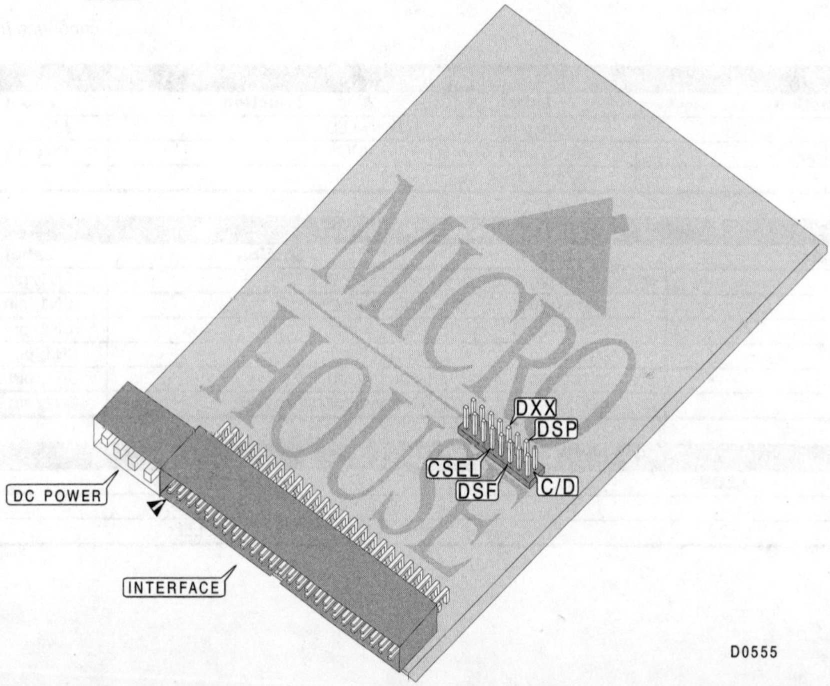

DRIVE CONFIGURATION	
Selection	**C/D**
➪ Single drive only	closed
Master drive in a two drive system	closed
Slave drive in a two drive system	open
Note: The rest of the jumpers are factory configured - do not alter.	

SAMSUNG ELECTRONICS, INC.

STG-31271A, STG-31601A

Interface IDE(AT)
Size 3.5 in. third height
View Bottom

DRIVE CONFIGURATION		
Setting	**JP1/Jumper 1**	**JP1/Jumper 2**
Single drive only	Closed	Open
Drive is slave in a two drive system	Open	Open
Master/Slave determined by Cable Select	Open	Closed

USER CONFIGURABLE SETTINGS		
Function	**Label**	**Position**
Jumper information not available	JP1/jumper 3	Unidentified
Jumper information not available	JP1/jumper 4	Unidentified

SAMSUNG ELECTRONICS, INC.

TBR-31081A

Interface IDE(AT)
Size 3.5 in. third height
View Bottom

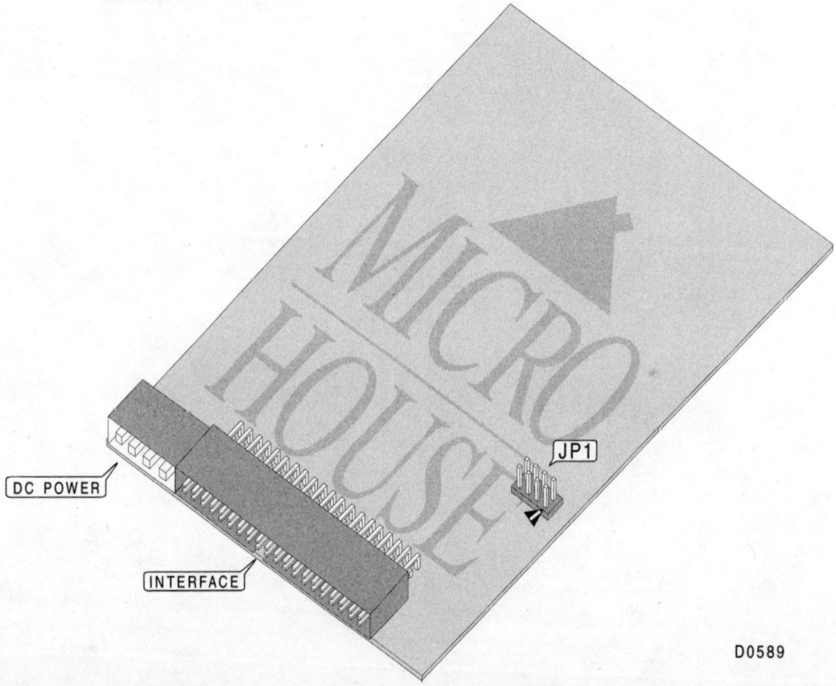

DRIVE CONFIGURATION		
Setting	**JP1/Jumper 1**	**JP1/Jumper 2**
Single drive only	Closed	Open
Drive is master when slave uses DASP signal	Closed	Open
Drive is master when slave does not use DASP signal	Closed	Closed
Drive is slave in a two drive system	Open	Open

USER CONFIGURABLE SETTINGS		
Function	**Label**	**Position**
Jumper information not available	JP1/jumper 3	Unidentified
Jumper information not available	JP1/jumper 4	Unidentified

SAMSUNG ELECTRONICS, INC.

WNR-31271A, WNR-31601A, WNR-32100A, WNR-32101A, WNR-32501A

Interface IDE(AT)
Size 3.5 in. quarter height
View Bottom

DRIVE CONFIGURATION		
Setting	**CD**	**CSEL**
Single drive only	Closed	Open
Drive is master in two drive system	Closed	Open
Drive is slave in two drive system	Open	Open
Master/Slave determined by Cable Select	Open	Closed

USER CONFIGURABLE SETTINGS		
Function	**Label**	**Position**
Jumper information not available	DCF	Unidentified
Jumper information not available	DXX	Unidentified

SEAGATE TECHNOLOGY INC.

MARATHON 130SL, 170SL, 210SL, 260SL, 340, 455, 520
ST9051A, ST9052A, ST9077A, ST9080A, ST9096A, ST9100A, ST9100AG, ST9140AG, ST9144A, ST9145A, ST9145AG, ST9150AG, ST9190AG, ST9235A, ST9235AG, ST9240A, ST9240AG, ST9295AG, ST9300AG, ST9385AG, ST9550AG, ST9655AG

Interface: IDE (44PIN)
Size: 2.5 in. fifth height (some models are 0.75 , some are 0.5 high)
View: Bottom

DRIVE CONFIGURATION		
Selection	**JP1**	**JP2**
➩ Single drive only	open	open
Master drive in a two drive system when slave is a Seagate drive	open	closed
Slave drive in a two drive system when master is a 2.5 Seagate drive	closed	open
Note: Do not jumper both JP1 & JP2 at the same time. This position is reserved for factory use.		

SEAGATE TECHNOLOGY, INC.

ST11200N, ST11200ND, ST11201N, ST11201ND, ST11700N, ST11700ND, ST11900N, ST11900ND, ST12400N, ST12400ND, ST1830N, ST1950N, ST1980N, ST1980ND, ST3500N, ST3600N, ST3600ND, ST3610N, ST3610ND, HAWK 1, HAWK 2(VER. 1)

Interface: SCSI-2
Size: 3.5 in. half height (ST1xxxN)
3.5 in. third height (ST3xxxN)
View: Bottom

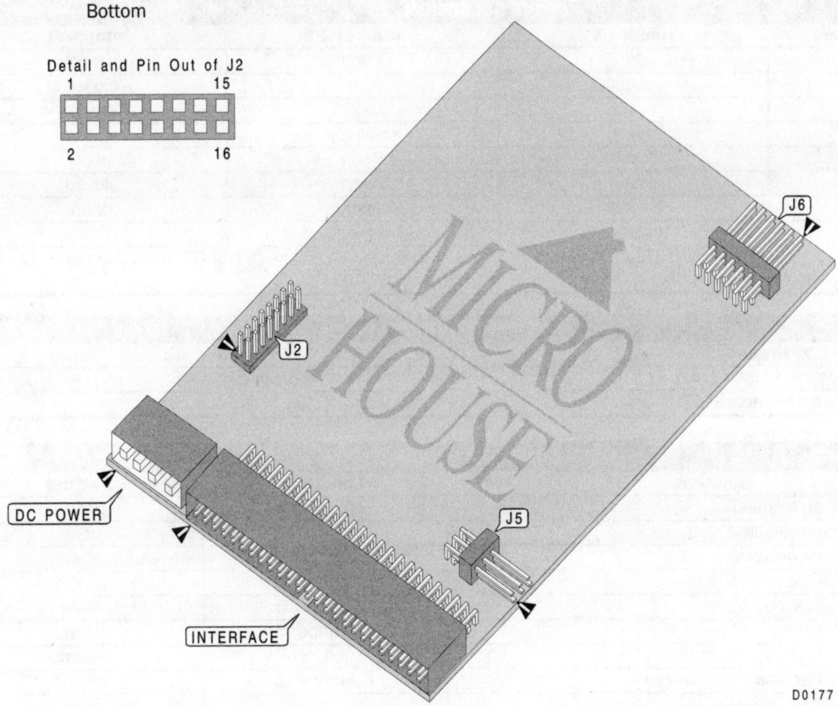

DRIVE SELECTION - J5			
Drive Number	**Jumper 1**	**Jumper 2**	**Jumper 3**
0	open	open	open
1	closed	open	open
2	open	closed	open
3	closed	closed	open
4	open	open	closed
5	closed	open	closed
6	open	closed	closed
7	closed	closed	closed

Continued on next page . . .

SEAGATE TECHNOLOGY, INC.

ST11200N, ST11200ND, ST11201N, ST11201ND, ST11700N, ST11700ND, ST11900N, ST11900ND, ST12400N, ST12400ND, ST1830N, ST1950N, ST1980N, ST1980ND, ST3500N, ST3600N, ST3600ND, ST3610N, ST3610ND, HAWK 1, HAWK 2 (VER. 1)

. . . continued from previous page

DRIVE SELECTION - J5			
Drive Number	**Jumper 1**	**Jumper 2**	**Jumper 3**
0	open	open	open
1	open	open	closed
2	open	closed	open
3	open	closed	closed
4	closed	open	open
5	closed	open	closed
6	closed	closed	open
7	closed	closed	closed

Note: This table applies to ST1980N, ST11200N, and ST3500N/ST3600N with PCB# 75782480 only. Use the table on the previous page for any other PCB#.

CONNECTIONS	
Purpose	**Location**
External Spindle Sync	J6/pins 5 & 6
Hard drive active LED	J6/pins 3 & 4

USER CONFIGURABLE SETTINGS		
Function	**Location**	**Setting**
➪ Drive termination enabled	J2/jumper 3	closed
Drive termiantion disabled	J2/jumper 3	open
➪ SCSI parity check disabled	J2/jumper 4	open
SCSI parity check enabled	J2/jumper 4	closed
➪ Write protect disabled	J2/jumper 5	open
Write protect enabled	J2/jumper 5	closed
➪ Spindle motor starts on power-up	J2/jumper 6	open
Spindle motor starts on command only	J2/jumper 6	closed
➪ Spindle motor delay disabled	J2/jumper 7	open
Spindle motor delay enabled (ID x 12 sec.)	J2/jumper 7	closed
➪ Factory configured-do not alter	J2/jumper 8	open

TERMINATION POWER SETTINGS			
Function	**J2/jumper 1**	**J2/jumper 2**	**J2/pins 2 & 4**
➪ TERMPWR supplied by drive to drive	open	Closed	N/A
TERMPWR supplied by drive to SCSI pin 26	closed	Open	N/A
TERMPWR supplied by drive to drive and SCSI pin 26	closed	Closed	N/A
TERMPWR is not connected to drive or SCSI bus	open	Open	N/A
TERMPWR supplied by SCSI pin 26 to drive	N/A	N/A	closed

MISCELLANEOUS TECHNICAL NOTES

Spindle motor starts on power-up and Spindle motor start delay enabled are mutually exclusive.

SCSI I.D. may also be set using J6/jumpers 4, 5, and 6. These Jumpers perform the same function and are mutually exclusive with J5/jumpers 1, 2, and 3.

SEAGATE TECHNOLOGY, INC.

BARRACUDA 1 (VER. 2), BARRACUDA 2 (VER. 3), ST11950W, ST11950WD, ST12550W, ST12550WD

Interface: SCSI-2
Size: 3.5 in. Half Height
View: Top

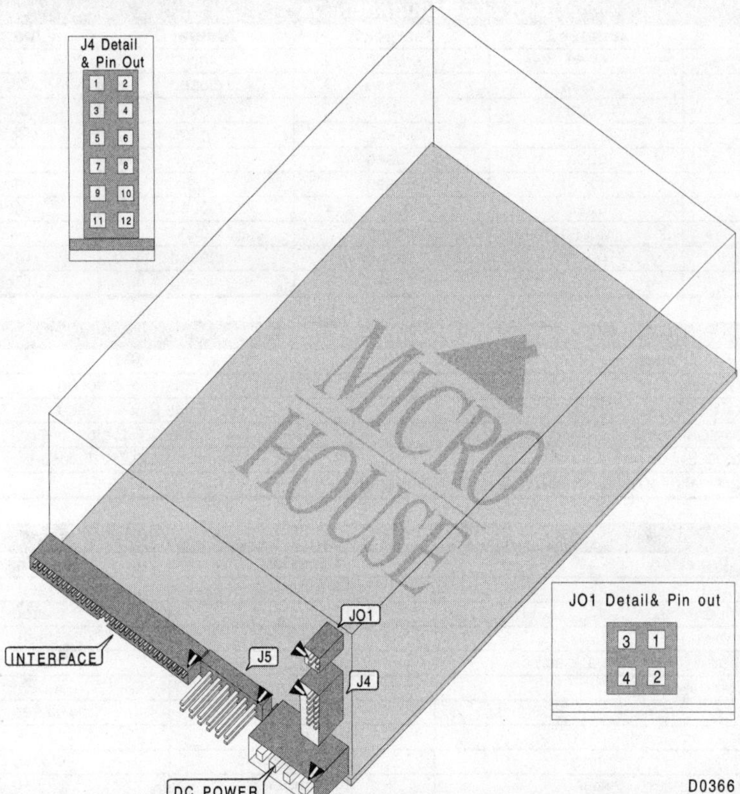

DRIVE CONFIGURATION - J5				
Drive Number	**Jumper 1**	**Jumper 2**	**Jumper 3**	**Jumper 4**
0	open	open	open	open
1	closed	open	open	open
2	open	closed	open	open
3	closed	closed	open	open
4	open	open	closed	open
5	closed	open	closed	open

Continued on next page . . .

SEAGATE TECHNOLOGY, INC.

BARRACUDA 1 (VER. 2), BARRACUDA 2 (VER. 3), ST11950W, ST11950WD, ST12550W, ST12550WD

. . . continued from previous page

DRIVE CONFIGURATION - J4				
Drive Number	**Jumper 1**	**Jumper 2**	**Jumper 3**	**Jumper 4**
6	open	closed	closed	Open
7	closed	closed	closed	Open
8	open	open	open	Closed
9	closed	open	open	Closed
10	open	closed	open	Closed
11	closed	closed	open	Closed
12	open	open	closed	Closed
13	closed	open	closed	Closed
14	open	closed	closed	Closed
15	closed	closed	closed	Closed

TERMINATION POWER SETTINGS	
Function	**J01**
➪ TERMPWR supplied by drive *	pins 1 & 2 closed
TERMPWR supplied by SCSI bus	pins 2 & 4 closed
TERMPWR supplied to SCSI bus *	pins 1 & 3 closed
TERMPWR supplied to SCSI bus and drive *	pins 1 & 3, 2 & 4 closed
Note: * These settings are valid for single-ended drives only (W models).	

USER CONFIGURABLE SETTINGS		
Function	**Location**	**Setting**
➪ Write protection disabled	J4/jumper 1	open
Write protection enabled	J4/jumper 1	closed
➪ Delay motor start disabled	J4/jumper 2	open
Delay motor start enabled (target ID x 10sec.)	J4/jumper 2	closed
➪ Spindle motor starts on power-up	J4/jumper 3	open
Target waits for Start Unit command from SCSI host	J4/jumper 3	closed
➪ Parity check enabled	J4/jumper 4	open
Parity check disabled	J4/jumper 4	closed
➪ Factory configured - do not alter	J4/jumper 5	closed
➪ Drive terminators disabled*	J4/jumper 6	open
Drive terminators enabled*	J4/jumper 6	closed
Note: * These settings are valid for single-ended drives only (W models).		

CONNECTIONS	
Function	**Location**
Remote LED connector	J5/jumper 5
Spindle sync	J5/jumper 6

SEAGATE TECHNOLOGY, INC.

ST15230NC

Interface: SCSI-2 Fast
Size: 3.5 in. half height
View: Bottom

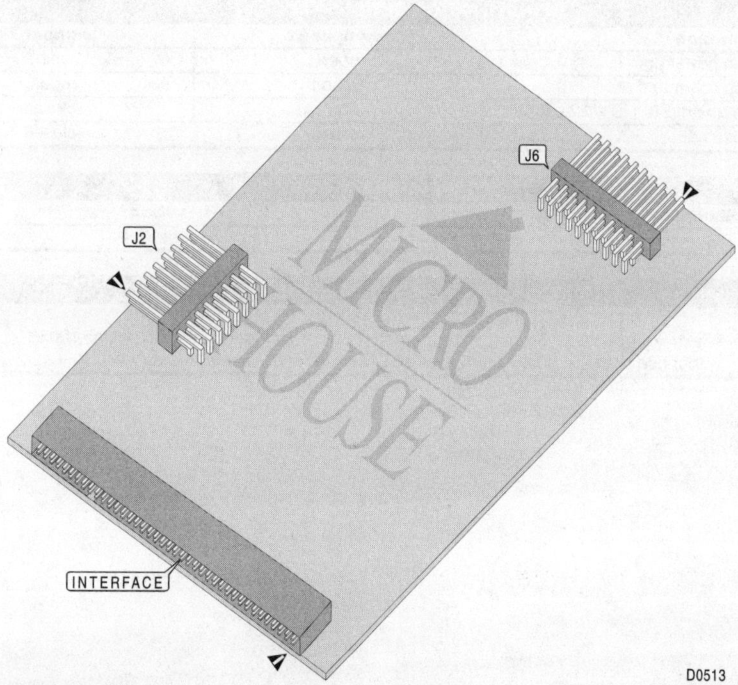

DRIVE SELECTION - J6			
SCSI ID Number	**Jumper 1**	**Jumper 2**	**Jumper 3**
➩ 0	open	open	open
1	closed	open	open
2	open	closed	open
3	closed	closed	open
4	open	open	closed
5	closed	open	closed
6	open	closed	closed
7	closed	closed	closed

USER CONFIGURABLE SETTINGS		
Function	**Location**	**Setting**
➩ Factory configured - do not alter	J2/jumper 1	open
➩ Factory configured - do not alter	J2/jumper 2	open
➩ Factory configured - do not alter	J2/jumper 3	open
➩ SCSI parity enabled	J2/jumper 4	closed
SCSI parity disabled	J2/jumper 4	open
➩ Write protect disabled	J2/jumper 5	open
Write protect enabled	J2/jumper 5	closed
➩ Factory configured - do not alter	J2/jumper 8	open
➩ Factory configured - do not alter	J6/jumper 5	open
➩ Factory configured - do not alter	J6/jumper 6	open

Continued on next page . . .

SEAGATE TECHNOLOGY, INC.

ST15230NC

. . . continued from previous page

SPINDLE MOTOR OPTIONS - J2		
Function	**Jumper 6**	**Jumper 7**
➪ Spindle motor starts on power up	open	open
Spindle motor starts on command only	closed	open
Spindle motor delay enabled (SCSI ID x 12 sec.)	open	closed
Spindle motor delay disabled	closed	closed

CONNECTIONS	
Function	**Location**
Remote LED connection	J6/jumper 4

MISCELLANEOUS TECHNICAL NOTES
J6/jumpers 7-10 come with cover installed - do not remove. SCSI ID, spindle motor options, sync. connector, remote LED are included in the 80 pin interface. Do not use both the 80 pin connector and J2 or J6 for the same function.

SEAGATE TECHNOLOGY, INC.

HAWK 2LP (VER. 2), HAWK 4 (VER. 1), ST15230W/WD, ST31230W/WD, ST32430W/WD

Interface: SCSI-2 Fast Wide
Size: 3.5 in. third height
View: Bottom

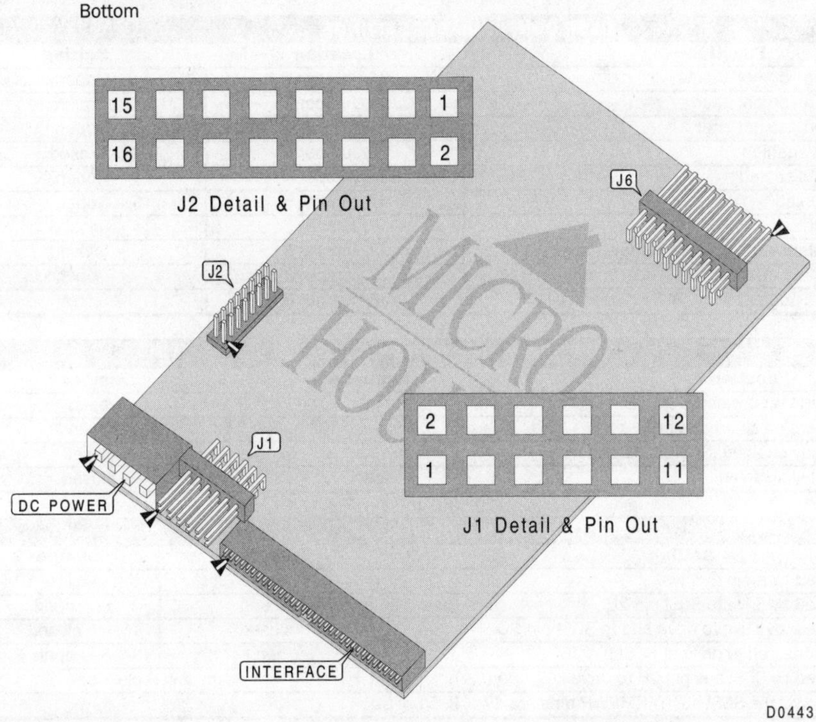

DRIVE SELECTION - J1				
SCSI ID Number	**Jumper 1**	**Jumper 2**	**Jumper 3**	**Jumper 4**
0	open	open	open	open
1	closed	open	open	open
2	open	closed	open	open
3	closed	closed	open	open
4	open	open	closed	open
5	closed	open	closed	open
6	open	closed	closed	open
7	closed	closed	closed	open
8	open	open	open	closed
9	closed	open	open	closed
10	open	closed	open	closed
11	closed	closed	open	closed
12	open	open	closed	closed
13	closed	open	closed	closed
14	open	closed	closed	closed
15	closed	closed	closed	closed

Continued on next page . . .

SEAGATE TECHNOLOGY, INC.

HAWK 2LP (VER. 2), HAWK 4 (VER. 1), ST15230W/WD, ST31230W/WD, ST32430W/WD

. . . continued from previous page

USER CONFIGURABLE SETTINGS		
Function	**Location**	**Setting**
➪ Factory configured - do not alter	J1/Jumper 5	open
➪ Factory configured - do not alter	J1/Jumper 6	open
➪ Factory configured - do not alter	J2/Jumper 3	open
➪ SCSI parity check enabled	J2/Jumper 4	closed
SCSI parity check disabled	J2/Jumper 4	open
➪ Write protect disabled	J2/Jumper 5	open
Write protect enabled	J2/Jumper 5	closed
➪ Terminating resistors enabled (W models only)	J2/Jumper 8	closed
Terminating resistors disabled (W models only)	J2/Jumper 8	open
➪ Factory configured - do not alter (WD models only)	J2/Jumper 8	N/A

SPINDLE MOTOR OPTIONS - J2		
Function	**Jumper 6**	**Jumper 7**
➪ Spindle motor starts on power up	open	open
Spindle motor starts on command only	closed	open
Spindle motor delay enabled (SCSI ID x 12 sec.)	open	closed
Spindle motor delay disabled	closed	closed

TERMINATION POWER - J2		
Setting	**Jumper 1**	**Jumper 2**
➪ TERMPWR supplied by drive to drive	open	closed
TERMPWR supplied by drive to SCSI pin 26	closed	open
TERMPWR supplied by drive to drive and to SCSI pin 26	closed	closed
TERMPWR not connected to drive or SCSI pin 26	open	open
TERMPWR supplied by SCSI bus pin 26 to drive (W models only)	pins 2 & 4 closed	
Note: For 15230W/WD the SCSI bus TERMPWR pins are 17, 18, 51 & 52.		

CONNECTIONS	
Function	**Location**
Remote LED connection	J1/pins 8 & 11

MISCELLANEOUS TECHNICAL NOTES
J1 & J6 are mutually exclusive. When configuring the drive with J1, do not alter J6. Same settings and pin-out apply for both jumpers. J6 jumpers 7-10 are shipped with cover installed. Do not remove cover or install jumpers.11

SEAGATE TECHNOLOGY, INC.

ST31250DC/WC, ST32550DC/WC

Interface:	SCSI-2 Fast
Size:	3.5 in. third height
View:	Bottom

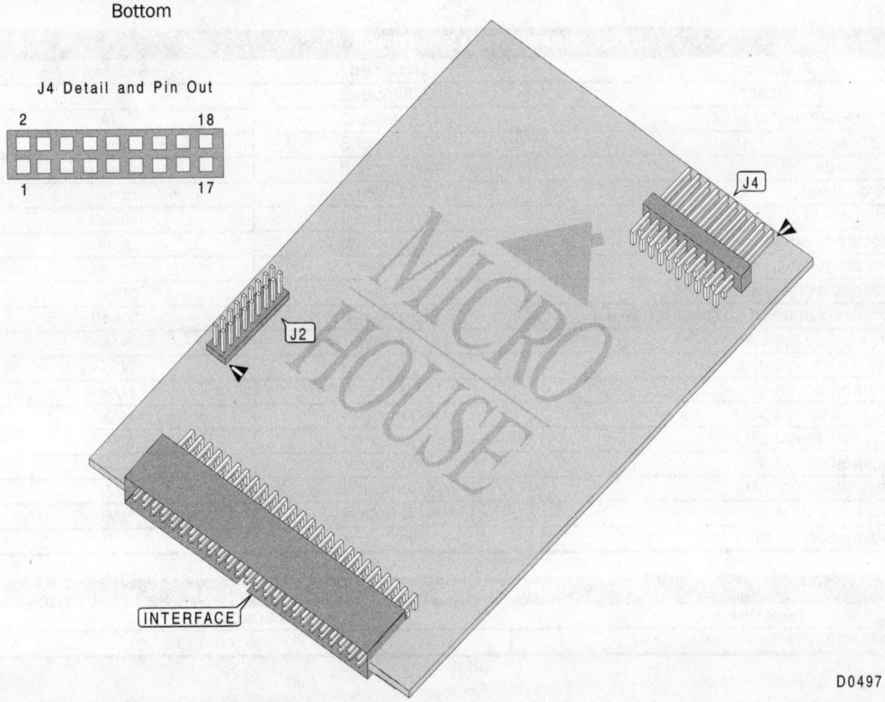

DRIVE SELECTION - J4				
SCSI ID Number	**Jumper 1**	**Jumper 2**	**Jumper 3**	**Jumper 4**
⇨ 0	open	open	open	open
1	closed	open	open	open
2	open	closed	open	open
3	closed	closed	open	open
4	open	open	closed	open
5	closed	open	closed	open
6	open	closed	closed	open
7	closed	closed	closed	open
8	open	open	open	closed
9	closed	Open	open	closed
10	open	Closed	open	closed
11	closed	Closed	open	closed
12	open	Open	closed	closed
13	closed	Open	closed	closed
14	open	Closed	closed	closed
15	closed	Closed	closed	closed

Continued on next page. . .

SEAGATE TECHNOLOGY, INC.

ST31250DC/WC, ST32550DC/WC

. . .continued from previous page

USER CONFIGURABLE SETTINGS		
Function	**Location**	**Setting**
➪ Factory configured-do not alter	J2/jumper 1	N/A
➪ Factory configured-do not alter	J2/jumper 2	N/A
➪ SCSI parity check enabled	J2/jumper 3	open
SCSI parity check disabled	J2/jumper 3	closed
➪ Write protect disabled	J2/jumper 4	open
Write protect enabled	J2/jumper 4	closed
➪ Spindle motor starts on power-up	J2/jumper 5	open
Spindle motor starts on command only	J2/jumper 5	closed
➪ Spindle motor delay disabled	J2/jumper 6	open
Spindle motor delay enabled (10 sec. X SCSI ID)	J2/jumper 6	closed
➪ Factory configured-do not alter	J2/jumper 7	N/A
➪ Factory configured-do not alter	J2/jumper 9	N/A
➪ Factory configured-do not alter	J2/jumper 8	N/A
➪ Factory configured-do not alter	J4/jumper 5	N/A
➪ Factory configured-do not alter	J4/jumper 6	N/A
➪ Factory configured-do not alter	J4/jumper 7	N/A
➪ Factory configured-do not alter	J4/jumper 8	N/A
➪ Factory configured-do not alter	J4/jumper 9	N/A
➪ Factory configured-do not alter	J4/jumper 10	N/A

CONNECTIONS	
Function	**Location**
Remote LED connection	J4/pins 8 & 11

SEAGATE TECHNOLOGY, INC.

BARRACUDA 2LP (VER. 4), ST31250WD, ST32550WD

Interface: SCSI-2 Fast
Size: 3.5 in. third height
View: Bottom

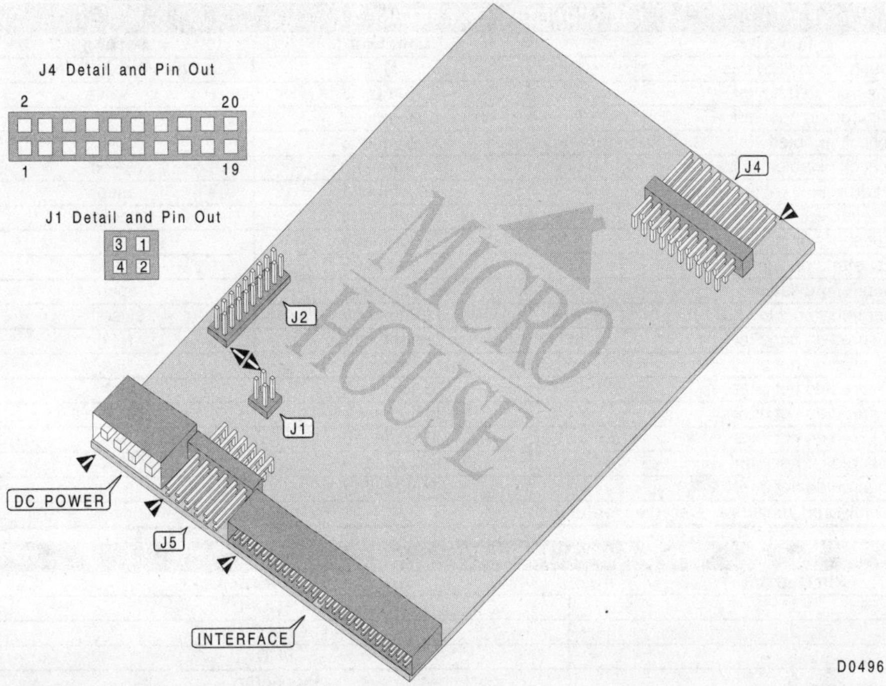

DRIVE SELECTION-J5				
Drive Number	**Jumper 1**	**Jumper 2**	**Jumper 3**	**Jumper 4**
⇨ 0	open	open	open	open
1	closed	open	open	open
2	open	closed	open	open
3	closed	closed	open	open
4	open	open	closed	open
5	closed	open	closed	open
6	open	closed	closed	open
7	closed	closed	closed	open
8	open	open	open	closed
9	closed	open	open	closed
10	open	closed	open	closed
11	closed	closed	open	closed
12	open	open	closed	closed
13	closed	open	closed	closed
14	open	closed	closed	closed
15	closed	closed	closed	closed

Continued on next page. . .

SEAGATE TECHNOLOGY, INC.

BARRACUDA 2LP (VER. 4), ST31250WD, ST32550WD

. . .continued from previous page

USER CONFIGURABLE SETTINGS		
Function	**Location**	**Setting**
➪ TERMPWR supplied to SCSI bus	J1	pins 1 & 3 closed
➪ Factory configured-do not alter	J2/jumper 1	N/A
➪ Factory configured-do not alter	J2/jumper 2	N/A
➪ SCSI parity check enabled	J2/jumper 3	open
SCSI parity check disabled	J2/jumper 3	closed
➪ Write protect disabled	J2/jumper 4	open
Write protect enabled	J2/jumper 4	closed
➪ Spindle motor starts on power-up	J2/jumper 5	open
Spindle motor starts on command only	J2/jumper 5	closed
➪ Spindle motor delay disabled	J2/jumper 6	open
Spindle motor delay enabled (10 sec. X SCSI ID)	J2/jumper 6	closed
➪ Factory configured-do not alter	J2/jumper 7	N/A
➪ Factory configured-do not alter	J2/jumper 8	N/A
➪ Factory configured-do not alter	J2/jumper 9	N/A
➪ Factory configured-do not alter	J4/jumper 7	N/A
➪ Factory configured-do not alter	J4/jumper 8	N/A
➪ Factory configured-do not alter	J4/jumper 9	N/A
➪ Factory configured-do not alter	J4/jumper 10	N/A
Note: Factory configured jumpers are reserved for future use.		

J4 PIN-OUT	
Pin Number	**Significance**
1	SCSI ID=0
2	Fault LED (+)
3	SCSI ID=1
4	Proprietary
5	SCSI ID=2
6	SSREF (+)
7	SCSI ID=3
8	Drive active LED (+)
9	Termination enable
10	GRD
11	+5v
12	Unused
Note: See the USER CONFIGURABLE SETTINGS table for the significance of the remaining J4 jumpers.	

CONNECTIONS	
Function	**Location**
Spindle sych. connector	J5/jumper 6
Drive active LED	J5/jumper 2

SEAGATE TECHNOLOGY, INC.

MEDALIST 1080 (VER. 1), ST3780A, ST31220A

Interface: IDE (AT)
Size: 3.5 in. third height
View: Bottom

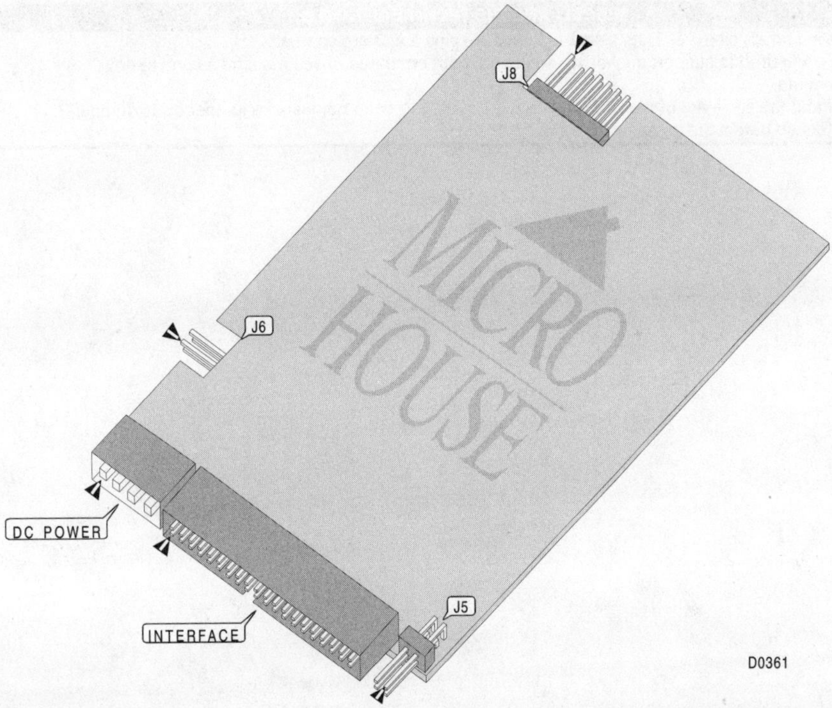

DRIVE CONFIGURATION J5		
Selection	**Jumper 1**	**Jumper 2**
➪ Single drive system	open	open
Drive is slave in a two drive system	closed	open
Drive is master in a two drive system	open	closed
Cable select *	closed	closed
Note: The * indicates you must also install a jumper on pins 6 & 8 of J8.		

USER CONFIGURABLE SETTINGS		
Function	**Location**	**Setting**
Dual drive emulation mode enabled	J6/jumper 1 & 2	closed
Dual drive emulation mode disabled	J6/jumper 1 & 2	open
Cable select enabled	J8/pins 6 & 8	closed

CONNECTIONS	
Function	**Location**
Remote LED connection	J8/jumper 7

Continued on next page . . .

SEAGATE TECHNOLOGY, INC.

MEDALIST 1080 (VER. 1), ST3780A, ST31220A

. . . continued from previous page

MISCELLANEOUS TECHNICAL NOTES
Note:Shunts installed on J5/pins 1 & 3, J8/pins 1 & 2, and J6/pins 1 & 3 are spares. When the drive is configured to support dual-drive emulation, both emulated drives respond as one to any power-management command. If both the master and the slave are of the ST31220 family drives, you need not install a jumper on J5/jumper 2. This is because the master can detect the presence of a slave with DASP.

WESTERN DIGITAL CORPORATION

AT CAVIAR DRIVES: WDAC1170, WDAC1210, WDAC1270, WDAC1365, WDAC140,WDAC1425, WDAC160, WDAC21000, WDAC2120, WDAC21200, WDAC21600, WDAC2170, WDAC2200, WDAC2250, WDAC2340, WDAC2420, WDAC2540, WDAC2635, WDAC2700, WDAC280, WDAC2850, WDAC31000, WDAC31200, WDAC31600, WDAC32100, WDAC33100, WDAC34000

Interface: IDE(AT)
EIDE (Models 21000, 21200, 21600, 2850, 31000, 31600, 32100, 33100, 34000)
Size: 3.5 in. third height
View: Top

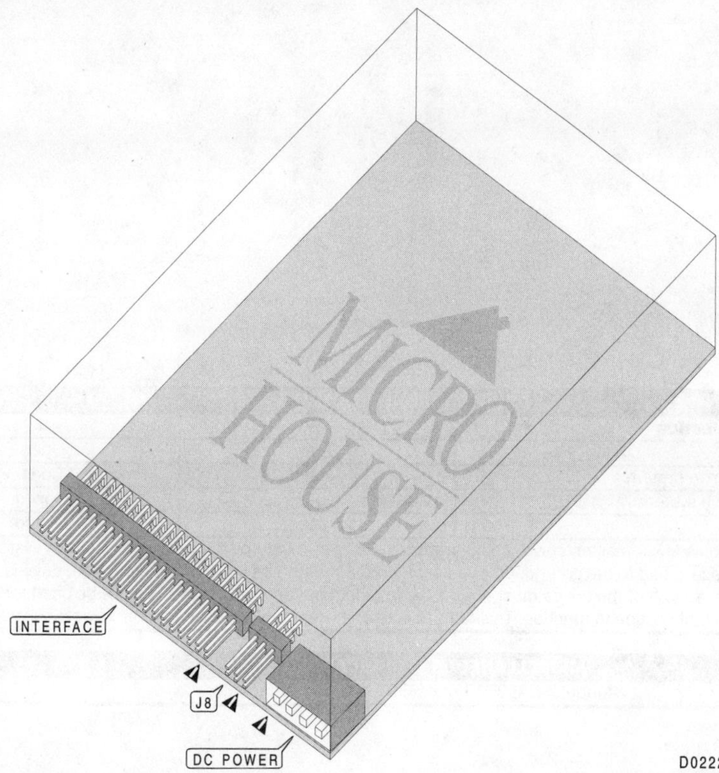

D0222

DRIVE CONFIGURATION			
Selection	**J8/Jumper 1**	**J8/Jumper 2**	**J8/Jumper 3**
➩ Single drive only	Open	open	open
Master drive in a two drive system	Open	open	closed
Slave drive in a two drive system	Open	closed	open
Cable Select	Closed	Open	Open

MISCELLANEOUS TECHNICAL NOTES
On older versions of some Caviar drives J8/jumper 1 must be closed to make it slave to a Conner Peripherals CP3022 or CP342. On newer versions leave J8/jumper 1 open. Caviar drives are shipped with a spare shunt jumpered across J8/jumpers 2 & 3.

WESTERN DIGITAL CORPORATION

WDAL1100, WDAL2200, WDAL2540

Interface: IDE (44PIN)
Size: 2.5 in. fifth height
View: Bottom

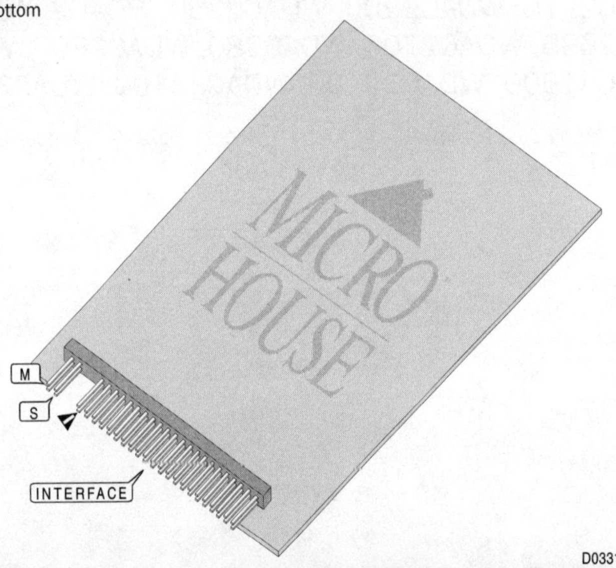

D0331

DRIVE CONFIGURATION		
Selection	**M**	**S**
➪ Single drive only	Open	Open
Master drive in a two drive system	Closed	Open
Slave drive in a two drive system	Open	Closed
Cable select option	Closed	Closed
Note: Cable select option allows master/slave configuration to be determined by the cabling of the host system. If pin 28 on the interface is tied to the ground, drive is considered a master, if pin 28 is not grounded, drive is considered a slave. Both of the drives must have a Cable Select option and pin 28 must not be used for any other signal in order for this option to function. This feature is useful when drive is used in a removable subsystem.		

MISCELLANEOUS TECHNICAL NOTES
Model WDAL2540 uses EIDE 44-pin interface.

Chapter 6

Hard Drive Specifications

Introduction

This chapter contains the specifications required when setting up a hard drive. It is alphabetically sorted by manufacturer. It contains the actual physical specifications and the IDE manufacturers' recommended logical parameters.

Changes in Make and Model

Several companies have changed names or have been bought out by other companies. If you can't find the model you are looking for, look through the other manufacturers to see if it is under another brand name. The model numbers usually stay unchanged when the drive changes to another manufacturer.

The following is a list of some brand names to cross reference when looking for a drive model:

- ATASI and PRIAM may be the same
- CDC, CONNER, IMPRIMIS and SEAGATE may be the same (see Seagate in this chapter for the conversion chart)
- MAXTOR and MINISCRIBE may be the same
- OTARI and DISCTRON may be the same
- PRIAM and VERTEX may be the same
- YE DATA and C.ITOH may be the same
- BRAND and NCL may be the same

There may be others not listed here, so look thoroughly through all brands to locate your drive.

Categories and Abbreviations

The following list defines the abbreviations used in the table headings that follow.

INT:	Interface type.
CAP:	Capacity in MB.
FF:	Form Factor denotes the diameter of the drive platter(s) in inches.
HGT:	The drive's height.
CODE:	Encoding scheme.
LZ:	Landing zone (or parking, shipping) cylinder. Not used on most of the newer drives.
WP:	Write Precompensation cylinder. Not used on most of the newer drives.
RWC:	Reduced Write Current cylinder. Not used on most of the newer drives.
MTBF:	Mean Time Between Failures, in thousands of hours.
PHDS:	Number of *physical* data heads. Servo heads are not included in this figure.
PCYL:	Number of *physical* user-accessible cylinders. *If the drive is an IDE drive and this number exceeds 1,024, a translation must be used.*
PSPT:	Number of *physical* sectors per track.
LHDS:	Number of *logical* data heads. Servo heads are not included in this figure.
LCYL:	Number of *logical* user-accessible cylinders.
LSPT:	Number of *logical* sectors per track.

In addition, the following abbreviations are used to save space within the table rows:

INT:	COMM=COMMODORE
	IDE-CPQ=IDE-COMPAQ
	U-SCSI=ULTRA-SCSI
	PCM=PCMCIA
	U=UNIDEN
FF:	FH=Full
	HH=Half
	3H=1"
	4H=less than 1"
CODE:	RLL 1=RLL 1,7
	RLL 2=RLL 2,7
	RLL 3=RLL 3,9

PSPT: VAR=Variable sectors per track.

Manufacturers Recommended IDE Parameters

Older BIOSes can only access 1,024 cylinders, and most IDE drives now use more. If this is the case, an IDE drive's physical parameters cannot be used to achieve a safe setup with full capacity. Most IDE drives manufactured after 1990 support a logical translation to alleviate this problem. Please see "The IDE Interface" in Chapter Three for a more detailed explanation of IDE/BIOS limits.

LHDS, LCYL, and LSPT are the figures you should enter in a user-definable BIOS Drive Type (usually Type 46 and/or Type 47). If these figures aren't listed for an IDE drive and the actual physical parameters cannot be used, you can use the values from the IDE Quick Reference Chart (chapter 7). Alternatively, you many need to upgrade the BIOS to a newer version that supports the IDE Identify Drive command if the drive supports that command. See "ATA Commands" in Chapter Two for a discussion of this command.

Drive Specifications

ADVANTAGE MEMORY CORPORATION

MODEL	INT	CAP	FF	HGT	CODE	LZ	WP	RWC	MTBF	PHDS	PCYL	PSPT	LHDS	LCYL	LSPT
POCKETDRIVE 170MB	PCM	170.8	2.50	5H	U	AUTO	N/A	N/A	250	0	0	0			
POCKETDRIVE 260MB	PCM	260.4	2.50	5H	U	AUTO	N/A	N/A	250	0	0	0			

ALPS ELECTRIC (USA), INC.

MODEL	INT	CAP	FF	HGT	CODE	LZ	WP	RWC	MTBF	PHDS	PCYL	PSPT	LHDS	LCYL	LSPT
DR311C901		0								0	0	0			
DR311C901A		0								0	0	0			
DR311D911A		0								0	0	0			
DR312C901	IDE(AT)	211.54	3.50	3H	RLL 1	AUTO	N/A	N/A	150	4	2105	0	16	992	26
DR312C911	IDE(AT)	211.5	3.50	3H	RLL 1	AUTO	N/A	N/A	150	4	2108	0	16	992	26
DR312D	SCSI-2	211.5	3.50	3H	RLL 1	AUTO	N/A	N/A	150	4	2108	0			
DR312D911A	SCSI-2	211.5	3.50	3H	RLL 1	AUTO	N/A	N/A	150	4	2108	0			
DRPO-20A	ST-506/412	16.37	3.50	HH	RLL 2		616	616	0	2	615	26			
DRPO-20D	ST-506/412	16.37	3.50	HH	RLL 2		616	616	0	2	615	26			

AMPEX CORPORATION

MODEL	INT	CAP	FF	HGT	CODE	LZ	WP	RWC	MTBF	PHDS	PCYL	PSPT	LHDS	LCYL	LSPT
PYXIS-07	ST-506/412	5.57	5.25	FH	MFM		132	132	0	2	320	17			
PYXIS-13	ST-506/412	11.14	5.25	FH	MFM		132	132	0	4	320	17			
PYXIS-20	ST-506/412	16.71	5.25	FH	MFM		132	132	0	6	320	17			
PYXIS-27	ST-506/412	22.28	5.25	FH	MFM		132	132	0	8	320	17			

AREAL TECHNOLOGY, INC.

MODEL	INT	CAP	FF	HGT	CODE	LZ	WP	RWC	MTBF	PHDS	PCYL	PSPT	LHDS	LCYL	LSPT
A120	IDE(44PIN)	136.9	2.50	5H	RLL 2	AUTO	N/A	N/A	100	4	1070	250	10	535	50
A130	IDE(44PIN)	131.48	2.50	4H	RLL 1	AUTO	N/A	N/A	150	2	1438	0	5	856	60
A170	IDE(44PIN)	172.03	2.50	4H	RLL 2	AUTO	N/A	N/A	100	4	1344	0	10	672	50
A175L	IDE(44PIN)	166.99	2.50	4H	RLL 1	AUTO	N/A	N/A	250	2	0	0	6	950	60
A180	IDE(44PIN)	182.7	2.50	5H	RLL 2	AUTO	N/A	N/A	100	4	1430	250	10	715	50
A260	IDE(44PIN)	262.96	2.50	4H	RLL 1	AUTO	N/A	N/A	150	4	1438	0	10	856	60
A265	IDE(44PIN)	262.97	2.50	4H	RLL 1	AUTO	N/A	N/A	150	4	0	0	10	856	60
A340	IDE(44PIN)	394.81	2.50	4H	RLL 1	AUTO	N/A	N/A	100	6	1501	0	12	1020	63
A345L	IDE(44PIN)	333.99	2.50	4H	RLL 1	AUTO	N/A	N/A	250	4	0	0	12	950	60
A520L	IDE(44PIN)	502.03	2.50	4H	RLL 1	AUTO	N/A	N/A	250	6	0	0	16	1020	63
A525	IDE(44PIN)	526.41	2.50	4H	RLL 1	AUTO	N/A	N/A	100	8	1501	0	16	1020	63
A700L	IDE(44PIN)	700.88	2.50	4H	RLL 1	AUTO	N/A	N/A	250	8	0	0	16	1424	63
A85	IDE(44PIN)	86.01	2.50	4H	RLL 2	AUTO	N/A	N/A	100	2	1344	0	10	672	25
A90	IDE(XT/AT)	90.78	2.50	4H	RLL 2	AUTO	N/A	N/A	100	2	1430	62			
MD-2050	SCSI	50.31	2.50	4H	RLL 2	AUTO	N/A	N/A	45	2	819	60			

AREAL TECHNOLOGY, INC.

MODEL	INT	CAP	FF	HGT	CODE	LZ	WP	RWC	MTBF	PHDS	PCYL	PSPT	LHDS	LCYL	LSPT
MD-2060	IDE(44PIN)	62.4	2.50	4H	RLL 2	AUTO	N/A	N/A	100	2	1024	119	7	1024	17
MD-2080	IDE(44PIN)	80.35	2.50	4H	RLL 2	AUTO	N/A	N/A	100	2	1330	59	14	665	17
MD-2085	IDE(44PIN)	85.9	2.50	4H	RLL 2	AUTO	N/A	N/A	100	2	1410	59	14	705	17

ATASI TECHNOLOGY, INC.

MODEL	INT	CAP	FF	HGT	CODE	LZ	WP	RWC	MTBF	PHDS	PCYL	PSPT	LHDS	LCYL	LSPT
3020	ST-506/412	16.84	5.25	FH	MFM		320	320	0	3	645	17			
3033	ST-506/412	28.07	5.25	FH	MFM		320	320	40	5	645	17			
3046	ST-506/412	39.29	5.25	FH	MFM	644	320	320	0	7	645	17			
3051	ST-506/412	42.89	5.25	FH	MFM		350	359	0	7	704	17			
3051+	ST-506/412	44.66	5.25	FH	MFM	732	350	359	0	7	733	17			
3053	ST-506/412	44.66	5.25	FH	MFM		368	350	0	7	733	17			
3075	ST-506/412	71.3	5.25	FH	MFM		1,025	1,025	40	8	1024	17			
3085	ST-506/412	71.3	5.25	FH	MFM	1024	512		40	8	1024	17			
519	ST-506/412	159.8	5.25	FH	MFM		N/A	N/A	40	15	1224	17			
519R	ST-506/412	244.4	5.25	FH	RLL 2		N/A	N/A	40	15	1224	26			
638	ESDI-10MHZ	338.68	5.25	FH	RLL	AUTO	N/A	N/A	40	15	1225	35			
676	ESDI-15MHZ	676.82	5.25	FH	RLL	AUTO	N/A	N/A	150	15	1632	54			
7120	SCSI	1046.39	5.25	FH	RLL	AUTO	N/A	N/A	150	15	1919	71			
738	SCSI	338.68	5.25	FH	RLL	AUTO	N/A	N/A	40	15	1225	35			
776	SCSI	676.82	5.25	FH	RLL	AUTO	N/A	N/A	150	15	1632	54			

ATTO TECHNOLOGY, INC.

MODEL	INT	CAP	FF	HGT	CODE	LZ	WP	RWC	MTBF	PHDS	PCYL	PSPT	LHDS	LCYL	LSPT
SCSI EXPANDER	SCSI-2	0	3.50	HH	U				200	0	0	0			
SILICONCACHE	SCSI	512	NONE	NO	NONE				150	0	0	0			
SILICONDISK PLUS	SCSI-2 F	512		FH	U	N/A	N/A	N/A	150	0	0	0			
SILICONDISK PRO	SCSI-2 F	512	NONE	NO	NONE				0	0	0	0			

AURA ASSOCIATES

MODEL	INT	CAP	FF	HGT	CODE	LZ	WP	RWC	MTBF	PHDS	PCYL	PSPT	LHDS	LCYL	LSPT
AU 63	PCM	63.92		–	RLL 1	AUTO			100	2	0	0			
AU 126	PCM	125.85		–	RLL 1	AUTO			100	4	0	0			

BASF

MODEL	INT	CAP	FF	HGT	CODE	LZ	WP	RWC	MTBF	PHDS	PCYL	PSPT	LHDS	LCYL	LSPT
6185	ST-506/412	22.97	5.25	FH	MFM		220	220	0	6	440	17			
6186	ST-506/412	15.31	5.25	FH	MFM		220	220	0	4	440	17			
6187	ST-506/412	7.65	5.25	FH	MFM		220	220	0	2	440	17			
6188-R1	ST-506/412	10.65	5.25	FH	MFM				0	2	612	17			
6188-R3	ST-506/412	21.3	5.25	FH	MFM				0	4	612	17			

BELFORT MEMORY INTERNATIONAL

MODEL	INT	CAP	FF	HGT	CODE	LZ	WP	RWC	MTBF	PHDS	PCYL	PSPT	LHDS	LCYL	LSPT
B5108A	IDE(AT)	1086	5.25	3H	RLL 1	AUTO	N/A	N/A	300	2	0	0	16	2100	63
B5128A	IDE(AT)	1220	5.25	3H	RLL 1	N/A	N/A	N/A	300	2	0	0			
B5150A	IDE(AT)	1430	5.25	3H	RLL 1	N/A	N/A	N/A	300	2	0	0			
B5256A	IDE(AT)	2441	5.25	3H	RLL 1	N/A	N/A	N/A	300	4	0	0			
B5300A	IDE(AT)	2861	5.25	3H	RLL 1	N/A	N/A	N/A	300	4	0	0			
B5450A	IDE(AT)	4291	5.25	3H	RLL 1	N/A	N/A	N/A	300	6	0	0			

BRAND TECHNOLOGIES

MODEL	INT	CAP	FF	HGT	CODE	LZ	WP	RWC	MTBF	PHDS	PCYL	PSPT	LHDS	LCYL	LSPT
BT8085	ST-506/412	71.3	5.25	FH	MFM	AUTO	N/A	N/A	50	8	1024	17			
BT8128	ST-506/412	109.05	5.25	FH	RLL	AUTO	N/A	N/A	50	8	1024	26			
BT8170E	ESDI-10MHZ	142.6	5.25	FH	RLL 2	AUTO	N/A	N/A	50	8	1024	34			
BT8170S	SCSI	142.6	5.25	FH	RLL 2	AUTO	N/A	N/A	50	8	1024	34			
BT9121A	IDE(AT)	107.45	3.50	HH	RLL 2	AUTO	N/A	N/A	50	5	1166	36	10	583	36
BT9121E	ESDI-10MHZ	107.45	3.50	HH	RLL 2	AUTO	N/A	N/A	50	5	1166	36			
BT9121S	SCSI	107.45	3.50	HH	RLL 2	AUTO	N/A	N/A	50	5	1166	36			
BT9170A	IDE(AT)	150.44	3.50	HH	RLL 2	AUTO	N/A	N/A	50	7	1166	36	14	583	36
BT9170E	ESDI-10MHZ	150.44	3.50	HH	RLL 2	AUTO	N/A	N/A	50	7	1166	36			
BT9170S	SCSI	150.44	3.50	HH	RLL 2	AUTO	N/A	N/A	50	7	1166	36			
BT9220A	IDE(AT)	200.39	3.50	HH	RLL 2	AUTO	N/A	N/A	50	9	1208	36	16	401	61
BT9220E	ESDI-10MHZ	200.39	3.50	HH	RLL 2	AUTO	N/A	N/A	50	9	1208	36			
BT9220S	SCSI	200.39	3.50	HH	RLL 2	AUTO	N/A	N/A	50	9	1208	36			
BT9400S	IDE(AT)	400	3.50	FH	RLL 1	AUTO	N/A	N/A	100	6	1800	0	16	801	61
BT9650A	IDE(AT)	650	3.50	FH	RLL 1	AUTO	N/A	N/A	100	6	1800	0	16	1024	63

BULL HN INFORMATION SYSTEMS, INC.

MODEL	INT	CAP	FF	HGT	CODE	LZ	WP	RWC	MTBF	PHDS	PCYL	PSPT	LHDS	LCYL	LSPT
D-530	ST-506/412	25.77	5.25	FH	MFM	987	988	988	0	3	987	17			
D-550	ST-506/412	42.95	5.25	FH	MFM	987	988	988	0	5	987	17			
D-570	ST-506/412	60.13	5.25	FH	MFM	987	988	988	0	7	987	17			
D-585	ST-506/412	71.04	5.25	FH	RLL	1166	1,166	1,166	0	7	1166	17			

C.ITOH ELECTRONICS

MODEL	INT	CAP	FF	HGT	CODE	LZ	WP	RWC	MTBF	PHDS	PCYL	PSPT	LHDS	LCYL	LSPT
YD-3042	SCSI	43.57	3.50	HH	RLL	AUTO	N/A	N/A	40	4	788	0			
YD-3081B	SCSI	45.45	3.50	HH	RLL 2	AUTO	N/A	N/A	30	2	1057	0			
YD-3082	SCSI	87.14	3.50	3H	RLL	AUTO	N/A	N/A	40	8	788	0			
YD-3082B	SCSI	90.91	3.50	HH	RLL 2	AUTO	N/A	N/A	30	4	1057	0			
YD-3083B	SCSI	136.37	3.50	HH	RLL 2	AUTO	N/A	N/A	30	6	1057	0			
YD-3084B	SCSI	181.83	3.50	HH	RLL 2	AUTO	N/A	N/A	30	8	1057	0			
YD-3161B	IDE(AT)	45.45	3.50	3H	RLL 2	AUTO	N/A	N/A	40	2	1057	0			
YD-3162B	IDE(AT)	90.91	3.50	3H	RLL 2	AUTO	N/A	N/A	40	4	1057	0			

C.ITOH ELECTRONICS

MODEL	INT	CAP	FF	HGT	CODE	LZ	WP	RWC	MTBF	PHDS	PCYL	PSPT	LHDS	LCYL	LSPT
YD-3181B	SCSI	45.45	3.50	3H	RLL 2	AUTO	N/A	N/A	40	2	1057	0			
YD-3182B	SCSI	90.91	3.50	3H	RLL 2	AUTO	N/A	N/A	40	4	1057	0			
YD-3530	ST-506/412	31.81		–	MFM		732	732	0	5	731	17			
YD-3540	ST-506/412	44.53	5.25	HH	MFM		732	732	0	7	731	0			

CALLUNA TECHNOLOGY, INC.

MODEL	INT	CAP	FF	HGT	CODE	LZ	WP	RWC	MTBF	PHDS	PCYL	PSPT	LHDS	LCYL	LSPT
CALLUNACARD 130MB	PCM	130	2.50	5H	RLL 1	AUTO	N/A	N/A	150	2	0	0	10	687	37
CALLUNACARD 170MB	PCM	170	2.50	5H	RLL 1	AUTO	N/A	N/A	150	4	0	0	6	893	62
CALLUNACARD 260MB	PCM	260	2.50	5H	RLL 1	AUTO	N/A	N/A	150	4	0	0	15	691	49
CALLUNACARD 340MB	PCM	340	2.50	5H	RLL 1	AUTO	N/A	N/A	150	4	0	0	12	893	62
CALLUNACARD 510MB	PCM	510	2.50	5H	RLL 1	AUTO	N/A	N/A	150	6	0	0	16	1005	62
CT170MC	PCM	170.13	2.50	5H	RLL 1	AUTO	N/A	N/A	150	4	1467	0			
CT260MC	PCM	260.35	2.50	5H	RLL 1	AUTO	N/A	N/A	150	4	1781	0			

CARDIFF PERIPHERALS CORPORATION

MODEL	INT	CAP	FF	HGT	CODE	LZ	WP	RWC	MTBF	PHDS	PCYL	PSPT	LHDS	LCYL	LSPT
F-3053	ST-506/412	44.56	3.50	HH	MFM				0	5	1024	17			
F-3080E	ST-506/412	68.15	3.50	HH	RLL 2	AUTO	N/A	N/A	0	5	1024	26			
F-3080S	SCSI	68.15	3.50	HH	RLL 2	AUTO	N/A	N/A	0	5	1024	26			
F-3127E	ESDI-10MHZ	91.75	3.50	HH	RLL 2	AUTO	N/A	N/A	0	5	1024	35			
F-3127S	SCSI	91.75	3.50	HH	RLL 2	AUTO	N/A	N/A	0	5	1024	35			

CENTURY DATA, INC.

MODEL	INT	CAP	FF	HGT	CODE	LZ	WP	RWC	MTBF	PHDS	PCYL	PSPT	LHDS	LCYL	LSPT
CAST-10203(E)	ESDI-10MHZ	56.44	5.25	FH	RLL 2	AUTO	N/A	N/A	0	3	1050	35			
CAST-10203(S)	SCSI	56.44	5.25	FH	RLL 2	AUTO	N/A	N/A	0	3	1050	35			
CAST-10304(E)	ESDI-10MHZ	75.26	5.25	FH	RLL 2	AUTO	N/A	N/A	0	4	1050	35			
CAST-10304(S)	SCSI	75.26	5.25	FH	RLL 2	AUTO	N/A	N/A	0	4	1050	35			
CAST-10305(E)	ESDI-10MHZ	94.08	5.25	FH	RLL 2	AUTO	N/A	N/A	0	5	1050	35			
CAST-10305(S)	SCSI	94.08	5.25	FH	RLL 2	AUTO	N/A	N/A	0	5	1050	35			
CAST-14404(E)	ESDI-10MHZ	113.97	5.25	HH	RLL 2	AUTO	N/A	N/A	0	4	1590	35			
CAST-14404(S)	SCSI	113.97	5.25	HH	RLL 2	AUTO	N/A	N/A	0	4	1590	35			
CAST-14405(E)	ESDI-10MHZ	142.46	5.25	HH	RLL 2	AUTO	N/A	N/A	0	5	1590	35			
CAST-14405(S)	SCSI	142.46	5.25	HH	RLL 2	AUTO	N/A	N/A	0	5	1590	35			
CAST-14406(E)	ESDI-10MHZ	170.95	5.25	HH	RLL 2	AUTO	N/A	N/A	0	6	1590	35			
CAST-14406(S)	SCSI	170.95	5.25	HH	RLL 2	AUTO	N/A	N/A	0	6	1590	35			
CAST-24509(E)	ESDI-10MHZ	257.88	5.25	FH	RLL 2	AUTO	N/A	N/A	0	9	1599	35			
CAST-24509(S)	SCSI	257.88	5.25	FH	RLL 2	AUTO	N/A	N/A	0	9	1599	35			
CAST-24611 (E)	ESDI-10MHZ	315.19	5.25	FH	RLL 2	AUTO	N/A	N/A	0	11	1599	35			
CAST-24611(S)	SCSI	315.19	5.25	FH	RLL 2	AUTO	N/A	N/A	0	11	1599	35			
CAST-24713 (E)	ESDI-10MHZ	372.5	5.25	FH	RLL 2	AUTO	N/A	N/A	0	13	1599	35			
CAST-24713(S)	SCSI	372.5	5.25	FH	RLL 2	AUTO	N/A	N/A	0	13	1599	35			

CMS ENHANCEMENTS, INC.

MODEL	INT	CAP	FF	HGT	CODE	LZ	WP	RWC	MTBF	PHDS	PCYL	PSPT	LHDS	LCYL	LSPT
16.0 GB 4MM DAT	SCSI-2	16000	3.50	HH	DAT				180	0	0	0			
2.0 GB 4MM DAT	SCSI-2	2000	3.50	HH	DAT				60	0	0	0			
2.5 GB 8MM DAT	SCSI	2500	5.25	HH	DAT				80	0	0	0			
5.0 GB 8MM DAT	SCSI-2	5000	5.25	HH	DAT				50	0	0	0			
8.0 GB 4MM DAT	SCSI-2	8000	3.50	HH	DAT				60	0	0	0			
B040A3	IDE(AT)	40	3.50	3H	RLL 2	AUTO	N/A	N/A	50	2	1045	40	5	977	17
B040A5(12-PIN)	IDE(AT)	42.82	3.50	3H	RLL 2	AUTO	N/A	N/A	150	2	820	0	6	820	17
B040A5(18-PIN)	IDE(AT)	42.82	3.50	3H	RLL 2	AUTO	N/A	N/A	150	2	820	0	8	820	17
B040S3	SCSI	42.02	3.50	3H	RLL 2	AUTO	N/A	N/A	50	2	1026	40			
B080A3 (ver 2)	IDE(AT)	84.1	3.50	3H	RLL 1	AUTO	N/A	N/A	100	5	1053	39	8	529	39
B080A3 (ver 1)	IDE(AT)	84.1	3.50	3H	RLL 1	530	529	N/A	100	4	1053	39	8	529	39
B080A5	IDE(AT)	89.1	3.50	3H	RLL 2	1025	1025	N/A	150	3	1024	17	10	1024	17
B100A5	IDE(AT)	106.9	3.50	3H	RLL 2	1025	1025	N/A	150	12	1024	17	12	1024	17
B100S3	SCSI	121.5	3.50	3H	RLL 2	AUTO	N/A	N/A	40	4	1524	9			
B120A2	IDE(AT)	126.4	3.50	3H	RLL 2	AUTO	N/A	N/A	100	4	1349	0	8	872	35
B120A4	IDE(AT)	127.9	3.50	3H	RLL 1	AUTO	0	N/A	250	2	1536	0	16	919	17
B120A5	IDE(AT)	130.6	3.50	3H	RLL 2	AUTO	1001	N/A	150	2	0	0	15	1001	17
B150A3	IDE(AT)	170	3.50	3H	RLL 1	AUTO	332	N/A	250	2	2111	67	16	322	63
B170A4	IDE(AT)	170.81	3.50	3H	RLL 1	AUTO	N/A	N/A	250	4	1536	0	15	1011	22
B200A5	IDE(AT)	213	3.50	3H	RLL 1	1024	1024	N/A	250	4	1862	62	12	1024	34
B200S3	SCSI	212	3.50	3H	RLL 2	AUTO	N/A	N/A	256	4	2124	49			
B340A2	IDE(AT)	341.2	3.50	3H	RLL 2	1010	1010	N/A	100	2	1010	55	12	1010	55
B340A5	IDE(AT)	341	3.50	3H	RLL 1	AUTO	768	N/A	250	3	1985	62	14	768	62
B425A5	IDE(AT)	425	3.50	3H	RLL 1	AUTO	1018	N/A	250	7	2128	0	14	1018	62
B500A3	IDE(AT)	528	3.50	3H	RLL 1	1024	1024	N/A	250	6	2243	90	16	1024	63
CQLTE-120	IDE(AT)	127	2.50	4H	RLL 2	AUTO	980	N/A	150	15	980	17	15	980	17
CQLTE-200	IDE(AT)	209	3.50	4H	RLL 2	985	985	N/A	150	13	985	32	13	985	32
D080M3	ST-506/412	83.97	3.50	HH	MFM				30	9	1072	17			
D20XT-OK	ST-506/412	21.41	3.50	HH	MFM				20	4	615	17			
D30XT-OK	ST-506/412	32.74	3.50	HH	RLL				20	4	615	26			
D40XT-WS	ST-506/412	42.51	3.50	HH	MFM				45	5	977	17			
F070M3-A	ST-506/412	85.3	5.25	FH	MFM	AUTO	N/A	N/A	25	8	1024	17			
F070M3-N	ST-506/412	85.3	3.50	FH	MFM	AUTO	N/A	N/A	25	8	1024	17			
F070M3-X	ST-506/412	85.3	5.25	FH	MFM	AUTO	N/A	N/A	25	8	1024	17			
F1.0S1	SCSI	1034	5.25	FH	RLL 2	AUTO	N/A	N/A	150	15	1925	71			
F115ESDI-T	ESDI-10MHZ	65.58	5.25	FH	RLL 2	AUTO	N/A	N/A	25	4	915	35			
F150AT-CA	ESDI-10MHZ	150	5.25	FH	RLL 2	AUTO	N/A	N/A	0	4	915	17			
F150EQ-WCA	ESDI-10MHZ	150	5.25	FH	RLL 2	AUTO	N/A	N/A	0	0	0	0			
F15AT-WCA	ESDI-10MHZ	150	5.25	FH	RLL 2	AUTO	N/A	N/A	40	0	0	0			
F2.1S1	SCSI	2100	5.25	FH	RLL 2	AUTO	N/A	N/A	256	21	2267	0			
F2.1S5	SCSI-2	2129	5.25	FH	RLL 2	AUTO	N/A	N/A	250	19	2624	0			
F320AT-CA	ESDI-10MHZ	320	5.25	FH	RLL 2	AUTO	N/A	N/A	40	0	0	0			
F660S1	SCSI	676.8	5.25	FH	RLL 2	AUTO	N/A	N/A	150	15	1632	54			

CMS ENHANCEMENTS, INC.

MODEL	INT	CAP	FF	HGT	CODE	LZ	WP	RWC	MTBF	PHDS	PCYL	PSPT	LHDS	LCYL	LSPT
F70ESDI-T	ESDI-10MHZ	73.13	5.25	FH	RLL	AUTO	N/A	N/A	25	7	583	35			
H330S1	SCSI	343.3	5.25	HH	RLL 2	AUTO	N/A	N/A	150	7	1774	54			
H330E1 EXPRESS	ESDI-15MHZ	344.49	5.25	HH	RLL 2	AUTO	N/A	N/A	150	7	1780	54			
H340E1 EXPRESS	ESDI-15MHZ	344.49	5.25	HH	RLL 2	AUTO	N/A	N/A	150	7	1780	54			
H660S1	SCSI-2	668	5.25	HH	RLL 2	AUTO	N/A	N/A	150	7	2112	0			
K020M4	ST-506/412	21.41	3.50	HH	MFM	AUTO	N/A	N/A	43	4	615	17			
K030R6-A	ST-506/412	32.74	3.50	HH	RLL	AUTO	N/A	N/A	20	4	615	26			
K030R6-N	ST-506/412	32.74	3.50	HH	RLL	AUTO	N/A	N/A	20	4	615	26			
K030R6-X	ST-506/412	32.74	3.50	HH	RLL	AUTO	N/A	N/A	20	4	615	26			
K040A4-A	IDE(AT)	42.54	5.25	FH	RLL 2	AUTO	N/A	N/A	40	4	670	31	5	977	17
K040A4-N	IDE(AT)	42.54	5.25	FH	RLL 2	AUTO	N/A	N/A	40	4	670	31	5	977	17
K040A4-X	IDE(AT)	42.54	5.25	FH	RLL 2	AUTO	N/A	N/A	40	4	670	31	5	977	17
K040S3-N	SCSI	42.65	3.50	3H	RLL 2	AUTO	N/A	N/A	20	2	1047	41			
K040S4-A	SCSI	41.3	3.50	HH	RLL 2	AUTO	N/A	N/A	40	4	656	31			
K040S4-N	SCSI	41.3	3.50	HH	RLL 2	AUTO	N/A	N/A	40	4	656	31			
K040S4-X	SCSI	41.3	3.50	HH	RLL 2	AUTO	N/A	N/A	40	4	656	31			
K065M2	ST-506/412	65	3.50	HH	MFM	AUTO	N/A	N/A	70	9	1024	17			
K080M2	ST-506/412	83.9	3.50	HH	MFM	AUTO	N/A	N/A	70	9	1024	17			
K1.0A1	IDE(AT)	976	3.50	HH	RLL 1	AUTO	1891	N/A	300	9	2372	0	16	1891	63
K120A4	IDE(AT)	130	3.50	HH	RLL 2	AUTO	N/A	N/A	150	7	0	0	15	1001	17
K150S1	SCSI-2	158.8	3.50	HH	RLL 2	AUTO	N/A	N/A	70	7	1024	17			
K2.1S5	SCSI-2	2135	3.50	HH	RLL 1	AUTO	N/A	N/A	500	19	2711	84			
K200A2	IDE(AT)	200	3.50	HH	RLL 2	AUTO	N/A	N/A	30	9	1216	34			
K20M25/30-OK	ST-506/412	21.41	3.50	HH	MFM				50	4	615	17			
K20M25/30-WS	ST-506/412	21.41	3.50	HH	MFM				20	4	615	17			
K30M25/30-OK	ST-506/412	32.74	3.50	HH	RLL				50	4	615	26			
K30M25/30-WS	ST-506/412	32.11	3.50	HH	MFM				20	6	615	17			
K340A5-CM (VER 1)	IDE(AT)	340	3.50	HH	U	AUTO	N/A	N/A	150	9	1474	0			
K340A5-CM (VER 2)	IDE(AT)	340	3.50	HH	NONE	AUTO	N/A	N/A	150	9	1474	0			
K380S1	SCSI-2	380	3.50	HH	RLL	AUTO	N/A	N/A	150	14	1120	6			
K40M25/30-WS	ST-506/412	42.51	3.50	HH	MFM				45	5	977	17			
K425A5 (VER 1)	IDE(AT)	426	3.50	HH	U	AUTO	N/A	N/A	150	9	1474	0	16	1024	63
K425A5 (VER 2)	IDE(AT)	426	3.50	HH	AUTO	AUTO	N/A	N/A	150	9	1474	0	16	1024	63
K500A1	IDE(AT)	528	3.50	HH	RLL 1	AUTO	1024	N/A	300	8	1760	0	16	1024	63
K80M25/30	ST-506/412	83.97	3.50	HH	MFM				30	9	1072	17			
PDT1300 DAT	SCSI-2	2000	5.25	HH	DAT				50	0	0	0			
PDT2600 DAT	SCSI-2	5700	5.25	HH	DAT				50	0	0	0			
PI120LP	SCSI	127	3.50	3H	RLL 1	AUTO	N/A	N/A	250	3	1528	0			
PI240LP	SCSI	245	3.50	3H	RLL 1	AUTO	N/A	N/A	250	4	1818	0			
PS EXPRESS 150	ESDI-10MHZ	150	5.25	FH	RLL	AUTO	N/A	N/A	40	7	583	35			
PS EXPRESS 320	ESDI-10MHZ	320	5.25	FH	RLL	AUTO	N/A	N/A	40	15	1631	0			
SCSI 105MB REMOVABLE	SCSI	105	3.50	3H	RLL 2	AUTO	N/A	N/A	100	2	4086	0			
SCSI 120 S4	SCSI	127	3.50	3H	RLL 1	AUTO	N/A	N/A	250	3	1528	0			
SCSI 128MB REMOVABLE	SCSI	128	3.50	HH	U	AUTO	N/A	N/A	30	0	0	25	16	919	17

CMS ENHANCEMENTS, INC.

MODEL	INT	CAP	FF	HGT	CODE	LZ	WP	RWC	MTBF	PHDS	PCYL	PSPT	LHDS	LCYL	LSPT
SCSI 2.0 GB S7	SCSI-2	2103	3.50	HH	RLL 1	AUTO	N/A	N/A	500	16	3042	59			
SCSI 240 S6	SCSI	245	5.25	3H	RLL 1	AUTO	N/A	N/A	300	4	1952	48			
SCSI 256MB REMOVABLE	SCSI	256	8.00	HH	U	AUTO	N/A	N/A	30	0	0	0	16	489	63
SCSI 340 S5	SCSI-2	340	3.50	3H	RLL 1	AUTO	N/A	N/A	250	3	1691	0			
SCSI 40 S4	SCSI	42	3.50	3H	RLL 1	AUTO	N/A	N/A	250	1	1536	54			
SCSI 525 S3	SCSI	545.9	3.50	3H	RLL 1	AUTO	N/A	N/A	256	6	2243	60			
SCSI 525 S4	SCSI-2	525	3.50	3H	RLL 1	AUTO	N/A	N/A	350	6	2448	0			
SCSI 80 S4	SCSI	85	3.50	3H	RLL 1	AUTO	N/A	N/A	250	2	1528	0			
SCSI-2 3.0 GB S1	SCSI-2	3022	5.25	FH	RLL 2	AUTO	N/A	N/A	300	21	2772	0			
SPRINTER 45	SCSI	44.4	5.25	HH	RLL 2	AUTO	N/A	N/A	80	2	1275	34			
ST-3290A	IDE(AT)	261.3	3.50	3H	RLL 1	AUTO	1001	N/A	250	4	1691	34	15	1001	34

COGITO

MODEL	INT	CAP	FF	HGT	CODE	LZ	WP	RWC	MTBF	PHDS	PCYL	PSPT	LHDS	LCYL	LSPT
CG-906	ST-506/412	5.32	5.25	HH	MFM		128	128	0	2	306	17			
CG-912	ST-506/412	10.65	5.25	HH	MFM		128	128	0	4	306	17			
CG-925	ST-506/412	21.3	5.25	HH	MFM		307	307	0	4	612	17			
PT-912	ST-506/412	10.65	5.25	HH	MFM		307	307	0	2	612	17			
PT-925	ST-506/412	21.3	5.25	HH	MFM		307	307	0	4	612	17			

COMPAQ COMPUTER CORPORATION

MODEL	INT	CAP	FF	HGT	CODE	LZ	WP	RWC	MTBF	PHDS	PCYL	PSPT	LHDS	LCYL	LSPT
113640-001	IDE-CPQ	43.13	3.50	HH	RLL	AUTO	N/A	N/A	40	2	1053	40			
113641-001	IDE-CPQ	112.45	3.50	HH	RLL	AUTO	N/A	N/A	40	8	832	33			
115145-001	IDE-CPQ	84.34	3.50	HH	RLL	AUTO	N/A	N/A	40	6	832	33			
115147-001	ESDI-15MHZ	325.02	5.25	HH	RLL	AUTO	N/A	N/A	60	7	1744	52			
115158-001	ESDI-15MHZ	651.35	5.25	FH	RLL	AUTO	N/A	N/A	40	15	1631	52			
115627-001	IDE-CPQ	112.45	3.50	HH	RLL	AUTO	N/A	N/A	40	8	832	33			
115830-001	ESDI-10MHZ	318.56	5.25	FH	RLL	AUTO	N/A	N/A	40	15	1220	34			
116562-001	IDE-CPQ	123.96	3.50	HH	RLL	AUTO	N/A	N/A	40	4	1552	39			
116565-001	IDE-CPQ	207.94	3.50	HH	RLL	AUTO	N/A	N/A	40	8	1336	38			
122136-001	IDE-CPQ	60.7	3.50	HH	RLL	AUTO	N/A	N/A	40	2	1520	39			
131067-001	IDE-CPQ	510.41	3.50	HH	RLL	AUTO	N/A	N/A	150	12	1806	46			
131362-001	ESDI-15MHZ	325.02	5.25	HH	RLL	AUTO	N/A	N/A	60	7	1744	52			
CFA-170A	IDE(AT)	171.6	3.50	3H	RLL 1	AUTO	N/A	N/A	250	0	0	0	16	332	63
CP-2088 HONSHU	IDE(AT)	85	2.50	3H	RLL 1	AUTO	N/A	N/A	150	4	0	0	8	548	38
CP-30061 HOPI	IDE(AT)	60.86	3.50	3H	RLL 1	AUTO	N/A	N/A	150	2	1524	39	4	762	39
CP-30061G HOPI	IDE(AT)	60.86	3.50	3H	RLL 1	AUTO	N/A	N/A	150	2	1524	39	4	762	39
CP-30081 HOPI	IDE(AT)	84.5	3.50	3H	RLL 2	AUTO	N/A	N/A	150	4	1058	39	8	526	39
CP-30081E JAGUAR	IDE(AT)	85.06	3.50	3H	RLL 1	AUTO	N/A	N/A	150	2	1806	46	8	526	39
CP-30101	IDE(AT)	121.72	3.50	3H	RLL 1	AUTO	N/A	N/A	150	4	1524	39	8	762	39
CP-30101G HOPI	IDE(AT)	121.72	3.50	3H	RLL 1	AUTO	N/A	N/A	150	4	1524	39	8	762	39

COMPAQ COMPUTER CORPORATION

MODEL	INT	CAP	FF	HGT	CODE	LZ	WP	RWC	MTBF	PHDS	PCYL	PSPT	LHDS	LCYL	LSPT
CP-30171E	IDE(AT)	170.13	3.50	3H	RLL 1	AUTO	N/A	N/A	150	4	1806	46	8	903	46
CP-3041	IDE(AT)	42.88	3.50	3H	RLL 2	AUTO	N/A	N/A	150	2	1047	40	5	980	17
CP-3111	IDE(AT)	112.45	3.50	HH	RLL	AUTO	N/A	N/A	0	8	832	33	8	832	33
CP-3181	IDE(AT)	84.34	3.50	HH	RLL 2	AUTO	N/A	N/A	150	6	832	33	6	832	33
CP-3201I	IDE(AT)	215.33	3.50	HH	RLL 2	AUTO	N/A	N/A	150	8	1348	39	16	683	38
CP-341	IDE(AT)	42.86	3.50	HH	RLL 2	AUTO	N/A	N/A	20	4	805	26	4	805	26
CP-341I	IDE(AT)	42.86	3.50	HH	RLL 2	AUTO	N/A	N/A	20	4	805	26	4	805	26
CP-3501	IDE(AT)	510.41	3.50	HH	RLL 2	AUTO	N/A	N/A	150	12	1806	46	16	989	63
CP-4041	IDE-CPQ	42.64	3.50	4H	RLL 2	AUTO	N/A	N/A	150	2	1096	38	5	980	17

COMPUTER MEMORIES, INC.

MODEL	INT	CAP	FF	HGT	CODE	LZ	WP	RWC	MTBF	PHDS	PCYL	PSPT	LHDS	LCYL	LSPT
5018H	ST-506/412	15	5.25	FH	MFM				0	2	0	17			
514	ST-506/412	58.55		–	MFM		1,225	1,225	0	7	961	17			
7660	ST-506/412	50.13	5.25	FH	MFM		450	961	0	6	960	17			
CM 3206	ST-506/412	10.65	5.25	FH	MFM				0	4	306	17			
CM 3412	ST-506/412	10.65	5.25	FH	MFM		256	306	0	4	306	17			
CM 3426	ST-506/412	21.41	5.25	FH	MFM		256	616	0	4	615	17			
CM 5205	ST-506/412	4.45	5.25	FH	MFM		128	128	0	2	256	17			
CM 5206	ST-506/412	5.32	5.25	FH	MFM		256	307	0	2	306	32			
CM 5410	ST-506/412	8.91	5.25	FH	MFM		128	128	0	4	256	17			
CM 5412	ST-506/412	10.65	5.25	FH	MFM		128	307	0	4	306	32			
CM 5616	ST-506/412	13.36	5.25	FH	MFM		257	257	0	6	256	17			
CM 5619	ST-506/412	15.98	5.25	FH	MFM		128	307	0	6	306	32			
CM 5826	ST-506/412	21.3	5.25	FH	MFM				0	8	306	17			
CM 6213	ST-506/412	11.14	5.25	FH	MFM		256	641	0	2	640	17			
CM 6413	ST-506/412	10.7	5.25	FH	MFM		256	615	0	2	615	17			
CM 6426	ST-506/412	22.28	5.25	FH	MFM		256	641	0	4	640	17			
CM 6426S	ST-506/412	22.28	5.25	FH	MFM		256	256	0	4	640	17			
CM 6640	ST-506/412	32.11	5.25	FH	MFM		256	616	0	6	615	17			
CM 7000	ST-506/412	44.66	5.25	FH	MFM		512	733	0	7	733	17			
CM 7030	ST-506/412	25.52	5.25	FH	MFM		512	733	0	4	733	17			
CM 7038	ST-506/412	31.9	5.25	FH	MFM		512	733	0	5	733	17			
CM 7053	ST-506/412	44.66	5.25	FH	MFM		512	733	0	7	733	17			
CM 7085	ST-506/412	71.3	5.25	FH	MFM		512	1,024	0	8	1024	17			
CM 7660	ST-506/412	50.13	5.25	FH	MFM		512	961	0	6	960	17			
CM 7880	ST-506/412	66.84	5.25	FH	MFM		450	961	0	8	960	17			

CONNER PERIPHERALS, INC.

MODEL	INT	CAP	FF	HGT	CODE	LZ	WP	RWC	MTBF	PHDS	PCYL	PSPT	LHDS	LCYL	LSPT
CFA-1080A	IDE(AT)	1000	3.50	3H	RLL 1	AUTO	N/A	N/A	250	0	0	0	16	2100	63
CFA-1275A	IDE(AT)	1278	3.50	3H	U	2480	N/A	N/A	300	6	2479	0	16	2479	63
CFA-1275S	SCSI-2	1278	3.50	3H	RLL 1	AUTO	N/A	N/A	300	6	0	0			

CONNER PERIPHERALS, INC.

MODEL	INT	CAP	FF	HGT	CODE	LZ	WP	RWC	MTBF	PHDS	PCYL	PSPT	LHDS	LCYL	LSPT
CFA-170A	IDE(AT)	163.4	3.50	3H	RLL 1	AUTO	N/A	N/A		0	0	0	16	332	63
CFA-170S	SCSI	170.13	3.50	3H	RLL 1	AUTO	N/A	N/A	150	4	1806	46			
CFA-2161A	IDE(AT)	2147	3.50	3H	RLL 1	AUTO	N/A	N/A	500	8	4474	0	16	4095	63
CFA-270A	IDE(AT)	270.5	3.50	3H	RLL 1	AUTO			0	2	2805	0	16	524	63
CFA-270S	SCSI-2	270.5	3.50	3H	RLL 1	AUTO			0	2	2805	0			
CFA-340A	IDE(AT)	343	3.50	3H	RLL 1	AUTO	N/A	N/A	250	4	2111	0	16	665	63
CFA-340S	SCSI-2	343	3.50	3H	RLL 1	AUTO	N/A	N/A	250	4	0	0			
CFA-425A	IDE(AT)	426	3.50	3H	U	863	N/A	N/A	300	2	862	0	16	826	63
CFA-540A	IDE(AT)	540.86	3.50	3H	RLL 1	AUTO	N/A	N/A	250	4	2111	0	16	1048	63
CFA-540S	SCSI-2	541	3.50	3H	RLL 1	AUTO	N/A	N/A	300	4	2805	0			
CFA-810A	IDE(AT)	810	3.50	3H	RLL 1	1572	N/A	N/A	300	0	0	0	16	1572	63
CFA-850A	IDE(AT)	850	3.50	3H	RLL 1	N/A	N/A	N/A	250	4	3640	111	16	1652	63
CFA-850S	SCSI-2	852	3.50	3H	RLL 1	AUTO	N/A	N/A	300	4	0	0			
CFL-350A	IDE(AT)	350	2.50	FH	RLL 1	905	N/A	N/A	300	4	2225	0	12	905	63
CFL-420A	IDE(AT)	422	2.50	5H	RLL 1	818	N/A	N/A	300	4	2393	0	16	818	63
CFN-170A	IDE(AT)	168.2	2.50	4H	RLL 1	AUTO	N/A	N/A	250	4	1339	0	16	326	63
CFN-170S	SCSI-2	168.2	2.50	4H	RLL 1	AUTO	N/A	N/A	250	4	1339	0			
CFN-250A	IDE(AT)	252.7	2.50	4H	RLL 1	AUTO			0	6	1339	0	16	489	63
CFN-250S	SCSI-2	252.7	2.50	4H	RLL 1	AUTO			0	6	1339	0			
CFN-340A	IDE(AT)	344.5	2.50	4H	RLL 1	AUTO	N/A	N/A	250	6	1598	0	16	667	63
CFN-340S	SCSI-2	344.5	2.50	4H	RLL 1	AUTO	N/A	N/A	250	6	1598	0			
CFP-1060B	SCSI-2 D	1062.44	3.50	3H	RLL 1				500	8	2757	0			
CFP-1060E	SCSI	1062.44	3.50	3H	RLL 1				500	8	2757	0			
CFP-1060S	SCSI-2	1062.44	3.50	3H	RLL 1				500	8	2757	0			
CFP-1060W	SCSI-2 W	1062.44	3.50	3H	RLL 1				500	8	2757	0			
CFP-1080S	SCSI-2 F	1080	3.50	3H	RLL 1	AUTO	N/A	N/A	500	6	0	0			
CFP-2105S	SCSI-2 W	2147	3.50	3H	U	AUTO	N/A	N/A	999	10	3892	0			
CFP-2105W	SCSI-2	2147	3.50	3H	U	AUTO	N/A	N/A	999	10	3892	0			
CFP-2107E	SCSI-2 FW	2110	3.50	3H	RLL 1	AUTO	N/A	N/A	999	10	3924	91			
CFP-2107S	SCSI-2 F	2110	3.50	3H	RLL 1	AUTO	N/A	N/A	999	10	3924	91			
CFP-2107W	SCSI-2 FW	2110	3.50	3H	RLL 1	AUTO	N/A	N/A	999	10	3924	91			
CFP-4207E	SCSI-2 FW	4295	3.50	3H	RLL 1	AUTO	N/A	N/A	999	20	3924	96			
CFP-4207S	SCSI-2 W	4295	3.50	3H	RLL 1	AUTO	N/A	N/A	999	20	3924	96			
CFP-4207W	SCSI-2 FW	4295	3.50	3H	RLL 1		N/A	N/A	999	20	3924	96			
CFS-1081A	IDE(AT)	1080	3.50	3H	U	2097	N/A	N/A	300	4	3924	0	16	2097	63
CFS-1275A	IDE(AT)	1278	3.50	HH	RLL 1	2479	0		250	0	0	0	16	2479	63
CFS-1275A (REV. A)	IDE(AT)	1275	3.50	HH	RLL 1	AUTO	N/A	N/A	250	6	3640	0	16	2477	63
CFS-1621A	IDE(AT)	1620	3.50	3H	U	3146	N/A	N/A	300	6	3924	0	16	3146	63
CFS-210A	IDE(AT)	213.23	3.50	3H	RLL 1	AUTO	N/A	N/A	250	0	0	0	16	685	38
CFS-270A	IDE(AT)	270.9	3.50	3H	RLL 1	AUTO	N/A	N/A	250	2	0	0	16	525	63
CFS-420A	IDE(AT)	426.8	3.50	3H	RLL 1	AUTO			250	4	2388	0	16	826	63
CFS-425A	IDE(AT)	425	3.50	3H	RLL 1	AUTO	N/A	N/A	250	2	0	0	16	826	63
CFS-540A	IDE(AT)	540	3.50	3H	RLL 1	AUTO	N/A	N/A	250	4	0	0	16	1050	63

CONNER PERIPHERALS, INC.

MODEL	INT	CAP	FF	HGT	CODE	LZ	WP	RWC	MTBF	PHDS	PCYL	PSPT	LHDS	LCYL	LSPT
CFS-541A	IDE(AT)	540	3.50	3H	RLL 1	1048	N/A	N/A	300	2	3924	0	16	1048	63
CFS-635A	IDE(AT)	635	3.50	3H	RLL 1	AUTO	N/A	N/A	300	3	3640	0	16	1238	63
CFS-850A	IDE(AT)	850	3.50	3H	RLL 1	AUTO	N/A	N/A	300	4	3640	0	16	1651;	63
CP-2020 KATO	SCSI	21.39	2.50	4H	RLL 2	AUTO	N/A	N/A	150	2	653	32			
CP-2022	IDE(AT)	21.39	2.50	4H	RLL 2	AUTO	N/A	N/A	0	2	653	32	2	653	32
CP-2024	IDE(XT/AT)	21.39	2.50	4H	RLL 2	AUTO	N/A	N/A	150	2	653	32	2	653	32
CP-2034	IDE(AT)	32.02	2.50	4H	RLL 2	AUTO	N/A	N/A	150	2	823	38	2	823	38
CP-2044	IDE(44PIN)	42.6	2.50	4H	RLL 2	980	N/A	N/A	150	4	548	38	5	980	17
CP-2044PK	IDE(44PIN)	42.6	2.50	4H	RLL 2	980	N/A	N/A	150	4	548	38	5	980	17
CP-2064	IDE(AT)	64.04	2.50	4H	RLL 2	AUTO	N/A	N/A	150	4	823	38	4	823	38
CP-2084	IDE(AT)	85.37	2.50	4H	RLL 1	AUTO	N/A	N/A	150	4	1097	38	8	548	38
CP-2124	IDE(AT)	121.6	2.50	4H	RLL 1	762	0		150	4	0	53	8	762	39
CP-2304	IDE(AT)	215.33	3.50	4H	RLL	AUTO	N/A	N/A	0	8	1348	39			
CP-3000	IDE(AT)	42.8	3.50	3H	RLL 2	AUTO	N/A	N/A	150	2	1045	40	5	980	17
CP-30060	SCSI	60.86	3.50	3H	RLL 1	AUTO	N/A	N/A	150	2	1524	39			
CP-30064	IDE(AT)	60.86	3.50	3H	RLL 1	AUTO	N/A	N/A	150	2	1524	39	4	762	39
CP-30064H	IDE(AT)	60.86	3.50	3H	RLL 1	AUTO	N/A	N/A	150	2	1524	39	4	762	39
CP-30069	MCA	60.86	3.50	3H	RLL 1	AUTO	N/A	N/A	150	2	1524	39			
CP-30080	SCSI	84.5	3.50	3H	RLL 2	AUTO	N/A	N/A	150	4	1058	39			
CP-30080E	SCSI	85.06	3.50	4H	RLL 1	AUTO	N/A	N/A	150	2	1806	46			
CP-30084	IDE(AT)	84.5	3.50	3H	RLL 2	AUTO	N/A	N/A	150	4	1058	39	8	526	39
CP-30084E	IDE(AT)	85.06	3.50	3H	RLL 1	AUTO	N/A	N/A	150	2	1806	46	4	903	46
CP-30100	SCSI	121.72	3.50	3H	RLL 1	AUTO	N/A	N/A	150	4	1524	39			
CP-30104	IDE(AT)	121.56	3.50	3H	RLL 1	AUTO	N/A	N/A	150	4	1522	39	8	762	39
CP-30104H	IDE(AT)	121.56	3.50	3H	RLL 1	AUTO	N/A	N/A	150	4	1522	39	8	762	39
CP-30109	MCA	121.56	3.50	3H	RLL 1	AUTO	N/A	N/A	150	4	1522	39			
CP-30124	IDE(AT)	125.02	3.50	3H	RLL 1	AUTO	N/A	N/A	250	2	1985	62	5	895	55
CP-30170	SCSI	170.13	3.50	3H	RLL 1	AUTO	N/A	N/A	150	4	1806	46			
CP-30170E	SCSI	170	3.50	3H	RLL 1	AUTO	N/A	N/A	150	4	1806	46			
CP-30174	IDE(AT)	171.6	3.50	3H	RLL 1	AUTO	N/A	N/A	250	0	0	0	16	332	63
CP-30174E	IDE(AT)	170.13	3.50	3H	RLL 1	AUTO	N/A	N/A	150	4	1806	46	8	903	46
CP-3020	SCSI	21.01	3.50	3H	RLL 2	AUTO	N/A	N/A	150	2	622	33			
CP-30200	SCSI-2	212.64	3.50	3H	RLL 2	AUTO	N/A	N/A	150	4	2119	49			
CP-30204	IDE(AT)	212.64	3.50	3H	RLL 2	AUTO	N/A	N/A	150	4	2119	49	16	683	38
CP-3022	IDE(AT)	21.49	3.50	3H	RLL 2	AUTO	N/A	N/A	150	2	636	33	2	636	33
CP-3024	IDE(AT)	21.49	3.50	3H	RLL 2	AUTO	N/A	N/A	150	2	636	33	2	636	33
CP-30254	IDE(AT)	251	3.50	3H	RLL 1	895	N/A	N/A	250	0	0	0	10	895	55
CP-30254H	IDE(AT)	251	3.50	3H	RLL 1	895	N/A	N/A	250	0	0	0	10	895	55
CP-30340	SCSI-2	343	3.50	3H	RLL 1	AUTO	N/A	N/A	250	4	0	0			
CP-30344	IDE(AT)	343	3.50	3H	RLL 1	665	N/A	N/A	300	0	0	0	16	665	63
CP-3040	SCSI	42.02	3.50	3H	RLL 2	AUTO	N/A	N/A	150	2	1026	40			
CP-3044	IDE(AT)	42.88	3.50	3H	RLL 2	AUTO	N/A	N/A	150	2	1047	40	5	980	17
CP-3046F	IDE(AT)	42.8	3.50	3H	U	977	300	N/A	100	2	1045	40	5	977	17
CP-30540	SCSI	0	3.50	3H	RLL 1				0	0	0	0			

CONNER PERIPHERALS, INC.

MODEL	INT	CAP	FF	HGT	CODE	LZ	WP	RWC	MTBF	PHDS	PCYL	PSPT	LHDS	LCYL	LSPT
CP-30544	IDE(AT)	545.51	3.50	3H	RLL 2	AUTO	N/A	N/A	150	12	1806	49	16	1057	63
CP-3100	SCSI	104.89	3.50	3H	RLL 2	AUTO	N/A	N/A	150	8	776	33			
CP-3102	IDE(AT)	104.89	3.50	HH	RLL 2	AUTO	N/A	N/A	150	8	776	33	8	776	33
CP-3104	IDE(AT)	104.89	3.50	HH	RLL 2	AUTO	N/A	N/A	150	8	776	33	8	776	33
CP-3114	IDE(AT)	112.45	3.50	HH	RLL	AUTO	N/A	N/A	0	8	832	33	8	832	33
CP-31370	SCSI-2	1300	3.50	3H	RLL	AUTO	N/A	N/A	0	14	2386	0			
CP-3150	SCSI	52.44	3.50	HH	RLL 2	AUTO	N/A	N/A	150	4	776	33			
CP-3180	SCSI	84.34	3.50	HH	RLL 2	AUTO	N/A	N/A	150	6	832	33			
CP-3184	IDE(AT)	84.34	3.50	HH	RLL 2	AUTO	N/A	N/A	150	6	832	33	6	832	33
CP-3200	SCSI	212.61	3.50	HH	RLL	AUTO	N/A	N/A	150	8	1366	38			
CP-3200F	SCSI	212.61	3.50	HH	RLL 2	AUTO	N/A	N/A	150	8	1366	38			
CP-3204	IDE(AT)	215.33	3.50	HH	RLL 2	AUTO	N/A	N/A	150	8	1348	39	16	683	38
CP-3204F	IDE(AT)	212.61	3.50	HH	RLL 2	AUTO	N/A	N/A	50	8	1366	38	16	683	38
CP-3209F	MCA	212.61	3.50	HH	RLL 2	AUTO	N/A	N/A	50	8	1366	38			
CP-3304	IDE(AT)	340.27	3.50	HH	RLL 1	AUTO	N/A	N/A	150	8	1806	46	16	659	63
CP-3360	SCSI-2	362.47	3.50	3H	RLL 2	AUTO	N/A	N/A	150	8	1806	49			
CP-3364	IDE(AT)	362.47	3.50	3H	RLL 2	AUTO	N/A	N/A	150	8	1806	49	16	702	63
CP-340	SCSI	41.95	3.50	HH	RLL 2	AUTO	N/A	N/A	20	4	788	26			
CP-341	IDE(AT)	0	3.50	HH	U	AUTO	N/A	N/A		0	0	0			
CP-341I	IDE(AT)	0	3.50	HH	U	AUTO	N/A	N/A		0	0	0			
CP-342	IDE(AT)	42.86	3.50	HH	RLL 2	AUTO	N/A	N/A	0	4	805	26	4	805	26
CP-344	IDE(AT)	42.86	3.50	HH	RLL 2	AUTO	N/A	N/A	20	4	805	26	4	805	26
CP-346	IDE(AT)	42.86	3.50	HH	RLL 2	AUTO	N/A	N/A	20	4	805	26	4	805	26
CP-3500	SCSI-2	510.41	3.50	HH	RLL 2	AUTO	N/A	N/A	150	12	1806	46			
CP-3504	IDE(AT)	509.38	3.50	HH	RLL 2	AUTO	N/A	N/A	150	12	1806	46	16	987	63
CP-3540	SCSI-2	543.7	3.50	3H	RLL 2	AUTO	N/A	N/A	150	12	1806	49			
CP-3544 SUMMIT	IDE(AT)	544.3	3.50	HH	RLL 2	AUTO	N/A	N/A	150	12	1808	49	16	1023	63
CP-4024	IDE(XT/AT)	21.58	3.50	4H	RLL 2	AUTO	N/A	N/A	150	2	620	34	2	620	34
CP-4044	IDE(XT/AT)	42.64	3.50	4H	RLL 2	AUTO	N/A	N/A	150	2	1096	38	5	980	17

CONTROL DATA CORPORATION

MODEL	INT	CAP	FF	HGT	CODE	LZ	WP	RWC	MTBF	PHDS	PCYL	PSPT	LHDS	LCYL	LSPT
9415-5 21 WREN I	ST-506/412	18.2	5.25	FH	MFM	698	0	0	30	3	697	17			
9415-5 25 WREN I	ST-506/412	24.26	5.25	FH	MFM	698	0	0	30	4	697	17			
9415-5 28 WREN I	ST-506/412	24.26	5.25	FH	MFM	698	0	0	30	4	697	17			
9415-5 36 WREN I	ST-506/412	30.33	5.25	FH	MFM	698	0	0	30	5	697	17			
9415-5 38 WREN I	ST-506/412	31.91	5.25	FH	MFM	698	0	0	30	5	977	17			
94155-021 WREN II	ST-506/412	18.2	5.25	FH	MFM		697	697	0	3	697	17			
94155-025 WREN II	ST-506/412	24.26	5.25	FH	MFM		128	697	0	4	697	17			
94155-036 WREN II	ST-506/412	31.9	5.25	FH	MFM		128	697	0	5	733	17			
94155-038 WREN II	ST-506/412	31.9	5.25	FH	MFM		0	734	0	5	733	17			
94155-048 WREN II	ST-506/412	40.25	5.25	FH	MFM	AUTO	128	926	40	5	925	17			
94155-051 WREN II	ST-506/412	43.04	5.25	FH	MFM		128	990	0	5	989	17			
94155-057 WREN II	ST-506/412	48.3	5.25	FH	MFM	AUTO	128	926	40	6	925	17			
94155-067 WREN II	ST-506/412	56.35	5.25	FH	MFM	AUTO	128	926	40	7	925	17			
94155-077 WREN II	ST-506/412	64.4	5.25	FH	MFM	AUTO	128	926	40	8	925	17			
94155-085 WREN II	ST-506/412	71.3	5.25	FH	MFM	AUTO	128	1,025	40	8	1024	17			
94155-085P WREN II	ST-506/412	71.3	5.25	FH	MFM	AUTO	128	1,025	40	8	1024	17			
94155-086 WREN II	ST-506/412	72.46	5.25	FH	MFM	AUTO	128	926	40	9	925	17			
94155-096 WREN II	ST-506/412	80.21	5.25	FH	MFM	AUTO			40	9	1024	17			
94155-135 WREN II	ST-506/412	115.01	5.25	FH	RLL 2	AUTO	128	961	40	9	960	26			
94156-048	ESDI-10MHZ	80.51	5.25	FH	RLL 2	AUTO	N/A	N/A	40	5	925	34			
94156-48 WREN II	ESDI-10MHZ	80.51	5.25	FH	RLL 2	AUTO	N/A	N/A	40	5	925	34			
94156-67 WREN II	ESDI-10MHZ	112.71	5.25	FH	RLL 2	AUTO	N/A	N/A	40	7	925	34			
94156-86 WREN II	ESDI-10MHZ	144.92	5.25	FH	RLL 2	AUTO	N/A	N/A	40	9	925	34			
94161-086 2X7 PIN	SCSI	86.82	5.25	FH	RLL 2	AUTO	N/A	N/A	100	5	969	35			
94161-086 3X5 PIN	SCSI	86.82	5.25	FH	RLL 2	AUTO	N/A	N/A	100	5	969	35			
94161-101 2X7 PIN	SCSI	84.34	5.25	FH	RLL 2	AUTO	N/A	N/A	100	5	969	34			
94161-101 3X5 PIN	SCSI	84.34	5.25	FH	RLL 2	AUTO	N/A	N/A	100	5	969	34			
94161-103 2X7 PIN	SCSI	104.18	5.25	FH	RLL 2	AUTO	N/A	N/A	100	6	969	35			
94161-103 3X5 PIN	SCSI	104.18	5.25	FH	RLL 2	AUTO	N/A	N/A	100	6	969	35			
94161-121 2X7 PIN	SCSI	121.55	5.25	FH	RLL 2	AUTO	N/A	N/A	100	7	969	35			
94161-121 3X5 PIN	SCSI	121.55	5.25	FH	RLL 2	AUTO	N/A	N/A	100	7	969	35			
94161-138 2X7 PIN	SCSI	138.91	5.25	FH	RLL 2	AUTO	N/A	N/A	100	8	969	35			
94161-138 3X5 PIN	SCSI	138.91	5.25	FH	RLL 2	AUTO	N/A	N/A	100	8	969	35			
94161-141 2X7 PIN	SCSI	121.5	5.25	FH	RLL 2	AUTO	N/A	N/A	100	7	969	35			
94161-141 3X5 PIN	SCSI	121.55	5.25	FH	RLL 2	AUTO	N/A	N/A	100	7	969	35			
94161-151 2X7 PIN	SCSI	156.28	5.25	FH	RLL 2	AUTO	N/A	N/A	100	9	969	35			
94161-151 3X5 PIN	SCSI	156.28	5.25	FH	RLL 2	AUTO	N/A	N/A	100	9	969	35			

CONTROL DATA CORPORATION

MODEL	INT	CAP	FF	HGT	CODE	LZ	WP	RWC	MTBF	PHDS	PCYL	PSPT	LHDS	LCYL	LSPT
94161-155 2X7 PIN	SCSI	151.81	5.25	FH	RLL 2	AUTO	N/A	N/A	100	9	969	34			
94161-155 3X5 PIN	SCSI	151.81	5.25	FH	RLL 2	AUTO	N/A	N/A	100	9	969	34			
94161-182 2X7 PIN	SCSI	156.28	5.25	FH	RLL 2	AUTO	N/A	N/A	100	9	969	35			
94161-182 3X5 PIN	SCSI	156.28	5.25	FH	RLL 2	AUTO	N/A	N/A	100	9	969	35			
94166-101 WREN III	ESDI-10MHZ	89.3	5.25	FH	RLL 2	AUTO	N/A	N/A	100	5	969	35			
94166-121 WREN III	ESDI-10MHZ	107.16	5.25	FH	RLL 2	AUTO	N/A	N/A	100	7	969	36			
94166-141 WREN III	ESDI-10MHZ	125.02	5.25	FH	RLL 2	AUTO	N/A	N/A	100	7	969	36			
94166-161 COMPAQ	ESDI-10MHZ	160.74	5.25	FH	RLL 2	AUTO	N/A	N/A	100	9	969	36			
94166-161 WREN III	ESDI-10MHZ	142.88	5.25	FH	RLL 2	AUTO	N/A	N/A	100	8	969	36			
94166-182 WREN III	ESDI-10MHZ	160.74	5.25	FH	RLL 2	AUTO	N/A	N/A	100	9	969	36			
94171-300 WREN IV	SCSI	300.42	5.25	FH	RLL 2	AUTO	N/A	N/A	0	9	1412	0			
94171-307 WREN IV	SCSI	300.42	5.25	FH	RLL 2	AUTO	N/A	N/A	0	9	1412	0			
94171-327 WREN IV	SCSI	300.42	5.25	FH	RLL 2	AUTO	N/A	N/A	0	9	1412	0			
94171-344 WREN IV	SCSI	323.08	5.25	FH	RLL 2	AUTO	N/A	N/A	0	9	1549	0			
94171-350 WREN IV	SCSI	307	5.25	FH	RLL 2	AUTO	N/A	N/A	100	9	1412	0			
94171-375 WREN IV	SCSI	330	5.25	FH	RLL 2	AUTO	N/A	N/A	0	9	1549	0			
94171-376 WREN IV	SCSI	323	5.25	FH	RLL 2	AUTO	N/A	N/A	100	9	1549	0			
94171-376D WREN IV	SCSI	323	5.25	FH	RLL 2	AUTO	N/A	N/A	100	9	1549	0			
94181-385D WREN V	SCSI	337	5.25	FH	RLL 2	AUTO	N/A	N/A	100	15	791	0			
94181-385H WREN V	SCSI	337	5.25	FH	RLL 2	AUTO	N/A	N/A	100	15	791	0			
94181-574 WREN V	SCSI	574	5.25	FH	RLL 2	AUTO	N/A	N/A	100	15	1549	0			
94181-702 WREN V	SCSI	601	5.25	FH	RLL 2	AUTO	N/A	N/A	100	15	1546	0			
94181-702D WREN V	SCSI	601	5.25	FH	RLL 2	AUTO	N/A	N/A	100	15	1546	0			
94186-265 WREN V	ESDI-10MHZ	227.72	5.25	FH	RLL 2	AUTO	N/A	N/A	100	9	1412	35			
94186-324 WREN V	ESDI-10MHZ	278.33	5.25	FH	RLL 2	AUTO	N/A	N/A	0	11	1412	35			
94186-383 WREN V	ESDI-10MHZ	338.1	5.25	FH	RLL 2	AUTO	N/A	N/A	100	13	1412	36			
94186-383H WREN V	ESDI-10MHZ	338.1	5.25	FH	RLL 2	AUTO	N/A	N/A	100	15	1224	36			
94186-442 WREN V	ESDI-10MHZ	379.54	5.25	FH	RLL 2	AUTO	N/A	N/A	100	15	1412	35			
94191-766 WREN VI	SCSI	676.82	5.25	FH	RLL 2	AUTO	N/A	N/A	100	15	1632	54			
94191-766D WREN VI	SCSI	676.82	5.25	FH	RLL 2	AUTO	N/A	N/A	100	15	1632	54			
94196-383 WREN VI	ESDI-10MHZ	338.1	5.25	FH	RLL 2	AUTO	N/A	N/A	100	7	1747	54			
94196-766 WREN VI	ESDI-10MHZ	676.82	5.25	FH	RLL 2	AUTO	N/A	N/A	100	15	1632	54			
94204-051 WREN II	IDE(AT)	43.04	5.25	HH	RLL 2	AUTO	N/A	N/A	40	5	989	26			
94204-065 WREN II	IDE(AT)	63.09	5.25	HH	RLL 2	AUTO	N/A	N/A	40	5	948	26			
94204-071 WREN II	IDE(AT)	63.09	5.25	HH	RLL 2	AUTO	N/A	N/A	40	5	1032	27			
94204-074 WREN II	IDE(AT)	63.09	5.25	HH	RLL 2	AUTO	N/A	N/A	40	5	948	26	5	948	26
94204-081 WREN II	IDE(AT)	71.33	5.25	HH	RLL 2	AUTO	N/A	N/A	40	5	1032	27	10	516	27
94205-030 WREN II	ST-506/412	25.82	5.25	FH	MFM	AUTO		989	40	3	989	17			
94205-041 WREN II	ST-506/412	43.04	5.25	HH	MFM	AUTO	128	990	40	4	989	17			
94205-051 WREN II	ST-506/412	43.04	5.25	HH	MFM	AUTO	128	990	40	5	989	17			
94205-053 WREN II	ST-506/412	43.04	5.25	HH	MFM	AUTO	128	990	40	5	1024	17			
94205-071 WREN II	ST-506/412	43.04	5.25	HH	RLL	AUTO	128	990	40	5	989	26			
94205-075 WREN II	ST-506/412	61.82	5.25	HH	RLL	AUTO	128	966	40	5	966	25			

CONTROL DATA CORPORATION

MODEL	INT	CAP	FF	HGT	CODE	LZ	WP	RWC	MTBF	PHDS	PCYL	PSPT	LHDS	LCYL	LSPT
94205-077 WREN II	ST-506/412	65.82	5.25	HH	RLL 2	AUTO			40	5	989	26			
94208-051 WREN II	IDE(AT)	42.82	5.25	HH	MFM	AUTO	128	N/A	40	5	966	25			
94208-062 WREN II	IDE(SDI)	61.82	5.25	HH	RLL	AUTO	128	N/A	40	5	966	25			
94208-075 WREN II	IDE(AT)	65.82	5.25	HH	RLL 2	AUTO	N/A	N/A	0	5	989	26			
94211-091 WREN II	SCSI	91.75	5.25	HH	RLL 2	AUTO	N/A	N/A	100	5	992	35			
94211-106 WREN III	SCSI	91.75	5.25	HH	RLL 2	AUTO	N/A	N/A	100	5	1024	35			
94211-106M WREN III	SCSI	91.75	5.25	HH	RLL 2	AUTO	N/A	N/A	40	5	1024	35			
94216-106 WREN III	ESDI-10MHZ	89.12	5.25	HH	RLL 2	AUTO	N/A	N/A	100	5	1024	34			
94221-125 WREN V	SCSI	107	5.25	HH	RLL 2	AUTO	N/A	N/A	100	3	1544	0			
94221-169 WREN V	SCSI	158.99	5.25	HH	RLL 2	AUTO	N/A	N/A	100	5	1310	0			
94221-184 WREN V	SCSI	158.99	5.25	HH	RLL 2	AUTO	N/A	N/A	100	5	1310	0			
94221-209 WREN V	SCSI	179	5.25	HH	RLL 2	AUTO	N/A	N/A	100	5	1544	0			
94241-383 WREN VI	SCSI	331	5.25	HH	RLL 2	AUTO	N/A	N/A	100	7	1261	0			
94241-502 WREN VI	SCSI	435	5.25	HH	RLL 2	AUTO	N/A	N/A	100	7	1765	0			
94241-502M WREN VI	SCSI-MAC	440	5.25	HH	RLL 2	AUTO	N/A	N/A	100	7	1765	0			
94244-274 WREN VI	IDE(AT)	232.56	5.25	HH	RLL 2	AUTO	N/A	N/A	100	5	1747	52	10	873	54
94244-383 WREN VI	IDE(AT)	338.1	5.25	HH	RLL 2	AUTO	N/A	N/A	100	7	1747	54	14	873	54
94246-182 WREN VI	ESDI-15MHZ	160.69	5.25	HH	RLL 2	AUTO	N/A	N/A	100	4	1453	54			
94246-383 WREN VI	ESDI-15MHZ	338.1	5.25	HH	RLL 2	AUTO	N/A	N/A	100	7	1747	54			
94311-136 SWIFT SL	SCSI	120	3.50	3H	RLL 2	AUTO	N/A	N/A	70	5	0	0			
94311-136S SWIFT SL	SCSI-2	114.92	3.50	3H	RLL 2	AUTO	N/A	N/A	70	5	1247	36			
94314-136 SWIFT SL	IDE(AT)	120	3.50	3H	RLL 2	AUTO	N/A	N/A	70	5	0	0			
94316-111 SWIFT	ESDI-10MHZ	98	3.50	HH	RLL 2	AUTO	N/A	N/A	70	5	0	0			
94316-136 SWIFT SL	ESDI-10MHZ	120	3.50	HH	RLL 2	AUTO	N/A	N/A	70	5	0	0			
94316-155 SWIFT	ESDI-10MHZ	138.31	3.50	HH	RLL 2	AUTO	N/A	N/A	70	7	1072	36			
94316-200 SWIFT	ESDI-10MHZ	177	3.50	HH	RLL 2	AUTO	N/A	N/A	70	5	0	0			
94351-090 SWIFT	SCSI	79.28	3.50	HH	RLL 2	AUTO	N/A	N/A	150	5	1068	29			
94351-111 SWIFT	SCSI	98.42	3.50	HH	RLL 2	AUTO	N/A	N/A	70	5	1068	36			
94351-126 SWIFT	SCSI	111	3.50	HH	RLL 2	AUTO	N/A	N/A	70	7	1068	29			
94351-128 SWIFT	SCSI	111	3.50	HH	RLL 2	AUTO	N/A	N/A	70	7	1068	29			
94351-133S SWIFT	SCSI-2	116.85	3.50	HH	RLL 2	AUTO	N/A	N/A	70	5	1268	36			
94351-155 SWIFT	SCSI	137.79	3.50	HH	RLL 2	AUTO	N/A	N/A	70	7	1068	36			
94351-155S SWIFT	SCSI-2	137.79	3.50	HH	RLL 2	AUTO	N/A	N/A	70	7	1068	36			
94351-160 SWIFT	SCSI	142.71	3.50	HH	RLL 2	AUTO	N/A	N/A	150	9	1068	29			
94351-172 SWIFT	SCSI	177.16	3.50	HH	RLL 2	AUTO	N/A	N/A	70	9	1068	36			
94351-186S SWIFT	SCSI-2	163.6	3.50	HH	RLL 2	AUTO	N/A	N/A	150	7	1268	36			
94351-200 SWIFT	SCSI	177.16	3.50	HH	RLL 2	AUTO	N/A	N/A	150	9	1068	36			
94351-200S SWIFT	SCSI-2	177.16	3.50	HH	RLL 2	AUTO	N/A	N/A	150	9	1068	36			
94351-230 SWIFT	SCSI	210.34	3.50	HH	RLL 2	AUTO	N/A	N/A	70	9	1268	36			
94351-230S SWIFT	SCSI-2	210.34	3.50	HH	RLL 2	AUTO	N/A	N/A	70	9	1268	36			
94354-090 SWIFT	IDE(AT)	79.58	3.50	HH	RLL 2	AUTO	N/A	N/A	150	5	1072	29	16	335	29
94354-111 SWIFT	IDE(AT)	98.79	3.50	HH	RLL 2	AUTO	N/A	N/A	70	5	1072	36	10	536	36
94354-126 SWIFT	IDE(AT)	111.41	3.50	HH	RLL 2	AUTO	N/A	N/A	150	7	1072	29	16	469	29
94354-133 SWIFT	IDE(AT)	117.22	3.50	HH	RLL 2	AUTO	N/A	N/A	70	5	1272	36	10	636	36

CONTROL DATA CORPORATION

MODEL	INT	CAP	FF	HGT	CODE	LZ	WP	RWC	MTBF	PHDS	PCYL	PSPT	LHDS	LCYL	LSPT
94354-155 SWIFT	IDE(AT)	138.31	3.50	HH	RLL 2	AUTO	N/A	N/A	70	7	1072	36	14	596	36
94354-160 SWIFT	IDE(AT)	143.25	3.50	HH	RLL 2	AUTO	N/A	N/A	150	9	1072	29	16	603	29
94354-186 SWIFT	IDE(AT)	164.11	3.50	HH	RLL 2	AUTO	N/A	N/A	150	7	1272	36	14	636	36
94354-200 SWIFT	IDE(AT)	177.83	3.50	HH	RLL 2	AUTO	N/A	N/A	150	9	1072	36	12	804	36
94354-239 SWIFT	IDE(AT)	211	3.50	HH	RLL 2	AUTO	N/A	N/A	70	9	1272	36	14	817	36
94355-055 SWIFT II	ST-506/412	46	3.50	HH	MFM	AUTO			70	5	0	17			
94355-100 SWIFT	ST-506/412	83.97	3.50	HH	MFM	AUTO	300	1,073	150	9	1072	17			
94355-150 SWIFT	ST-506/412	128.43	3.50	HH	RLL 2	AUTO	300	1,073	150	9	1072	26			
94356-111 SWIFT	ESDI-10MHZ	98.79	3.50	HH	RLL 2	AUTO	N/A	N/A	150	5	1072	36			
94356-155 SWIFT	ESDI-10MHZ	138.31	3.50	HH	RLL 2	AUTO	N/A	N/A	70	7	1072	36			
94356-200 SWIFT	ESDI-10MHZ	177.83	3.50	HH	RLL 2	AUTO	N/A	N/A	70	9	1072	36			
94601-12D WREN VII	SCSI D	1035	5.25	FH	RLL 2	AUTO	N/A	N/A	150	15	1931	0			
94601-12G WREN VII	SCSI	1035	5.25	FH	RLL 2	AUTO	N/A	N/A	150	15	1931	0			
94601-12GM WREN VII	SCSI-MAC	1035	5.25	FH	RLL 2	AUTO	N/A	N/A	150	15	1931	0			
94601-767H WREN VII	SCSI-MAC	676	5.25	FH	RLL 2	AUTO	N/A	N/A	100	15	1356	0			
97155-036	ST-506/412	30	8.00	FH	MFM				70	0	0	17			
97501-15G ELITE	SCSI-2	1500	5.25	FH	RLL 2	AUTO	N/A	N/A	100	17	0	0			
BJ7D5A/77731600	ST-506/412	18.2	5.25	FH	MFM		128		0	3	697	17			
BJ7D5A/77731601	ST-506/412	18.2	5.25	FH	MFM		128		0	3	697	17			
BJ7D5A/77731602	ST-506/412	30.33	5.25	FH	MFM		128		0	5	697	17			
BJ7D5A/77731603	ST-506/412	30.33	5.25	FH	MFM		128		0	5	697	17			
BJ7D5A/77731604	ST-506/412	36	5.25	FH	MFM		128		0	5	697	0			
BJ7D5A/77731605	ST-506/412	30.33	5.25	FH	MFM		128		0	5	697	17			
BJ7D5A/77731606	ST-506/412	27	5.25	FH	MFM		128		0	0	0	17			
BJ7D5A/77731607	ST-506/412	18.2	5.25	FH	MFM		128		0	3	697	17			
BJ7D5A/77731608	ST-506/412	30.33	5.25	FH	MFM		128		0	5	697	17			
BJ7D5A/77731609	ST-506/412	30.33	5.25	FH	MFM		128		0	5	697	17			
BJ7D5A/77731610	ST-506/412	18.2	5.25	FH	MFM		128		0	3	697	17			
BJ7D5A/77731611	ST-506/412	30.33	5.25	FH	MFM		128		0	5	697	17			
BJ7D5A/77731612	ST-506/412	24.26	5.25	FH	MFM		128		0	4	697	17			
BJ7D5A/77731613	ST-506/412	31.9	5.25	FH	MFM		128		0	5	733	17			
BJ7D5A/77731614	ST-506/412	24.26	5.25	FH	MFM		128		0	4	697	17			
BJ7D5A/77731615	ST-506/412	24.26	5.25	FH	MFM		128		0	4	697	17			
BJ7D5A/77731616	ST-506/412	31.9	5.25	FH	MFM		128		0	5	733	17			
BJ7D5A/77731617	ST-506/412	30.33	5.25	FH	MFM		128		0	5	697	17			
BJ7D5A/77731618	ST-506/412	30.33	5.25	FH	MFM		128		0	5	697	17			
BJ7D5A/77731619	ST-506/412	30.33	5.25	FH	MFM		128		0	5	697	17			
BJ7D5A/77731620	ST-506/412	30.33	5.25	FH	MFM		128		0	5	697	17			

CORE INTERNATIONAL, INC.

MODEL	INT	CAP	FF	HGT	CODE	LZ	WP	RWC	MTBF	PHDS	PCYL	PSPT	LHDS	LCYL	LSPT
3SHC230	SCSI	230	3.50	FH	RLL	AUTO	N/A	N/A	150	5	0	0			
AT-115F	ESDI-10MHZ	160.74	5.25	FH	RLL 2	AUTO	N/A	N/A	0	9	969	54			

CORE INTERNATIONAL, INC.

MODEL	INT	CAP	FF	HGT	CODE	LZ	WP	RWC	MTBF	PHDS	PCYL	PSPT	LHDS	LCYL	LSPT
AT-145	ST-506/412	58.97	5.25	FH	MFM				0	7	968	17			
AT-150	ESDI-10MHZ	156.28	5.25	FH	RLL 2	AUTO	N/A	N/A	99	9	969	35			
AT-30	ST-506/412	31.9	5.25	FH	MFM				0	5	733	17			
AT-30R	ST-506/412	48.78	5.25	FH	RLL 2				0	5	733	26			
AT-32	ST-506/412	31.9	5.25	HH	MFM				0	5	733	17			
AT-32R	ST-506/412	48.78	5.25	HH	RLL 2				0	5	733	26			
AT-40	ST-506/412	40.21	5.25	FH	MFM				0	5	924	17			
AT-40R	ST-506/412	61.5	5.25	FH	RLL 2				0	5	924	26			
ATPLUS-20	ST-506/412	21.41	5.25	HH	MFM				50	4	615	17			
ATPLUS-43	ST-506/412	42.99	5.25	HH	MFM				50	5	988	17			
ATPLUS-43R	ST-506/412	65.76	5.25	HH	RLL 2				50	5	988	26			
ATPLUS-44	ST-506/412	44.66	3.50	HH	MFM				50	7	733	17			
ATPLUS-44R	ST-506/412	68.3	3.50	HH	RLL 2				50	7	733	26			
ATPLUS-72	ST-506/412	72.38	5.25	FH	MFM				50	9	924	17			
ATPLUS-72R	ST-506/412	110.7	5.25	FH	RLL 2				50	9	924	26			
ATPLUS-80	ST-506/412	80.21	3.50	HH	MFM				50	9	1024	17			
ATPLUS-80R	ST-506/412	122.68	3.50	HH	RLL 2				50	9	1024	26			
HC1000-FH	SCSI	1005.58	5.25	FH	RLL 2	AUTO	N/A	N/A	150	16	1918	64			
HC1000-FH	ESDI-24MHZ	1056.76	5.25	FH	RLL 2	AUTO	N/A	N/A	99	15	1787	77			
HC150-FH	ESDI-10MHZ	151.81	5.25	FH	RLL 2	AUTO	N/A	N/A	100	9	969	34			
HC150-FH	SCSI	156.28	5.25	FH	RLL 2	AUTO	N/A	N/A	150	9	969	35			
HC150-HH	ESDI-10MHZ	156.8	5.25	HH	RLL 2	AUTO	N/A	N/A	100	7	1250	35			
HC175-FH	ESDI-10MHZ	172.89	5.25	FH	RLL 2	AUTO	N/A	N/A	150	9	1072	35			
HC200	IDE(AT)	200	5.25	FH	RLL	AUTO	N/A	N/A	150	8	0	0	12	986	33
HC260-FH	ESDI-10MHZ	260.62	5.25	FH	RLL 2	AUTO	N/A	N/A	0	12	1212	35			
HC310-5FH VER 1	ESDI-10MHZ	338.41	5.25	FH	RLL 2	AUTO	N/A	N/A	150	15	1224	36			
HC310-5FH VER 2	ESDI-10MHZ	319.54	5.25	FH	RLL 2	AUTO	N/A	N/A	100	13	1412	34			
HC310-5FH VER 3	ESDI-10MHZ	319.54	5.25	FH	RLL 2	AUTO	N/A	N/A	100	13	1412	34			
HC310-FH (SCSI)	SCSI	331.9	5.25	FH	RLL 2	AUTO	N/A	N/A	150	8	1447	56			
HC310-HH VER 1	ESDI-15MHZ	325.58	5.25	HH	RLL 2	AUTO	N/A	N/A	100	7	1747	52			
HC310-HH VER 2	ESDI-15MHZ	325.58	5.25	HH	RLL 2	AUTO	N/A	N/A	100	7	1747	52			
HC315-FH	ESDI-15MHZ	337.83	5.25	FH	RLL 2	AUTO	N/A	N/A	150	8	1447	57			
HC380-FH	ESDI-10MHZ	379.54	5.25	FH	RLL 2	AUTO	N/A	N/A	150	15	1412	35			
HC40-FH	ESDI-10MHZ	40.42	5.25	FH	RLL 2	AUTO	N/A	N/A	150	4	564	35			
HC650-FH (SCSI)	SCSI	666.73	5.25	FH	RLL 2	AUTO	N/A	N/A	150	14	1661	56			
HC650-FH VER 1	ESDI-15MHZ	679.09	5.25	FH	RLL 2	AUTO	N/A	N/A	100	15	1661	53			
HC650-FH VER 2	ESDI-15MHZ	676.09	5.25	FH	RLL 2	AUTO	N/A	N/A	100	15	1661	53			
HC650-FH VER 3	ESDI-15MHZ	676.82	5.25	FH	RLL 2	AUTO	N/A	N/A	150	15	1632	54			
HC655-FH	ESDI-15MHZ	675.66	5.25	FH	RLL 2	AUTO	N/A	N/A	150	16	1447	57			
HC90	ESDI-10MHZ	86.82	5.25	HH	RLL 2	AUTO	N/A	N/A	0	5	969	35			
MC120	MCA	120.58	5.25	FH	RLL	AUTO	N/A	N/A	45	8	920	32			
MC60	MCA	60.81	5.25	FH	RLL	AUTO	N/A	N/A	45	4	928	32			
OPTIMA 30	ST-506/412	31.9	5.25	HH	MFM				0	5	733	17			
OPTIMA 30R	ST-506/412	48.78	5.25	HH	RLL 2				0	5	733	26			

CORE INTERNATIONAL, INC.

MODEL	INT	CAP	FF	HGT	CODE	LZ	WP	RWC	MTBF	PHDS	PCYL	PSPT	LHDS	LCYL	LSPT
OPTIMA 40	ST-506/412	41.9	5.25	HH	MFM				35	5	963	17			
OPTIMA 40R	ST-506/412	64.09	5.25	HH	RLL 2				35	5	963	26			
OPTIMA 70	ST-506/412	71.91	5.25	FH	MFM				35	9	918	17			
OPTIMA 70R	ST-506/412	109.98	5.25	FH	RLL 2				35	9	918	26			
OPTIMA 80	ST-506/412	80.21	3.50	HH	MFM				35	9	1024	17			
OPTIMA 80R	ST-506/412	122.68	3.50	HH	RLL 2				35	9	1024	26			

DAEYOUNG ELECTRONICS

MODEL	INT	CAP	FF	HGT	CODE	LZ	WP	RWC	MTBF	PHDS	PCYL	PSPT	LHDS	LCYL	LSPT
DX-3120A	IDE(AT)	120	3.50	3H	RLL 2	AUTO	N/A	N/A	50	3	1040	27	8	866	34

DIGITAL EQUIPMENT CORPORATION

MODEL	INT	CAP	FF	HGT	CODE	LZ	WP	RWC	MTBF	PHDS	PCYL	PSPT	LHDS	LCYL	LSPT
DP3055	SCSI-2	551	3.50	3H	RLL 1	AUTO	N/A	N/A	700	4	3115	0			
DP3055W	SCSI-2 W	551	3.50	3H	RLL 1	AUTO	N/A	N/A	700	4	3115	0			
DP3110	SCSI-2	1103	3.50	3H	RLL 1	AUTO	N/A	N/A	700	8	3115	0			
DP3110W	SCSI-2 W	1103	3.50	3H	RLL 1	AUTO	N/A	N/A	700	8	3115	0			
DSP3053L	SCSI-2 W	535	3.50	3H	RLL 1	AUTO	N/A	N/A	500	4	3117	0			
DSP3053LD	SCSI-2 D	535	3.50	3H	RLL 1	AUTO	N/A	N/A	500	4	3117	0			
DSP3053LR	SCSI-2 D80	535	3.50	3H	RLL 1	AUTO	N/A	N/A	500	4	3117	0			
DSP3053LW	SCSI-2 W	535	3.50	3H	RLL 1	AUTO	N/A	N/A	500	4	3117	0			
DSP3053LWD	SCSI-2 WD	535	3.50	3H	RLL 1	AUTO	N/A	N/A	500	4	3117	0			
DSP3085	SCSI-2	852	3.50	HH	NONE	AUTO	N/A	N/A	0	14	0	57			
DSP3105	SCSI-2	1050	3.50	HH	NONE	AUTO	N/A	N/A	250	14	0	57			
DSP3105D	SCSI-2 D	1050	3.50	HH	NONE	AUTO	N/A	N/A	250	14	0	57			
DSP3107L	SCSI-2 D	1070	3.50	3H	RLL 1	AUTO	N/A	N/A	500	8	3117	0			
DSP3107LD	SCSI-2 D	1070	3.50	3H	RLL 1	AUTO	N/A	N/A	500	8	3117	0			
DSP3107LR	SCSI-2 D80	1070	3.50	3H	RLL 1	AUTO	N/A	N/A	500	8	3117	0			
DSP3107LW	SCSI-2 W	1070	3.50	3H	RLL 1	AUTO	N/A	N/A	500	8	3117	0			
DSP3107LWD	SCSI-2 WD	1070	3.50	3H	RLL 1	AUTO	N/A	N/A	500	8	3117	0			
DSP3133L	SCSI-2	1338	3.50	3H	RLL 1	AUTO	N/A	N/A	500	10	3117	0			
DSP3133LD	SCSI-2 D	1338	3.50	3H	RLL 1	AUTO	N/A	N/A	500	10	3117	0			
DSP3133LR	SCSI-2 F	1337.5	3.50	3H	RLL 1	N/A	N/A	N/A	500	10	3117	0			
DSP3133LW	SCSI-2 WD	1338	3.50	3H	RLL 1	AUTO	N/A	N/A	500	10	3117	0			
DSP3133LWD	SCSI-2 W	1338	3.50	3H	RLL 1	AUTO	N/A	N/A	500	10	3117	0			
DSP3160	SCSI-2	1600	3.50	3H	RLL 1	AUTO	N/A	N/A	350	16	2599	0			
DSP3160W	SCSI-2 W	1600	3.50	3H	RLL 1	AUTO	N/A	N/A	350	16	2599	0			
DSP3210	SCSI-2	2148	3.50	3H	RLL 1	AUTO	N/A	N/A	500	16	3045	0			
DSP3210W	SCSI-2 W	2148	3.50	3H	RLL 1	AUTO	N/A	N/A	500	16	3045	0			
DSP3210WD	SCSI-2 WD	2148	3.50	3H	RLL 1	AUTO	N/A	N/A	500	16	3045	0			
RZ26	SCSI-2	1050	3.50			AUTO	N/A	N/A		0	2570	0			
RZ27	SCSI-2	1600	3.50			AUTO	N/A	N/A		0	2599	0			

DIGITAL EQUIPMENT CORPORATION

MODEL	INT	CAP	FF	HGT	CODE	LZ	WP	RWC	MTBF	PHDS	PCYL	PSPT	LHDS	LCYL	LSPT
RZ28	SCSI-2	2100	3.50			AUTO	N/A	N/A		0	3062	0			
RZ35	SCSI-2	852	3.50			AUTO	N/A	N/A		0	2086	0			
RZ55	SCSI	0		--	RLL	AUTO	N/A	N/A	0	0	0	0			
RZ55E	SCSI	0		--	RLL	AUTO	N/A	N/A	0	0	0	0			

DIGITAL STORAGE SOLUTIONS

MODEL	INT	CAP	FF	HGT	CODE	LZ	WP	RWC	MTBF	PHDS	PCYL	PSPT	LHDS	LCYL	LSPT
DSS-1750	SCSI	1750	3.50	3H	RLL 1	AUTO	N/A	N/A	300	0	0	0			

DISCTEC

MODEL	INT	CAP	FF	HGT	CODE	LZ	WP	RWC	MTBF	PHDS	PCYL	PSPT	LHDS	LCYL	LSPT
RHD-120	IDE(AT)	129.8	3.50	3H	RLL	AUTO	N/A	N/A	100	0	0	0			
RHD-180	IDE(AT)	183	3.50	3H	RLL	AUTO	N/A	N/A	100	0	0	0			
RHD-20	IDE(AT)	21.41	3.50	3H	RLL	AUTO	N/A	N/A	50	0	0	0			
RHD-210	IDE(AT)	209.7	3.50	3H	RLL	AUTO	N/A	N/A	150	0	0	0			
RHD-60	IDE(AT)	62.91	3.50	3H	RLL	AUTO	N/A	N/A	100	0	0	0			
RHD-80	IDE(AT)	81.33	3.50	3H	RLL	AUTO	N/A	N/A	150	0	0	0			

DISCTRON, INC.

MODEL	INT	CAP	FF	HGT	CODE	LZ	WP	RWC	MTBF	PHDS	PCYL	PSPT	LHDS	LCYL	LSPT
D-503	ST-506/412	2.66	5.25	FH	MFM				0	2	153	17			
D-504	ST-506/412	3.74	5.25	FH	MFM				0	2	215	17			
D-506	ST-506/412	5.32	5.25	FH	MFM				0	4	153	17			
D-507	ST-506/412	5.32	5.25	FH	MFM		128	128	0	2	306	17			
D-509	ST-506/412	7.48	5.25	FH	MFM		128	128	0	4	215	17			
D-512	ST-506/412	10.65	5.25	FH	MFM				0	8	153	17			
D-513	ST-506/412	11.22	5.25	FH	MFM		128	128	0	6	215	17			
D-514	ST-506/412	10.65	5.25	FH	MFM		128	128	0	4	306	17			
D-518	ST-506/412	14.97	5.25	FH	MFM		128	128	0	8	215	17			
D-519	ST-506/412	15.98	5.25	FH	MFM		128	128	0	6	306	17			
D-526	ST-506/412	21.3	5.25	FH	MFM		128	128	0	8	306	17			

DMA SYSTEMS

MODEL	INT	CAP	FF	HGT	CODE	LZ	WP	RWC	MTBF	PHDS	PCYL	PSPT	LHDS	LCYL	LSPT
306	ST-506/412	10.65	5.25	HH	MFM		400	612	0	2	612	17			

ELCOH

MODEL	INT	CAP	FF	HGT	CODE	LZ	WP	RWC	MTBF	PHDS	PCYL	PSPT	LHDS	LCYL	LSPT
DISCACHE10	ST-506/412	11.14	5.25	FH	MFM		321	321	0	4	320	17			
DISCACHE20	ST-506/412	22.28	5.25	FH	MFM		321	321	0	8	320	17			

EPSON, INC.

MODEL	INT	CAP	FF	HGT	CODE	LZ	WP	RWC	MTBF	PHDS	PCYL	PSPT	LHDS	LCYL	LSPT
HD-560	ST-506/412	21.41	5.25	HH	MFM		300	615	0	4	615	17			
HD-830	ST-506/412	10.65	5.25	HH	MFM				0	2	612	17			
HD-850	ST-506/412	10.65	5.25	HH	MFM				0	4	306	17			
HD-860	ST-506/412	21.3	5.25	HH	MFM				0	4	612	17			
HMD-710	ST-506/412	10.7	5.25	HH	MFM				20	2	612	32			
HMD-720	ST-506/412	21.41	5.25	HH	MFM				20	4	612	32			
HMD-726A	SCSI	20.78	3.50	HH	RLL 2	AUTO			20	4	615	33			
HMD-755	ST-506/412	21.41	5.25	HH	RLL 2				20	2	615	34			
HMD-765	ST-506/412	41.82	5.25	HH	RLL 2				20	4	615	34			

FUJI ELECTRIC CORPORATION

MODEL	INT	CAP	FF	HGT	CODE	LZ	WP	RWC	MTBF	PHDS	PCYL	PSPT	LHDS	LCYL	LSPT
FK301-13	ST-506/412	10.65	3.50	HH	MFM				45	4	306	17			
FK302-13	ST-506/412	10.65	3.50	HH	MFM		307	613	0	2	612	17			
FK302-26	ST-506/412	21.3	3.50	HH	MFM		307	613	0	4	612	17			
FK302-39	ST-506/412	31.96	3.50	HH	MFM		307	613	0	6	612	17			
FK303-52	ST-506/412	42.82	3.50	HH	MFM				20	8	615	17			
FK305-26	ST-506/412	21.41	3.50	HH	MFM		616		0	4	615	17			
FK305-39	ST-506/412	32.11	3.50	HH	MFM		616		20	6	615	17			
FK305-58	ST-506/412	32.11	3.50	HH	MFM		616		20	6	615	17			
FK305-58R	ST-506/412	49.12	3.50	HH	RLL 2		616		20	6	615	26			
FK308S-39R	SCSI	32.74	3.50	HH	RLL 2	AUTO	N/A	N/A	20	4	615	26			
FK308S-58R	SCSI	49.12	3.50	HH	RLL 2	AUTO	N/A	N/A	0	6	615	26			
FK309-26	ST-506/412	21.41	3.50	HH	MFM		616		20	4	615	17			
FK309-39R	ST-506/412	32.74	3.50	HH	RLL 2		616		20	4	615	26			
FK309S-50R	SCSI	41.56	3.50	HH	RLL 2	AUTO	N/A	N/A	20	4	615	33			

FUJITSU CORPORATION

MODEL	INT	CAP	FF	HGT	CODE	LZ	WP	RWC	MTBF	PHDS	PCYL	PSPT	LHDS	LCYL	LSPT
DYNADISQ 260 M2681TA	IDE(AT)	260	3.50	–	U	AUTO	N/A	N/A	250	3	0	0	11	977	48
DYNADISQ 260 M2681TA	SCSI	525	3.50	–	U	AUTO	N/A	N/A	250	6	0	0	16	1024	63
DYNADISQ 350 M2682TA	IDE(AT)	350	3.50	–	U	AUTO	N/A	N/A	250	4	0	0	11	992	63
DYNADISQ 350 M2682TA	SCSI	350	3.50	–	U	AUTO	N/A	N/A	250	4	0	0	11	992	63
DYNADISQ 525 M2684TA	IDE(AT)	525	3.50	–	U	AUTO	N/A	N/A	250	6	0	0	16	1024	63
DYNADISQ 525 M2684TA	SCSI	525	3.50	–	U	AUTO	N/A	N/A	250	6	0	0	16	1024	63
M1603S	SCSI-2	540	3.50	3H	RLL 1	AUTO	N/A	N/A	800	3	3457	94			
M1603SAU	SCSI-2	540	3.50	3H	RLL 1	AUTO	N/A	N/A		2	0	0			
M1603TA	IDE(AT)	544	3.50	3H	RLL 1	AUTO	N/A	N/A	250	3	3457	0	16	1055	63
M1606S	SCSI-2	1080	3.50	3H	RLL 1	AUTO	N/A	N/A	800	6	3457	94			
M1606SAU	SCSI-2	1080	3.50	3H	RLL 1	AUTO	N/A	N/A	800	3	0	0			
M1606TA	IDE(AT)	1089	3.50	3H	RLL 1	AUTO	N/A	N/A	250	6	3457	0	16	2111	63
M1612TAU	IDE(AT)	546.68	3.50	3H	PRML	AUTO	N/A	N/A	300	2	4133	0	16	1057	63

FUJITSU CORPORATION

MODEL	INT	CAP	FF	HGT	CODE	LZ	WP	RWC	MTBF	PHDS	PCYL	PSPT	LHDS	LCYL	LSPT
M1614TAU	IDE(AT)	1091	3.50	3H	PRML	AUTO	N/A	N/A	300	4	4133	0	16	2114	63
M1623TAU	IDE(AT)	1702	3.5	TH	PRMI	AUTO	N/A	N/A	300	0	0	0	16	3298	63
M1624TAU	IDE(AT)	2269	3.50	3H	PRML	AUTO	N/A	N/A	300	4	6288	0	16	4092	63
M1636TAU	IDE(AT)	1286	3.50	3H	PRML	AUTO	N/A	N/A	300	2	6288	0	16	2490	63
M1638TAU	IDE(AT)	2572	3,50	3H	PRML	AUTO	N/A	N/A	300	4	6288	0	16	4982	63
M2225D	ST-506/412	20.15	3.50	HH	MFM				30	4	615	32			
M2225D2	ST-506/412	20.15	3.50	HH	MFM				30	4	615	32			
M2225DR	ST-506/412	32.74	3.50	HH	RLL 2				30	4	615	26			
M2226D	ST-506/412	30.22	3.50	HH	MFM				30	6	615	32			
M2226D2	ST-506/412	30.22	3.50	HH	MFM				30	6	615	32			
M2226DR	ST-506/412	49.12	3.50	HH	RLL 2				30	6	615	26			
M2227D	ST-506/412	40.3	3.50	HH	MFM				30	8	615	32			
M2227D2	ST-506/412	40.3	3.50	HH	MFM				30	8	615	32			
M2227DR	ST-506/412	65.49	3.50	HH	RLL 2				30	8	615	26			
M2230	ST-506/412	5.24	5.25	FH	MFM		128	320	0	2	320	0			
M2230AS	ST-506/412	5.57	5.25	FH	MFM		128	320	0	2	320	17			
M2230AT	ST-506/412	5.57	5.25	HH	MFM		128	320	0	2	320	32			
M2231	ST-506/412	5.32	5.25	FH	MFM				0	2	306	17			
M2233	ST-506/412	10.48	5.25	FH	MFM		128	320	0	4	320	0			
M2233AS	ST-506/412	11.14	5.25	FH	MFM		128	320	0	4	320	17			
M2233AT	ST-506/412	20.97	5.25	HH	MFM		18	320	0	4	320	32			
M2234	ST-506/412	15.73	5.25	FH	MFM		128	320	0	6	320	0			
M2234AS	ST-506/412	16.71	5.25	FH	MFM		128	320	0	6	320	17			
M2235	ST-506/412	22.28	5.25	FH	MFM		128	320	0	8	320	0			
M2235AS	ST-506/412	22.28	5.25	FH	MFM		128	320	0	8	320	17			
M2241AS	ST-506/412	24.7	5.25	FH	MFM		375		20	4	754	16			
M2241AS2	ST-506/412	24.7	5.25	FH	MFM		375		20	4	754	32			
M2242AS	ST-506/412	43.23	5.25	FH	MFM		754	754	30	7	754	16			
M2242AS2	ST-506/412	43.23	5.25	FH	MFM				30	7	754	32			
M2243AS	ST-506/412	67.94	5.25	FH	MFM		754	754	30	11	754	16			
M2243AS2	ST-506/412	67.94	5.25	FH	MFM				0	11	754	32			
M2243R	ST-506/412	110.51	5.25	HH	RLL 2				0	7	1186	26			
M2243T	ST-506/412	72.26	5.25	HH	MFM				0	7	1186	17			
M2244E	ESDI-10MHZ	134.84	5.25	FH	RLL 2	AUTO	N/A	N/A	30	5	823	64			
M2244S	SCSI	73.74	5.25	FH	RLL 2	AUTO	N/A	N/A	35	5	823	65			
M2244SA	SCSI	73.74	5.25	FH	RLL 2	AUTO	N/A	N/A	35	5	823	35			
M2244SB	SCSI	73.74	5.25	FH	RLL 2	AUTO	N/A	N/A	35	5	823	19			
M2245E	ESDI-10MHZ	188.77	5.25	FH	RLL 2	AUTO	N/A	N/A	30	7	823	64			
M2245S	SCSI	103.23	5.25	FH	RLL 2	AUTO	N/A	N/A	35	7	823	64			
M2245SA	SCSI	103.23	5.25	FH	RLL 2	AUTO	N/A	N/A	35	7	823	35			
M2245SB	SCSI	103.23	5.25	FH	RLL 2	AUTO	N/A	N/A	35	7	823	19			
M2246E	ESDI-10MHZ	147.48	5.25	FH	RLL 2	AUTO	N/A	N/A	30	10	823	0			
M2246S	SCSI	147.48	5.25	FH	RLL 2	AUTO	N/A	N/A	35	10	823	65			
M2246SA	SCSI	147.48	5.25	FH	RLL 2	AUTO	N/A	N/A	35	10	823	35			

FUJITSU CORPORATION

MODEL	INT	CAP	FF	HGT	CODE	LZ	WP	RWC	MTBF	PHDS	PCYL	PSPT	LHDS	LCYL	LSPT
M2246SB	SCSI	147.48	5.25	FH	RLL 2	AUTO	N/A	N/A	35	10	823	19			
M2247E	ESDI-10MHZ	142.5	5.25	FH	RLL 1	AUTO	N/A	N/A	30	7	1243	64			
M2247S	SCSI	289.56	5.25	FH	RLL 1	AUTO	N/A	N/A	30	7	1243	65			
M2247SA	SCSI	155.92	5.25	FH	RLL 1	AUTO	N/A	N/A	30	7	1243	35			
M2247SB	SCSI	84.64	5.25	FH	RLL 1	AUTO	N/A	N/A	30	7	1243	19			
M2248E	ESDI-10MHZ	224	5.25	FH	RLL 1	AUTO	N/A	N/A	30	11	1243	64			
M2248S	SCSI	455.03	5.25	FH	RLL 1	AUTO	N/A	N/A	30	11	1243	65			
M2248SA	SCSI	245.02	5.25	FH	RLL 1	AUTO	N/A	N/A	30	11	1243	35			
M2248SB	SCSI	133.01	5.25	FH	RLL 1	AUTO	N/A	N/A	30	11	1243	19			
M2249E	ESDI-10MHZ	305.5	5.25	FH	RLL 1	AUTO	N/A	N/A	30	15	1243	64			
M2249S	SCSI	620.5	5.25	FH	RLL 1	AUTO	N/A	N/A	30	15	1243	65			
M2249SA	SCSI	334.11	5.25	FH	RLL 1	AUTO	N/A	N/A	30	15	1243	35			
M2249SB	SCSI	181.37	5.25	FH	RLL 1	AUTO	N/A	N/A	30	15	1243	19			
M2261E	ESDI-15MHZ	366.72	5.25	FH	RLL 1	AUTO	N/A	N/A	200	8	1658	54			
M2261H	SCSI	346.34	5.25	FH	RLL 1	AUTO	N/A	N/A	200	8	1658	94			
M2261HA	SCSI	346.34	5.25	FH	RLL 1	AUTO	N/A	N/A	200	8	1658	51			
M2261HB	SCSI	346.34	5.25	FH	RLL 1	AUTO	N/A	N/A	200	8	1658	27			
M2261S	SCSI	346.34	5.25	FH	RLL 1	AUTO	N/A	N/A	200	8	1658	94			
M2261SA	SCSI	346.34	5.25	FH	RLL 1	AUTO	N/A	N/A	200	8	1658	51			
M2261SB	SCSI	346.34	5.25	FH	RLL 1	AUTO	N/A	N/A	200	8	1658	27			
M2262E	ESDI-15MHZ	504.24	5.25	FH	RLL 1	AUTO	N/A	N/A	200	11	1658	54			
M2262H	SCSI	476.23	5.25	FH	RLL 1	AUTO	N/A	N/A	200	11	1658	94			
M2262HA	SCSI	476.23	5.25	FH	RLL 1	AUTO	N/A	N/A	200	11	1658	51			
M2262HB	SCSI	476.23	5.25	FH	RLL 1	AUTO	N/A	N/A	200	11	1658	27			
M2262S	SCSI	476.23	5.25	FH	RLL 1	AUTO	N/A	N/A	200	11	1658	94			
M2262SA	SCSI	476.23	5.25	FH	RLL 1	AUTO	N/A	N/A	200	11	1658	51			
M2262SB	SCSI	476.23	5.25	FH	RLL 1	AUTO	N/A	N/A	200	11	1658	27			
M2263E	ESDI-15MHZ	687.6	5.25	FH	RLL 1	AUTO	N/A	N/A	200	15	1658	54			
M2263H	SCSI	674.87	5.25	FH	RLL 1	AUTO	N/A	N/A	200	15	1658	94			
M2263HA	SCSI	674.87	5.25	FH	RLL 1	AUTO	N/A	N/A	200	15	1658	53			
M2263HB	SCSI	674.87	5.25	FH	RLL 1	AUTO	N/A	N/A	200	15	1658	27			
M2263S	SCSI	674.87	5.25	FH	RLL 1	AUTO	N/A	N/A	200	15	1658	53			
M2263SA	SCSI	674.87	5.25	FH	RLL 1	AUTO	N/A	N/A	200	15	1658	53			
M2263SB	SCSI	674.87	5.25	FH	RLL 1	AUTO	N/A	N/A	200	15	1658	53			
M2266E	ESDI-15MHZ	674.87	5.25	FH	RLL 1	AUTO	N/A	N/A	200	15	1658	53			
M2266H	SCSI	953.1	5.25	FH	RLL 1	AUTO	N/A	N/A	200	15	1658	0			
M2266HA	SCSI	1079.1	5.25	FH	RLL 1	AUTO	N/A	N/A	200	15	1658	0			
M2266HB	SCSI	1140.2	5.25	FH	RLL 1	AUTO	N/A	N/A	200	15	1658	0			
M2266S	SCSI	953.1	5.25	FH	RLL 1	AUTO	N/A	N/A	200	15	1658	0			
M2266SA	SCSI	1079.1	5.25	FH	RLL 1	AUTO	N/A	N/A	200	15	1658	0			
M2266SB	SCSI	1140.24	5.25	FH	RLL 1	AUTO	N/A	N/A	200	15	1658	0			
M2611ES	SCSI	46.44	3.50	HH	RLL 1	AUTO	N/A	N/A	50	2	1334	68			
M2611ESA	SCSI	46.44	3.50	HH	RLL 1	AUTO	N/A	N/A	50	2	1334	34			

FUJITSU CORPORATION

MODEL	INT	CAP	FF	HGT	CODE	LZ	WP	RWC	MTBF	PHDS	PCYL	PSPT	LHDS	LCYL	LSPT
M2611ESB	SCSI	46.44	3.50	HH	RLL 1	AUTO	N/A	N/A	50	2	1334	17			
M2611ET(JUMPER)	IDE(AT)	45.07	3.50	HH	RLL 1	AUTO	N/A	N/A	50	2	1334	33	4	667	33
M2611H	SCSI	46.44	3.50	HH	RLL 1	AUTO	N/A	N/A	50	2	1334	34			
M2611S	SCSI	46.44	3.50	HH	RLL 1	AUTO	N/A	N/A	50	2	1334	68			
M2611SA	SCSI	46.44	3.50	HH	RLL 1	AUTO	N/A	N/A	50	2	1334	34			
M2611SB	SCSI	46.44	3.50	HH	RLL 1	AUTO	N/A	N/A	50	2	1334	17			
M2611T(JUMPER)	IDE(AT)	45.07	3.50	HH	RLL 1	AUTO	N/A	N/A	50	2	1334	33	4	667	33
M2611T(SWITCH)	IDE(AT)	45.07	3.50	HH	RLL 1	AUTO	N/A	N/A	50	2	1334	33			
M2612ES	SCSI	92.88	3.50	HH	RLL 1	AUTO	N/A	N/A	50	4	1334	68			
M2612ESA	SCSI	92.88	3.50	HH	RLL 1	AUTO	N/A	N/A	50	4	1334	34			
M2612ESB	SCSI	92.88	3.50	HH	RLL 1	AUTO	N/A	N/A	50	4	1334	17			
M2612ET(JUMPER)	IDE(AT)	90.15	3.50	HH	RLL 1	AUTO	N/A	N/A	50	4	1334	33	8	667	33
M2612ET(SWITCH)	IDE(AT)	90.15	3.50	HH	RLL 1	AUTO	N/A	N/A	50	4	1334	33			
M2612S	SCSI	92.88	3.50	HH	RLL 1	AUTO	N/A	N/A	50	4	1334	68			
M2612SA	SCSI	92.88	3.50	HH	RLL 1	AUTO	N/A	N/A	50	4	1334	34			
M2612SB	SCSI	92.88	3.50	HH	RLL 1	AUTO	N/A	N/A	50	4	1334	17			
M2612T(JUMPER)	IDE(AT)	90.15	3.50	HH	RLL 1	AUTO	N/A	N/A	50	4	1334	33	8	667	33
M2612T(SWITCH)	IDE(AT)	90.15	3.50	HH	RLL 1	AUTO	N/A	N/A	50	4	1334	33			
M2612TA	IDE(AT)	545			PRML	AUTO	N/A	N/A	300	2	0	0	16	1057	63
M2613ES	SCSI	139.33	3.50	HH	RLL 1	AUTO	N/A	N/A	50	6	1334	68			
M2613ESA	SCSI	139.33	3.50	HH	RLL 1	AUTO	N/A	N/A	50	6	1334	34			
M2613ESB	SCSI	139.33	3.50	HH	RLL 1	AUTO	N/A	N/A	50	6	1334	17			
M2613ET(JUMPER)	IDE(AT)	135.23	3.50	HH	RLL 1	AUTO	N/A	N/A	50	6	1334	33	12	667	33
M2613ET(SWITCH)	IDE(AT)	135.23	3.50	HH	RLL 1	AUTO	N/A	N/A	50	6	1334	33			
M2613S	SCSI	139.33	3.50	HH	RLL 1	AUTO	N/A	N/A	50	6	1334	68			
M2613SA	SCSI	139.33	3.50	HH	RLL 1	AUTO	N/A	N/A	50	6	1334	34			
M2613SB	SCSI	139.33	3.50	HH	RLL 1	AUTO	N/A	N/A	50	6	1334	17			
M2613T(JUMPER)	IDE(AT)	135.23	3.50	HH	RLL 1	AUTO	N/A	N/A	50	6	1334	33	12	667	33
M2613T(SWITCH)	IDE(AT)	135.23	3.50	HH	RLL 1	AUTO	N/A	N/A	50	6	1334	33			
M2614ES	SCSI	185.77	3.50	HH	RLL 1	AUTO	N/A	N/A	50	8	1334	68			
M2614ESA	SCSI	185.77	3.50	HH	RLL 1	AUTO	N/A	N/A	50	8	1334	34			
M2614ESB	SCSI	185.77	3.50	HH	RLL 1	AUTO	N/A	N/A	50	8	1334	17			
M2614ET(JUMPER)	IDE(AT)	180.31	3.50	HH	RLL 1	AUTO	N/A	N/A	50	8	1334	33	16	667	33
M2614ET(SWITCH)	IDE(AT)	180.31	3.50	HH	RLL 1	AUTO	N/A	N/A	50	8	1334	33	16	667	33
M2614S	SCSI	185.77	3.50	HH	RLL 1	AUTO	N/A	N/A	50	8	1334	68			
M2614SA	SCSI	185.77	3.50	HH	RLL 1	AUTO	N/A	N/A	50	8	1334	34			
M2614SB	SCSI	185.77	3.50	HH	RLL 1	AUTO	N/A	N/A	50	8	1334	17			
M2614T(JUMPER)	IDE(AT)	180.31	3.50	HH	RLL 1	AUTO	N/A	N/A	50	8	1334	33	16	667	33
M2614T(SWITCH)	IDE(AT)	180.31	3.50	HH	RLL 1	AUTO	N/A	N/A	50	8	1334	33			
M2614TA	IDE(AT)	1091	3.5	TH	PRML	AUTO	N/A	N/A		4	0	0	16	2114	63
M2616ET(JUMPER)	IDE(AT)	104.21	3.50	HH	RLL 1	AUTO	N/A	N/A	50	4	1542	33	8	771	33
M2616ET(SWITCH)	IDE(AT)	104.21	3.50	HH	RLL 1	AUTO	N/A	N/A	50	4	1542	33	8	771	33
M2616SA	SCSI	104.21	3.50	HH	RLL 1	AUTO	N/A	N/A	50	4	1542	33			
M2616T(JUMPER)	IDE(AT)	104.21	3.50	HH	RLL 1	AUTO	N/A	N/A	50	4	1542	33	8	771	33

FUJITSU CORPORATION

MODEL	INT	CAP	FF	HGT	CODE	LZ	WP	RWC	MTBF	PHDS	PCYL	PSPT	LHDS	LCYL	LSPT
M2616T(SWITCH)	IDE(AT)	104.21	3.50	HH	RLL 1	AUTO	N/A	N/A	50	4	1542	33	8	771	33
M2621S	SCSI-2	234.12	3.50	HH	RLL 1	AUTO	N/A	N/A	200	5	1429	64			
M2622F	SCSI-2	330.17	3.50	HH	RLL 1	AUTO	N/A	N/A	200	7	1429	0			
M2622FA	SCSI-2	330.17	3.50	HH	RLL 1	AUTO	N/A	N/A	200	7	1429	0			
M2622FB	SCSI-2	330.17	3.50	HH	RLL 1	AUTO	N/A	N/A	200	7	1429	0			
M2622HA	SCSI	330	3.50	–	U					7	1429	0			
M2622S	SCSI-2	327.77	3.50	HH	RLL 1	AUTO	N/A	N/A	200	7	1429	64			
M2622SA	SCSI-2	327.77	3.50	HH	RLL 1	AUTO	N/A	N/A	200	7	1429	64			
M2622T	IDE(AT)	327.77	3.50	HH	RLL 1	AUTO	N/A	N/A	200	7	1429	64	10	1013	63
M2623F	SCSI-2	425.13	3.50	HH	RLL 1	AUTO	N/A	N/A	200	9	1429	0			
M2623FA	SCSI-2	425.13	3.50	HH	RLL 1	AUTO	N/A	N/A	200	9	1429	0			
M2623FB	SCSI-2	425.13	3.50	HH	RLL 1	AUTO	N/A	N/A	200	9	1429	0			
M2623HA	SCSI	425	3.50	–	U					9	1429	0			
M2623S	SCSI-2	421.42	3.50	HH	RLL 1	AUTO	N/A	N/A	200	9	1429	64			
M2623SA	SCSI-2	421.42	3.50	HH	RLL 1	AUTO	N/A	N/A	200	9	1429	64			
M2623T	IDE(AT)	421.42	3.50	HH	RLL 1	AUTO	N/A	N/A	200	9	1429	63	13	1002	63
M2624F	SCSI-2	520.1	3.50	HH	RLL 1	AUTO	N/A	N/A	200	11	1429	0			
M2624FA	SCSI-2	520.1	3.50	HH	RLL 1	AUTO	N/A	N/A	200	11	1429	0			
M2624FB	SCSI-2	520.1	3.50	HH	RLL 1	AUTO	N/A	N/A	200	11	1429	0			
M2624HA	SCSI	520.1	3.50	–	U					7	1429	0			
M2624S	SCSI-2	515.08	3.50	HH	RLL 1	AUTO	N/A	N/A	200	11	1429	64			
M2624SA	SCSI-2	515.08	3.50	HH	RLL 1	AUTO	N/A	N/A	200	11	1429	64			
M2624T	IDE(AT)	515.08	3.50	HH	RLL 1	AUTO	N/A	N/A	200	11	1429	64	16	995	63
M2635S	SCSI-2	160	2.50	4H	RLL	AUTO	N/A	N/A	150	4	1569	0			
M2635T	IDE(AT)	160	2.50	4H	RLL	AUTO	N/A	N/A	150	4	1569	0	8	620	63
M2637S	SCSI-2	240	2.50	4H	RLL	AUTO	N/A	N/A	150	6	1569	0			
M2637SA	SCSI-2	240.1	3.50	3H	RLL 1	AUTO	N/A	N/A	0	0	0	0			
M2637T	IDE(AT)	240	2.50	4H	RLL	AUTO	N/A	N/A	150	6	1569	0	8	930	63
M2651HA	SCSI-2	1313	5.25	FH	RLL 1	AUTO	N/A	N/A	200	15	1944	88			
M2651SA	SCSI-2	1313	5.25	FH	RLL 1	AUTO	N/A	N/A	200	15	1944	88			
M2652HA	SCSI-2	1628.28	5.25	FH	RLL 1	AUTO	N/A	N/A	200	20	1893	84			
M2652SA	SCSI-2	1628.28	5.25	FH	RLL 1	AUTO	N/A	N/A	200	20	1893	84			
M2653HA	SCSI-2 D	1404.39	5.25	FH	RLL	AUTO	N/A	N/A	300	15	2078	88			
M2653SA	SCSI-2	1404.39	5.25	FH	RLL	AUTO	N/A	N/A	300	15	2078	88			
M2654HA	SCSI-2 D	2061.71	5.25	FH	RLL	AUTO	N/A	N/A	300	21	2179	88			
M2654SA	SCSI-2 D	2061.71	5.25	FH	RLL	AUTO	N/A	N/A	300	21	2179	88			
M2681S	SCSI	264.24	3.50	HH	RLL	AUTO	N/A	N/A	300	3	2378	0			
M2681T	IDE(AT)	264	3.50	HH	RLL	AUTO	N/A	N/A	300	0	0	0	11	977	48
M2681TA	IDE(AT)	264	3.50			AUTO	N/A	N/A		0	0	0	11	977	48
M2682S	SCSI	353.54	3.50	HH	RLL	AUTO	N/A	N/A	300	4	2378	0			
M2682T	IDE(AT)	352	3.50	HH	RLL	AUTO	N/A	N/A	300	0	0	0	11	992	63
M2682TA	IDE(AT)	0	3.50			AUTO	N/A	N/A		0	0	0	16	1024	63
M2684S	SCSI	532.14	3.50	HH	RLL	AUTO	N/A	N/A	300	6	2378	0			

FUJITSU CORPORATION

MODEL	INT	CAP	FF	HGT	CODE	LZ	WP	RWC	MTBF	PHDS	PCYL	PSPT	LHDS	LCYL	LSPT
M2684T	IDE(AT)	528	3.50	HH	RLL	AUTO	N/A	N/A	300	0	0	0	16	1024	VAR
M2691EH	SCSI	756	3.50	HH	RLL 1	AUTO	N/A	N/A		0	0	0			
M2691EHA	SCSI-2 D	648	3.50	HH	RLL 1	AUTO	N/A	N/A	300	9	1818	0			
M2691EQ	SCSI	756.44	3.50	HH	RLL 1	AUTO		N/A	0	9	1819	0			
M2691ER	SCSI D	756.44	3.50	HH	RLL 1	AUTO	N/A	N/A	0	9	1819	0			
M2691ES	SCSI	756	3.50	HH	RLL 1	AUTO	N/A	N/A		0	0	0			
M2691ESA	SCSI-2	648	3.50	HH	RLL 1	AUTO	N/A	N/A	300	9	1818	0			
M2692EH	SCSI	924	3.50	HH	RLL 1	AUTO	N/A	N/A		0	0	0			
M2692EHA	SCSI-2 D	793	3.50	HH	RLL 1	AUTO	N/A	N/A	300	11	1818	0			
M2692EQ	SCSI	924.54	3.50	HH	RLL 1	AUTO	N/A	N/A	0	11	1819	0			
M2692ER	SCSI D	924.54	3.50	HH	RLL 1	AUTO	N/A	N/A	0	11	1819	0			
M2692ES	SCSI	924	3.50	HH	RLL 1	AUTO	N/A	N/A		0	0	0			
M2692ESA	SCSI-2	793	3.50	HH	RLL 1	AUTO	N/A	N/A	300	11	1818	0			
M2693EH	SCSI	1092	3.50	HH	RLL 1	AUTO	N/A	N/A		0	0	0			
M2693EHA	SCSI-2 D	938	3.50	HH	RLL 1	AUTO	N/A	N/A	300	13	1818	0			
M2693EQ	SCSI	1092.64	3.50	HH	RLL 1	AUTO	N/A	N/A	0	13	1819	0			
M2693ER	SCSI D	1092.64	3.50	HH	RLL 1	AUTO	N/A	N/A	0	13	1819	0			
M2693ES	SCSI	1092	3.50	HH	RLL 1	AUTO	N/A	N/A		0	0	0			
M2693ESA	SCSI-2	938	3.50	HH	RLL 1	AUTO	N/A	N/A	300	13	1818	0			
M2694EH	SCSI	1260	3.50	HH	RLL 1					0	0	0			
M2694EHA	SCSI-2 D	1083	3.50	HH	RLL 1	AUTO	N/A	N/A	300	15	1818	0			
M2694EQ	SCSI	1260.74	3.50	HH	RLL 1	AUTO	N/A	N/A	0	15	1819	0			
M2694ER	SCSI D	1260.74	3.50	HH	RLL 1	AUTO	N/A	N/A	0	15	1819	0			
M2694ES	SCSI-2	1083.9	3.50	HH	RLL 1	AUTO	N/A	N/A	300	15	1819	0			
M2694ESA	SCSI-2	1083	3.50	HH	RLL 1	AUTO	N/A	N/A	300	15	1818	0			
M2694HA	SCSI	1083	3.50	–	U					15	1819	0			
M2694SA	SCSI	1083	3.50	–	U					15	1819	0			
M2706SA	SCSI-2	532	2.50	5H	PRML	AUTO	N/A	AUTO	300	6	2306	0			
M2915QA	SCSI-2 W	2176	3.50	HH	PRML	AUTO	N/A	N/A	500	15	3012	0			
M2915SA	SCSI-2	2176	3.50	HH	RLL	AUTO	N/A	N/A	500	15	3184	89		3012	
M2932QA	SCSI-2 W	2177	3.50	HH	PRML	AUTO	N/A	AUTO	800	9	3422	0			
M2932RA	SCSI-2 WD	2177	3.50	HH	PRML	AUTO	N/A	AUTO	800	9	3422	0			
M2932SA	SCSI-2	2177	3.50	HH	PRML	AUTO	N/A	AUTO	800	9	3422	0			
M2934QA	SCSI-2 WD	4355	3.50	HH	PRML	AUTO	N/A	AUTO	800	18	3422	0			
M2934RA	SCSI-2	4355	3.50	HH	PRML	AUTO	N/A	AUTO	800	18	3422	0			
M2934SA	SCSI-2	4355	3.50	HH	PRML	AUTO	N/A	AUTO	800	18	3422	0			
M2949SYU	SCSI-2	9100	3.50	HH	RLL	AUTO	N/A	N/A		0	0	0			
M2952Q	U-SCSI W	4350	3.50	3H	PRML	AUTO	N/A	N/A	1000	0	0	0			
M2952R	U-SCSI W	4350	3.50	3H	PRML	AUTO	N/A	N/A	1000	0	0	0			
M2952S	SCSI-2	2200	3.50	3H	PRML	AUTO	N/A	N/A		0	0	0			
M2952SYU	SCSI-2	2200	3.50	3H	PRML	AUTO	N/A	N/A		0	0	0			
M2954Q	U-SCSI W	2200	3.50	3H	PRML	AUTO	N/A	N/A	1000	0	0	0			
M2954R	U-SCSI W	2200	3.50	3H	PRML	AUTO	N/A	N/A	1000	0	0	0			
M2954S	SCSI-2	4400	3.50	3H	PRML	AUTO	N/A	N/A		0	0	0			

FUJITSU CORPORATION

MODEL	INT	CAP	FF	HGT	CODE	LZ	WP	RWC	MTBF	PHDS	PCYL	PSPT	LHDS	LCYL	LSPT
M2954SYU	SCSI-2	4350	3.50	3H	PRML	AUTO	N/A	N/A		0	0	0			
MPA3017AT	IDE(AT)	1750	3.50	3H	U	AUTO	N/A	N/A	250	6	3	87	16	3390	63
MPA3026AT	IDE(AT)	1750	3.50	3H	U	AUTO	N/A	N/A	250	3	8713	132	16	5086	63
MPA3035AT	IDE(AT)	3500	3.50	3H	U	AUTO	N/A	N/A	250	3	8713	132	16	6780	63
MPA3043AT	IDE(AT)	4370	3.50	3H	U	AUTO	N/A	N/A	250	3	8713	132	16	9042	63
MPA3052AT	IDE(AT)	5250	3.50	3H	U	AUTO	N/A	N/A	250	3	8713	132	16	10850	63

HEWLETT-PACKARD COMPANY

MODEL	INT	CAP	FF	HGT	CODE	LZ	WP	RWC	MTBF	PHDS	PCYL	PSPT	LHDS	LCYL	LSPT
D1296A	ST-506/412	21.41	5.25	HH	MFM	670	300	N/A	100	4	615	17			
D1297A	ST-506/412	42.82	5.25	HH	MFM	AUTO	N/A	N/A	100	6	820	17			
D1660A	ESDI-15MHZ	330	5.25	FH	RLL 2	AUTO	N/A	N/A	150	8	1457	57			
D1661A	ESDI-15MHZ	670	5.25	FH	RLL 2	AUTO	N/A	N/A	150	16	1457	57			
D1674A	ESDI-15MHZ	103	5.25	FH	RLL 2				0	0	0	0			
D1675A	ESDI-15MHZ	155	5.25	FH	RLL 2				0	0	0	0			
D1676A	ESDI-15MHZ	310	5.25	FH	RLL 2				0	0	0	0			
D1697A	IDE(AT)	240	3.50	3H	RLL 1	722	-1	N/A	250	4	1800	0	8	930	63
HP 97501A		10	3.50	HH	MFM				0	2	698	28	8	142	17
HP 97501B		20.02	3.50	HH	MFM				0	2	1400	28	8	288	17
HP 97530S	SCSI	204	5.25	FH	RLL 2	AUTO	N/A	N/A	0	6	0	0			
HP 97532D	SCSI D	107.67	5.25	FH	RLL 2	AUTO	N/A	N/A	99	4	1643	64			
HP 97532E	ESDI-10MHZ	107.67	5.25	FH	RLL 2	AUTO	N/A	N/A	99	4	1643	64			
HP 97532S	SCSI	107.67	5.25	FH	RLL 2	AUTO	N/A	N/A	99	4	1643	64			
HP 97532T	SCSI-2	107.67	5.25	FH	RLL 2	AUTO	N/A	N/A	99	4	1643	64			
HP 97533D	SCSI D	161.51	5.25	FH	RLL 2	AUTO	N/A	N/A	99	6	1643	64			
HP 97533E	ESDI-10MHZ	161.51	5.25	FH	RLL 2	AUTO	N/A	N/A	99	6	1643	64			
HP 97533S	SCSI	161.51	5.25	FH	RLL 2	AUTO	N/A	N/A	99	6	1643	64			
HP 97533T	SCSI-2	161.51	5.25	FH	RLL 2	AUTO	N/A	N/A	99	6	1643	64			
HP 97536D	SCSI D	323.02	5.25	FH	RLL 2	AUTO	N/A	N/A	99	12	1643	64			
HP 97536E	ESDI-10MHZ	323.02	5.25	FH	RLL 2	AUTO	N/A	N/A	99	12	1643	64			
HP 97536S	SCSI	323.02	5.25	FH	RLL 2	AUTO	N/A	N/A	99	12	1643	64			
HP 97536T	SCSI-2	323.02	5.25	FH	RLL 2	AUTO	N/A	N/A	99	12	1643	64			
HP 97544D	SCSI D	331.9	5.25	HH	RLL 2	AUTO	N/A	N/A	150	8	1447	56			
HP 97544E	ESDI-15MHZ	337.83	5.25	FH	RLL 2	AUTO	N/A	N/A	150	8	1447	57			
HP 97544P	SCSI-2	331.9	5.25	FH	RLL 2	AUTO	N/A	N/A	150	8	1447	56			
HP 97544S	SCSI	331.9	5.25	HH	RLL 2	AUTO	N/A	N/A	150	8	1447	56			
HP 97544T	SCSI-2	331.9	5.25	FH	RLL 2	AUTO	N/A	N/A	150	8	1447	56			
HP 97548D	SCSI D	663.81	5.25	HH	RLL 2	AUTO	N/A	N/A	150	16	1447	56			
HP 97548E	ESDI-15MHZ	675.66	5.25	FH	RLL 2	AUTO	N/A	N/A	150	16	1447	57			
HP 97548P	SCSI-2	663.81	5.25	FH	RLL 2	AUTO	N/A	N/A	150	16	1447	56			
HP 97548S	SCSI	663.81	5.25	HH	RLL 2	AUTO	N/A	N/A	150	16	1447	56			
HP 97548T	SCSI-2	663.81	5.25	FH	RLL 2	AUTO	N/A	N/A	150	16	1447	56			
HP 97549P	SCSI-2	1001.91	5.25	FH	RLL 2	AUTO	N/A	N/A	150	16	1911	64			

HEWLETT-PACKARD COMPANY

MODEL	INT	CAP	FF	HGT	CODE	LZ	WP	RWC	MTBF	PHDS	PCYL	PSPT	LHDS	LCYL	LSPT
HP 97549T	SCSI-2	1001.91	5.25	FH	RLL 2	AUTO	N/A	N/A	150	16	1911	64			
HP 97556E	ESDI-24MHZ	688.14	5.25	FH	RLL 2	AUTO	N/A	N/A	150	11	1697	72			
HP 97556P	SCSI-2	677.19	5.25	FH	RLL 2	AUTO	N/A	N/A	150	11	0	72			
HP 97556S	SCSI-2	677.19	5.25	FH	RLL 2	AUTO	N/A	N/A	150	11	1670	72			
HP 97556T	SCSI-2	677.19	5.25	FH	RLL 2	AUTO	N/A	N/A	150	11	1670	72			
HP 97558E	ESDI-24MHZ	1084.9	5.25	FH	RLL 2	AUTO	N/A	N/A	150	15	1962	72			
HP 97558P	SCSI-2	1069.97	5.25	FH	RLL 2	AUTO	N/A	N/A	150	15	1935	72			
HP 97558S	SCSI-2	1069.97	5.25	FH	RLL 2	AUTO	N/A	N/A	150	15	1935	72			
HP 97558T	SCSI-2	1069.97	5.25	FH	RLL 2	AUTO	N/A	N/A	150	15	1935	72			
HP 97560E	ESDI-24MHZ	1374.21	5.25	FH	RLL 2	AUTO	N/A	N/A	150	19	1962	72			
HP 97560P	SCSI-2	1355.3	5.25	FH	RLL 2	AUTO	N/A	N/A	150	19	1935	72			
HP 97560S	SCSI-2	1355.3	5.25	FH	RLL 2	AUTO	N/A	N/A	150	19	1935	72			
HP 97560T	SCSI-2	1355.3	5.25	FH	RLL 2	AUTO	N/A	N/A	150	19	1935	72			
HP C2233	SCSI-2	234.67	3.50	HH	RLL 1	AUTO	N/A	N/A	150	5	1546	0			
HP C2233 OPT. 060	IDE(AT)	238.68	3.50	HH	RLL 1	AUTO	N/A	N/A	150	5	1546	0	16	462	63
HP C2234	SCSI-2	328.53	3.50	HH	RLL 1	AUTO	N/A	N/A	150	7	1546	0			
HP C2234 OPT. 060	IDE(AT)	334.16	3.50	HH	RLL 1	AUTO	N/A	N/A	150	7	1546	0	16	647	63
HP C2235	SCSI-2	422.4	3.50	HH	RLL 1	AUTO	N/A	N/A	150	9	1546	0			
HP C2235 OPT. 060	IDE(AT)	429.64	3.50	HH	RLL 1	AUTO	N/A	N/A	150	9	1546	0	16	832	63
HP C2244	SCSI-2	566	3.50	HH	RLL 1	AUTO	N/A	N/A	300	7	2051	79			
HP C2245	SCSI-2	728	3.50	HH	RLL 1	AUTO	N/A	N/A	300	9	2051	79			
HP C2246	SCSI-2	890	3.50	HH	RLL 1	AUTO	N/A	N/A	300	11	2051	79			
HP C2247	SCSI-2	1052.09	3.50	HH	RLL 1	AUTO	N/A	N/A	150	13	1981	0			
HP C2490 D	SCSI-2 D	2170	3.50	HH	RLL 1	AUTO	N/A	N/A	500	9	3610	0			
HP C2490 SE	SCSI-2	0	3.50	HH	RLL 1				500	0	0	0			
HP C2490A	SCSI-2	2100	3.50	HH	RLL 1	AUTO	N/.A	N/A	500	18	2630	90			
HP C2490A 001	SCSI-2	2100	3.50	HH	RLL 1	AUTO	N/.A	N/A	500	18	2630	90			
HP C2490A 002	SCSI-2 D	2170	3.50	3H	RLL 1				500	9	3610	0			
HP C3007 (12 PINSET)	SCSI-2	41.92	5.25	FH		AUTO	N/A	N/A	0	4	622	33			
HP C3007 (9 PINSET)	SCSI-2	41.92	5.25	FH		AUTO	N/A	N/A	0	4	622	33			
HP C3009 (12 PINSET)	SCSI-2	41.92	5.25	FH		AUTO	N/A	N/A	0	4	622	33			
HP C3009 (9 PINSET)	SCSI-2	41.92	5.25	FH		AUTO	N/A	N/A	0	4	622	33			
HP C3010 (12 PINSET)	SCSI-2	41.92	5.25	FH		AUTO	N/A	N/A	0	4	622	33			
HP C3010 (9 PINSET)	SCSI-2	41.92	5.25	FH		AUTO	N/A	N/A	0	4	622	33			
HP C3013A OPT. 001	IDE(AT)	21.42	1.30	4H	RLL	AUTO	N/A	N/A	300	3	700	0	4	615	17
HP C3013A OPT. 003	PCM	21.42	1.30	4H	RLL	AUTO	N/A	N/A	300	3	700	0			
HP C3014A OPT. 001	IDE(AT)	42.83	1.30	4H	RLL	AUTO	N/A	N/A	300	4	786	0	6	820	17
HP C3014A OPT. 003	PCM	42.83	1.30	4H	RLL	AUTO	N/A	N/A	300	4	786	0			
HP C3323 D	SCSI-2 D	1050	3.50	3H	RLL 1	AUTO	N/A	N/A	500	7	2982	100			
HP C3323 SE	SCSI-2	1050	3.50	3H	RLL 1	AUTO	N/A	N/A	500	7	2982	100			
HP C3323A	SCSI-2	1050	3.50	3H	RLL 1	AUTO	N/A	N/A	500	7	2982	100			
HP C3323A 001	SCSI-2	1050	3.50	3H	RLL 1	AUTO	N/A	N/A	500	7	2982	100			
HP C3323A 002	SCSI-2 D	1050	3.50	3H	RLL 1	AUTO	N/A	N/A	500	7	2982	100			
HP C3324A 001	SCSI-2	1050	3.50	3H	RLL 1	AUTO	N/A	N/A	500	7	2982	100			

HEWLETT-PACKARD COMPANY

MODEL	INT	CAP	FF	HGT	CODE	LZ	WP	RWC	MTBF	PHDS	PCYL	PSPT	LHDS	LCYL	LSPT
HP C3324A 002	SCSI-2	1050	3.50	3H	RLL 1	AUTO	N/A	N/A	800	5	3610	100			
HP C3324A 012	SCSI-2 WD	170.8	3.50	3H	PRML	AUTO	N/A	N/A	800	0	0	0			
HP C3324A 041	SCSI SCA	1200	3.50	3H	PRML	AUTO	N/A	N/A	800	5	3610	0			
HP C3324A 043	SCSI SCA	1200	3.50	3H	PRML	AUTO	N/A	N/A	800	5	3610	0			
HP C3324A 044	SCSI SCA	2170	3.50	3H	PRML	AUTO	N/A	N/A	800	9	3610	0			
HP C3325A 001	SCSI-2	1050	3.50	3H	RLL 1	AUTO	N/A	N/A	500	7	2982	100			
HP C3325A 002	SCSI-2 D	1050	3.50	3H	RLL 2	AUTO	N/A	N/A	800	9	3610	100			
HP C3325A 012	SCSI-2 WD	170.8	3.50	3H	PRML	AUTO	N/A	N/A	800	0	0	0			
HP C3325A 041	SCSI SCA	1200	3.50	3H	PRML	AUTO	N/A	N/A	800	5	3610	0			
HP C3325A 043	SCSI SCA	1200	3.50	3H	PRML	AUTO	N/A	N/A	800	5	3610	0			
HP C3325A 044	SCSI SCA	2170	3.50	3H	PRML	AUTO	N/A	N/A	800	9	3610	0			
HP C3724D	SCSI-2 D	1200	3.50	3H	RLL 2	AUTO	N/A	N/A	800	5	3610	0			
HP C3724S	SCSI-2	1200	3.50	3H	PRML	AUTO	N/A	N/A	800	5	3610	0			
HP C3724SC	SCSI SCA	1200	3.50	3H	PRML	AUTO	N/A	N/A	800	5	3610	0			
HP C3724TC	SCSI SCA	1200	3.50	3H	PRML	AUTO	N/A	N/A	800	5	3610	0			
HP C3724W	SCSI-2 WD	1200	3.50	3H	PRML	AUTO	N/A	N/A	800	5	3610	0			
HP C3724WC	SCSI SCA	1200	3.50	3H	PRML	AUTO	N/A	N/A	800	5	3610	0			
HP C3725D	SCSI-2 D	2170	3.50	3H	RLL 2	AUTO	N/A	N/A	800	9	3610	0			
HP C3725S	SCSI-2	2170	3.50	3H	PRML	AUTO	N/A	N/A	800	9	3610	0			
HP C3725SC	SCSI SCA	2170	3.50	3H	PRML	AUTO	N/A	N/A	800	9	3610	0			
HP C3725TC	SCSI SCA	2170	3.50	3H	PRML	AUTO	N/A	N/A	800	9	3610	0			
HP C3725W	SCSI-2 WD	2170	3.50	3H	PRML	AUTO	N/A	N/A	800	9	3610	0			
HP C3725WC	SCSI SCA	2170	3.50	3H	PRML	AUTO	N/A	N/A	800	9	3610	0			
HP C5270A	IDE(AT)	1084	3.50	3H		AUTO	N/A	N/A		0	0	0	16	2105	63
HP C5271A	IDE(AT)	1626	3.50	3H		AUTO	N/A	N/A		0	0	0	16	3158	63
HP C5272A	IDE(AT)	1336	3.50	3H		AUTO	N/A	N/A		0	0	0	16	2595	63
HP C5273A	IDE(AT)	2004	3.50	3H		AUTO	N/A	N/A		0	0	0	16	3893	63
HP C5280A	IDE(AT)	1084	3.50	3H		AUTO	N/A	N/A		0	0	0	16	2105	63
HP C5281A	IDE(AT)	1626	3.50	3H		AUTO	N/A	N/A		0	0	0	16	3158	63
HP C5435A (SURESTORE 1300V)	IDE(AT)	1336	3.50	3H	RLL 1					4	0	0			
HP C5436A (SURESTORE 2000V)	IDE(AT)	2004	3.50	3H	RLL 1	N/A				6	0	0			

HITACHI CORPORATION

MODEL	INT	CAP	FF	HGT	CODE	LZ	WP	RWC	MTBF	PHDS	PCYL	PSPT	LHDS	LCYL	LSPT
DK 211A-51	IDE(AT)	510	2.50	4H	RLL 1	AUTO	N/A	N/A	300	6	2244	0	16	1021	61
DK 211A-54	IDE(AT)	680	2.50	4H	RLL 1	AUTO	N/A	N/A	300	6	2244	0	16	1047	63
DK 211A-68	IDE(AT)	680	2.50	4H	RLL 1	AUTO	N/A	N/A	300	8	2094	0	16	1384	60
DK 212A-10	IDE(AT)	1080	2.50	4H	RLL 1	AUTO	N/A	N/A	300	8	2602	0	16	2095	63
DK 212A-81	IDE(AT)	810	2.50	4H	RLL 1	AUTO	N/A	N/A	300	6	2602	0	16	1571	63
DK 213A-13	IDE(AT)	1350	2.50	5H		AUTO	N/A	N/A		0	0	0	16	2625	63
DK 213A-18	IDE(AT)	1800	2.50	5H		AUTO	N/A	N/A		0	0	0	16	3491	63

HITACHI CORPORATION

MODEL	INT	CAP	FF	HGT	CODE	LZ	WP	RWC	MTBF	PHDS	PCYL	PSPT	LHDS	LCYL	LSPT
DK 221A-34	IDE(AT)	340	2.50	5H	RLL 1	AUTO	N/A	N/A	300	4	2094	0	16	692	60
DK 222A-27	IDE(AT)	540	2.50	5H	RLL 1	AUTO	N/A	N/A	300	2	2602	0	16	550	60
DK 222A-54	IDE(AT)	540	2.50	5H	RLL 1	AUTO	N/A	N/A	300	4	2602	0	16	1047	63
DK 223A-11	IDE(AT)	1080	2.50	5H		AUTO	N/A	N/A		0	0	0	16	2094	63
DK 223A-81	IDE(AT)	810	2.50	5H		AUTO	N/A	N/A		0	0	0	16	1575	63
DK 224A-14	IDE(AT)	1440	2.50	5H		AUTO	N/A	N/A		0	0	0	16	2792	63
DK 225A-14	IDE(AT)	1440	2.50	5H		AUTO	N/A	N/A		0	0	0	16	2792	63
DK 225A-21	IDE(AT)	2160	2.50	5H		AUTO	N/A	N/A	300	16	4188	63			
DK 226A-32	IDE(AT)	3240	2.50	5H					300	16	6282	63			
DK 301-1	ST-506/412	10.65	3.50	HH	MFM				0	4	306	17			
DK 301-2	ST-506/412	15.98	3.50	HH	MFM				0	6	306	17			
DK 312C-20	SCSI	209.34	3.50	HH	RLL 2	AUTO	N/A	N/A	150	10	1076	38			
DK 312C-25	SCSI	251.21	3.50	HH	RLL 2	AUTO	N/A	N/A	100	12	1076	38			
DK 314C-41	SCSI	418.96	3.50	HH	RLL 2	AUTO	N/A	N/A	150	14	1169	50			
DK 315C-10	SCSI	1054	3.50	HH	RLL 1	AUTO	N/A	N/A	200	11	2469	59			
DK 315C-10D	SCSI	1054	3.50	HH	RLL 1	AUTO	N/A	N/A	200	11	2469	59			
DK 315C-10M	SCSI	10.7	3.50	HH	MFM		616	616	0	2	615	17			
DK 315C-11	SCSI	1105	3.50	HH	RLL 1	AUTO	N/A	N/A	200	15	2480	58			
DK 315C-14	SCSI	1437	3.50	HH	RLL 1	AUTO	N/A	N/A	200	15	2464	57			
DK 315C-14D	SCSI	1437	3.50	HH	RLL 1	AUTO	N/A	N/A	200	15	2464	57			
DK 318H-91	U-SCSI	9100	3.50	–					0	0	0	0			
DK 318H-91WC	SCSI	9100	3.50	–					0	0	0	0			
DK 318H-91WS	SCSI W	4370	3.50	–					0	0	0	0			
DK 325C	SCSI	573	3.50	3H	RLL 1	AUTO	N/A	N/A	200	6	2458	57			
DK 325C-57	SCSI	573	3.50	3H	RLL 1	AUTO	N/A	N/A	200	6	2458	57			
DK 326C-10	SCSI-2	1052	3.50	3H	RLL 1	AUTO	NA	N/A	400	7	3320	0			
DK 326C-10WD	SCSI-2	1052	3.50	3H	RLL 1	AUTO	N/A	N/A	400	7	3320	0			
DK 326C-6	SCSI-2	601	3.50	3H	RLL 1	AUTO	NA	N/A	400	4	3315	0			
DK 326C-6WD	SCSI-2	601	3.50	HH	RLL 1	N/A	N/A	N/A	400	5	3202	0			
DK 328C-10	SCSI-2	1050	3.50	3H	RLL 1	AUTO	N/A	N/A	800	3	0	0			
DK 328C-10WD	SCSI-2 WD	1050	3.50	3H	RLL 1	AUTO	N/A	N/A	800	3	0	0			
DK 328C-10WS	SCSI-2 W	1050	3.50	3H	RLL 1	AUTO	N/A	N/A	800	3	0	0			
DK 328C-21	SCSI-2	2100	3.50	3H	RLL 1	AUTO	N/A	N/A	800	5	0	0			
DK 328C-21WD	SCSI-2 WD	2100	3.50	3H	RLL 1	AUTO	N/A	N/A	800	5	0	0			
DK 328C-21WS	SCSI-2 W	2100	3.50	3H	RLL 1	AUTO	N/A	N/A	800	5	0	0			
DK 328C-43	SCSI-2	4300	3.50	3H	RLL 1	AUTO	N/A	N/A	800	10	0	0			
DK 328C-43WD	SCSI-2 WD	4300	3.50	3H	RLL 1	AUTO	N/A	N/A	800	10	0	0			
DK 328C-43WS	SCSI-2 W	4300	3.50	3H	RLL 1	AUTO	N/A	N/A	800	10	0	0			
DK 328H-43	U-SCSI	9100	3.50	–					0	0	0	0			
DK 328H-43WC	SCSI	9100	3.50	–					0	0	0	0			
DK 328H-43WS	SCSI W	4370	3.50	–					0	0	0	0			
DK 505-2	ST-506/412	23.04	5.25	HH	MFM				20	4	615	32			
DK 511-3	ST-506/412	28.63	5.25	FH	MFM		128		20	5	699	32			
DK 511-5	ST-506/412	40.08	5.25	FH	MFM		128		20	7	699	32			

HITACHI CORPORATION

MODEL	INT	CAP	FF	HGT	CODE	LZ	WP	RWC	MTBF	PHDS	PCYL	PSPT	LHDS	LCYL	LSPT
DK 511-8	ST-506/412	134.84	5.25	FH	MFM				20	10	823	32			
DK 512-12	ESDI-10MHZ	94.4	5.25	FH	RLL 2	AUTO	N/A	N/A	20	7	823	0			
DK 512-17	ESDI-10MHZ	134.8	5.25	FH	RLL 2	AUTO	N/A	N/A	20	10	823	0			
DK 512-8	ESDI-10MHZ	67.4	5.25	FH	RLL 2	AUTO	N/A	N/A	20	5	823	0			
DK 512C-12	SCSI	188.77	5.25	FH	RLL 2	AUTO	N/A	N/A	20	7	823	64			
DK 512C-17	SCSI	269.68	5.25	FH	RLL 2	AUTO	N/A	N/A	20	10	823	64			
DK 512C-8	SCSI	134.84	5.25	FH	RLL 2	AUTO	N/A	N/A	20	5	823	64			
DK 514-38	ESDI-15MHZ	330.1	5.25	FH	RLL 2	AUTO	N/A	N/A	30	14	903	51			
DK 514C-38 +A/D2	SCSI	330.1	5.25	FH	RLL 2	AUTO	N/A	N/A	30	14	903	51			
DK 514C-38 -A/D1	SCSI	330.1	5.25	FH	RLL 2	AUTO	N/A	N/A	30	14	903	51			
DK 514C-38 REV. A/D2	SCSI	321.8	5.25	FH	RLL 2	AUTO	N/A	N/A	30	14	898	50			
DK 515-12	ESDI-20MHZ	1229	5.25	FH	RLL 2	AUTO	N/A	N/A	150	15	0	69			
DK 515-78	ESDI-20MHZ	673.13	5.25	FH	RLL 2	AUTO	N/A	N/A	150	14	1361	69			
DK 515C-78	SCSI	673.13	5.25	FH	RLL 2	AUTO	N/A	N/A	150	14	1361	69			
DK 515C-78D	SCSI	673.13	5.25	FH	RLL 2	AUTO	N/A	N/A	150	14	1361	69			
DK 516-12	ESDI-24MHZ	1056.76	5.25	FH	RLL 2	AUTO	N/A	N/A	100	15	1787	77			
DK 516-15	ESDI-24MHZ	1321.68	5.25	FH	RLL 2	AUTO	N/A	N/A	150	15	2235	77			
DK 516C-16	SCSI-2	1351.15	5.25	FH	RLL 2	AUTO	N/A	N/A	100	15	2172	81			
DK 516C-16D	SCSI-2	1342	5.25	FH	RLL 2	AUTO	N/A	N/A	150	15	2172	81			
DK 517C-26	SCSI	0	5.25	FH	NONE				0	0	0	0			
DK 517C-37	SCSI-2	2873	5.25	FH	RLL	AUTO	N/A	N/A	200	21	3299	0			
DK 517C-37D	SCSI-2	2873	5.25	FH	RLL 1	AUTO	N/A	N/A	400	21	3299	81			
DK 521-5	ST-506/412	42.98	5.25	HH	MFM				0	6	823	17			
DK 522-10	ESDI-10MHZ	91.01	5.25	HH	RLL 2	AUTO	N/A	N/A	30	6	823	36			
DK 522C-10	SCSI	88.05	5.25	HH	RLL 2	AUTO	N/A	N/A	30	6	819	35			
DK 524C-20	SCSI	168.9	5.25	HH	RLL 2	AUTO	N/A	N/A	40	6	1100	50			
DR 311D	SCSI-2	105.8	3.50	3H	RLL 1	AUTO	N/A	N/A	150	2	2108	0			

HYOSUNG COMPUTER

MODEL	INT	CAP	FF	HGT	CODE	LZ	WP	RWC	MTBF	PHDS	PCYL	PSPT	LHDS	LCYL	LSPT
HC 8085	ST-506/412	71.3	5.25	FH	RLL	AUTO	N/A	N/A	28	8	1024	17			
HC 8128	ST-506/412	109.05	5.25	FH	RLL	AUTO	N/A	N/A	28	8	1024	26			
HC 8170E	ESDI-10MHZ	150.99	5.25	FH	RLL	AUTO	N/A	N/A	28	8	1024	36			

IBM CORPORATION

MODEL	INT	CAP	FF	HGT	CODE	LZ	WP	RWC	MTBF	PHDS	PCYL	PSPT	LHDS	LCYL	LSPT
0664-CSH	SCSI-2	4000	5.25	FH		AUTO	N/A	N/A		0	0	0			
0664-ESH	SCSI-2 W	4000	5.25	FH		AUTO	N/A	N/A		0	0	0			
115MB	ESDI-10MHZ	118.05	5.25	–	RLL	AUTO			0	7	915	36			
120MB	ST-506/412	120.58	5.25	FH	MFM				0	8	920	32			
20MB	ST-506/412	21.3	5.25	FH	MFM	663	306		0	4	612	17			
20MB PS/2	ST-506/412	21.3	3.50	HH	MFM		128		0	4	612	17			
30MB	ST-506/412	31.48	5.25	FH	MFM	663	300		0	4	615	25			
314MB	ESDI-10MHZ	319.87	5.25	FH	RLL	AUTO	N/A	N/A	0	15	1225	34			
44MB	ST-506/412	44.66	5.25	FH	MFM	733	300		0	7	733	17			
60MB	ST-506/412	60.86	5.25	FH	MFM				0	6	762	26			
70MB	ESDI-10MHZ	75.22	5.25	FH	RLL	AUTO	N/A	N/A	0	7	583	36			
DALA-3540	IDE(AT)	540	3.50			AUTO	N/A	N/A	300	0	0	0			
DALS-3540		0								0	0	0			
DAQA-32160	IDE(AT)	2160	3.50	3H		AUTO	N/A	N/A		4	0	0	16	4200	63
DAQA-33240	IDE(AT)	2160	3.50	3H		AUTO	N/A	N/A		6	0	0	16	6296	63
DBOA-2360	IDE(AT)	360	2.50	4H		AUTO	N/A	N/A	300	0	0	0			
DBOA-2528	IDE(AT)	528	2.50	4H		AUTO	N/A	N/A	300	0	0	0			
DBOA-2540	IDE(AT)	540	2.50	4H		AUTO	N/A	N/A	300	0	0	0			
DBOA-2720	IDE(AT)	720	2.50	4H		AUTO	N/A	N/A	300	0	0	0			
DCAA-33610	IDE(AT)	3610	3.50	3H	PRML	AUTO	N/A	N/A		5	0	0	16	7000	63
DCAA-34330	IDE(AT)	4330	3.50	3H	PRML	AUTO	N/A	N/A		6	0	0	16	8400	63
DCAS-32160 (50-PIN)	U-SCSI	2160	3.50	3H		AUTO	N/A	N/A		0	0	0			
DCAS-32160 (68-PIN)	U-SCSI W	2160	3.50	3H		AUTO	N/A	N/A		0	0	0			
DCAS-32160 (80-PIN)	U-SCSI W	2160	3.50	3H		AUTO	N/A	N/A		0	0	0			
DCAS-34330 (50-PIN)	U-SCSI	4330	3.50	3H		AUTO	N/A	N/A		0	0	0			
DCAS-34330 (68-PIN)	U-SCSI W	4330	3.50	3H		AUTO	N/A	N/A		0	0	0			
DCAS-34330 (80-PIN)	U-SCSI W	4330	3.50	3H		AUTO	N/A	N/A		0	0	0			
DCHS-34550 (50-PIN)	SCSI-2 F	4550	3.50	3H	PRML	AUTO	N/A	N/A		0	0	0			
DCHS-34550 (68-PIN)	SCSI-2 FW	4550	3.50	3H	PRML	AUTO	N/A	N/A		0	0	0			
DCHS-34550 (80-PIN)	U-SCSI	4550	3.50	3H	PRML	AUTO	N/A	N/A		0	0	0			
DCHS-39100 (50-PIN)	SCSI-2 F	9111	3.50	3H	PRML	AUTO	N/A	N/A		0	0	0			
DCHS-39100 (68-PIN)	SCSI-2 FW	9111	3.50	3H	PRML	AUTO	N/A	N/A		0	0	0			
DCHS-39100 (80-PIN)	U-SCSI	9111	3.50	3H	PRML	AUTO	N/A	N/A		0	0	0			
DCRA-22160	IDE(AT)	2160	2.50	5H	U	AUTO	N/A	N/A		0	0	0	16	4200	63
DDLA-21215	IDE(AT)	1215	2.50	5H						0	0	0			
DDLA-21620	IDE(AT)	1620	2.50	5H						0	0	0			
DESKSTAR (VER. 2)	IDE(AT)	0	3.50	3H		AUTO	N/A	N/A	300	0	0	0	16	954	36
DESKSTAR (VER. 3)	IDE(AT)	540	3.50			AUTO	N/A	N/A	300	0	0	0			
DESKSTAR (VER. 4)	IDE(AT)	0	3.50	3H		AUTO	N/A	N/A		3	0	0	16	2480	63
DESKSTAR (VER. 1)	SCSI-2 F	281	3.50	3H					300	2	0	0			
DESKSTAR (VER. 2)		0								0	0	0			
DESKSTAR 540MB (06H8558,9)	SCSI-2 F	540	3.50	3H	PRML	AUTO	N/A	N/A	350	2	4899	0			
DESKSTAR-3	IDE(AT)	0	3.50	3H		AUTO	N/A	N/A		4	0	0	16	4200	63

IBM CORPORATION

MODEL	INT	CAP	FF	HGT	CODE	LZ	WP	RWC	MTBF	PHDS	PCYL	PSPT	LHDS	LCYL	LSPT
DESKSTAR-4	IDE(AT)	0	3.50	3H	PRML	AUTO	N/A	N/A		5	0	0	16	7000	63
DESKSTAR-5	IDE(AT)	6480	3.50	FH						8	0	0			
DESKSTAR-5	IDE(AT)	6480	3.50	FH						8	0	0			
DESKSTAR-8	IDE(AT)	8041	3.50	FH						16	16351	63			
DESKSTAR-XP	IDE(AT)	0	3.50			AUTO	N/A	N/A	300	0	0	0			
DESKSTAR-XP	SCSI-2 F	0	3.50	3H	U	AUTO	N/A	N/A	0	2	0	0			
DFHS-31080	SCSI-2 F	1120	3.50	3H		AUTO	N/A	N/A		0	0	0			
DFHS-31080	U-SCSI FW	1120	3.50	3H	PRML	AUTO	N/A	N/A	100	0	0	0			
DFHS-31080	U-SCSI FWD	1120	3.50	3H	PRML	AUTO	N/A	N/A	100	0	0	0			
DFHS-31080 (80-PIN)	SCSI SCA	1120	3.50	3H	PRML				100	0	0	0			
DFHS-32160	SCSI-2 F	2250	3.50	3H		AUTO	N/A	N/A	1000	0	0	0			
DFHS-32160	U-SCSI FW	2250	3.50	3H	PRML	AUTO	N/A	N/A	100	0	0	0			
DFHS-32160	U-SCSI FWD	2250	3.50	3H	PRML	AUTO	N/A	N/A	100	0	0	0			
DFHS-32160 (80-PIN)	SCSI SCA	2250	3.50	3H	PRML				100	0	0	0			
DFHS-34320	SCSI-2 F	4510	3.50	3H		AUTO	N/A	N/A		0	0	0			
DFHS-34320	U-SCSI FW	4510	3.50	3H		AUTO	N/A	N/A		0	0	0			
DFHS-34320	U-SCSI FWD	4510	3.50	3H		AUTO	N/A	N/A		0	0	0			
DFHS-34320 (80-PIN)	SCSI SCA	4510	3.50	3H	PRML				100	0	0	0			
DFMS-31080	SCSI-2 F	1320	3.50	3H		AUTO	N/A	N/A		0	0	0			
DFMS-31080	U-SCSI FW	1320	3.50	3H	PRML	AUTO	N/A	N/A	100	0	0	0			
DFMS-31080	U-SCSI FWD	1320	3.50	3H	PRML	AUTO	N/A	N/A	100	0	0	0			
DFMS-31080 (80-PIN)	SCSI SCA	1120	3.50	3H	PRML				100	0	0	0			
DFMS-32160	SCSI-2 F	2320	3.50	3H		AUTO	N/A	N/A		0	0	0			
DFMS-32160	U-SCSI FW	2320	3.50	3H	PRML	AUTO	N/A	N/A	100	0	0	0			
DFMS-32160	U-SCSI FWD	2320	3.50	3H	PRML	AUTO	N/A	N/A	100	0	0	0			
DFMS-32160 (80-PIN)	SCSI SCA	2320	3.50	3H	PRML	AUTO	N/A	N/A	100	0	0	0			
DFMS-32600	SCSI-2 F	2650	3.50	3H						0	0	0			
DFMS-32600	U-SCSI FW	2650	3.50	3H	PRML				100	0	0	0			
DFMS-32600 (80-PIN)	SCSI SCA	2650	3.50	3H	PRML				100	0	0	0			
DFMS-32600	U-SCSI FWD	2650	3.50	3H	PRML				100	0	0	0			
DFMS-34320	SCSI-2 F	4320	3.50	3H		AUTO	N/A	N/A		0	0	0			
DFMS-34320	U-SCSI FW	4320	3.50	3H	PRML	AUTO	N/A	N/A	100	0	0	0			
DFMS-34320	U-SCSI FWD	4320	3.50	3H	PRML	AUTO	N/A	N/A	100	0	0	0			
DFMS-34320 (80-PIN)	SCSI SCA	4510	3.50	3H	PRML				100	0	0	0			
DFMS-35250	SCSI-2 F	5310	3.50	3H		AUTO	N/A	N/A		0	0	0			
DFMS-35250	U-SCSI FW	5310	3.50	3H	PRML	AUTO	N/A	N/A	100	0	0	0			
DFMS-35250	U-SCSI FWD	5310	3.50	3H	PRML	AUTO	N/A	N/A		0	0	0			
DFMS-35250 (80-PIN)	SCSI SCA	5310	3.50	3H	PRML				100	0	0	0			
DHAA-2270		0								0	0	0			
DHAA-2405	IDE(44PIN)	405.13	2.50	4H		AUTO	N/A	N/A	300	0	0	0	16	785	63
DHAA-2540	IDE(44PIN)	540.35	2.50	4H		AUTO	N/A	N/A	300	0	0	0	16	1047	63
DHAA-2810	IDE(AT)	810	2.50	4H		AUTO	N/A	N/A	300	0	0	0			
DHAS-2270	SCSI-2	270	2.50	5H						0	0	0			

IBM CORPORATION

MODEL	INT	CAP	FF	HGT	CODE	LZ	WP	RWC	MTBF	PHDS	PCYL	PSPT	LHDS	LCYL	LSPT
DHAS-2405	SCSI-2	405	2.50	5H						0	0	0			
DHAS-2540	SCSI-2	540	2.50	5H						0	0	0			
DHEA-34330 (VER. 1.7)	IDE(AT)	4033	2.50		U	AUTO	N/A	N/A	250	16	8400	63			
DHEA-34331	IDE(AT)	4026	3.50	FH						16	8184	63			
DHEA-34860 (VER. 1.7)	IDE(AT)	4086	2.50		U	AUTO	N/A	N/A	250	16	8400	63			
DHEA-36480	IDE(AT)	6480	3.50	FH						8	0	0			
DHEA-36480 (VER. 1.7)	IDE(AT)	6480	3.50	FH						8	0	0			
DHEA-36481	IDE(AT)	6041	3.50	FH						16	12592	63			
DHEA-38451	IDE(AT)	8041	3.50	FH						16	16351	63			
DJAA-31270	IDE(AT)	1270	3.50	3H		AUTO	N/A	N/A		3	0	0	16	2480	63
DJAA-31700	IDE(AT)	1700	3.50	3H		AUTO	N/A	N/A		4	0	0	16	3308	63
DLGA-22690	IDE(AT)	2690	2.50	5H		AUTO	N/A	N/A		0	0	0	16	5216	63
DLGA-23080	IDE(AT)	3080	2.50	5H		AUTO	N/A	N/A		0	0	0	16	5968	63
DMCA-21080	IDE(AT)	1080	2.50	5H						0	0	0	16	2100	63
DMCA-21440	IDE(AT)	1440	2.50	5H						0	0	0	16	2800	63
DMCA-22160	IDE(AT)	2160	2.50	5H					300	6	4928	0	16	4200	
DORS-32160 (50-PIN)	U-SCSI	2160	3.50	3H		AUTO	N/A	N/A	800	0	0	0	5		
DORS-32160 (68-PIN)	U-SCSI W	2160	3.50	3H		AUTO	N/A	N/A	800	0	0	0	5		
DORS-32160 (80-PIN)	U-SCSI W	2160	3.50	3H		AUTO	N/A	N/A	800	0	0	0	5		
DPEA-30540	IDE(AT)	540	3.50			AUTO	N/A	N/A	300	0	0	0			
DPEA-30810	IDE(AT)	810	3.50	3H		AUTO	N/A	N/A	300	0	0	0			
DPEA-31080	IDE(AT)	1080	3.50			AUTO	N/A	N/A	300	0	0	0			
DPES-30540	SCSI-2 F	540	3.50	3H	U	AUTO	N/A	N/A	0	2	0	0			
DPES-30810	SCSI-2 F	810	3.50	3H	U	AUTO	N/A	N/A	0	3	0	0			
DPES-31080	SCSI-2 F	1080	3.50	3H	U	AUTO	N/A	N/A	0	4	0	0			
DPLA-24480	IDE(AT)	4480	2.50	FH						15	9264	63			
DPLA-25120	IDE(AT)	4480	2.50	FH						15	10592	63			
DPRA-20810	IDE(AT)	810	2.50	4H		AUTO	N/A	N/A	300	0	0	0			
DPRA-21215	IDE(AT)	1215	2.50	4H		AUTO	N/A	N/A	300	0	0	0			
DPRS-20810	SCSI-2	810	2.50	5H		AUTO	N/A	N/A		6	3481	0	0		0
DPRS-21215	SCSI-2	1215	2.50	5H		AUTO	N/A	N/A		6	3481	0	0		0
DSAA-3270	IDE(AT)	281.34	3.50	3H		AUTO	N/A	N/A	300	0	0	0	16	954	36
DSAA-3360	IDE(AT)	365.29	3.50	3H		AUTO	N/A	N/A	300	0	0	0	16	929	48
DSAA-3540 VER. 1	IDE(AT)	548.09	3.50	3H		AUTO	N/A	N/A	300	0	0	0	16	1062	63
DSAA-3540 VER. 2	IDE(AT)	528.48	3.50	3H		AUTO	N/A	N/A	300	0	0	0	16	1024	63
DSAA-3720	IDE(AT)	730.79	3.50	3H		AUTO	N/A	N/A	300	0	0	0	16	1416	63
DSAS-3270	SCSI-2 F	281	3.50	3H					300	2	0	0			
DSAS-3360	SCSI-2 F	365	3.50	3H					300	2	0	0			
DSAS-3540	SCSI-2 F	548	3.50	3H					300	3	0	0			
DSAS-3720	SCSI-2 F	730	3.50	3H					300	4	0	0			
DSOA-20540	IDE(44PIN)	540	2.50	5H		AUTO	N/A	N/A		0	0	0	16	1050	63
DSOA-20810	IDE(44PIN)	810	2.50	5H		AUTO	N/A	N/A		0	0	0	16	1575	63
DSOA-21080	IDE(44PIN)	1080	2.50	5H		AUTO	N/A	N/A		0	0	0	16	2100	63
DTCA-23240	IDE(AT)	3240	2.50	FH						16	6304	63			

IBM CORPORATION

MODEL	INT	CAP	FF	HGT	CODE	LZ	WP	RWC	MTBF	PHDS	PCYL	PSPT	LHDS	LCYL	LSPT
DTCA-24090	IDE(AT)	4090	2.50	FH						16	7944	63			
DTNA-21800	IDE(AT)	1800	2.50	5H	PRML	AUTO	N/A	N/A		0	0	0	16	3500	63
DTNA-22160	IDE(AT)	2160	2.50	5H	PRML	AUTO	N/A	N/A		0	0	0	16	4200	63
DVAA-2810	IDE(AT)	810	2.50	5H	U	AUTO	N/A	N/A	300	16	1571	63			
DVAS-2810	SCSI-2	810	2.50	5H		AUTO	N/A	N/A		6	2788	0	0		0
DYKA-22160	IDE(AT)	2130	2.50	FH						16	4200	63			
DYKA-23240	IDE(AT)	3240	2.50	FH						16	6304	63			
DYLA-26480	IDE(AT)	6480	2.50	FH						15	16944	63			
DYLA-28100	IDE(AT)	8100	2.50	FH						15	16944	63			
H2172-A2	IDE(44PIN)	172.16	2.50	4H		AUTO	N/A	N/A	0	0	0	0	10	989	34
H2172-S2	SCSI-2	172	2.50	3H		AUTO	N/A	N/A	300	0	0	0			
H2258-A3	IDE(44PIN)	258.24	2.50	4H		AUTO	N/A	N/A	0	0	0	0	15	989	34
H2258-S3	SCSI-2	258	2.50	3H		AUTO	N/A	N/A	300	0	0	0			
H2344-A4	IDE(44PIN)	344.33	2.50	4H		AUTO	N/A	N/A	0	0	0	0	15	915	49
H2344-S4	SCSI-2	344	2.50	3H		AUTO	N/A	N/A	300	0	0	0			
H3133-A2	IDE(AT)	133.56	3.50	3H		AUTO	N/A	N/A	250	0	0	0	15	1023	17
H3171-A2	IDE(AT)	171.29	3.50	3H		AUTO	N/A	N/A	250	0	0	0	10	984	34
H3256-A3	IDE(AT)	257.16	3.50	3H		AUTO	N/A	N/A	250	0	0	0	16	872	36
H3342-A4	IDE(AT)	342.88	3.50	3H		AUTO	N/A	N/A	250	0	0	0	16	872	48
PS/2 MODEL 25 20MB	ST-506/412	20	5.25	–					0	0	0	0			
TRAVELSTAR (VER. 2)	IDE(AT)	810	2.50	5H	U	AUTO	N/A	N/A	300	16	1571	63			
TRAVELSTAR		0								0	0	0			
TRAVELSTAR (VER. 2)	SCSI-2	810	2.50	5H		AUTO	N/A	N/A		6	2788	0	0		0
TRAVELSTAR	SCSI-2	0	2.50	5H						0	0	0			
TRAVELSTAR-2LP	IDE(44PIN)	0	2.50	5H		AUTO	N/A	N/A		0	0	0	16	1575	63
TRAVELSTAR-2XP	IDE(AT)	0	2.50	5H	U	AUTO	N/A	N/A		0	0	0	16	4200	63
TRAVELSTAR-3GN	IDE(AT)	8100	2.50	FH						15	16944	63			
TRAVELSTAR-3LP	IDE(AT)	0	2.50	5H						0	0	0	16	2100	63
TRAVELSTAR-3XP	IDE(AT)	0	2.50	5H		AUTO	N/A	N/A		0	0	0	16	5216	63
TRAVELSTAR-4GT	IDE(AT)	4480	2.50	FH						15	10592	63			
TRAVELSTAR-4LP	IDE(AT)	0	2.50	5H	PRML	AUTO	N/A	N/A		0	0	0	16	3500	63
TRAVELSTAR-5GS	IDE(AT)	4480	2.50	FH						15	10592	63			
TRAVELSTAR-8GS	IDE(AT)	8100	2.50	FH						15	16944	63			
TRAVELSTAR-LP	IDE(AT)	0	2.50	4H		AUTO	N/A	N/A	300	0	0	0			
TRAVELSTAR-VP	IDE(AT)	0	2.50	5H						0	0	0			
TRAVELSTAR-XP	IDE(AT)	0	2.50	4H		AUTO	N/A	N/A	300	0	0	0			
TRAVELSTAR-XP	SCSI-2	0	2.50	5H		AUTO	N/A	N/A		6	3481	0	0		0
TYPE 0661 MODEL 371	SCSI-2	326.51	3.50	HH	RLL	AUTO	N/A	N/A	300	14	949	48			
TYPE 0661 MODEL 467	SCSI-2	412.53	3.50	HH	RLL	AUTO	N/A	N/A	300	14	1199	48			
TYPE 0662 MODEL A10 (SPITFIRE)	IDE(AT)	1051.8	3.50	3H		AUTO	N/A	N/A	500	0	0	0	16	2038	63
TYPE 0662 MODEL S12 (SPITFIRE)	SCSI-2	F10503.50	FH	U	AUTO	N/A	N/A	800	16	2038	63				

IBM CORPORATION

MODEL	INT	CAP	FF	HGT	CODE	LZ	WP	RWC	MTBF	PHDS	PCYL	PSPT	LHDS	LCYL	LSPT
TYPE 0662 MODEL S1D (SPITFIRE)	SCSI-2 F	10503.50	FH	U	AUTO	N/A	N/A	800	16	2038	63				
TYPE 0662 MODEL SW1 (SPITFIRE)	SCSI-2 W	1050	3.50	FH	U	AUTO	N/A	N/A	800	16	2038	63			
TYPE 0662 MODEL SWD (SPITFIRE)	SCSI-2 W	1050	3.50	FH	U	AUTO	N/A	N/A	800	16	2038	63			
TYPE 0663 MODEL E12	SCSI	1044	3.50	HH	RLL	AUTO	N/A	N/A	500	13	2469	0			
TYPE 0663 MODEL E15	SCSI	1206	3.50	HH	RLL	AUTO	N/A	N/A	500	15	2469	0			
TYPE 0663 MODEL H11	SCSI	0	3.50	HH	RLL				400	0	0	0			
TYPE 0663 MODEL H12	SCSI	1039.61	3.50	HH	RLL	AUTO	N/A	N/A	400	15	2051	66			
TYPE 0663 MODEL L08	SCSI	623.76	3.50	HH	RLL	AUTO	N/A	N/A	400	9	2051	66			
TYPE 0663 MODEL L11	SCSI	900.99	3.50	HH	RLL	AUTO	N/A	N/A	400	13	2051	66			
TYPE 0663 MODEL L12	SCSI	1039.61	3.50	HH	RLL	AUTO	N/A	N/A	880	15	2051	66			
TYPE 0664 CSH/ESH	SCSI	4000	5.25	FH	RLL	AUTO	N/A	N/A	375	2	2857	0			
TYPE 0664 MODEL M1H (50-PIN)	SCSI-2 F	2000	3.50	HH		AUTO	N/A	N/A		6	2857	0			
TYPE 0664 MODEL M1H (68-PIN)	SCSI-2 FW	2000	3.50	HH		AUTO	N/A	N/A		6	2857	0			
TYPE 0664 MODEL N1H (50-PIN)	SCSI-2 F	2000	3.50	HH		AUTO	N/A	N/A		6	2857	0			
TYPE 0664 MODEL N1H (68-PIN)	SCSI-2 FW	2000	3.50	HH		AUTO	N/A	N/A		6	2857	0			
TYPE 0665 MODEL 38	ST-506/412	31.9	5.25	FH	MFM	AUTO	N/A	N/A	0	5	733	17			
TYPE 0665 MODEL 53	ST-506/412	44.66	5.25	FH	MFM	AUTO	N/A	N/A	0	7	733	17			
TYPE 0667 MODEL 61	ESDI-10MHZ	52.14	5.25	FH	RLL	AUTO	N/A	N/A	0	5	582	35			
TYPE 0667 MODEL 85	ESDI-10MHZ	73	5.25	FH	RLL	AUTO	N/A	N/A	0	7	582	64			
TYPE 0669 MODEL 61	ESDI-10MHZ	48.42	5.25	FH	RLL	AUTO	N/A	N/A	0	5	582	35			
TYPE 0669 MODEL 85	ESDI-10MHZ	67.79	5.25	FH	RLL	AUTO	N/A	N/A	0	7	582	64			
TYPE 0671 MODEL S11	SCSI	234.38	5.25	FH	RLL	AUTO	N/A	N/A	0	11	1224	34			
TYPE 0671 MODEL S15	SCSI	319.61	5.25	FH	RLL	AUTO	N/A	N/A	0	15	1224	34			
TYPE 0681 MODEL 500	SCSI	476.26	5.25	FH	RLL	AUTO	N/A	N/A	150	11	1458	58			
TYPE 0681 MODEL1000	SCSI	865.93	5.25	FH	RLL	AUTO	N/A	N/A	150	20	1458	58			
ULTRASTAR 2ES (50-PIN)	U-SCSI	0	3.50	3H		AUTO	N/A	N/A		0	0	0			
ULTRASTAR 2ES (68-PIN)	U-SCSI W	0	3.50	3H		AUTO	N/A	N/A		0	0	0			
ULTRASTAR 2ES (80-PIN)	U-SCSI W	2160	3.50	3H		AUTO	N/A	N/A		0	0	0			
ULTRASTAR 2XP (50 PIN)	SCSI-2 F	4550	3.50	3H	PRML	AUTO	N/A	N/A		0	0	0			
ULTRASTAR 2XP (68-PIN)	SCSI-2 FW	4550	3.50	3H	PRML	AUTO	N/A	N/A		0	0	0			
ULTRASTAR 2XP (80-PIN)	U-SCSI	4550	3.50	3H	PRML	AUTO	N/A	N/A		0	0	0			
ULTRASTAR ES (50-PIN)	U-SCSI	2160	3.50	3H		AUTO	N/A	N/A	800	0	0	0	5		
ULTRASTAR ES (68-PIN)	U-SCSI W	2160	3.50	3H		AUTO	N/A	N/A	800	0	0	0	5		
ULTRASTAR ES (80-PIN)	U-SCSI W	0	3.50	3H		AUTO	N/A	N/A	800	0	0	0	5		
WD-12	ST-506/412	10	5.25	FH	MFM		296	296	0	4	306	17			
WD-2120	MCA	127.28	3.50	3H	RLL	AUTO	N/A	N/A	150	4	1243	50			
WD-240	MCA	43.65	2.50	4H	RLL	AUTO	N/A	N/A	150	2	1122	38			
WD-25	ST-506/412	20	5.25	FH	MFM		296	296	0	0	0	17			
WD-3158	MCA	120.58	3.50	FH	RLL	AUTO	N/A	N/A	45	8	920	32			

IBM CORPORATION

MODEL	INT	CAP	FF	HGT	CODE	LZ	WP	RWC	MTBF	PHDS	PCYL	PSPT	LHDS	LCYL	LSPT
WD-3160	MCA	163.09	3.50	HH	RLL	AUTO	N/A	N/A	110	8	1021	39			
WD-380	MCA	81.54	3.50	HH	RLL	AUTO	N/A	N/A	110	4	1021	39			
WD-387	MCA	60.81	3.50	HH	RLL	AUTO	N/A	N/A	45	4	928	32			
WD-L40	MCA	41.45	3.50	HH	RLL	AUTO	N/A	N/A	90	2	1038	39			
WD-L40S	MCA	41.45	3.50	HH	RLL	AUTO	N/A	N/A	90	2	1038	39			
WDA-2120	IDE(AT)	126.51	3.50	3H	RLL 1	AUTO	N/A	N/A	150	4	1243	50	15	969	17
WDA-240	IDE(AT)	43.1	2.50	4H	RLL 1	AUTO	N/A	N/A	150	2	1122	38	8	619	17
WDA-260	IDE(AT)	63.64	3.50	3H	RLL	AUTO	N/A	N/A	150	2	1243	50			
WDA-280	IDE(AT)	86.08	2.50	3H	RLL 1	AUTO	N/A	N/A	150	4	1122	38	10	989	17
WDA-3160	IDE(AT)	81.54	3.50	HH	RLL	AUTO	N/A	N/A	110	4	1021	39			
WDA-380	IDE(AT)	81.54	3.50	HH	RLL	AUTO	N/A	N/A	110	4	1021	39			
WDA-L160	IDE(AT)	171.29	3.50	3H	RLL 1	AUTO	N/A	N/A	150	4	1923	44	10	984	34
WDA-L40	IDE(AT)	41.53	3.50	3H	RLL 2	AUTO	N/A	N/A	90	2	1040	39			
WDA-L42	IDE(AT)	42.61	3.50	3H	RLL 2	AUTO	N/A	N/A	90	2	1067	39			
WDA-L80	IDE(AT)	85.64	3.50	3H	RLL 1	AUTO	N/A	N/A	150	2	1923	44	10	984	17
WDA-S260	IDE(AT)	63.3	3.50	4H	RLL 1	AUTO	N/A	N/A	150	2	1243	50	8	909	17
WDS-240	SCSI-2	43.65	2.50	5H	RLL	AUTO	N/A	N/A	150	2	1122	38			
WDS-280	SCSI-2	87	2.50	5H		AUTO	N/A	N/A		4	1122	38			
WDS-3100	SCSI-2	108.27	3.50	4H	RLL	AUTO	N/A	N/A	150	2	2009	0			
WDS-3160	SCSI-2	163.09	3.50	HH	RLL	AUTO	N/A	N/A	110	8	1021	39			
WDS-3200	SCSI-2	216.53	3.50	4H	RLL	AUTO	N/A	N/A	150	4	2009	0			
WDS-380	SCSI-2	81.54	3.50	HH	RLL	AUTO	N/A	N/A	110	4	1021	39			
WDS-L160	SCSI	0	3.50	5H		AUTO	N/A	N/A	150	2	0	112			
WDS-L40	SCSI-2	41.45	3.50	FH	RLL	AUTO	N/A	N/A	90	2	1038	39			
WDS-L42	SCSI-2	42.57	3.50	3H	RLL 1	AUTO	N/A	N/A	80	2	1066	39			
WDS-L80	SCSI	0	3.50	5H		AUTO	N/A	N/A	150	2	0	112			

INFOSYSTEMS, LTD.

MODEL	INT	CAP	FF	HGT	CODE	LZ	WP	RWC	MTBF	PHDS	PCYL	PSPT	LHDS	LCYL	LSPT
IS-70270	IDE(AT)	270	3.50	6	RLL 1	AUTO	N/A	N/A	300	0	0	0	16	540	63
IS-71070	IDE(AT)	1070	3.50	6	RLL 1	AUTO	N/A	N/A	300	0	0	0	16	2140	63

INTEGRAL PERIPHERALS

MODEL	INT	CAP	FF	HGT	CODE	LZ	WP	RWC	MTBF	PHDS	PCYL	PSPT	LHDS	LCYL	LSPT
1820 MUSTANG	IDE(44PIN)	21.41	1.80	5H	RLL 1	AUTO	N/A	N/A	100	2	0	0	4	615	17
1820 MUSTANG (REV. B)	IDE(44PIN)	21.4	1.80	5H	RLL 1	AUTO	N/A	N/A	100	2	0	0	4	615	17
1842 STINGRAY	IDE(AT)	42.51	1.80	5H	RLL 1	AUTO	N/A	N/A	100	3	0	0	5	977	17
1862 MAVERICK	IDE(AT)	64.04	1.80	4H	RLL 1	AUTO	N/A	N/A	100	3	0	0			
8105PA VIPER (REV. A)	PCM	105.4	1.80	5H	RLL 1	AUTO	N/A	N/A	250	4	0	0	16	757	17
8105PA VIPER (REV. C)	PCM	105.4	1.80	5H	RLL 1	AUTO	N/A	N/A	250	4	0	0	16	757	17
8170PA VIPER (REV. B)	PCM	170.8	1.80	5H	RLL 1	AUTO	N/A	N/A	250	4	0	0	10	981	34
8260PA VIPER (REV. A)	PCM	260.7	1.80	5H	PRML	AUTO	N/A	N/A	250	4	0	0	16	936	34
8340PA VIPER	PCM	341.1	1.80	5H	PRML	AUTO	N/A	N/A	250	4	0	0	16	771	54

INTERNATIONAL MEMORIES, INC. (IMI)

MODEL	INT	CAP	FF	HGT	CODE	LZ	WP	RWC	MTBF	PHDS	PCYL	PSPT	LHDS	LCYL	LSPT
5006	ST-506/412	5.32	5.25	FH	MFM		214	307	0	2	306	17			
5006H	ST-506/412	5	5.25	FH	MFM				10	2	0	32			
5007	ST-506/412	5.32	5.25	FH	MFM				0	2	306	17			
5012	ST-506/412	10.65	5.25	FH	MFM		214	307	0	4	306	17			
5012H	ST-506/412	15	5.25	FH	MFM				0	0	0	17			
5018	ST-506/412	15.98	5.25	FH	MFM		214	307	0	6	306	17			
7720	ST-506/412	20	8.00	FH	MFM				0	0	0	17			
7740	ST-506/412	40	8.00	FH	MFM				0	0	0	17			

IOMEGA CORPORATION

MODEL	INT	CAP	FF	HGT	CODE	LZ	WP	RWC	MTBF	PHDS	PCYL	PSPT	LHDS	LCYL	LSPT
BETA 20I	SCSI	0	5.25	HH					0	0	0	0			
BETA 20L	SCSI	0		--					0	0	0	0			
BETA 44	SCSI	0		--					0	0	0	0			
JAZ SCSI	SCSI-2	1020	2.50	3H	U					0	0	0			
ZIP 100 INTERNAL	IDE(AT)	100	3.50	3H	U					0	0	0	12	512	32

J&A COMPUTER CO., LTD.

MODEL	INT	CAP	FF	HGT	CODE	LZ	WP	RWC	MTBF	PHDS	PCYL	PSPT	LHDS	LCYL	LSPT
D5882	SCSI	675	5.25	FH	RLL 1				50	15	1633	53			

JCT

MODEL	INT	CAP	FF	HGT	CODE	LZ	WP	RWC	MTBF	PHDS	PCYL	PSPT	LHDS	LCYL	LSPT
100	ST-506/412	3.93	5.25	HH	MFM				0	2	226	17			
1000	COMM	5	5.25	HH	MFM				0	0	0	17			
1005	COMM	7	5.25	HH	MFM				0	0	0	17			
1010	COMM	14	5.25	HH	MFM				0	0	0	17			
105	ST-506/412	5.32	5.25	HH	MFM				0	2	306	17			
110	ST-506/412	14	5.25	HH	MFM				0	0	0	17			
120	ST-506/412	20	5.25	HH	MFM				0	0	0	17			

JTS CORPORATION

MODEL	INT	CAP	FF	HGT	CODE	LZ	WP	RWC	MTBF	PHDS	PCYL	PSPT	LHDS	LCYL	LSPT
C2000-3AF	IDE(AT)	2000	3.50	4H	U	AUTO		N/A		16	3882	63			
P1000-2AF	IDE(AT)	1000	3.50	5H	RLL 1	AUTO	N/A	N/A	250	0	0	0	16	1942	63
P1200-2AF	IDE(AT)	1200	3.50	5H	RLL 1	AUTO	N/A	N/A	250	4	0	0	16	2332	63
P1600-3AF	IDE(AT)	1600	3.50	5H	RLL 1	AUTO	N/A	N/A	250	0	0	0	16	3108	63
P3540AF	IDE(AT)	540.5	3.50	5H	RLL 1	AUTO		N/A	250	0	0	0	16	1049	63
P3850AF	IDE(AT)	850.5	3.50	5H	RLL 1	AUTO		N/A	250	0	0	0	16	1648	63

JVC

MODEL	INT	CAP	FF	HGT	CODE	LZ	WP	RWC	MTBF	PHDS	PCYL	PSPT	LHDS	LCYL	LSPT
JD-3812MOZA	JVC(26PIN)	10.61	3.50	HH	MFM	N/A	N/A	N/A	30	2	610	17			
JD-3824TA	JVC(26PIN)	21.43	3.50	HH	RLL 2	N/A	N/A	N/A	30	2	436	48			
JD-3848HA	JVC(26PIN)	42.76	3.50	HH	RLL 2	N/A	N/A	N/A	30	4	436	48			
JD-E2042M	IDE(AT)	42.84	2.50	4H	RLL 1	AUTO	N/A	N/A	130	2	973	43			
JD-E2085M	IDE(AT)	85.68	2.50	4H	RLL 1	AUTO	N/A	N/A	130	4	973	43			
JD-E2825P(A)	IDE(AT)	21.41	2.50	4H	RLL 2	AUTO	N/A	N/A	30	2	581	36			
JD-E2825P(S)	SCSI	21.41	2.50	4H	RLL 2	AUTO	N/A	N/A	30	2	581	36			
JD-E2825P(X)	IDE(XT)	21.41	2.50	4H	RLL 2	AUTO	N/A	N/A	30	2	581	36			
JD-E2850P(A)	IDE(AT)	42.52	2.50	4H	RLL 2	AUTO	N/A	N/A	130	3	791	35			
JD-E2850P(S)	SCSI	42.52	2.50	4H	RLL 2	AUTO	N/A	N/A	130	3	0	35			
JD-E3896V(A)	IDE(AT)	84.73	3.50	3H	RLL 2	AUTO	N/A	N/A	30	4	862	48			
JD-E3896V(S)	SCSI	84.73	3.50	3H	RLL 2	AUTO	N/A	N/A	30	4	862	48			
JD-E3896V(X)	IDE(XT)	84.73	3.50	3H	RLL 2	AUTO	N/A	N/A	30	4	862	48			

KALOK CORPORATION

MODEL	INT	CAP	FF	HGT	CODE	LZ	WP	RWC	MTBF	PHDS	PCYL	PSPT	LHDS	LCYL	LSPT
KL 3100	IDE(AT)	105.2	3.50	HH	RLL 2	AUTO	N/A	N/A	100	6	820	0	6	979	35
KL 3120	IDE(AT)	120.6	3.50	HH	RLL 2	AUTO	N/A	N/A	100	6	820	0	6	981	40
KL 320	ST-506/412	21.41	3.50	HH	MFM				43	4	615	17			
KL 330	ST-506/412	32.74	3.50	HH	RLL 2				43	4	615	26			
KL 332	MCA	32.74	3.50	HH	RLL				0	4	615	26			
KL 340	ST-506/412	42.82	3.50	HH	MFM				50	6	820	17			
KL 341	SCSI	42.91	3.50	HH	RLL 2	AUTO	N/A	N/A	40	4	676	31			
KL 342	MCA	42.91	3.50	HH	RLL 2	AUTO	N/A	N/A	40	4	676	31			
KL 343	IDE(AT)	42.91	3.50	HH	RLL 2	AUTO	N/A	N/A	100	4	676	31			
KL 360	ST-506/412	65.49	3.50	HH	RLL 2				50	6	820	26			
KL 381	SCSI	84	3.50	HH	RLL 2	AUTO	N/A	N/A	50	6	815	0			
KL 383	IDE(AT)	84	3.50	HH	RLL 2	AUTO	N/A	N/A	50	6	815	0	6	815	33
P5-125(A)	IDE(AT)	125.8	3.50	5H	RLL 1	AUTO	N/A	N/A	100	2	2048	0			
P5-125(S)	SCSI-2	125.8	3.50	5H	RLL 1	AUTO	N/A	N/A	100	2	2048	0			
P5-250(A)	IDE(AT)	251.9	3.50	5H	RLL 1	AUTO	N/A	N/A	100	4	2048	0			
P5-250(S)	SCSI-2	251.9	3.50	5H	RLL 1	AUTO	N/A	N/A	100	4	2048	0			

KYOCERA ELECTRONICS, INC.

MODEL	INT	CAP	FF	HGT	CODE	LZ	WP	RWC	MTBF	PHDS	PCYL	PSPT	LHDS	LCYL	LSPT
KC 20A	ST-506/412	21.44	3.50	HH	MFM	644	0	N/A	40	4	616	17			
KC 20B	ST-506/412	21.41	3.50	HH	MFM	664	0	N/A	40	4	615	17			
KC 30A	ST-506/412	32.8	3.50	HH	RLL 2	664	0	N/A	40	4	616	26			
KC 30B	ST-506/412	32.74	3.50	HH	RLL 2	664	0	N/A	40	4	615	26			
KC 40GA	IDE(AT)	39.55	3.50	HH	RLL 2	AUTO	N/A	N/A	40	2	1073	0	4	577	33
KC 80GA	IDE(AT)	78.81	3.50	HH	RLL 2	AUTO	N/A	N/A	40	4	1069	0	8	577	33

LAPINE

MODEL	INT	CAP	FF	HGT	CODE	LZ	WP	RWC	MTBF	PHDS	PCYL	PSPT	LHDS	LCYL	LSPT
3522	ST-506/412	10.65	3.50	HH	MFM			307	0	4	306	17			
LT 10	ST-506/412	10.7	3.50	HH	MFM			616	0	2	615	17			
LT 20	ST-506/412	21.41	3.50	HH	MFM			616	0	4	615	17			
LT 200	ST-506/412	21.37	3.50	HH	MFM			615	0	4	614	17			
LT 2000	ST-506/412	21.37	3.50	HH	MFM			615	0	4	614	17			
LT 300	ST-506/412	32.69	3.50	HH	RLL 2			615	0	4	614	26			
LT 3065	ST-506/412	10.65	3.50	HH	RLL 2		128	306	0	4	306	17			
LT 3512	ST-506/412	10.65	3.50	HH	RLL 2		128	306	0	4	306	17			
LT 3522	ST-506/412	10.65	3.50	HH	RLL 2		128	306	0	4	306	17			
LT 3532	ST-506/412	32.69	3.50	HH	RLL 2			615	0	4	614	26			
TITAN 20	ST-506/412	21.41	3.50	HH	MFM				0	4	615	17			
TITAN 30	ST-506/412	32.74	3.50	HH	RLL				0	4	615	26			

MAXTOR CORPORATION

MODEL	INT	CAP	FF	HGT	CODE	LZ	WP	RWC	MTBF	PHDS	PCYL	PSPT	LHDS	LCYL	LSPT
250837	IDE(44PIN)	837	2.50	5H	RLL 1	AUTO	N/A	N/A	300	5	3196	0	16	1621	63
250840	IDE(AT)	840	2.50	5H	RLL 1	AUTO	N/A	N/A	300	5	3232	0			
25084A	IDE(AT)	84	2.50	3H	RLL	AUTO	N/A	N/A	150	0	0	0	16	569	18
251005	IDE(44PIN)	11005	2.50	5H	RLL 1	AUTO	N/A	N/A	300	6	3196	0	16	1945	63
251010	IDE(AT)	1010	2.50	5H	RLL 1	AUTO	N/A	N/A	300	6	3232	0			
25125S	SCSI	128.2	2.50	4H	RLL 1	AUTO	N/A	N/A	150	4	1092	0			
25128A	IDE(AT)	128.2	2.50	4H	RLL 1	AUTO	N/A	N/A	150	4	1092	0	14	1024	17
251340	IDE(44PIN)	1340	2.50	5H	RLL 1	AUTO	N/A	N/A	300	8	3196	0	16	2594	63
251350	IDE(AT)	1350	2.50	5H	RLL 1	AUTO	N/A	N/A	300	8	3232	0			
25252A	IDE(AT)	252	2.50	3H	RLL	AUTO	N/A	N/A	150	0	0	0	16	569	54
25252S	SCSI	252	2.50	4H	U	AUTO	N/A	N/A	0	6	1418	0			
2585A	IDE(AT)	85.41	2.50	4H	RLL 1	AUTO	N/A	N/A	150	4	1092	0	10	981	17
2585S	SCSI	85.41	2.50	4H	RLL 1	AUTO	N/A	N/A	150	4	1092	0			
7040A	IDE(AT)	43.13	3.50	3H	RLL 1	AUTO	N/A	N/A	150	2	1170	36	5	981	17
7040A (VER. 2)	IDE(AT)	41	3.50	FH		AUTO	N/A	N/A	150	4	1170	1			
7040S	SCSI	42.57	3.50	3H	RLL 1	AUTO	N/A	N/A	40	2	1155	36			
7060A	IDE(AT)	65.2	3.50	3H	RLL 1	AUTO	N/A	N/A	150	2	1516	42	7	1024	17
7060S	SCSI	64.42	3.50	3H	RLL 1	AUTO	N/A	N/A	150	2	1498	42			
7080A	IDE(AT)	85.15	3.50	3H	RLL 1	AUTO	N/A	N/A	150	4	1155	36	10	981	17
7080A (VER. 2)	IDE(AT)	82	3.50	FH		AUTO	N/A	N/A	150	4	1170	36			
7080S	SCSI	85.15	3.50	3H	RLL 1	AUTO	N/A	N/A	150	4	1155	36			
71000A	IDE(AT)	1002	3.50	3H	RLL 1	AUTO	N/A	N/A	300	3	0	0			
71050A	IDE(AT)	1050	3.50	3H	U	AUTO	N/A	N/A	0	0	0	0	16	2045	63
71084A	IDE(AT)	1	3.50	4H	RLL 1	AUTO	N/A	N/A	300	4	3	0	16	2,105	63
71084AP	IDE(AT)	1	3.50	4H	RLL 1	AUTO	N/A	N/A	300	4	3	0	16	2,105	63
7120A	IDE(AT)	130.4	3.50	3H	RLL 1	AUTO	N/A	N/A	150	4	1516	42	6	771	55
7120S	SCSI	128.85	3.50	3H	RLL 1	AUTO	N/A	N/A	150	4	1498	42			
71260A	IDE(AT)	1260	3.50	3H	U	AUTO	N/A	N/A	0	0	0	0	16	2448	63
71260AP	IDE(AT)	1260	3.50	3H	U	AUTO	N/A	N/A	0	0	0	0	16	2448	63

MAXTOR CORPORATION

MODEL	INT	CAP	FF	HGT	CODE	LZ	WP	RWC	MTBF	PHDS	PCYL	PSPT	LHDS	LCYL	LSPT
7131A	IDE(AT)	130	3.50	3H	RLL 1	AUTO	N/A	N/A	150	0	0	0	8	1002	32
71336A	IDE(AT)	1336	3.50	3H	RLL 1	AUTO	N/A	N/A	300	4	4702	0	16	2595	63
71336AP	IDE(AT)	1336	3.50	3H	RLL 1	AUTO	N/A	N/A	300	4	4702	0	16	2595	63
71350A	IDE(AT)	1350	3.50	3H	RLL 1	AUTO	N/A	N/A	300	0	0	0	16	2624	63
71350AP	IDE(AT)	1350	3.50	3H	RLL 1	AUTO	N/A	N/A	300	0	0	0	16	2624	63
7135AV	IDE(AT)	135	3.50	HH	U		N/A	N/A	0	0	0	0	13	966	21
71626A	IDE(AT)	1626	3.50	3H	RLL 1	AUTO	N/A	N/A	300	6	3721	0	16	3158	63
71626AP	IDE(AT)	1	3.50	3H	RLL 1	AUTO	N/A	N/A	300	6	3	0	16	3158	63
71670A	IDE(AT)	1670	3.50	3H	RLL 1	AUTO	N/A	N/A	300	5	4702	0	16	3244	63
71670AP	IDE(AT)	1670	3.50	3H	RLL 1	AUTO	N/A	N/A	300	5	4702	0	16	3244	63
71687A		0								0	0	0			
71687AP		0								0	0	0			
7170A	IDE(AT)	170	3.50	3H	RLL 1	AUTO	N/A	N/A	150	0	0	0	15	866	26
7171A	IDE(AT)	171	3.50	3H	RLL 1	AUTO	N/A	N/A	150	0	0	0	15	866	26
72004A	IDE(AT)	2004	3.50	3H	RLL 1	AUTO	N/A	N/A	300	6	4702	0	16	3893	63
72004AP	IDE(AT)	2004	3.50	3H	RLL 1	AUTO	N/A	N/A	300	6	4702	0	16	3893	63
72025AP	IDE(AT)	2025	3.50	3H	RLL 1	AUTO	N/A	N/A	300	0	0	0	16	3936	63
7213A	IDE(AT)	212.78	3.50	3H	RLL 1	AUTO	N/A	N/A	150	4	1690	0	16	683	38
7213S	SCSI	212.78	3.50	3H	RLL 1	AUTO	N/A	N/A	150	4	1690	0			
7245A	IDE(AT)	245.57	3.50	3H	RLL 1	AUTO	N/A	N/A	150	4	1944	0	16	967	31
7245S	SCSI	245	3.50	3H	RLL 1	AUTO	N/A	N/A	150	4	1944	0			
72577AP	IDE(AT)	2577	3.50	3H	RLL 1	AUTO	N/A	N/A	300	0	0	0	16	4996	63
72700AP	IDE(AT)	2700	3.50	3H	RLL 1	AUTO	N/A	N/A	300	0	0	0	16	5248	63
7270AV	IDE(AT)	270	3.50	HH	U		N/A	N/A	0	0	0	0	11	959	50
7270S	SCSI	270	3.50	3H	RLL 1	AUTO	N/A	N/A	300	4	1629	0			
7273A	IDE(AT)	275.39	3.50	3H	RLL 1	AUTO	N/A	N/A	150	2	2771	0	16	967	31
7273A (NEWER VER)	IDE(AT)	273	3.50	3H	RLL 1	AUTO	N/A	N/A	300	2	2793	0	16	1012	33
7290A	IDE(AT)	290	3.50	HH	RLL 1	AUTO	N/A	N/A	150	0	0	0	9	998	63
7290S	SCSI	290	3.50	3H	RLL 1	AUTO	N/A	N/A	0	4	1765	0			
7345A	IDE(AT)	345	3.50	3H	RLL 1	AUTO	N/A	N/A	150	4	0	0	15	790	57
7345S	SCSI	345		–	RLL 1	AUTO	N/A	N/A	0	4	2219	0			
7405A	IDE(AT)	405	3.50	3H	U	AUTO	N/A	N/A	150	0	0	0	16	989	50
7405AV	IDE(AT)	405	3.50	HH	U		N/A	N/A	0	0	0	0	16	989	50
7420AV	IDE(AT)	420	3.50	HH	U		N/A	N/A	0	0	0	0	16	986	52
7425AV	IDE(AT)	425	3.50	3H	U	AUTO	N/A	N/A	0	0	0	0	16	1002	52
7540A	IDE(AT)	528.4	3.50	3H	RLL 1	AUTO	N/A	N/A	150	4	2771	0	16	1024	63
7540AV	IDE(AT)	540	3.50	HH	U		N/A	N/A	0	0	0	0	16	1046	63
7541A	IDE(AT)	541	3.50	4H	RLL 1	AUTO	N/A	N/A	300	2	3	0	16	1,052	63
7541AP	IDE(AT)	541	3.50	4H	RLL 1	AUTO	N/A	N/A	300	2	3	0	16	1,052	63
7546A	IDE(AT)	528.48	3.50	3H	RLL 1	AUTO	N/A	N/A	150	4	2771	0	16	1024	63
7546A (NEWER VER)	IDE(AT)	547	3.50	3H	RLL 1	AUTO	N/A	N/A	300	4	2793	0	16	1060	63
7668A	IDE(AT)	668	3.50	3H	RLL 1	AUTO	N/A	N/A	300	2	4702	0	16	1297	63
7668AP	IDE(AT)	668	3.50	3H	RLL 1	AUTO	N/A	N/A	300	2	4702	0	16	1297	63

MAXTOR CORPORATION

MODEL	INT	CAP	FF	HGT	CODE	LZ	WP	RWC	MTBF	PHDS	PCYL	PSPT	LHDS	LCYL	LSPT
7850AV	IDE(AT)	850	3.50	3H	U	AUTO	N/A	N/A	0	0	0	0	16	1648	63
8050S	SCSI	50	3.50	HH	RLL 2	AUTO	N/A	N/A	30	4	809	26			
8051A	IDE(AT)	42.72	3.50	HH	RLL 2	AUTO	N/A	N/A	150	4	745	28	4	745	28
8051S	SCSI	44.9	3.50	HH	RLL 2	AUTO	N/A	N/A	150	4	783	0			
80875A	IDE(AT)	875	3.50	3H		AUTO	N/A	N/A		0	0	0	16	1700	63
80875A2	IDE(AT)	875	3.50			AUTO	N/A	N/A	0	0	0	0			
81080A3	IDE(AT)	1080	3.50			AUTO	N/A	N/A	0	0	0	0			
81081A2	IDE(AT)	1081	3.50	3H						0	0	0	16	2100	63
81275A3	IDE(AT)	1275	3.50			AUTO	N/A	N/A	0	0	0	0			
81280A	IDE(AT)	213	2.50		U					0	0	0	16	684	38
81280A2	IDE(AT)	1280	3.50	3H		AUTO	N/A	N/A		0	0	0	16	2481	63
81312A	IDE(AT)	1312	3.50	3H		AUTO	N/A	N/A		0	0	0	16	2548	63
81312A3	IDE(AT)	1312	3.50			AUTO	N/A	N/A	0	0	0	0			
81620A3	IDE(AT)	1620	3.50	3H						0	0	0	16	3250	63
81630A4	IDE(AT)	1630	3.50			AUTO	N/A	N/A	150	0	0	0			
81750A	IDE(AT)	1750	3.50	3H		AUTO	N/A	N/A		0	0	0	16	3400	63
81750A2	IDE(AT)	1750	3.50			AUTO	N/A	N/A	0	0	0	0			
81750A4	IDE(AT)	1750	3.50			AUTO	N/A	N/A	0	0	0	0			
81750D2	IDE(AT)	1750	3.50			AUTO	N/A	N/A	0	0	0	0			
82100A4	IDE(AT)	2100	3.50	3H						0	0	0	16	4092	63
82160D2	IDE(AT)	2160	3.50			AUTO	N/A	N/A	0	0	0	0			
82187A	IDE(AT)	2187	3.50	3H		AUTO	N/A	N/A		0	0	0	16	4248	63
82187A5	IDE(AT)	2187	3.50			AUTO	N/A	N/A	0	0	0	0			
82400A4	IDE(AT)	2400	3.50		RLL 8	AUTO	N/A	N/A	0	0	0	0			
82559A4	IDE(AT)	2559	3.50		RLL 8	AUTO	N/A	N/A	0	0	0	0			
82560A	IDE(AT)	2560	3.50	3H	RLL					0	0	0	16	4962	63
82560A3	IDE(AT)	2560	3.50			AUTO	N/A	N/A	0	0	0	0			
82560A4	IDE(AT)	2560	3.50	3H						0	0	0	16	4962	63
82560D3	IDE(AT)	3240	3.50			AUTO	N/A	N/A	0	0	0	0			
82577A6	IDE(AT)	2577	3.50			AUTO	N/A	N/A	0	0	0	0			
82580A5	IDE(AT)	2580	3.50	3H						0	0	0	16	5004	63
82625A	IDE(AT)	2625	3.50	3H		AUTO	N/A	N/A		0	0	0	16	5100	63
82625A6	IDE(AT)	2625	3.50			AUTO	N/A	N/A	0	0	0	0			
83062A	IDE(AT)	3062	3.50	3H		AUTO	N/A	N/A		0	0	0	16	5948	63
83062A7	IDE(AT)	3062	3.50			AUTO	N/A	N/A	0	0	0	0			
83200A5	IDE(AT)	3200	3.50			AUTO	N/A	N/A	0	0	0	0			
83200A6	IDE(AT)	3200	3.50			AUTO	N/A	N/A	0	0	0	0			
83200A8	IDE(AT)	3200	3.50			AUTO	N/A	N/A	0	0	0	0			
83201A6	IDE(AT)	3201	3.50	3H						0	0	0	16	6218	63
83202A6	IDE(AT)	3202	3.50			AUTO	N/A	N/A	0	0	0	0			
83209A5	IDE(AT)	3209	3.50			AUTO	N/A	N/A	0	0	0	0			
83240A4	IDE(AT)	3240	3.50		RLL 8	AUTO	N/A	N/A	0	0	0	0			
83240D3	IDE(AT)	3240	3.50			AUTO	N/A	N/A	0	0	0	0			
83240D4	IDE(AT)	3240	3.50			AUTO	N/A	N/A	0	0	0	0			

MAXTOR CORPORATION

MODEL	INT	CAP	FF	HGT	CODE	LZ	WP	RWC	MTBF	PHDS	PCYL	PSPT	LHDS	LCYL	LSPT
83500A	IDE(AT)	3500	3.50	3H		AUTO	N/A	N/A		0	0	0	16	6800	63
83500A4	IDE(AT)	3500	3.50			AUTO	N/A	N/A	0	0	0	0			
83500A8	IDE(AT)	3500	3.50			AUTO	N/A	N/A	0	0	0	0			
83500D4	IDE(AT)	3500	3.50			AUTO	N/A	N/A	0	0	0	0			
83840A	IDE(AT)	3840	3.50	3H	RLL					0	0	0	16	7441	63
83840A6	IDE(AT)	3840	3.50	3H						0	0	0	16	7443	63
84000A6	IDE(AT)	4000	3.50	3H						0	0	0	16	7763	63
84004A8	IDE(AT)	4004	3.50			AUTO	N/A	N/A	0	0	0	0			
84200A8	IDE(AT)	4200	3.50			AUTO	N/A	N/A	0	0	0	0			
84320A5	IDE(AT)	4320	3.50			AUTO	N/A	N/A	0	0	0	0			
84320A8	IDE(AT)	4320	3.50	3H						0	0	0	16	8400	63
84320D4	IDE(AT)	4320	3.50			AUTO	N/A	N/A	0	0	0	0			
84320D5	IDE(AT)	4320	3.50			AUTO	N/A	N/A	0	0	0	0			
85120A	IDE(AT)	5120	3.50	3H	RLL					0	0	0	16	9924	63
85120A6	IDE(AT)	5120	3.50			AUTO	N/A	N/A	0	0	0	0			
85120A8	IDE(AT)	5120	3.50	3H						0	0	0	16	9924	63
85121A8	IDE(AT)	5121	3.50			AUTO	N/A	N/A	0	0	0	0			
85250D6	IDE(AT)	5250	3.50			AUTO	N/A	N/A	0	0	0	0			
86480A8	IDE(AT)	6480	3.50			AUTO	N/A	N/A	0	0	0	0			
86480D6	IDE(AT)	6480	3.50			AUTO	N/A	N/A	0	0	0	0			
86480D8	IDE(AT)	6480	3.50			AUTO	N/A	N/A	0	0	0	0			
87000A8	IDE(AT)	700	3.50			AUTO	N/A	N/A	0	0	0	0			
87000D8	IDE(AT)	7000	3.50			AUTO	N/A	N/A	0	0	0	0			
88400D8	IDE(AT)	8400	3.50			AUTO	N/A	N/A	0	0	0	0			
EXT-4175	ESDI-10MHZ	149.15	5.25	FH					20	7	1224	34			
EXT-4280	ESDI-10MHZ	149.1	5.25	FH					20	11	1224	34			
EXT-4380	ESDI-10MHZ	319.61	5.25	FH	RLL	AUTO	N/A	N/A	20	15	1224	34			
LXT-100S	SCSI	93.07	3.50	HH	RLL 1	AUTO	N/A	N/A	150	8	733	31			
LXT-200A	IDE(AT)	207	3.50	HH	RLL 1	AUTO	N/A	N/A	150	7	1320	0	15	816	32
LXT-200S OPTION A	SCSI	207	3.50	HH	RLL 2	AUTO	N/A	N/A	150	7	1320	0			
LXT-200S OPTION B	SCSI	207	3.50	HH	RLL 2	AUTO	N/A	N/A	150	7	1320	0			
LXT-213A	IDE(AT)	212	3.50	HH	RLL 1	AUTO	N/A	N/A	150	7	1320	0	16	683	38
LXT-213S OPTION A	SCSI	212	3.50	HH	RLL 1	AUTO	N/A	N/A	150	7	1320	0			
LXT-213S OPTION B	SCSI	212	3.50	HH	RLL 1	AUTO	N/A	N/A	150	7	1320	0			
LXT-340A	IDE(AT)	340	3.50	HH	RLL 2	AUTO	N/A	N/A	150	7	1560	0	16	654	63
LXT-340S OPTION A	SCSI	340	3.50	HH	RLL 2	AUTO	N/A	N/A	150	7	1560	0			
LXT-340S OPTION B	SCSI	340	3.50	HH	RLL 2	AUTO	N/A	N/A	150	7	1560	0			
LXT-437A	IDE(AT)	437	3.50	HH	RLL 2	AUTO	N/A	N/A	150	9	1560	0	16	842	63
LXT-437S OPTION A	SCSI	437	3.50	HH	RLL 2	AUTO	N/A	N/A	150	9	1560	0			
LXT-437S OPTION B	SCSI	437	3.50	HH	RLL 2	AUTO	N/A	N/A	150	9	1560	0			
LXT-50S	SCSI	46.53	3.50	HH	RLL 1	AUTO	N/A	N/A	40	4	733	31			
LXT-535A	IDE(AT)	528.48	3.50	HH	RLL 2	AUTO	N/A	N/A	150	11	1560	0	16	1024	63
LXT-535S OPTION A	SCSI	535	3.50	HH	RLL 2	AUTO	N/A	N/A	150	11	1560	0			

MAXTOR CORPORATION

MODEL	INT	CAP	FF	HGT	CODE	LZ	WP	RWC	MTBF	PHDS	PCYL	PSPT	LHDS	LCYL	LSPT
LXT-535S OPTION B	SCSI	535	3.50	HH	RLL 2	AUTO	N/A	N/A	150	11	1560	0			
MX9217SDN	SCSI-2 D	2170	3.50	3H	PRML	AUTO	N/A	N/A	800	9	3703	0			
MX9217SDW	SCSI-2 WD	2170	3.50	3H	PRML	AUTO	N/A	N/A	800	9	3703	0			
MX9217SSN	SCSI-2	2170	3.50	3H	PRML	AUTO	N/A	N/A	800	9	3703	0			
MX9217SSW	SCSI SCA	2170	3.50	3H	PRML	AUTO	N/A	N/A	800	9	3703	0			
MXT-1240S	SCSI-2	1240	3.50	3H	RLL 1	AUTO	N/A	N/A	300	0	0	0			
MXT-540A	IDE(AT)	528.48	3.50	3H	RLL 1	AUTO	N/A	N/A	300	0	0	0	16	1024	63
MXT-540AL	IDE(AT)	528.5	3.50	3H	RLL 1	AUTO	N/A	N/A	300	7	2234	0	16	1024	63
MXT-540SL	SCSI-2	540	3.50	3H	RLL 1	AUTO	N/A	N/A	300	0	0	0			
PANTHER P0-12S	SCSI-2	1051	5.25	FH	RLL	AUTO	N/A	N/A	100	15	1795	0			
PANTHER P1-08E	ESDI-24MHZ	696.4	5.25	FH	RLL	AUTO	N/A	N/A	100	9	1778	85			
PANTHER P1-08S	SCSI-2	696.4	5.25	FH	RLL	AUTO	N/A	N/A	100	9	1778	85			
PANTHER P1-12E	ESDI-24MHZ	1051.43	5.25	FH	AUTO	N/A	N/A	100K	0	15	1778	77			
PANTHER P1-12S	SCSI-2	1005.48	5.25	FH	RLL	AUTO	N/A	N/A	100	19	1216	85			
PANTHER P1-13E	ESDI-24MHZ	1160.67	5.25	FH	AUTO	N/A	N/A	N/A	100	15	1778	85			
PANTHER P1-16E	ESDI-24MHZ	1331.82	5.25	FH	RLL	AUTO	N/A	N/A	100	19	1778	77			
PANTHER P1-17E	ESDI-24MHZ	1470.19	5.25	FH	RLL	AUTO	N/A	N/A	100	19	1778	85			
PANTHER P1-17S	SCSI-2	1470.19	5.25	FH	RLL	AUTO	N/A	N/A	100	19	1778	85			
RXT-800 SERIES	SCSI	0		–	RLL				0	0	0	0			
XT 8800E	ESDI-10MHZ	694.68	5.25	FH	RLL 1	AUTO	N/A	N/A	150	15	1274	71			
XT-1050	ST-506/412	39.25	5.25	FH	MFM				150	5	902	17			
XT-1065	ST-506/412	55.93	5.25	FH	MFM		919	919	150	7	918	17			
XT-1085	ST-506/412	71.3	5.25	FH	MFM		N/A	N/A	150	8	1024	17			
XT-1105	ST-506/412	87.89	5.25	FH	MFM		919	919	150	11	918	17			
XT-1120R	ST-506/412	104.85	5.25	FH	RLL 2		N/A	N/A	70	8	1024	25			
XT-1140	ST-506/412	119.85	5.25	FH	MFM		N/A	N/A	150	15	918	17			
XT-1240R	ST-506/412	196.6	5.25	FH	RLL 2		N/A	N/A	150	15	1024	25			
XT-2085	ST-506/412	74.57	5.25	FH	MFM		1,225	1,225	150	7	1224	17			
XT-2140	ST-506/412	117.19	5.25	FH	MFM		1,225	1,225	150	11	1224	17			
XT-2190	ST-506/412	159.8	5.25	FH	MFM		N/A	N/A	150	15	1224	17			
XT-3170 NON-SHROUDE	SCSI	141	5.25	FH	RLL	AUTO	N/A	N/A	20	9	1224	25			
XT-3170 SHROUDED	SCSI	141	5.25	FH	RLL	AUTO	N/A	N/A	20	9	1224	25			
XT-3280 NON-SHROUDE	SCSI	235	5.25	FH	RLL	AUTO	N/A	N/A	20	15	1224	25			
XT-3280 SHROUDED	SCSI	235	5.25	FH	RLL	AUTO	N/A	N/A	20	15	1224	25			
XT-3380	SCSI	319.61	5.25	FH	RLL	AUTO	N/A	N/A	20	15	1224	34			
XT-4170E	ESDI-10MHZ	153.53	5.25	FH	RLL 1	AUTO	N/A	N/A	30	7	1224	36			
XT-4170S PCB REV. 3	SCSI	157.92	5.25	FH	RLL 1	AUTO	N/A	N/A	150	7	1224	36			
XT-4170S PCB REV. 1	SCSI	157.92	5.25	FH	RLL 1	AUTO	N/A	N/A	150	7	1224	36			
XT-4230E	ESDI-10MHZ	197.4	5.25	FH	RLL 1	AUTO	N/A	N/A	150	9	1224	36			
XT-4280E	ESDI-10MHZ	241.27	5.25	FH	RLL	AUTO	N/A	N/A	30	11	1224	35			
XT-4280S PCB REV. 3	SCSI	248.16	5.25	FH	RLL	AUTO	N/A	N/A	30	11	1224	36			
XT-4280S PCB REV. 1	SCSI	248.16	5.25	FH	RLL	AUTO	N/A	N/A	30	11	1224	36			
XT-4380E	ESDI-10MHZ	329.01	5.25	FH	RLL 1	AUTO	N/A	N/A	30	15	1224	36			
XT-4380EF	ESDI-10MHZ	329.01	5.25	FH	RLL 1	AUTO	N/A	N/A	30	15	1224	36			

MAXTOR CORPORATION

MODEL	INT	CAP	FF	HGT	CODE	LZ	WP	RWC	MTBF	PHDS	PCYL	PSPT	LHDS	LCYL	LSPT
XT-4380F PCB REV.3	SCSI	338.41	5.25	FH	RLL 1	AUTO	N/A	N/A	150	15	1224	36			
XT-4380S PCB REV. 3	SCSI	338.41	5.25	FH	RLL 1	AUTO	N/A	N/A	150	15	1224	36			
XT-4380S PCB REV. 1	SCSI	338.41	5.25	FH	RLL 1	AUTO	N/A	N/A	150	15	1224	36			
XT-8380DS	SCSI	360	5.25	FH	U	AUTO	N/A	N/A	150	8	1632	0			
XT-8380E VER. 1	ESDI-15MHZ	360.97	5.25	FH	RLL 1	AUTO	N/A	N/A	150	8	1632	54			
XT-8380E VER. 2	ESDI-15MHZ	360.97	5.25	FH	RLL 1	AUTO	N/A	N/A	150	8	1632	54			
XT-8380E VER. 3	ESDI-15MHZ	360.97	5.25	FH	RLL 1	AUTO	N/A	N/A	150	8	1632	54			
XT-8380E VER. 4	ESDI-15MHZ	360.97	5.25	FH	RLL 1	AUTO	N/A	N/A	150	8	1632	54			
XT-8380E VER. 5	ESDI-15MHZ	360.97	5.25	FH	RLL 1	AUTO	N/A	N/A	150	8	1632	54			
XT-8380E VER. 6	ESDI-15MHZ	360.97	5.25	FH	RLL 1	AUTO	N/A	N/A	150	8	1632	54			
XT-8380EH VER. 1	ESDI-15MHZ	360.97	5.25	FH	RLL 1	AUTO	N/A	N/A	150	8	1632	54			
XT-8380EH VER. 2	ESDI-15MHZ	360.97	5.25	FH	RLL 1	AUTO	N/A	N/A	150	8	1632	54			
XT-8380EH VER. 3	ESDI-15MHZ	360.97	5.25	FH	RLL 1	AUTO	N/A	N/A	150	8	1632	54			
XT-8380EH VER. 4	ESDI-15MHZ	360.97	5.25	FH	RLL 1	AUTO	N/A	N/A	150	8	1632	54			
XT-8380EH VER. 5	ESDI-15MHZ	360.97	5.25	FH	RLL 1	AUTO	N/A	N/A	150	8	1632	54			
XT-8380EH VER. 6	ESDI-15MHZ	360.97	5.25	FH	RLL 1	AUTO	N/A	N/A	150	8	1632	54			
XT-8380S	SCSI	360.97	5.25	FH	RLL 1	AUTO	N/A	N/A	150	8	1632	54			
XT-8380SH	SCSI	360.3	5.25	FH	RLL 1	AUTO	N/A	N/A	150	8	1632	54			
XT-8610E VER. 1	ESDI-15MHZ	541.45	5.25	FH	RLL 1	AUTO	N/A	N/A	150	12	1632	54			
XT-8610E VER. 2	ESDI-15MHZ	541.45	5.25	FH	RLL 1	AUTO	N/A	N/A	150	12	1632	54			
XT-8610E VER. 3	ESDI-15MHZ	541.45	5.25	FH	RLL 1	AUTO	N/A	N/A	150	12	1632	54			
XT-8610E VER. 4	ESDI-15MHZ	541.45	5.25	FH	RLL 1	AUTO	N/A	N/A	150	12	1632	54			
XT-8610E VER. 5	ESDI-15MHZ	541.45	5.25	FH	RLL 1	AUTO	N/A	N/A	150	12	1632	54			
XT-8610E VER. 6	ESDI-15MHZ	541.45	5.25	FH	RLL 1	AUTO	N/A	N/A	150	12	1632	54			
XT-8702S	SCSI	616.68	5.25	FH	RLL 1	AUTO	N/A	N/A	150	15	1487	54			
XT-8760DS	SCSI	676	5.25	FH	U	AUTO	N/A	N/A	150	15	1632	0			
XT-8760E VER. 1	ESDI-15MHZ	676.82	5.25	FH	RLL 1	AUTO	N/A	N/A	150	15	1632	54			
XT-8760E VER. 2	ESDI-15MHZ	676.82	5.25	FH	RLL 1	AUTO	N/A	N/A	150	15	1632	54			
XT-8760E VER. 3	ESDI-15MHZ	676.82	5.25	FH	RLL 1	AUTO	N/A	N/A	150	15	1632	54			
XT-8760E VER. 4	ESDI-15MHZ	676.82	5.25	FH	RLL 1	AUTO	N/A	N/A	150	15	1632	54			
XT-8760E VER. 5	ESDI-15MHZ	676.82	5.25	FH	RLL 1	AUTO	N/A	N/A	150	15	1632	54			
XT-8760E VER. 6	ESDI-15MHZ	676.82	5.25	FH	RLL 1	AUTO	N/A	N/A	150	15	1632	54			
XT-8760EH VER. 1	ESDI-15MHZ	676.82	5.25	FH	RLL 1	AUTO	N/A	N/A	150	15	1632	54			
XT-8760EH VER. 2	ESDI-15MHZ	676.82	5.25	FH	RLL 1	AUTO	N/A	N/A	150	15	1632	54			
XT-8760EH VER. 3	ESDI-15MHZ	676.82	5.25	FH	RLL 1	AUTO	N/A	N/A	150	15	1632	54			
XT-8760EH VER. 4	ESDI-15MHZ	676.82	5.25	FH	RLL 1	AUTO	N/A	N/A	150	15	1632	54			
XT-8760EH VER. 5	ESDI-15MHZ	676.82	5.25	FH	RLL 1	AUTO	N/A	N/A	150	15	1632	54			
XT-8760EH VER. 6	ESDI-15MHZ	676.82	5.25	FH	RLL 1	AUTO	N/A	N/A	150	15	1632	54			
XT-8760S	SCSI	676.82	5.25	FH	RLL 1	AUTO	N/A	N/A	150	15	1632	54			
XT-8760SH	SCSI	675.6	5.25	FH	RLL 1	AUTO	N/A	N/A	150	15	1632	54			
XT-8800E	ESDI-24MHZ	694	5.25	FH	RLL 1	AUTO	N/A	N/A	150	15	1274	71			
XT81000E	ESDI-24MHZ	889.89	5.25	FH	RLL 1	AUTO	N/A	N/A	150	15	1632	71			

MEGA DRIVE SYSTEMS

MODEL	INT	CAP	FF	HGT	CODE	LZ	WP	RWC	MTBF	PHDS	PCYL	PSPT	LHDS	LCYL	LSPT
P105	SCSI	103.3	3.50	HH	RLL 2	AUTO	N/A	N/A	50	6	1019	33			
P120	SCSI	120	3.50	HH	RLL 1	AUTO	N/A	N/A	50	5	1123	0			
P170	SCSI	168	3.50	HH	RLL 1	AUTO	N/A	N/A	50	7	1123	0			
P210	SCSI	210	3.50	HH	RLL 1	AUTO	N/A	N/A	50	7	1156	0			
P320	SCSI	320	3.50	HH	RLL 1	AUTO	N/A	N/A	150	15	886	0			
P42	SCSI	42.27	3.50	HH	RLL 2	AUTO	N/A	N/A	50	3	834	33			
P425	SCSI	426	3.50	HH	RLL 1	AUTO	N/A	N/A	75	9	1512	0			
P84	SCSI	84.54	3.50	HH	RLL 2	AUTO	N/A	N/A	50	6	834	33			

MEMOREX TELEX CORPORATION

MODEL	INT	CAP	FF	HGT	CODE	LZ	WP	RWC	MTBF	PHDS	PCYL	PSPT	LHDS	LCYL	LSPT
321	ST-506/412	5.57	5.25	FH	MFM		128	321	0	2	320	17			
322	ST-506/412	11.14	5.25	FH	MFM		128	321	0	4	320	17			
323	ST-506/412	16.71	5.25	FH	MFM		128	321	0	6	320	17			
324	ST-506/412	22.28	5.25	FH	MFM		128	321	0	8	320	17			
450	ST-506/412	10.65	5.25	FH	MFM		350	321	0	2	612	17			
512	ST-506/412	25.09	5.25	FH	MFM		480	321	0	3	961	17			
513	ST-506/412	41.82	5.25	FH	MFM		480	321	0	5	961	17			
514	ST-506/412	58.55	5.25	FH	MFM		480	961	0	7	961	0			

MEMTECH TECHNOLOGY

MODEL	INT	CAP	FF	HGT	CODE	LZ	WP	RWC	MTBF	PHDS	PCYL	PSPT	LHDS	LCYL	LSPT
AT2500-12	IDE(AT)	12	2.50	5H					128	16	44	32			
AT2500-192	IDE(AT)	192	2.50	5H					128	16	761	32			
AT2500-24	IDE(AT)	24	2.50	5H					128	16	92	32			
AT2500-48	IDE(AT)	48	2.50	5H					128	16	188	32			
AT2500-96	IDE(AT)	96	2.50	5H					128	16	380	32			
SC35	SCSI	864	3.50	HH						0	0	32			

MICROPOLIS CORPORATION

MODEL	INT	CAP	FF	HGT	CODE	LZ	WP	RWC	MTBF	PHDS	PCYL	PSPT	LHDS	LCYL	LSPT
1302	ST-506/412	20.4	5.25	FH	MFM	830	N/A	N/A	20	3	830	17			
1303	ST-506/412	34	5.25	FH	MFM	830	N/A	N/A	20	5	830	17			
1304	ST-506/412	40.8	5.25	FH	MFM	830	N/A	N/A	20	6	830	17			
1323	ST-506/412	35.6	5.25	FH	MFM	AUTO	N/A	N/A	35	4	1024	17			
1323A	ST-506/412	44.5	5.25	FH	MFM	AUTO	N/A	N/A	35	5	1024	17			
1324	ST-506/412	53.4	5.25	FH	MFM	1024	N/A	N/A	35	6	1024	17			
1324A	ST-506/412	62.3	5.25	FH	MFM	1024	N/A	N/A	35	7	1024	17			
1325	ST-506/412	71.3	5.25	FH	MFM	1024	N/A	N/A	35	8	1024	17			
1333	ST-506/412	35.6	5.25	FH	MFM	1024	N/A	N/A	25	4	1024	17			
1333A	ST-506/412	44.5	5.25	FH	MFM	1024	N/A	N/A	25	5	1024	17			
1334	ST-506/412	53.4	5.25	FH	MFM	1024	N/A	N/A	25	6	1024	17			
1334A	ST-506/412	62.3	5.25	FH	MFM	1024	N/A	N/A	25	7	1024	17			
1335	ST-506/412	71.3	5.25	FH	MFM	1024	N/A	N/A	25	8	1024	17			
1352	ESDI-10MHZ	37.74	5.25	FH	RLL 2	AUTO	N/A	N/A	0	2	1024	36			
1352A	ESDI-10MHZ	56.62	5.25	FH	RLL 2	AUTO	N/A	N/A	0	3	1024	36			
1353	ESDI-10MHZ	75.49	5.25	FH	RLL 2	AUTO	N/A	N/A	150	4	1024	36			
1353A	ESDI-10MHZ	94.37	5.25	FH	RLL 2	AUTO	N/A	N/A	150	5	1024	36			
1354	ESDI-10MHZ	113.24	5.25	FH	RLL 2	AUTO	N/A	N/A	150	6	1024	36			
1354A	ESDI-10MHZ	132.12	5.25	FH	RLL 2	AUTO	N/A	N/A	150	7	1024	36			
1355	ESDI-10MHZ	150.99	5.25	FH	RLL 2	AUTO	N/A	N/A	150	8	1024	36			
1373	SCSI	72.97	5.25	FH	RLL 2	AUTO	N/A	N/A	30	4	1018	35			
1373A	SCSI	91.21	5.25	FH	RLL 2	AUTO	N/A	N/A	30	5	1018	35			
1374	SCSI	109.45	5.25	FH	RLL 2	AUTO	N/A	N/A	30	6	1018	35			
1374A	SCSI	127.69	5.25	FH	RLL 2	AUTO	N/A	N/A	30	7	1018	35			
1375	SCSI	145.94	5.25	FH	RLL 2	AUTO	N/A	N/A	30	8	1018	35			
1516-10S	ESDI-24MHZ	678.29	5.25	FH	RLL 2	AUTO	N/A	N/A	0	10	1840	72			
1517-13	ESDI-24MHZ	922.52	5.25	FH	RLL 2	AUTO	N/A	N/A	0	13	1925	72			
1518	ESDI-10MHZ	1338	3.50			AUTO	N/A	N/A	15000	0	2100	0			
1518-14	ESDI-24MHZ	993.48	5.25	FH	RLL 1	AUTO	N/A	N/A	0	14	1925	72			
1518-15	ESDI-24MHZ	1354.75	5.25	FH	RLL 1	AUTO	N/A	N/A	150	15	2100	84			
1528-15	SCSI-2	1354.75	5.25	FH	RLL 1	AUTO	N/A	N/A	150	15	2100	84			
1538-15	ESDI-20MHZ	871.61	5.25	FH	RLL 1	AUTO	N/A	N/A	150	15	1669	68			
1548	SCSI-2	1748	5.25	FH	RLL 1	AUTO	N/A	N/A	150	15	2089	0			
1548-15	SCSI-2	1748	5.25	FH	RLL 1	AUTO	N/A	N/A	150	15	2089	0			
1554-07	ESDI-10MHZ	157.92	5.25	FH	RLL 2	AUTO	N/A	N/A	150	7	1224	36			
1555-08	ESDI-10MHZ	180.48	5.25	FH	RLL 2	AUTO	N/A	N/A	150	8	1224	36			
1555-09	ESDI-10MHZ	203.04	5.25	FH	RLL 2	AUTO	N/A	N/A	150	9	1224	36			
1556-10	ESDI-10MHZ	225.6	5.25	FH	RLL 2	AUTO	N/A	N/A	150	10	1224	36			
1556-11	ESDI-10MHZ	248.16	5.25	FH	RLL 2	AUTO	N/A	N/A	150	11	1224	36			
1557-12	ESDI-10MHZ	270.72	5.25	FH	RLL 2	AUTO	N/A	N/A	150	12	1224	36			
1557-13	ESDI-10MHZ	293.28	5.25	FH	RLL 2	AUTO	N/A	N/A	150	13	1224	36			
1558	ESDI-10MHZ	338.41	5.25	FH	RLL 2	AUTO	N/A	N/A	150	15	1224	36			
1558-14	ESDI-10MHZ	315.85	5.25	FH	RLL 2	AUTO	N/A	N/A	150	12	1224	36			
1558-15	ESDI-10MHZ	338.41	5.25	FH	RLL 2	AUTO	N/A	N/A	150	15	1224	36			

MICROPOLIS CORPORATION

MODEL	INT	CAP	FF	HGT	CODE	LZ	WP	RWC	MTBF	PHDS	PCYL	PSPT	LHDS	LCYL	LSPT
1564-07	ESDI-15MHZ	315.85	5.25	FH	RLL 2	AUTO	N/A	N/A	150	7	1632	54			
1565-08	ESDI-15MHZ	360.97	5.25	FH	RLL 2	AUTO	N/A	N/A	150	8	1632	54			
1565-09	ESDI-15MHZ	406.09	5.25	FH	RLL 2	AUTO	N/A	N/A	150	9	1632	54			
1566-10	ESDI-15MHZ	451.21	5.25	FH	RLL 2	AUTO	N/A	N/A	150	10	1632	54			
1566-11	ESDI-15MHZ	496.33	5.25	FH	RLL 2	AUTO	N/A	N/A	150	11	1632	54			
1567-12	ESDI-15MHZ	541.45	5.25	FH	RLL 2	AUTO	N/A	N/A	150	12	1632	54			
1567-13	ESDI-15MHZ	586.57	5.25	FH	RLL 2	AUTO	N/A	N/A	150	13	1632	54			
1568-14	ESDI-15MHZ	631.7	5.25	FH	RLL 2	AUTO	N/A	N/A	150	14	1632	54			
1568-15	ESDI-15MHZ	676.82	5.25	FH	RLL 2	AUTO	N/A	N/A	150	15	1632	54			
1574-07 VERSION 1	SCSI	157.92	5.25	FH	RLL 2	AUTO	N/A	N/A	150	7	1224	36			
1574-07 VERSION 2	SCSI	157.92	5.25	FH	RLL 2	AUTO	N/A	N/A	150	7	1224	36			
1575-08 VERSION 1	SCSI	180.48	5.25	FH	RLL 2	AUTO	N/A	N/A	150	8	1224	36			
1575-08 VERSION 2	SCSI	180.48	5.25	FH	RLL 2	AUTO	N/A	N/A	150	8	1224	36			
1575-09 VERSION 1	SCSI	203.04	5.25	FH	RLL 2	AUTO	N/A	N/A	150	9	1224	36			
1575-09 VERSION 2	SCSI	203.04	5.25	FH	RLL 2	AUTO	N/A	N/A	150	9	1224	36			
1576-10 VERSION 1	SCSI	225.6	5.25	FH	RLL 2	AUTO	N/A	N/A	150	10	1224	36			
1576-10 VERSION 2	SCSI	225.6	5.25	FH	RLL 2	AUTO	N/A	N/A	150	10	1224	36			
1576-11 VERSION 1	SCSI	248.16	5.25	FH	RLL 2	AUTO	N/A	N/A	150	11	1224	36			
1576-11 VERSION 2	SCSI	248.16	5.25	FH	RLL 2	AUTO	N/A	N/A	150	11	1224	36			
1577-12 VERSION 1	SCSI	270.72	5.25	FH	RLL 2	AUTO	N/A	N/A	150	12	1224	36			
1577-12 VERSION 2	SCSI	270.72	5.25	FH	RLL 2	AUTO	N/A	N/A	150	12	1224	36			
1577-13 VERSION 1	SCSI	293.28	5.25	FH	RLL 2	AUTO	N/A	N/A	150	13	1224	36			
1577-13 VERSION 2	SCSI	293.28	5.25	FH	RLL 2	AUTO	N/A	N/A	150	13	1224	36			
1578-14 VERSION 1	SCSI	315.85	5.25	FH	RLL 2	AUTO	N/A	N/A	150	14	1224	36			
1578-14 VERSION 2	SCSI	315.85	5.25	FH	RLL 2	AUTO	N/A	N/A	150	14	1224	36			
1578-15 VERSION 1	SCSI	338.41	5.25	FH	RLL 2	AUTO	N/A	N/A	150	15	1224	36			
1578-15 VERSION 2	SCSI	338.41	5.25	FH	RLL 2	AUTO	N/A	N/A	150	15	1224	36			
1586-11	SCSI	495.12	5.25	FH	RLL 2	AUTO	N/A	N/A	150	11	1628	54			
1587-12	SCSI	540.13	5.25	FH	RLL 2	AUTO	N/A	N/A	150	12	1628	54			
1587-13	SCSI	585.14	5.25	FH	RLL 2	AUTO	N/A	N/A	150	13	1628	54			
1588-14	SCSI	630.15	5.25	FH	RLL 2	AUTO	N/A	N/A	150	14	1628	54			
1588-15	SCSI	676.82	5.25	FH	RLL 2	AUTO	N/A	N/A	150	15	1632	54			
1588T-15	SCSI	676.82	5.25	FH	RLL 2	AUTO	N/A	N/A	150	15	1632	54			
1596-10S	SCSI	676.08	5.25	FH	RLL 2	AUTO	N/A	N/A	150	10	1834	72			
1597-13	SCSI	919.64	5.25	FH	RLL 2	AUTO	N/A	N/A	150	13	1919	72			
1598-14	SCSI	976.63	5.25	FH	RLL 2	AUTO	N/A	N/A	150	14	1919	71			
1598-15	SCSI-2	1051.29	5.25	FH	RLL 2	AUTO	N/A	N/A	150	15	1928	71			
1624	SCSI	668	5.25	FH	UNIDEN	AUTO	N/A	N/A	150	7	2089	0			
1624-07	SCSI-2	667	5.25	FH	RLL	AUTO	N/A	N/A	150	7	2099	0			
1653-04	ESDI-10MHZ	92.08	5.25	HH	RLL 2	AUTO	N/A	N/A	150	4	1249	36			
1653-05	ESDI-10MHZ	115.1	5.25	HH	RLL 2	AUTO	N/A	N/A	150	5	1249	36			
1654-06	ESDI-10MHZ	138.12	5.25	HH	RLL 2	AUTO	N/A	N/A	150	6	1249	36			
1654-07	ESDI-10MHZ	161.15	5.25	HH	RLL 2	AUTO	N/A	N/A	150	7	1249	36			
1663-04	ESDI-15MHZ	196.85	5.25	HH	RLL 2	AUTO	N/A	N/A	150	4	1780	54			

MICROPOLIS CORPORATION

MODEL	INT	CAP	FF	HGT	CODE	LZ	WP	RWC	MTBF	PHDS	PCYL	PSPT	LHDS	LCYL	LSPT
1663-05	ESDI-15MHZ	246.06	5.25	HH	RLL 2	AUTO	N/A	N/A	150	5	1780	54			
1664-06	ESDI-15MHZ	295.28	5.25	HH	RLL 2	AUTO	N/A	N/A	150	6	1780	54			
1664-07	ESDI-15MHZ	344.49	5.25	HH	RLL 2	AUTO	N/A	N/A	150	7	1780	54			
1673-04	SCSI	92.08	5.25	HH	RLL 2	AUTO	N/A	N/A	150	4	1249	36			
1673-05	SCSI	115.1	5.25	HH	RLL 2	AUTO	N/A	N/A	150	5	1249	36			
1673-05 MAC	SCSI-MAC	114.73	5.25	HH	RLL	AUTO	N/A	N/A	150	5	1245	36			
1674-06	SCSI	138.12	5.25	HH	RLL 2	AUTO	N/A	N/A	150	6	1249	36			
1674-07	SCSI	161.15	5.25	HH	RLL 2	AUTO	N/A	N/A	150	7	1249	36			
1674-07 MAC	SCSI-MAC	160.63	5.25	HH	RLL	AUTO	N/A	N/A	150	7	1245	36			
1683-04	SCSI	196.41	5.25	HH	RLL 2	AUTO	N/A	N/A	150	4	1776	54			
1683-04 MAC	SCSI-MAC	196.41	5.25	HH	RLL	AUTO	N/A	N/A	150	4	1776	54			
1683-05	SCSI	245.51	5.25	HH	RLL 2	AUTO	N/A	N/A	150	5	1776	54			
1684	SCSI	340	5.25	FH	UNIDEN	AUTO	N/A	N/A	150	7	1774	54			
1684-06	SCSI	294.61	5.25	HH	RLL 2	AUTO	N/A	N/A	150	6	1776	54			
1684-07	SCSI	343.33	5.25	HH	RLL 2	AUTO	N/A	N/A	150	7	1774	54			
1684-07 MAC	SCSI-MAC	343.71	5.25	HH	RLL	AUTO	N/A	N/A	150	7	1776	54			
1743-05	IDE(AT)	81.71	3.50	HH	RLL 2	AUTO	N/A	N/A	0	5	1140	0			
1744-06	IDE(AT)	98.05	3.50	HH	RLL 2	AUTO	N/A	N/A	0	6	1140	0			
1744-07	IDE(AT)	114.4	3.50	HH	RLL 2	AUTO	N/A	N/A	0	7	1140	0			
1745-08	IDE(AT)	130.74	3.50	HH	RLL 2	AUTO	N/A	N/A	0	8	1140	0			
1745-09	IDE(AT)	147.08	3.50	HH	RLL 2	AUTO	N/A	N/A	0	9	1140	0			
1773-05	SCSI	81.71	3.50	HH	RLL 2	AUTO	N/A	N/A	0	5	1140	0			
1774-06	SCSI	98.05	3.50	HH	RLL 2	AUTO	N/A	N/A	0	6	1140	0			
1774-07	SCSI	114.4	3.50	HH	RLL 2	AUTO	N/A	N/A	0	7	1140	0			
1775-08	SCSI	130.74	3.50	HH	RLL 2	AUTO	N/A	N/A	0	8	1140	0			
1775-09	SCSI	147.08	3.50	HH	RLL 2	AUTO	N/A	N/A	0	9	1140	0			
1908	SCSI-2	1408	5.25	HH	UNIDEN	AUTO	N/A	N/A	150	15	2100	0			
1908-15	SCSI-2	1408	5.25	FH	RLL	AUTO	N/A	N/A	150	15	2089	0			
1924	SCSI-2 F	2100	5.25	HH	UNIDEN	AUTO	N/A	N/A	250	21	2267	0			
1924 DIFFERENTIAL	SCSI-2 D	2100	5.25	HH	UNIDEN	AUTO	N/A	N/A	250	21	2267	0			
1924-21	SCSI-2	2100	5.25	FH	RLL	AUTO	N/A	N/A	250	21	2280	0			
1926	SCSI-2	2158	5.25	–	NONE	AUTO	N/A	N/A	300	15	2772	0			
1926-15	SCSI-2	2158	5.25	–	NONE	AUTO	N/A	N/A	300	15	2772	0			
1936	SCSI-2	3000	3.50	HH	UNIDEN	AUTO	N/A	N/A	250	21	2772	0			
1936-21	SCSI-2	3022	5.25	–	RLL 2	AUTO	N/A	N/A	300	21	2772	0			
1936AV	SCSI-2	3000	3.50	HH	UNIDEN	AUTO	N/A	N/A	250	21	2772	0			
1936D	SCSI-2 D	3000	3.50	HH	UNIDEN	AUTO	N/A	N.A	250	21	2772	0			
1991	SCSI-2 F	9091	5.25	FH	RLL 1	AUTO	N/A	N/A	650	21	4446	0			
1991AV	SCSI-2 F	9091	5.25	FH	RLL 1	AUTO	N/A	N/A	650	21	4446	0			
1991W	SCSI-2 FW	9091	5.25	FH	UNIDEN	AUTO	N/A	N/A	650	21	4477	0			
1991WAV	SCSI-2 FW	9091	5.25	FH	UNIDEN	AUTO	N/A	N/A	650	21	4477	0			
2105	SCSI	560	3.50	HH	UNIDEN	AUTO	N/A	N/A	300	8	1744	0			
2105-8	SCSI-2	560	3.50	HH	RLL 1	AUTO	N/A	N/A	300	8	1760	0			
2105A	IDE(AT)	650	3.50	HH	RLL	AUTO	N/A	N/A	300	8	1760	0	16	1084	63

MICROPOLIS CORPORATION

MODEL	INT	CAP	FF	HGT	CODE	LZ	WP	RWC	MTBF	PHDS	PCYL	PSPT	LHDS	LCYL	LSPT
2108	SCSI	693	3.50	HH	UNIDEN	AUTO	N/A	N/A	300	10	1744	0			
2108-6	SCSI-2	666	3.50	HH	RLL 1	AUTO	N/A	N/A	300	10	1745	0			
2112	SCSI	1050	3.50	HH	UNIDEN	AUTO	N/A	N/A	300	15	1744	0			
2112-15 REV. 1	SCSI-2 F	1050	3.50	HH	RLL 1	AUTO			300	15	1747	0			
2112-15 REV. 2	SCSI-2	1050	3.50	FH	RLL 1	AUTO	N/A	N/A	300	15	1747	0			
2112A	IDE(AT)	1050	3.50	HH	RLL	AUTO	N/A	N/A	300	15	1760	0	64	991	32
2112D	SCSI-2 D	1050	3.50	HH	UNIDEN	AUTO	N/A	N/A	300	15	1744	0			
2112DW	SCSI-2 F	1214	3.50			AUTO	N/A	N/A	30000	15	1760	0			
2200WD	SCSI-2	2044	3.50			AUTO	N/A	N/A	30000	15	2372	0			
2205-15	IDE(AT)	528.48	3.50	HH	NONE	AUTO	N/A	N/A	300	5	2372	0	16	1024	63
2205-5	SCSI-2	584	3.50	HH	RLL 1	AUTO	N/A	N/A	300	5	2372	0			
2205A	IDE(AT)	528.48	3.50	HH	NONE	AUTO	N/A	N/A	300	5	2372	0	16	1024	63
2207-6	SCSI-2	700	3.50	HH	RLL 1	AUTO	N/A	N/A	300	6	2372	0			
2210-9	SCSI-2	1050	3.50	HH	RLL 1	AUTO	N/A	N/A	300	9	2372	0			
2210A	IDE(AT)	976	3.50	HH	AUTO	AUTO	N/A	N/A	300	9	2372	0	16	1891	63
2210AV	SCSI-2	1000	3.50	HH	UNIDEN	AUTO	N/A	N/A	300	15	2360	0			
2217-15	SCSI-2	1765	3.50	FH	RLL 1	AUTO	N/A	N/A	300	15	2372	0			
2217-AV	SCSI-2	1700	3.50	HH	NONE	AUTO	N/A	N/A	300	15	2372	0			
2217A	IDE(AT)	1626	3.50	–	NONE	AUTO	N/A	N/A	300	15	2372	0	16	3152	63
3221	SCSI-2 F	2050	3.50	HH	UNIDEN	AUTO	N/A	N/A	650	0	3	0			
3243	SCSI-2 F	4294	3.50	HH	RLL 1	4125			650	10	3957	0			
3243AV	SCSI-2 F	4294	3.50	HH	RLL 1	4125			650	10	3957	0			
3243S	SCSI	4294	3.50	HH	UNIDEN	AUTO	N/A	N/A	650	0	3957	0			
3243W	SCSI-2 FW	4294	3.50	HH	RLL 1	AUTO	N/A	N/A	650	10	3957	0			
3243WAV	SCSI-2 FW	4294	3.50	HH	UNIDEN	AUTO	N/A	N/A	650	10	3957	0			
3243WD	SCSI-2 FWD	4294	3.50	FH	UNIDEN	AUTO	N/A	N/A	650	10	3957	0			
3243WDAV	SCSI-2 FWD	4294	3.50	FH	UNIDEN	AUTO	N/A	N/A	650	10	3957	0			
3387NS	U-SCSI	8700	3.50	HH		AUTO	N/A	N/A	650	0	4811	0			
3387SS	U-SCSI	8700	3.50	HH		AUTO	N/A	N/A	650	0	4811	0			
3387WS	U-SCSI	8700	3.50	HH		AUTO	N/A	N/A	650	0	4811	0			
3391AV	U-SCSI	9103	3.50	HH		AUTO	N/A	N/A	650	0	4735	0			
3391NS	U-SCSI	9103	3.50	3H		AUTO	N/A	N/A	650	0	4735	0			
3391SD	U-SCSI	9103	3.50	3H		AUTO	N/A	N/A	650	0	0	0			
3391SS	U-SCSI	9103	3.50	3H		AUTO	N/A	N/A	650	0	0	0			
3391WAV	U-SCSI	9103	3.50	HH		AUTO	N/A	N/A	650	0	4811	0			
3391WD	U-SCSI WD	9103	3.50	3H					650	0	4735	0			
3391WS	U-SCSI W	9103	3.50	3H		AUTO	N/A	N/A	650	0	0	0			
3418NS	U-SCSI	18250	3.50			AUTO	N/A	N/A	10000	0	7308	0			
3418WS	U-SCSI	18250	3.50			AUTO	N/A	N/A	10000	0	7308	0			
3420AV	U-SCSI	20000	3.50			AUTO	N/A	N/A	10000	0	7308	0			
3420WAV	U-SCSI	20000	3.50			AUTO	N/A	N/A	10000	0	7308	0			
4110 REV. A	SCSI-2	1050	3.50	HH	RLL	AUTO	N/A	N/A	300	9	2372	0			
4110 REV. B	SCSI-2 F	1056	3.50	3H	UNIDEN	AUTO	N/A	N/A	500	16	2048	0			
4110A	IDE(AT)	1057	3.50	3H	UNIDEN	AUTO	N/A	N/A	500	0	0	0	16	2048	63

MICROPOLIS CORPORATION

MODEL	INT	CAP	FF	HGT	CODE	LZ	WP	RWC	MTBF	PHDS	PCYL	PSPT	LHDS	LCYL	LSPT
4110AV	SCSI-2 F	1000	3.50	3H	UNIDEN	AUTO	N/A	N/A	500	16	2048	0			
4210	SCSI-2 F	1060	3.50	3H		AUTO	N/A	N/A	650	0	3290	0			
4210AV	SCSI-2 F	1060	3.50	3H		AUTO	N/A	N/A	650	0	3956	0			
4210W	SCSI-2 FW	1	3.50	HH	UNIDEN	AUTO	N/A	N/A	650	10	4	0			
4210WAV	SCSI-2 FW	1060	3.50	HH	UNIDEN	AUTO	N/A	N/A	650	10	4150	0			
4221	SCSI-2 F	2.05	3.50	3H	UNIDEN	AUTO	N/A	N/A	650	10	4150	0			
4221AV	SCSI-2 F	2.05	3.50	3H	UNIDEN	AUTO	N/A	N/A	650	10	4150	0			
4221W	SCSI-2 FW	2	3.50	HH	UNIDEN	AUTO	N/A	N/A	650	10	4150	0			
4221WAV	SCSI-2 FW	2050	3.50	HH	UNIDEN	AUTO	N/A	N/A	650	10	4150	0			
4221WD	SCSI-2 FWD	2050	3.50	HH	UNIDEN	AUTO	N/A	N/A	650	10	4150	0			
4221WDAV	SCSI-2 FWD	2050	3.50	HH	UNIDEN	AUTO	N/A	N/A	650	10	4150	0			
4341NS	U-SCSI	4130	3.50			AUTO	N/A	N/A	10000	0	4811	0			
4341SS	U-SCSI	4130	3.50	3H		AUTO	N/A	N/A	650	0	4811	0			
4341WS	U-SCSI	4130	3.50	3H		AUTO	N/A	N/A	650	0	4811	0			
4345AV	U-SCSI	4550	3.50	3H		AUTO	N/A	N/A	650	0	4811	0			
4345NS	U-SCSI	4550	3.50	3H		AUTO	N/A	N/A	650	0	4811	0			
4345SD	U-SCSI	455	3.50	3H		AUTO	N/A	N/A	650	0	0	0			
4345SS	U-SCSI	4550	3.50	3H		AUTO	N/A	N/A	650	0	0	0			
4345WAV	U-SCSI	4550	3.50	3H		AUTO	N/A	N/A	650	0	4811	0			
4345WD	U-SCSI	4550	3.50	3H		AUTO	N/A	N/A	650	0	0	0			
4345WS	U-SCSI W	4550	3.50	3H		AUTO	N/A	N/A	650	0	0	0			
4421 (TZ0030-01-7)	SCSI-2 F	2147	3.50	3H	UNIDEN	AUTO	N/A	N/A	650	7	4050	0			
4421 (TZ0030-02-5)	SCSI-2 F	2147	3.50	3H	UNIDEN	AUTO	N/A	N/A	650	7	4050	0			
4421AV	U-SCSI	2050	3.50	HH		AUTO	N/A	N/A	650	0	4050	0			
4525A (VER. 1)	IDE(AT)	2500	3.50	3H		AUTO	N/A	N/A	400	4	6807	0	16	4846	63
4525A (VER. 2)	IDE(AT)	2500	3.50	3H		AUTO	N/A	N/A	400	4	6807	0	16	4846	63
4540A (VER. 1)	IDE(AT)	4000	3.50	3H		AUTO	N/A	N/A	400	6	6807	0	16	7752	63
4540A (VER. 2)	IDE(AT)	4000	3.50	3H		AUTO	N/A	N/A	400	6	6807	0	16	7752	63
4550A (VER. 1)	IDE(AT)	5000	3.50	3H		AUTO	N/A	N/A	400	8	6807	0	16	9692	63
4550A (VER. 2)	IDE(AT)	5000	3.50	3H		AUTO	N/A	N/A	400	8	6807	0	16	9692	63
4691AV	U-SCSI	9100	3.50			AUTO	N/A	N/A	10000	0	7308	512			
4691NS	U-SCSI	9100	3.50			AUTO	N/A	N/A	10000	0	7308	0			
4691WAV	U-SCSI	9100	3.50			AUTO	N/A	N/A	10000	0	7308	0			
4691WS	U-SCSI	9100	3.50			AUTO	N/A	N/A	10000	0	7308	0			
4743NS	U-SCSI	4300	3.50	3H		AUTO	N/A	N/A	400	0	6807	0			
4743SS	U-SCSI	4300	3.50	3H		AUTO	N/A	N/A	400	0	6807	0			
4743WS	U-SCSI	4300	3.50	3H		AUTO	N/A	N/A	400	0	6807	0			

MICROSCIENCE INTERNATIONAL CORPORATION

MODEL	INT	CAP	FF	HGT	CODE	LZ	WP	RWC	MTBF	PHDS	PCYL	PSPT	LHDS	LCYL	LSPT
FH-21200	ESDI-24MHZ	1062.23	5.25	FH	RLL 2	AUTO	N/A	N/A	100	15	1921	72			
FH-21600	ESDI-24MHZ	1418.05	5.25	FH	RLL 2	AUTO	N/A	N/A	100	15	2147	86			
FH-2414	ESDI-15MHZ	366.72	5.25	FH	RLL 2	AUTO	N/A	N/A	100	8	1658	54			
FH-2777	ESDI-15MHZ	687.6	5.25	FH	RLL 2	AUTO	N/A	N/A	100	15	1658	54			

MICROSCIENCE INTERNATIONAL CORPORATION

MODEL	INT	CAP	FF	HGT	CODE	LZ	WP	RWC	MTBF	PHDS	PCYL	PSPT	LHDS	LCYL	LSPT
FH-31200	SCSI	1062.23	5.25	FH	RLL 2	AUTO	N/A	N/A	100	15	1921	72			
FH-31600	SCSI	1418.05	5.25	FH	RLL 2	AUTO	N/A	N/A	100	15	2147	86			
FH-3414	SCSI	366.72	5.25	FH	RLL 2	AUTO	N/A	N/A	100	8	1658	54			
FH-3777	SCSI	687.6	5.25	FH	RLL 2	AUTO	N/A	N/A	100	15	1658	54			
HH-1050	ST-506/412	44.56	5.25	HH	MFM		1,025	1,025	140	5	1024	17			
HH-1060	ST-506/412	65.53	5.25	HH	RLL 2		1,025	1,025	140	5	1024	25			
HH-1075	ST-506/412	62.39	5.25	HH	MFM		1,025	1,025	0	7	1024	17			
HH-1080	ST-506/412	68.15	5.25	HH	RLL 2				50	5	1024	26			
HH-1090	ST-506/412	80.05	5.25	HH	MFM		1,315	1,315	40	7	1314	17			
HH-1095	ST-506/412	95.42	5.25	HH	RLL 2		1,025	1,025	0	7	1024	26			
HH-1120	ST-506/412	122.44	5.25	HH	RLL 2		1,315	1,315	40	7	1314	26			
HH-21200	ESDI-10MHZ	128.45	5.25	HH	RLL 2	AUTO	N/A	N/A	40	7	1024	35			
HH-21600	ESDI-10MHZ	160.06	5.25	HH	RLL 2	AUTO	N/A	N/A	40	7	1276	35			
HH-2414	ESDI-15MHZ	366.72	5.25	HH	RLL 2	AUTO	N/A	N/A	100	8	1658	54			
HH-2777	ESDI-15MHZ	687.6	5.25	HH	RLL 2	AUTO	N/A	N/A	100	15	1658	54			
HH-312	ST-506/412	10.65	5.25	HH	MFM		307	307	0	4	306	17			
HH-3120	SCSI	121.09	5.25	HH	RLL 2	AUTO	N/A	N/A	40	5	1314	36			
HH-315	ST-506/412	10.65	5.25	HH	MFM		307	307	0	4	306	17			
HH-3160	SCSI	169.53	5.25	HH	RLL 2	AUTO	N/A	N/A	40	7	1314	36			
HH-325	ST-506/412	21.3	5.25	HH	MFM		613	613	0	4	612	17			
HH-330	ST-506/412	32.58	5.25	HH	RLL 2		613	613	0	4	612	26			
HH-4050	ST-506/412	44.56	3.50	HH	MFM		1,025	1,025	36	5	1024	17			
HH-4060	ST-506/412	68.15	3.50	HH	RLL 2				36	5	1024	26			
HH-4070	ST-506/412	62.39	3.50	HH	MFM				36	7	1024	17			
HH-4090	ST-506/412	95.42	3.50	HH	RLL 2				36	7	1024	26			
HH-5040	ESDI-10MHZ	45.96	3.50	HH	RLL 2	AUTO	N/A	N/A	36	3	855	35			
HH-5070	ESDI-10MHZ	76.6	3.50	HH	RLL 2	AUTO	N/A	N/A	36	5	855	35			
HH-5070-20	ESDI-10MHZ	86.01	3.50	HH	RLL 2	AUTO	N/A	N/A	36	5	960	35			
HH-5100	ESDI-10MHZ	107.25	3.50	HH	RLL 2	AUTO	N/A	N/A	60	7	855	35			
HH-5100-20	ESDI-10MHZ	120.42	3.50	HH	RLL 2	AUTO	N/A	N/A	60	7	960	35			
HH-5160	ESDI-10MHZ	159.43	3.50	HH	RLL 2	AUTO	N/A	N/A	36	7	1271	35			
HH-6100	SCSI	110.31	3.50	HH	RLL 2	AUTO	N/A	N/A	36	7	855	36			
HH-612	ST-506/412	10.65	5.25	HH	MFM		307	307	0	4	306	17			
HH-625	ST-506/412	21.3	5.25	HH	MFM		613	613	0	4	612	17			
HH-7040	IDE(AT)	47.27	3.50	HH	RLL 2	AUTO	N/A	N/A	36	3	855	36			
HH-7070-20	IDE(AT)	86.01	3.50	HH	RLL 2	960	960	N/A	36	5	960	35	5	960	35
HH-7100-00	IDE(AT)	107.25	3.50	HH	RLL 2	855	855	N/A	36	7	855	35	7	855	35
HH-7100-20	IDE(AT)	120.42	3.50	HH	RLL 2	960	960	N/A	60	7	960	35	7	960	35
HH-7100-21	IDE(AT)	121.31	3.50	HH	RLL 2	992	992	N/A	60	5	1077	44	7	960	35
HH-712	ST-506/412	10.65	5.25	HH	MFM		613	613	0	2	612	17			
HH-712A	ST-506/412	10.65	5.25	HH	MFM				0	2	612	17			
HH-7200	IDE(AT)	201.37	3.50	HH	RLL 2	977	977	N/A	60	7	1277	44	9	992	44
HH-725	ST-506/412	21.3	5.25	HH	MFM		613	613	0	4	612	17			
HH-738	ST-506/412	32.58	5.25	HH	RLL 2		613	613	0	4	612	26			

MICROSCIENCE INTERNATIONAL CORPORATION

MODEL	INT	CAP	FF	HGT	CODE	LZ	WP	RWC	MTBF	PHDS	PCYL	PSPT	LHDS	LCYL	LSPT
HH-7400	IDE(AT)	420	3.50	HH	RLL	AUTO	N/A	N/A	100	8	1904	0			
HH-8040	IDE(AT)	42.88	3.50	3H	RLL 2	AUTO	N/A	N/A	36	2	1047	40	5	977	17
HH-8040/MLC 48-000	IDE(AT)	41.94	3.50	3H	RLL 2	AUTO	N/A	N/A	36	2	1024	40			
HH-8080	IDE(AT)	85.09	3.50	3H	RLL	AUTO	N/A	N/A	100	2	1768	47			
HH-8200	IDE(AT)	210	3.50	3H	RLL	AUTO	N/A	N/A	100	4	1904	0			
HH-825	ST-506/412	21.41	5.25	HH	MFM		616	616	0	4	615	17			
HH-830	ST-506/412	32.74	5.25	HH	RLL 2		616	616	0	4	615	26			

MINISCRIBE

MODEL	INT	CAP	FF	HGT	CODE	LZ	WP	RWC	MTBF	PHDS	PCYL	PSPT	LHDS	LCYL	LSPT
1012	ST-506/412	10.65	5.25	FH	MFM	10	128	307	8	4	306	17			
2006	ST-506/412	5.32	5.25	FH	MFM	336	128	307	10	2	306	17			
2012	ST-506/412	10.65	5.25	FH	MFM	336	128	307	10	4	306	17			
3006	ST-506/412	5.32	5.25	HH	MFM	306	128	307	10	2	306	17			
3012	ST-506/412	10.65	5.25	HH	MFM	656	128	613	10	2	612	17			
3053	ST-506/412	44.56	5.25	HH	MFM	AUTO	512	1,024	30	5	1024	0			
3085	ST-506/412	71.28	5.25	HH	MFM	AUTO	512	1,170	35	7	1170	0			
3085E	ESDI-10MHZ	70.22	5.25	HH	RLL	AUTO	N/A	N/A	0	3	1270	36			
3085S	SCSI	69.39	5.25	HH	RLL	AUTO	N/A	N/A	0	3	1255	36			
3130E	ESDI-10MHZ	115.2	5.25	HH	RLL 2	AUTO	512	1,251	35	5	1250	36			
3130S	SCSI	115.66	5.25	HH	RLL	AUTO	512	1,256	0	5	1255	36			
3180E	ESDI-10MHZ	156.8	5.25	HH	RLL 2	AUTO	512	1,251	35	7	1250	35			
3180S	SCSI	161.28	5.25	HH	RLL 2	AUTO	512	1,256	35	7	1250	36			
3180SM	SCSI-MAC	161.28	5.25	HH	RLL	AUTO	N/A	N/A	35	7	1250	36			
3212	ST-506/412	10.65	5.25	HH	MFM	656	128	613	11	2	612	17			
3212P (PLUS)	ST-506/412	10.65	5.25	HH	MFM	656	128	613	0	2	612	17			
3412	ST-506/412	10.65	5.25	HH	MFM	336	128	307	11	4	306	17			
3425	ST-506/412	21.41	5.25	HH	MFM	656	128	616	20	4	615	17			
3425P (PLUS)	ST-506/412	21.41	5.25	HH	MFM	656	128	616	20	4	615	17			
3425S	SCSI	21.3	5.25	HH	RLL	656	N/A	N/A	20	4	612	17			
3438	ST-506/412	32.74	5.25	HH	RLL 2	656	128	616	20	4	615	26			
3438P (PLUS)	ST-506/412	32.74	5.25	HH	RLL 2	656	128	616	20	4	615	26			
3650	ST-506/412	42.24	5.25	HH	MFM	852	128	809	25	6	809	17			
3650F	ST-506/412	42.24	5.25	HH	MFM	852	128	810	0	6	809	17			
3650R	ST-506/412	64.61	5.25	HH	RLL 2	852	128	809	25	6	809	26			
3675	ST-506/412	64.61	5.25	HH	RLL 2	852	128	809	25	6	809	26			
4010	ST-506/412	8.35	5.25	FH	MFM	520	128	480	10	2	480	17			
4020	ST-506/412	16.71	5.25	FH	MFM	520	128	480	10	4	480	17			
5330	ST-506/412	25.06	5.25	FH	MFM		128	481	0	6	480	17			
5338	ST-506/412	31.96	5.25	FH	MFM		128	613	0	6	612	17			
5440	ST-506/412	33.42	5.25	FH	MFM		128	481	0	8	480	17			
5451	ST-506/412	42.61	5.25	FH	MFM		128	613	0	8	612	17			
6032	ST-506/412	26.73	5.25	FH	MFM	AUTO	512	1,024	25	3	1024	17			
6053	ST-506/412	44.56	5.25	FH	MFM	AUTO	512	1,024	25	5	1024	17			

MINISCRIBE

MODEL	INT	CAP	FF	HGT	CODE	LZ	WP	RWC	MTBF	PHDS	PCYL	PSPT	LHDS	LCYL	LSPT
6074	ST-506/412	62.39	5.25	FH	MFM		512	1,025	0	7	1024	17			
6079	ST-506/412	68.15	5.25	FH	RLL 2	AUTO	551	102	25	5	1024	26			
6085	ST-506/412	71.3	5.25	FH	MFM	AUTO	512	1,024	25	8	1024	17			
6128	ST-506/412	109.05	5.25	FH	RLL 2	AUTO	512	1,024	25	8	1024	26			
6170E	ESDI-10MHZ	142.6	5.25	FH	RLL	AUTO	N/A	N/A	0	8	1024	34			
6212	ST-506/412	10.65	5.25	FH	MFM		128	613	0	2	612	17			
7040A	IDE(AT)	42.57	3.50	3H	RLL 1	AUTO	N/A	N/A	150	2	1155	0	5	981	17
7040S	SCSI	42.57	3.50	3H	RLL 1	AUTO	N/A	N/A	40	2	1155	0			
7060A	IDE(AT)	65.2	3.50	3H	RLL 1	AUTO	N/A	N/A	150	2	1516	0	7	1024	17
7060S	SCSI	65.2	3.50	3H	RLL 1	AUTO	N/A	N/A	150	2	1516	0			
7080A	IDE(AT)	85.15	3.50	3H	RLL 1	AUTO	N/A	N/A	150	4	1155	0	10	981	17
7080S	SCSI	85.15	3.50	3H	RLL 1	AUTO	N/A	N/A	150	4	1155	0			
7120A	IDE(AT)	130.4	3.50	3H	RLL 1	AUTO	N/A	N/A	150	4	1516	0	14	1024	17
7120S	SCSI	130.4	3.50	3H	RLL 1	AUTO	N/A	N/A	150	4	1516	0			
7426	ST-506/412	21.3	3.50	HH	MFM		128	613	0	4	612	17			
8051A	IDE(AT)	42.72	3.50	HH	RLL 2	AUTO	N/A	N/A	150	4	745	0	4	745	28
8051S	SCSI	42.72	3.50	HH	RLL	AUTO	N/A	N/A	150	4	745	28			
8212	ST-506/412	10.7	3.50	HH	MFM	664	128	616	20	2	615	17			
8225	ST-506/412	20.52	3.50	HH	RLL 2	810	128	772	20	2	771	26			
8225AT	IDE(AT)	20.52	3.50	HH	RLL 2	820	N/A	N/A	30	2	771	26			
8225S	SCSI	20.52	3.50	HH	RLL 2	820	N/A	N/A	30	2	771	26			
8225XT	IDE(XT)	20.52	3.50	HH	RLL 2	820	N/A	N/A	30	2	771	26			
8412	ST-506/412	10.72	3.50	HH	MFM	336	128	616	20	4	308	17			
8425	ST-506/412	21.41	3.50	HH	MFM	664	128	616	20	4	615	17			
8425F	ST-506/412	21.41	3.50	HH	MFM	664	128	616	20	4	615	17			
8425S	SCSI	21.41	3.50	HH	RLL	664	N/A	N/A	20	4	615	17			
8425XT	IDE(XT)	21.41	3.50	HH	RLL	664	N/A	N/A	20	4	615	17			
8434F	ST-506/412	32.74	3.50	HH	RLL		128	616	0	4	615	26			
8438	ST-506/412	32.74	3.50	HH	RLL 2	664	128	616	20	4	615	26			
8438F	ST-506/412	32.74	3.50	HH	RLL 2	664	128	616	20	4	615	26			
8438P (PLUS)	ST-506/412	32.74	3.50	HH	RLL 2	664	128	615	20	4	615	26			
8438XT	IDE(XT)	32.74	3.50	HH	RLL	664	N/A	N/A	0	4	615	26			
8450	ST-506/412	41.05	3.50	HH	RLL 2	810	128	772	20	4	771	26			
8450AT	IDE(AT)	42.72	3.50	HH	RLL 2	820	N/A	N/A	30	4	745	28			
8450XT	IDE(XT)	42.86	3.50	HH	RLL 2	820	N/A	N/A	30	4	805	26			
9000E	ESDI-10MHZ	338.41	5.25	FH	RLL	AUTO	N/A	N/A	30	15	1224	36			
9000S	SCSI	347	5.25	FH	RLL	AUTO	N/A	N/A	30	15	1224	0			
9230E	ESDI-10MHZ	203.04	5.25	FH	RLL	AUTO	N/A	N/A	0	9	1224	36			
9230S	SCSI	203.04	5.25	FH	RLL	AUTO	N/A	N/A	50	9	1224	36			
9380E	ESDI-10MHZ	329.01	5.25	FH	RLL 2	AUTO	N/A	N/A	50	15	1224	35			
9380S	SCSI	327.39	5.25	FH	RLL 2	AUTO	N/A	N/A	50	15	1218	35			
9380SM	SCSI-MAC	319	5.25	FH	RLL	AUTO	N/A	N/A	50	15	1218	0			
9424E	ESDI-15MHZ	360	5.25	FH	RLL	AUTO	N/A	N/A	0	8	1661	53			

MINISCRIBE

MODEL	INT	CAP	FF	HGT	CODE	LZ	WP	RWC	MTBF	PHDS	PCYL	PSPT	LHDS	LCYL	LSPT
9424S	SCSI	355	5.25	FH	RLL	AUTO	N/A	N/A	0	8	1661	0			
9780E	ESDI-15MHZ	676.09	5.25	FH	RLL 1	AUTO	N/A	N/A	50	15	1661	53			
9780S	SCSI	676.09	5.25	FH	RLL	AUTO	N/A	N/A	50	15	1661	53			

MITSUBISHI ELECTRONICS

MODEL	INT	CAP	FF	HGT	CODE	LZ	WP	RWC	MTBF	PHDS	PCYL	PSPT	LHDS	LCYL	LSPT
M2860-2	ST-506/412	50	8.00	FH	MFM				0	0	0	17			
M2860-3	ST-506/412	85	8.00	FH	MFM				0	0	0	17			
MF504B	ST-506/412	21	8.00	FH	MFM				0	0	0	17			
MR521	ST-506/412	10.65	5.25	HH	MFM				0	2	612	17			
MR522	ST-506/412	21.3	5.25	HH	MFM				0	4	612	17			
MR533	ST-506/412	25.35	5.25	HH	MFM	971			0	3	971	17			
MR535	ST-506/412	42.51	5.25	HH	MFM		300	300	30	5	977	17			
MR535-V00	ST-506/412	42.51	5.25	HH	MFM		300	300	30	5	977	17			
MR535R	ST-506/412	65.02	5.25	HH	RLL 2				30	5	977	26			

MMI

MODEL	INT	CAP	FF	HGT	CODE	LZ	WP	RWC	MTBF	PHDS	PCYL	PSPT	LHDS	LCYL	LSPT
M 106	ST-506/412	5.32	3.50	HH	MFM		128		0	2	306	17			
M 112	ST-506/412	10.65	3.50	HH	MFM		128		0	4	306	17			
M 125	ST-506/412	21.3	3.50	HH	MFM		128		0	8	306	17			
M 212	ST-506/412	10.65	5.25	HH	MFM		128		0	4	306	17			
M 225	ST-506/412	21.3	5.25	HH	MFM		128		0	8	306	17			
M 306	ST-506/412	5.32	5.25	HH	MFM		128		0	2	306	17			
M 312	ST-506/412	10.65	5.25	HH	MFM		128		0	4	306	17			
M 325	ST-506/412	21.3	5.25	HH	MFM		128		0	8	306	17			

NCL AMERICA

MODEL	INT	CAP	FF	HGT	CODE	LZ	WP	RWC	MTBF	PHDS	PCYL	PSPT	LHDS	LCYL	LSPT
9170E	ESDI-10MHZ	150.44	3.50	HH	RLL 2	AUTO			50	7	1166	36			
9170S	SCSI	150.44	3.50	HH	RLL 2	AUTO			50	7	1166	36			
9220A	IDE(AT)	200.55	3.50	HH	RLL 2	AUTO			50	9	1209	36	16	401	61
9220E	ESDI-10MHZ	200.72	3.50	HH	RLL 2	AUTO			50	9	1210	36			
9220S	SCSI	200.72	3.50	HH	RLL 2	AUTO			50	9	1210	36			
NDC 5125	IDE(AT)	150.31	3.50	HH	RLL 2	AUTO			50	7	1165	36	10	583	36

NCR CORPORATION

MODEL	INT	CAP	FF	HGT	CODE	LZ	WP	RWC	MTBF	PHDS	PCYL	PSPT	LHDS	LCYL	LSPT
6091-5101	SCSI	323	5.25	–	RLL 2	AUTO	N/A	N/A	0	9	0	0			
6091-5301	SCSI	675	5.25	–	RLL 2	AUTO	N/A	N/A	0	15	0	0			
H6801-STD1-03-17	ST-506/412	53.12	3.50	HH	MFM		650		20	7	872	17			
H6801-STD1-07-17	IDE(AT)	45.33	3.50	HH	RLL 2	AUTO	N/A	N/A	20	3	868	34			

NCR CORPORATION

MODEL	INT	CAP	FF	HGT	CODE	LZ	WP	RWC	MTBF	PHDS	PCYL	PSPT	LHDS	LCYL	LSPT
H6801-STD1-10-17	IDE(AT)	104.89	3.50	HH	RLL 2	AUTO	N/A	N/A	150	8	776	33			
H6801-STD1-12-17	IDE(AT)	42.88	3.50	3H	RLL 2	AUTO	N/A	N/A	150	2	1047	40			
H6801-STD1-46-46	ST-506/412	21.41	3.50	HH	MFM	664	128	616	20	4	615	17			
H6801-STD1-47-46	ST-506/412	71.3	5.25	FH	MFM	AUTO	128	1,025	40	8	1024	17			
H6801-STD1-49-46	ESDI-10MHZ	121.55	5.25	FH	RLL 2	AUTO	N/A	N/A	100	7	969	35			

NEC TECHNOLOGIES, INC.

MODEL	INT	CAP	FF	HGT	CODE	LZ	WP	RWC	MTBF	PHDS	PCYL	PSPT	LHDS	LCYL	LSPT
D1528 M/N	SCSI	1350	3.50		RLL 1	AUTO	N/A	N/A	300	0	0	0			
D3126	ST-506/412	21.41	3.50	HH	MFM		256	616	20	4	615	17			
D3126H	ST-506/412	20	5.25			664	256	N/A		0	0	0	4	615	
D3142	ST-506/412	44.7	3.50	HH	MFM		N/A	N/A	30	8	642	17			
D3146H	ST-506/412	42.82	3.50	HH	MFM				20	8	615	17			
D3661	ESDI-10MHZ	118.05	3.50	HH	RLL 2	AUTO	N/A	N/A	30	7	915	36			
D3711	IDE(AT)	172.8	3.50	3H	RLL 1	AUTO	N/A	N/A	250	0	0	0	16	335	63
D3713	IDE(AT)	345.6	3.50	3H	RLL 1	AUTO	N/A	N/A	250	0	0	0	16	670	63
D3715	IDE(AT)	270	3.50	3H	RLL 1	AUTO	N/A	N/A	250	0	2924	0	16	524	63
D3717	IDE(AT)	540	3.50	3H	RLL 1	AUTO	N/A	N/A	250	0	0	0	16	1048	63
D3723	IDE(AT)	365	3.50	3H	RLL 1	AUTO	N/A	N/A	300	0	0	0	16	708	63
D3724	IDE(AT)	426.8	3.50	3H	PRML	AUTO	N/A	N/A	300	2	3493	0	16	827	63
D3725	IDE(AT)	730	3.50	3H	RLL 1	AUTO	N/A	N/A	300	0	0	0	16	1416	63
D3726	IDE(AT)	853.6	3.50	3H	PRML	AUTO	N/A	N/A	300	4	3493	0	16	1654	63
D3727	IDE(AT)	1083	3.50	3H	RLL 1	AUTO	N/A	N/A	300	0	0	0	16	2100	63
D3735	IDE(AT)	45.51	3.50	3H	RLL 1	AUTO	N/A	N/A	30	2	1084	41	4	542	41
D3741	IDE(AT)	45.04	3.50	HH	RLL	AUTO	N/A	N/A	0	8	423	26	7	733	17
D3743	IDE(AT)	540	3.50	3H	PRML	AUTO	N/A	N/A	300	2	3678	0	16	1048	63
D3745	IDE(AT)	1080	3.50	3H	PRML	AUTO	N/A	N/A	300	4	3678	0	16	2096	63
D3747	IDE(AT)	1620	3.50	3H	PRML	AUTO	N/A	N/A	300	6	3678	0	16	3144	63
D3755	IDE(AT)	104.96	3.50	3H	RLL 1	AUTO	N/A	N/A	30	4	1250	41	8	625	41
D3756	IDE(AT)	104.96	3.50	HH	RLL 1	AUTO	N/A	N/A	50	4	1250	41	8	625	41
D3761	IDE(AT)	118.05	3.50	HH	RLL 2	AUTO	N/A	N/A	30	7	915	36	7	915	35
D3781	IDE(AT)	425	3.50	HH	RLL 1	AUTO	N/A	N/A	50	9	1464	63	9	1464	63
D3815	SCSI-2	270.4	3.50	3H	RLL 1	AUTO	N/A	N/A	0	0	2924	0	16	524	63
D3817	SCSI-2	540.8	3.50	3H	RLL 1	AUTO	N/A	N/A	0	0	2924	0	16	1048	63
D3823	SCSI-2	365	3.50	3H	RLL 1	AUTO	N/A	N/A	300	0	0	0	16	708	63
D3825	SCSI-2	730	3.50	3H	RLL 1	AUTO	N/A	N/A	300	0	0	0	16	1416	63
D3827	SCSI-2	1083	3.50	3H	RLL 1	AUTO	N/A	N/A	300	0	0	0	16	2100	63
D3835	SCSI	45.09	3.50	3H	RLL 1	AUTO	N/A	N/A	30	2	1074	41			
D3843	SCSI-2	540	3.50	3H	PRML	AUTO	N/A	N/A	300	0	0	0	16	1048	63
D3845	SCSI-2	1080	3.50	3H	PRML	AUTO	N/A	N/A	300	4	3678	0	16	2080	63
D3847	SCSI-2	1620	3.50	3H	PRML	AUTO	N/A	N/A	300	6	3678	0	16	3144	63
D3855	SCSI	104.96	3.50	3H	RLL 1	AUTO	N/A	N/A	30	4	1250	41			
D3861	SCSI	114.77	3.50	HH	RLL 2	AUTO	N/A	N/A	30	7	915	35			
D3881	SCSI	425	3.50	HH	RLL 1	AUTO	N/A	N/A	50	9	1464	63			

NEC TECHNOLOGIES, INC.

MODEL	INT	CAP	FF	HGT	CODE	LZ	WP	RWC	MTBF	PHDS	PCYL	PSPT	LHDS	LCYL	LSPT
D3896	SCSI-2	2160	3.50	3H	RLL 1	AUTO	N/A	N/A	800	9	3928	0			
D5114	ST-506/412	5.32	5.25	–	MFM				0	2	306	17			
D5124	ST-506/412	10.75	5.25	HH	MFM		310	310	0	4	309	17			
D5126	ST-506/412	21.3	5.25	HH	MFM			613	0	4	612	17			
D5126H	ST-506/412	21.3	5.25	HH	MFM			613	0	4	612	17			
D5127	ST-506/412	60.45	5.25	HH	RLL 2	AUTO	N/A	616	0	4	615	17			
D5146	ST-506/412	42.82	5.25	HH	MFM			616	0	8	615	17			
D5146H	ST-506/412	42.82	5.25	HH	MFM			616	0	8	615	17			
D5452	ESDI-10MHZ	67.42	5.25	HH	RLL 2	AUTO	N/A	N/A	0	10	823	34			
D5652	ESDI-10MHZ	143.26	5.25	HH	RLL 2	AUTO	N/A	N/A	0	10	823	34			
D5655	ESDI-10MHZ	153.53	5.25	HH	RLL 2	AUTO	N/A	N/A	30	7	1224	35			
D5662	ESDI-10MHZ	329.01	5.25	FH	RLL 2	AUTO	N/A	N/A	45	15	1224	35			
D5682	ESDI-15MHZ	664.69	5.25	FH	RLL 1	AUTO	N/A	N/A	50	15	1633	53			
D5862	SCSI	328.2	5.25	FH	RLL 1	AUTO	N/A	N/A	50	15	1221	35			
D5882	SCSI	664.69	5.25	FH	RLL 1	AUTO	N/A	N/A	50	15	1633	53			
D5892	SCSI	1403.82	5.25	FH	RLL 1	AUTO	N/A	N/A	100	19	1678	86			
DSE1340A	IDE(AT)	1340	3.50	3H	PRML	AUTO	N/A	N/A	300	4	4323	0	16	2600	63
DSE1700A	IDE(AT)	1700	3.50	3H	PRML	AUTO	N/A	N/A	300	0	0	0	16	3306	63
DSE2010A	IDE(AT)	2010	3.50	3H	PRML	AUTO	N/A	N/A	300	6	4323	0	16	3900	63
DSE2100A	IDE(AT)	2100	3.50	3H	PRML	AUTO	N/A	N/A	300	0	0	0	16	4092	63
DSE2550A	IDE(AT)	2550	3.50	3H	PRML	AUTO	N/A	N/A	300	0	0	0	16	4960	63
LPS120AT	IDE(AT)	120	3.50	3H						0	0	0	5	901	53
LPS240AT	IDE(AT)	240	3.50	3H						0	0	0	13	723	51
LXT535	SCSI	535	3.50		RLL 1	AUTO	N/A	N/A	300	0	0	0			

NEI

MODEL	INT	CAP	FF	HGT	CODE	LZ	WP	RWC	MTBF	PHDS	PCYL	PSPT	LHDS	LCYL	LSPT
RD 3127	ST-506/412	10.65	5.25	HH	MFM				0	2	612	17			
RD 3255	ST-506/412	21.3	5.25	HH	MFM				0	4	612	17			
RD 4127	ST-506/412	10.65	5.25	FH	MFM				0	4	306	17			
RD 4255	ST-506/412	21.3	5.25	FH	MFM				0	8	306	17			

NEWBURY DATA

MODEL	INT	CAP	FF	HGT	CODE	LZ	WP	RWC	MTBF	PHDS	PCYL	PSPT	LHDS	LCYL	LSPT
NDR 1065	ST-506/412	55.93	5.25	FH	MFM				0	7	918	17			
NDR 1085	ST-506/412	71.3	5.25	FH	MFM				0	8	1024	17			
NDR 1105	ST-506/412	87.89	5.25	FH	MFM				0	11	918	17			
NDR 1140	ST-506/412	119.85	5.25	FH	MFM				0	15	918	17			
NDR 2085	ST-506/412	74.57	5.25	FH	MFM		1,224	1,224	0	7	1224	17			
NDR 2140	ST-506/412	117.19	5.25	FH	MFM		1,224	1,224	0	11	1224	17			
NDR 2190	ST-506/412	119.85	5.25	FH	MFM		1,224	1,224	0	15	918	17			
NDR 3170S	SCSI	146.64	5.25	FH	RLL 2	AUTO	N/A	N/A	0	9	1224	26			
NDR 320	ST-506/412	21.41	5.25	FH	MFM	615			0	4	615	17			

NEWBURY DATA

MODEL	INT	CAP	FF	HGT	CODE	LZ	WP	RWC	MTBF	PHDS	PCYL	PSPT	LHDS	LCYL	LSPT
NDR 3280S	SCSI	244.4	5.25	FH	RLL 2	AUTO	N/A	N/A	0	15	1224	26			
NDR 3380S	SCSI	319.61	5.25	FH	RLL 2	AUTO	N/A	N/A	50	15	1224	34			
NDR 340	ST-506/412	42.82	3.50	HH	MFM				0	8	615	17			
NDR 4175	ESDI-10MHZ	157.92	5.25	FH	RLL 2	AUTO	N/A	N/A	0	7	1224	36			
NDR 4380	ESDI-10MHZ	338.41	5.25	FH	RLL 2	AUTO	N/A	N/A	0	15	1224	36			
NDR 4380S	SCSI	319.61	5.25	FH	RLL	AUTO	N/A	N/A	0	15	1224	34			
PENNY 340	ST-506/412	42.82	5.25	HH	MFM		615	615	0	8	615	17			

NPL

MODEL	INT	CAP	FF	HGT	CODE	LZ	WP	RWC	MTBF	PHDS	PCYL	PSPT	LHDS	LCYL	LSPT
4064	ST-506/412	5	5.25	FH	MFM				0	0	0	17			
4127	ST-506/412	10	5.25	FH	MFM				0	0	0	17			
4191S	ST-506/412	15	5.25	FH	MFM				0	0	0	17			
4255	ST-506/412	20	5.25	FH	MFM				0	0	0	17			
NP 02-13	ST-506/412	11.14	5.25	HH	MFM	320	0	N/A	0	4	320	17			
NP 02-26A/26S	ST-506/412	22.28	5.25	HH	MFM	640	0	N/A	0	4	640	17			
NP 02-52A	ST-506/412	44.56	5.25	HH	MFM	640	640	N/A	0	8	640	17			
NP 03-20	ST-506/412	15.98	3.50	HH	MFM	306	0	N/A	0	6	306	17			
NP 03-6	ST-506/412	5.32	5.25	–	MFM				0	2	306	17			
NP 04-13T	ST-506/412	10	5.25	FH	MFM				0	0	0	17			
NP 04-55	ST-506/412	45.93	5.25	HH	MFM	754	0	N/A	0	7	754	17			
NP 04-85	ST-506/412	72.19	5.25	HH	MFM	754	0	N/A	0	11	754	17			

OKIDATA

MODEL	INT	CAP	FF	HGT	CODE	LZ	WP	RWC	MTBF	PHDS	PCYL	PSPT	LHDS	LCYL	LSPT
OD 526	ST-506/412	22.28	5.25	HH	RLL 2				0	4	640	17			
OD 540	ST-506/412	33.42	5.25	HH	RLL 2				0	6	640	17			

OLIVETTI

MODEL	INT	CAP	FF	HGT	CODE	LZ	WP	RWC	MTBF	PHDS	PCYL	PSPT	LHDS	LCYL	LSPT
HD 662/11	ST-506/412	10.65	5.25	HH	MFM				0	2	612	17			
HD 662/12	ST-506/412	21.3	5.25	HH	MFM				0	4	612	17			
XM 3220	ST-506/412	21.3	3.50	HH	MFM	656	128	N/A	0	4	612	17			
XM 5210	ST-506/412	10	5.25	HH	MFM				0	0	0	17			
XM 5220/2	ST-506/412	21.3	5.25	FH	MFM	656	128		0	4	612	17			
XM 5221	ST-506/412	21.41	5.25	HH	MFM	700	256	N/A	0	4	615	17			
XM 5340	ST-506/412	42.82	5.25	HH	MFM	819	256	256	0	6	820	17			
XM 5360	ST-506/412	42.82	5.25	HH	RLL	819	128	128	0	6	820	17			

OPTIMA TECHNOLOGY

MODEL	INT	CAP	FF	HGT	CODE	LZ	WP	RWC	MTBF	PHDS	PCYL	PSPT	LHDS	LCYL	LSPT
CONCORDE 1050	SCSI	1050	5.25	–	RLL 2	AUTO	N/A	N/A	50	15	0	0			

OPTIMA TECHNOLOGY

MODEL	INT	CAP	FF	HGT	CODE	LZ	WP	RWC	MTBF	PHDS	PCYL	PSPT	LHDS	LCYL	LSPT
CONCORDE 635	SCSI	640	5.25	–	RLL 2	AUTO	N/A	N/A	150	14	0	0			
DISKOVERY 420	SCSI	416	5.25	–	RLL 2	AUTO	N/A	N/A	100	8	0	0			
MINIPAK 100	SCSI	104	3.50	HH	RLL 2	AUTO	N/A	N/A	30	4	0	0			
MINIPAK 200	SCSI	209	3.50	HH	RLL 2	AUTO	N/A	N/A	40	8	0	0			
MINIPAK 300	SCSI	320	3.50	HH	RLL 2	AUTO	N/A	N/A	150	0	0	0			

ORCA TECHNOLOGY

MODEL	INT	CAP	FF	HGT	CODE	LZ	WP	RWC	MTBF	PHDS	PCYL	PSPT	LHDS	LCYL	LSPT
320A	IDE(AT)	370	3.50	HH	RLL 2	AUTO	N/A	N/A	100	9	0	0			
320S	SCSI	370	3.50	HH	RLL 2	AUTO	N/A	N/A	100	9	0	0			
400A	IDE(AT)	470	3.50	HH	RLL 2	AUTO	N/A	N/A	100	9	0	0			
400S	SCSI	470	3.50	HH	RLL 2	AUTO	N/A	N/A	100	9	0	0			
760E	ESDI-15MHZ	760	5.25	–	RLL 2	AUTO	N/A	N/A	50	15	0	0			
760S	SCSI	760	5.25	–	RLL 2	AUTO	N/A	N/A	50	15	0	0			
OT5H 53-MFM	ST-506/412	44.56	5.25	HH	MFM	1024			40	5	1024	17			

OTARI

MODEL	INT	CAP	FF	HGT	CODE	LZ	WP	RWC	MTBF	PHDS	PCYL	PSPT	LHDS	LCYL	LSPT
C-214	ST-506/412	10.65	5.25	FH	MFM		128	128	0	4	306	17			
C-507	ST-506/412	5.32	5.25	FH	MFM		128	128	0	2	306	17			
C-514	ST-506/412	10.65	5.25	FH	MFM		128	128	0	4	306	17			
C-519	ST-506/412	15.98	5.25	FH	MFM		128	128	0	6	306	17			
C-526	ST-506/412	21.3	5.25	FH	MFM		128	128	0	8	306	17			

PANASONIC INDUSTRIAL COMPANY

MODEL	INT	CAP	FF	HGT	CODE	LZ	WP	RWC	MTBF	PHDS	PCYL	PSPT	LHDS	LCYL	LSPT
JU-116	ST-506/412	21.41	3.50	HH	MFM		616	616	0	4	615	17			
JU-128	ST-506/412	44.66	3.50	HH	MFM		734	734	0	7	733	17			

PERIPHERALS ENTERPRISE CO., LTD.

MODEL	INT	CAP	FF	HGT	CODE	LZ	WP	RWC	MTBF	PHDS	PCYL	PSPT	LHDS	LCYL	LSPT
PL-100 TURBO	SCSI	105	3.50	HH	RLL 2	AUTO	N/A	N/A	60	4	0	0			
PL-200 TURBO	SCSI	210	3.50	HH	RLL 2	AUTO	N/A	N/A	50	7	0	0			
PL-320 TURBO	SCSI	320	3.50	HH	RLL 2	AUTO	N/A	N/A	100	14	0	0			
PT-225	ST-506/412	21.41	3.50	HH	MFM				0	4	615	17			
PT-234	ST-506/412	28.54	3.50	HH	MFM				0	4	820	17			
PT-238A	IDE(AT)	32.74	3.50	HH	RLL 2	AUTO	N/A	N/A	0	4	615	26			
PT-238R	ST-506/412	32.74	3.50	HH	RLL 2				0	4	615	26			
PT-238S	SCSI	32.74	3.50	HH	RLL 2	AUTO	N/A	N/A	0	4	615	26			
PT-251A	IDE(AT)	43.66	3.50	HH	RLL 2	AUTO	N/A	N/A	0	4	820	26			
PT-251R	ST-506/412	43.66	3.50	HH	RLL 2				0	4	820	26			
PT-251S	SCSI	43.66	3.50	HH	RLL 2	AUTO	N/A	N/A	0	4	820	26			

PERIPHERALS ENTERPRISE CO., LTD.

MODEL	INT	CAP	FF	HGT	CODE	LZ	WP	RWC	MTBF	PHDS	PCYL	PSPT	LHDS	LCYL	LSPT
PT-338	ST-506/412	32.11	3.50	HH	MFM				0	6	615	17			
PT-351	ST-506/412	42.82	3.50	HH	MFM				0	6	820	17			
PT-357A	IDE(AT)	49.12	3.50	HH	RLL 2	AUTO	N/A	N/A	0	6	615	26			
PT-357R	ST-506/412	49.12	3.50	HH	RLL 2				0	6	615	26			
PT-357S	SCSI	49.12	3.50	HH	RLL 2	AUTO	N/A	N/A	0	6	615	26			
PT-376A	IDE(AT)	65.49	3.50	HH	RLL 2	AUTO	N/A	N/A	0	6	820	26			
PT-376R	ST-506/412	65.49	3.50	HH	RLL 2				0	6	820	26			
PT-376S	SCSI	65.49	3.50	HH	RLL 2	AUTO	N/A	N/A	0	6	820	26			
PT-4102A	IDE(AT)	87.32	3.50	HH	RLL 2	AUTO	N/A	N/A	80	8	820	26			
PT-4102R	ST-506/412	87.32	3.50	HH	RLL 2	AUTO			80	8	820	26			

PLUS DEVELOPMENT CORPORATION

MODEL	INT	CAP	FF	HGT	CODE	LZ	WP	RWC	MTBF	PHDS	PCYL	PSPT	LHDS	LCYL	LSPT
HARDCARD 20	IDE(AT)	21.41	CARD	3H	RLL 2	AUTO	N/A	N/A	60	4	615	17			
HARDCARD II-80	IDE(AT)	80.51	CARD	3H	RLL 2	AUTO	N/A	N/A	0	10	925	17			
HARDCARD II-XL105	IDE(AT)	105.23	CARD	3H	RLL 2	AUTO	N/A	N/A	0	15	806	17			
HARDCARD II-XL50	IDE(AT)	52.31	CARD	3H	RLL 2	AUTO	N/A	N/A	0	10	601	17			
IMPULSE 105AT/LP	IDE(AT)	105.24	3.50	3H	RLL 2	AUTO			60	4	1219	0	16	755	17
IMPULSE 105S	SCSI-2	105.24	3.50	HH	RLL 2	AUTO			50	4	1219	0			
IMPULSE 105S/LP	SCSI-2	104.85	3.50	3H	RLL 2	AUTO			60	4	1219	0			
IMPULSE 120AT	IDE(AT)	120.74	3.50	HH	RLL 1	AUTO			50	5	1123	42	9	814	32
IMPULSE 120S	SCSI-2	120.74	3.50	HH	RLL 1	AUTO			50	5	1123	42			
IMPULSE 170AT	IDE(AT)	169.04	3.50	HH	RLL 1	AUTO			50	7	1123	42	10	966	34
IMPULSE 170S	SCSI-2	169.04	3.50	HH	RLL 1	AUTO			50	7	1123	42			
IMPULSE 210AT	IDE(AT)	210	3.50	HH	RLL 1	AUTO			50	7	1156	0	13	873	36
IMPULSE 210S	SCSI-2	210	3.50	HH	RLL 1	AUTO			50	7	1156	0			
IMPULSE 330AT	IDE(AT)	331.2	3.50	HH	RLL 1	AUTO			75	7	1512	0			
IMPULSE 330S	SCSI-2	331.2	3.50	HH	RLL 1	AUTO			75	7	1512	0			
IMPULSE 40AT	IDE(AT)	41.99	3.50	HH	RLL 2	AUTO	N/A	N/A	50	5	965	17	5	968	17
IMPULSE 40S	SCSI-2	42	3.50	HH	RLL 2	AUTO			50	3	0	0			
IMPULSE 425AT	SCSI-2	425.8	3.50	HH	RLL 1	AUTO			75	0	0	0			
IMPULSE 425S	SCSI-2	425.8	3.50	HH	RLL 1	AUTO			75	0	0	0			
IMPULSE 52AT/LP	IDE(AT)	52.29	3.50	3H	RLL 2	AUTO			60	8	751	17	8	751	17
IMPULSE 52S/LP	SCSI-2	52.2	3.50	3H	RLL 2	AUTO			60	2	0	0			
IMPULSE 80AT	IDE(AT)	83.99	3.50	HH	RLL 2	AUTO	N/A	N/A	50	10	965	17	6	611	17
IMPULSE 80AT/LP	IDE(AT)	85.78	3.50	3H	RLL 2	AUTO			60	16	616	17	6	611	17
IMPULSE 80S	SCSI-2	84	3.50	HH	RLL 2	AUTO			50	6	0	0			
IMPULSE 80S/LP	SCSI-2	85.7	3.50	3H	RLL 2	AUTO			60	4	0	0			

PRAIRIETEK

MODEL	INT	CAP	FF	HGT	CODE	LZ	WP	RWC	MTBF	PHDS	PCYL	PSPT	LHDS	LCYL	LSPT
PRAIRIE 120	IDE(XT/AT)	21.41	2.50	4H	RLL 2	AUTO	N/A	N/A	20	2	615	34			
PRAIRIE 140	IDE(XT/AT)	42.82	2.50	4H	RLL 2	AUTO	N/A	N/A	20	4	615	34			

PRAIRIETEK

MODEL	INT	CAP	FF	HGT	CODE	LZ	WP	RWC	MTBF	PHDS	PCYL	PSPT	LHDS	LCYL	LSPT
PRAIRIE 220A	IDE(AT)	21.3	2.50	3H	RLL 2	AUTO	N/A	N/A	20	2	612	34			
PRAIRIE 220S	SCSI	21.3	2.50	3H	RLL 2	AUTO	N/A	N/A	20	2	612	34			
PRAIRIE 240	IDE(XT/AT)	42.82	2.50	3H	RLL 2	AUTO	N/A	N/A	20	4	615	34			
PRAIRIE 242A	IDE(XT/AT)	42.82	2.50	3H	RLL 2	AUTO	N/A	N/A	20	4	615	34			
PRAIRIE 242S	SCSI	42.82	2.50	3H	RLL 2	AUTO	N/A	N/A	20	4	615	34			
PRAIRIE 282A	IDE(XT/AT)	82	2.50	3H	RLL 2	AUTO	N/A	N/A	20	4	0	34			
PRAIRIE 282S	SCSI	82	2.50	3H	RLL 2	AUTO	N/A	N/A	20	4	0	34			

PRIAM CORPORATION

MODEL	INT	CAP	FF	HGT	CODE	LZ	WP	RWC	MTBF	PHDS	PCYL	PSPT	LHDS	LCYL	LSPT
502	ST-506/412	46	5.25	FH	MFM		756	756	0	7	755	17			
504	ST-506/412	46	5.25	FH	MFM		756	756	0	7	755	17			
514	ST-506/412	117.19	5.25	FH	MFM				0	11	1224	0			
519	ST-506/412	159.8	5.25	FH	MFM		1,225	1,225	0	15	1224	0			
617	ESDI-10MHZ	153.66	5.25	FH	RLL 2	AUTO	N/A	N/A	0	7	1225	35			
628	ESDI-10MHZ	241.47	5.25	FH	RLL 2	AUTO	N/A	N/A	0	11	1225	35			
638	ESDI-10MHZ	329.28	5.25	FH	RLL 2	AUTO	N/A	N/A	0	15	1225	35			
717	SCSI	153.66	5.25	FH	RLL 2	AUTO	N/A	N/A	0	7	1225	35			
728	SCSI	241.47	5.25	FH	RLL 2	AUTO	N/A	N/A	0	11	1225	35			
738	SCSI	329.28	5.25	FH	RLL 2	AUTO	N/A	N/A	0	15	1225	0			
ID 120-E	ESDI-10MHZ	121.11	5.25	HH	RLL 2	AUTO	N/A	N/A	0	7	1024	33			
ID 130-PS041	ST-506/412	133.69	5.25	FH	MFM				40	15	1024	17			
ID 150	ESDI-10MHZ	160.06	5.25	HH	RLL 2	AUTO	N/A	N/A	0	7	1276	35			
ID 160-E	ESDI-10MHZ	158.05	5.25	FH	RLL 2	AUTO	N/A	N/A	0	7	1225	36			
ID 160-PS071	ESDI-10MHZ	158.05	5.25	FH	RLL 2	AUTO	N/A	N/A	0	7	1225	36			
ID 160-PSCSI	SCSI	159.55	5.25	FH	RLL 2	AUTO	N/A	N/A	0	7	1225	36			
ID 160-S	SCSI	159.55	5.25	FH	RLL 2	AUTO	N/A	N/A	0	7	1225	36			
ID 20	ST-506/412	25.61	5.25	FH	MFM				40	3	981	17			
ID 200L	IDE(AT)	201	3.50	HH	RLL 1	AUTO	N/A	N/A	100	7	1318	0			
ID 250-E	ESDI-10MHZ	248.37	5.25	FH	RLL	AUTO	N/A	N/A	0	11	1225	36			
ID 250-PS071	ESDI-10MHZ	248.37	5.25	FH	RLL	AUTO	N/A	N/A	0	11	1225	36			
ID 330-PS071	ESDI-10MHZ	330.39	5.25	FH	RLL 2	AUTO	N/A	N/A	0	15	1195	36			
ID 340H	SCSI	344.49	5.25	HH	RLL 2	AUTO	N/A	N/A	150	7	1780	54			
ID 45H-AT	ST-506/412	44.56	5.25	HH	MFM				40	5	1018	17			
ID 60AT	ST-506/412	59	5.25	FH	MFM	AUTO	N/A	N/A	150	5	1018	0			
ID 62-PS041	ST-506/412	62.39	5.25	FH	MFM				40	7	1024	17			
ID 75-Q	ST-506/412	73.98	5.25	FH	RLL	AUTO			40	5	1156	25			
ID/ED 40	ST-506/412	42.69	5.25	FH	MFM	981			40	5	981	17			
ID/ED 45-AT-D2	ST-506/412	44.3	5.25	FH	MFM				40	5	1018	17			
ID/ED 60	ST-506/412	59.77	5.25	FH	MFM	981			150	7	981	17			
ID/ED 62-AT-D2	ST-506/412	62.02	5.25	FH	MFM				40	7	1018	17			
ID/ED 100-Q	ST-506/412	108.65	5.25	HH	RLL 2	AUTO	N/A	N/A	40	7	1166	26			
ID/ED 130-AT-D2	ST-506/412	159.8	5.25	FH	MFM	AUTO	N/A	N/A	0	15	1224	17			
ID/ED 230-Q	ST-506/412	233.08	5.25	FH	RLL				40	15	1214	25			

PRIAM CORPORATION

MODEL	INT	CAP	FF	HGT	CODE	LZ	WP	RWC	MTBF	PHDS	PCYL	PSPT	LHDS	LCYL	LSPT
ID/ED 240	ST-506/412	243.6	5.25	FH	RLL 2	AUTO			40	15	1220	26			
ID/ED 250-PSCSI	SCSI	248.37	5.25	FH	RLL	AUTO	N/A	N/A	0	11	1225	36			
ID/ED 250-S	SCSI	248.37	5.25	FH	RLL	AUTO	N/A	N/A	0	11	1225	36			
ID/ED 330-E	ESDI-10MHZ	338.68	5.25	FH	RLL 2	AUTO	N/A	N/A	0	15	1225	36			
ID/ED 330-PSCSI	SCSI	338.68	5.25	FH	RLL 2	AUTO	N/A	N/A	0	15	1225	36			
ID/ED 330-S	SCSI	338.68	5.25	FH	RLL 2	AUTO	N/A	N/A	0	15	1225	36			
ID/ED 660	SCSI	675.16	5.25	FH	RLL	AUTO	N/A	N/A	150	15	1628	54			
ID/ED1000	SCSI	1046.39	5.25	FH	RLL	AUTO	N/A	N/A	150	15	1919	71			
INNERSPACE INTERNAL DISK SYSTEM	ST-506/412	32.74	3.50	HH	RLL 2				0	4	820	26			
S14	ST-506/412	117.19	5.25	–	MFM		1,224	1,224	0	11	1224	17			
S15	ST-506/412	159.8	5.25	–	MFM		1,224	1,224	0	15	1224	17			
TYPE A	ST-506/412	0	5.25	HH					0	0	0	0			
V130	ST-506/412	25.77	5.25	FH	MFM		988	988	0	3	987	17			
V150	ST-506/412	42.95	5.25	FH	MFM		988	988	0	5	987	17			
V160	ST-506/412	50.74	5.25	FH	MFM		1,167	1,167	0	5	1166	17			
V170	ST-506/412	60.13	5.25	FH	MFM		988	988	0	7	987	17			
V170R	ST-506/412	91.97	5.25	FH	RLL 2				0	7	987	26			
V185	ST-506/412	71.04	5.25	FH	MFM		1,167	1,167	0	7	1166	17			
V519	ST-506/412	159.8	5.25	FH	MFM	1223			0	15	1224	17			

QUANTUM CORPORATION

MODEL	INT	CAP	FF	HGT	CODE	LZ	WP	RWC	MTBF	PHDS	PCYL	PSPT	LHDS	LCYL	LSPT
ATLAS II XP32181S	U-SCSI	2180	3.50	3H	RLL 1	AUTO	N/A	N/A	1000	6	5952	0			
ATLAS II XP32181SCA	SCSI SCA	2180	3.50	3H	RLL 1	AUTO	N/A	N/A	1000	6	5952	0			
ATLAS II XP32181W	U-SCSI W	2180	3.50	3H	RLL 1	AUTO	N/A	N/A	1000	6	5952	0			
ATLAS II XP32181WD	U-SCSI WD	2180	3.50	3H	RLL 1	AUTO	N/A	N/A	1000	6	5952	0			
ATLAS II XP32275D		0								0	0	0			
ATLAS II XP34361S	U-SCSI	4360	3.50	3H	RLL 1	AUTO	N/A	N/A	1000	10	5952	0			
ATLAS II XP34361SCA	SCSI SCA	4360	3.50	3H	RLL 1	AUTO	N/A	N/A	1000	10	5952	0			
ATLAS II XP34361W	U-SCSI W	4360	3.50	3H	RLL 1	AUTO	N/A	N/A	1000	10	5952	0			
ATLAS II XP34361WD	U-SCSI WD	4360	3.50	3H	RLL 1	AUTO	N/A	N/A	1000	10	5952	0			
ATLAS II XP34550D		0								0	0	0			
ATLAS II XP39100D		0								0	0	0			
ATLAS II XP39100SCA	SCSI SCA	8678	3.50	FH	RLL 1	AUTO	N/A	N/A	1000	20	5952	0			
ATLAS XP31070S	SCSI-2 F	1075	3.50	3H	RLL 1	AUTO	N/A	N/A	800	5	3832	0			
ATLAS XP31070SCA	SCSI SCA	1075	3.50	3H	UNIDEN					5	3832	0			
ATLAS XP31070W	SCSI-2 FW	1075	3.50	3H	RLL 1	AUTO	N/A	N/A	800	5	3832	0			
ATLAS XP31070WD	SCSI-2 FWD	1075	3.50	3H	RLL 1	AUTO	N/A	N/A	800	5	3832	0			
ATLAS XP32150S	SCSI-2 F	2150	3.50	3H	RLL 1	AUTO	N/A	N/A	800	10	3832	0			
ATLAS XP32150W	SCSI-2 FW	2150	3.50	3H	RLL 1	AUTO	N/A	N/A	800	10	3832	0			
ATLAS XP32150WD	SCSI-2 FWD	2150	3.50	3H	RLL 1	AUTO	N/A	N/A	800	10	3832	0			
ATLAS XP34300S	SCSI-2 F	4300	3.50	HH	RLL 1	AUTO	N/A	N/A	800	20	3832	0			
ATLAS XP34300W	SCSI-2 FW	4300	3.50	HH	RLL 1	AUTO	N/A	N/A	800	20	3832	0			

QUANTUM CORPORATION

MODEL	INT	CAP	FF	HGT	CODE	LZ	WP	RWC	MTBF	PHDS	PCYL	PSPT	LHDS	LCYL	LSPT
ATLAS XP34300WD	SCSI-2 FWD	4300	3.50	HH	RLL 1	AUTO	N/A	N/A	800	20	3832	0			
BIGFOOT 1.2AT	IDE(AT)	1280	5.25	4H	PRML	AUTO	N/A	N/A	300	0	0	0	16	2492	63
BIGFOOT 1275	IDE(AT)	1275	5.25	4H	PRML	2492	N/A	N/A		2	5738	0	16	2492	63
BIGFOOT 1275AT	IDE(AT)	1275	5.25	4H	PRML	2492	N/A	N/A		2	5738	0	16	2492	63
BIGFOOT 2.5AT	IDE(AT)	2500	5.25	4H	PRML	AUTO	N/A	N/A	300	0	0	0	16	4994	63
BIGFOOT 2550	IDE(AT)	2500	5.25	4H	PRML	4994	N/A	N/A		4	5738	0	16	4994	63
BIGFOOT 2550AT	IDE(AT)	2500	5.25	4H	PRML	4994	N/A	N/A		4	5738	0	16	4994	63
BIGFOOT CY 2.1AT	IDE(AT)	170.8	3.50	5	UNIDEN	AUTO	N/A	N/A	250	16	4092	63			
BIGFOOT CY 4.3AT	IDE(AT)	4300	3.50	5	UNIDEN	AUTO	N/A	N/A	250	15	8960	63			
BIGFOOT CY 6.4AT	IDE(AT)	6400	3.50	5	UNIDEN	AUTO	N/A	N/A	250	15	13446	63			
CAPELLA VP31110	SCSI-2 F	1108	3.50	3H	RLL 1	AUTO	N/A	N/A	800	4	4165	0			
CAPELLA VP31110S	SCSI-2 F	0	3.50	HH	RLL 1	AUTO	N/A	N/A	300	0	0	0			
CAPELLA VP31110W	SCSI-2 FW	1108	3.50	3H	RLL 1	AUTO	N/A	N/A	800	4	4165	0			
CAPELLA VP31110WD	SCSI-2 FWD	1108	3.50	3H	RLL 1	AUTO	N/A	N/A	800	4	4165	0			
CAPELLA VP32210	SCSI-2 F	2216	3.50	3H	RLL 1	AUTO	N/A	N/A	800	8	4165	0			
CAPELLA VP32210S	SCSI-2 F	0	3.50	HH	RLL 1				300	0	0	0			
CAPELLA VP32210W	SCSI-2 FW	2216	3.50	3H	RLL 1	AUTO	N/A	N/A	800	8	4165	0			
CAPELLA VP32210WD	SCSI-2 FWD	2216	3.50	3H	RLL 1	AUTO	N/A	N/A	800	8	4165	0			
DAYTONA 127AT	IDE(AT)	127	2.50	3H	RLL 1	AUTO	N/A	N/A	350	0	0	0	9	677	41
DAYTONA 127S	SCSI	127	2.50	5H	RLL 1	AUTO	N/A	N/A	350	2	0	0			
DAYTONA 170AT	IDE(AT)	170	2.50	3H	RLL 1	AUTO	N/A	N/A	350	0	0	0	10	538	62
DAYTONA 170S	SCSI	170	2.50	5H	RLL 1	AUTO	N/A	N/A	350	3	0	0			
DAYTONA 256AT	IDE(AT)	256	2.50	3H	RLL 1	AUTO	N/A	N/A	350	0	0	0	11	723	63
DAYTONA 256S	SCSI	256	2.50	5H	RLL 1	AUTO	N/A	N/A	350	4	0	0			
DAYTONA 341AT	IDE(AT)	341	2.50	3H	RLL 1	AUTO	N/A	N/A	350	0	0	0	15	1011	44
DAYTONA 341S	SCSI	341	2.50	4H	RLL 1	AUTO	N/A	N/A	350	6	0	0			
DAYTONA 514AT	IDE(AT)	514	2.50	3H	RLL 1	AUTO	N/A	N/A	350	0	0	0	16	997	63
DAYTONA 514S	SCSI	514	2.50	4H	RLL 1	AUTO	N/A	N/A	350	8	0	0			
DSP3053L	SCSI-2 W	535	3.50	3H	RLL 1	AUTO	N/A	N/A	500	4	3117	0			
DSP3053LD	SCSI-2 D	535	3.50	3H	RLL 1	AUTO	N/A	N/A	500	4	3117	0			
DSP3085	SCSI-2	852	3.50	HH	NONE	AUTO	N/A	N/A	0	14	0	57			
DSP3105	SCSI-2	1050	3.50	HH	NONE	AUTO	N/A	N/A	250	14	0	57			
DSP3105D	SCSI-2	1050	3.50	HH	NONE	AUTO	N/A	N/A	250	14	0	57			
DSP3107L	SCSI-2 D	1070	3.50	3H	RLL 1	AUTO	N/A	N/A	500	8	3117	0			
DSP3107LD	SCSI-2 D	1070	3.50	3H	RLL 1	AUTO	N/A	N/A	500	8	3117	0			
DSP3133L	SCSI-2	1338	3.50	3H	RLL 1	AUTO	N/A	N/A	500	10	3117	0			
DSP3133LD	SCSI-2 D	1338	3.50	3H	RLL 1	AUTO	N/A	N/A	500	10	3117	0			
DSP3160	SCSI-2	1600	3.50	3H	RLL 1	AUTO	N/A	N/A	350	16	2599	0			
DSP3210	SCSI-2	2148	3.50	3H	RLL 1	AUTO	N/A	N/A	500	16	3045	0			
EMPIRE 1080S	U-SCSI	1232	3.50	HH	RLL 1	AUTO	N/A	N/A	500	8	2874	0			
EMPIRE 1400S	U-SCSI	1400	3.50	HH	PRML	AUTO	N/A	N/A	500	8	3115	0			
EMPIRE 1440S	U-SCSI	1440	3.50	HH	RLL 1	AUTO	N/A	N/A	500	8	0	0			
EMPIRE 2100S	U-SCSI	2100	3.50	HH	PRML	AUTO	N/A	N/A	500	12	3115	0			
EMPIRE 2160S	U-SCSI	2160	3.50	HH	RLL 1	AUTO	N/A	N/A	500	12	0	0			

QUANTUM CORPORATION

MODEL	INT	CAP	FF	HGT	CODE	LZ	WP	RWC	MTBF	PHDS	PCYL	PSPT	LHDS	LCYL	LSPT
EMPIRE 540S	U-SCSI	540	3.50	HH	RLL 1	AUTO	N/A	N/A	500	4	2874	0			
EPS 510	SCSI-2 F	107	5.25	FH	UNIDEN					0	0	0			
EPS 530	SCSI-2 F	267	5.25	FH	UNIDEN					0	0	0			
EPS 540	SCSI-2 F	428	5.25	FH	UNIDEN					0	0	0			
EPS 580	SCSI-2 F	856	5.25	FH	UNIDEN					0	0	0			
EUROPA 1080AT	IDE(AT)	1.08	2.50	4H	PRML	AUTO	N/A	N/A	350	8	2	0	15	2,352	60
EUROPA 540AT	IDE(AT)	540	2.50	5H	PRML	AUTO	N/A	N/A	350	4	2	0	15	1,176	60
EUROPA 810AT	IDE(AT)	810	2.50	4H	PRML	AUTO	N/A	N/A	350	6	2	0	15	1,764	60
FIREBALL 1080AT	IDE(AT)	1089	3.50	3H	PRML	2112	N/A	N/A	500	4	3835	0	16	2112	63
FIREBALL 1080S	U-SCSI	1092.7	3.50	3H	PRML	AUTO	N/A	N/A	500	4	3835	0			
FIREBALL 1280AT	IDE(AT)	1282	3.50	3H	PRML	AUTO	N/A	N/A	500	4	4142	0	16	2484	63
FIREBALL 1280S	U-SCSI	1282	3.50	3H	PRML	AUTO	N/A	N/A	500	4	4142	0			
FIREBALL 540AT	IDE(AT)	544	3.50	3H	PRML	1056	N/A	N/A	500	2	3835	0	16	1056	63
FIREBALL 540S	U-SCSI	545.4	3.50	3H	PRML	AUTO	N/A	N/A	500	2	3835	0			
FIREBALL 640AT	IDE(AT)	642	3.50	3H	PRML	AUTO	N/A	N/A	500	2	4142	0	16	1244	63
FIREBALL 640S	SCSI-2	852	3.50	3H	RLL 1	AUTO	N/A	N/A	300	4	3653	0	16	1647	63
FIREBALL ST 2.1S	U-SCSI	2111	3.50	0					0	6	820	17			
FIREBALL ST 3.2AT	U-SCSI	3228	3.50	0					0	16	6256	63			
FIREBALL ST 3.2S	U-SCSI	3228	3.50	0					0	16	6256	63			
FIREBALL ST 4.3AT	U-SCSI	4310	3.50	0					0	9	14848	63			
FIREBALL ST 4.3S	U-SCSI	4310	3.50	0					0	9	14848	63			
FIREBALL ST 6.4AT	U-SCSI	6448	3.50	0					0	15	13328	63			
FIREBALL ST 6.4S	U-SCSI	6448	3.50	0					0	15	13328	63			
FIREBALL TM 1.0AT	IDE(AT)	1080	3.50	3H	PRML	AUTO	N/A	N/A	400	2	6825	0			
FIREBALL TM 1.0S	U-SCSI	1082	3.50	3H		AUTO	N/A	N/A		0	0	0			
FIREBALL TM 2.1AT	IDE(AT)	2168	3.50	3H	PRML	AUTO	N/A	N/A	400	4	6825	0			
FIREBALL TM 2.1S	U-SCSI	2168	3.50	3H		AUTO	N/A	N/A		0	0	0			
FIREBALL TM 2.5AT		0								0	0	0			
FIREBALL TM 3.2AT	IDE(AT)	3254	3.50	3H	PRML	AUTO	N/A	N/A	400	6	6825	0			
FIREBALL TM 3.2S	U-SCSI	3254	3.50	3H		AUTO	N/A	N/A		0	0	0			
FIREBALL TM 3.8AT		0								0	0	0			
GODRIVE 120AT	IDE(AT)	126.62	2.50	3H	RLL 1	AUTO	N/A	N/A	150	4	1097	0	13	731	26
GODRIVE 120S	SCSI	126.62	2.50	3H	RLL 1	AUTO	N/A	N/A	150	4	1097	0			
GODRIVE 40AT	IDE(AT)	41.15	2.50	3H	RLL 1	AUTO	N/A	N/A	80	2	957	42	6	820	17
GODRIVE 40S	SCSI	42.9	2.50	3H	RLL 1	AUTO	N/A	N/A	80	2	870	0			
GODRIVE 60AT	IDE(AT)	63.03	2.50	3H	RLL 1	AUTO	N/A	N/A	150	2	1097	0	7	1024	17
GODRIVE 60S	SCSI	63	2.50	5H	RLL 1	AUTO	N/A	N/A	150	2	0	0			
GODRIVE 80AT	IDE(AT)	82.31	2.50	3H	RLL 1	AUTO	N/A	N/A	80	4	957	42	9	1024	17
GODRIVE 80S	SCSI	86.3	2.50	3H	RLL 1	AUTO	N/A	N/A	80	4	870	0			
GODRIVE GLS 127AT	IDE(AT)	127.9	2.50	3H	RLL 1	AUTO	N/A	N/A	350	3	1395	0	9	677	41
GODRIVE GLS 127S	SCSI	127.91	2.50	3H	RLL 1	AUTO	N/A	N/A	350	3	1395	0			
GODRIVE GLS 170AT	IDE(AT)	170.78	2.50	3H	RLL 1	AUTO	N/A	N/A	350	4	1395	0	10	538	62
GODRIVE GLS 170S	SCSI	170.78	2.50	3H	RLL 1	AUTO	N/A	N/A	350	4	1395	0			
GODRIVE GLS 256AT	IDE(AT)	256.53	2.50	3H	RLL 1	AUTO	N/A	N/A	350	6	1395	0	11	723	63

QUANTUM CORPORATION

MODEL	INT	CAP	FF	HGT	CODE	LZ	WP	RWC	MTBF	PHDS	PCYL	PSPT	LHDS	LCYL	LSPT
GODRIVE GLS 256S	SCSI	256.53	2.50	3H	RLL 1	AUTO	N/A	N/A	350	6	1395	0			
GODRIVE GLS 85AT	IDE(AT)	85.02	2.50	3H	RLL 1	AUTO	N/A	N/A	350	2	1395	0	10	722	23
GODRIVE GLS 85S	SCSI	85.03	2.50	3H	RLL 1	AUTO	N/A	N/A	350	2	1395	0			
GODRIVE GRS 160AT	IDE(AT)	169.87	2.50	3H	RLL 1	AUTO	N/A	N/A	150	4	1376	0	9	1024	36
GODRIVE GRS 160S	SCSI-2	168.16	2.50	3H	RLL 1	AUTO	N/A	N/A	150	4	1376	0			
GODRIVE GRS 80AT	IDE(AT)	84.08	2.50	3H	RLL 1	AUTO	N/A	N/A	150	2	1376	0	5	966	34
GODRIVE GRS 80S	SCSI-2	84.1	2.50	3H	RLL 1	AUTO	N/A	N/A	150	2	1376	0			
GRAND PRIX XP32151S	SCSI	2152	3.50	HH	UNIDEN	AUTO	N/A	N/A	800	10	0	0			
GRAND PRIX XP32151SCA	SCSI	3215	3.50	HH	UNIDEN	AUTO	N/A	N/A		0	0	0			
GRAND PRIX XP34301S	SCSI	4306	3.50	HH	UNIDEN	AUTO	N/A	N/A	800	20	0	0			
GRAND PRIX XP34301SCA	SCSI	3430	3.50	HH	UNIDEN	AUTO	N/A	N/A		0	0	0			
LIGHTNING 365AT	IDE(AT)	365.7	3.50	HH	RLL 1	AUTO	N/A	N/A	300	0	0	0	12	976	61
LIGHTNING 365S	SCSI-2	366.5	3.50	HH	RLL 1	AUTO	N/A	N/A	300	2	0	0			
LIGHTNING 540AT	IDE(AT)	541.3	3.50	HH	RLL 1	AUTO	N/A	N/A	300	0	0	0	16	1120	59
LIGHTNING 540S	SCSI-2	550.7	3.50	HH	RLL 1	AUTO	N/A	N/A	300	2	0	0			
LIGHTNING 730AT	IDE(AT)	730.8	3.50	HH	RLL 1	AUTO	N/A	N/A	300	0	0	0	16	1416	63
LIGHTNING 730S	SCSI-2	733.1	3.50	HH	RLL 1	AUTO	N/A	N/A	300	4	0	0			
MAVERICK 270AT	IDE(AT)	270.66	3.50	3H	RLL 1	944	N/A	N/A	300	2	2740	0	14	944	40
MAVERICK 270S	SCSI	270.7	3.50	3H	RLL 1	AUTO	N/A	N/A	300	2	0	0			
MAVERICK 540AT	IDE(AT)	541.32	3.50	3H	RLL 1	1120	N/A	N/A	300	4	2740	0	16	1048	63
MAVERICK 540S	SCSI	541.5	3.50	3H	RLL 1	AUTO	N/A	N/A	300	4	0	0			
PIONEER SG 1080AT	IDE(AT)	1000	3.50	3H					300	0	0	0			
PIONEER SG 2110AT	IDE(AT)	2100	3.50	3H					300	0	0	0			
PRODRIVE 1050S	SCSI	1054.71	3.50	HH	RLL 1	AUTO	N/A	N/A	150	12	2448	0			
PRODRIVE 105AT	IDE(AT)	105.24	3.50	HH	RLL 2	AUTO	N/A	N/A	60	4	1219	0	16	755	17
PRODRIVE 105S	SCSI	104.99	3.50	HH	RLL 2	AUTO	N/A	N/A	50	6	1019	0			
PRODRIVE 1080S	U-SCSI	1232	3.50	HH	RLL 1	AUTO	N/A	N/A	500	8	2874	0			
PRODRIVE 120AT	IDE(AT)	120.04	3.50	HH	RLL 1	AUTO	N/A	N/A	50	5	1123	0	9	814	32
PRODRIVE 120S	SCSI	120.04	3.50	HH	RLL 1	AUTO	N/A	N/A	50	5	1123	0			
PRODRIVE 1225S	SCSI	1230.5	3.50	HH	RLL 1	AUTO	N/A	N/A	150	14	2448	0			
PRODRIVE 160AT	IDE(AT)	168.39	3.50	HH	RLL 1	AUTO	N/A	N/A	80	4	839	0			
PRODRIVE 170AT	IDE(AT)	168.52	3.50	HH	RLL 1	AUTO	N/A	N/A	50	7	1123	0	10	968	34
PRODRIVE 170S	SCSI	169.52	3.50	HH	RLL 1	AUTO	N/A	N/A	50	7	1123	0			
PRODRIVE 1800S	SCSI-2	1800	3.50	HH	RLL 1	AUTO	N/A	N/A	350	14	2959	0			
PRODRIVE 210AT	IDE(AT)	209.19	3.50	HH	RLL 1	AUTO	N/A	N/A	50	7	1156	0	13	873	36
PRODRIVE 210S	SCSI	209.19	3.50	HH	RLL 1	AUTO	N/A	N/A	50	7	1156	0			
PRODRIVE 40AT	IDE(AT)	42.27	3.50	HH	RLL 2	AUTO	N/A	N/A	50	3	834	0	5	968	17
PRODRIVE 40S	SCSI	42.27	3.50	HH	RLL 2	AUTO	N/A	N/A	50	3	834	0			
PRODRIVE 425	IDE(AT)	426.57	3.50	HH	RLL 1	AUTO	N/A	N/A	150	9	1512	0	16	1021	51
PRODRIVE 425AT	IDE(AT)	426	3.50	HH	RLL 1	N/A	N/A	N/A	150	9	1520	44	16	1021	51
PRODRIVE 425IAT	IDE(AT)	426.57	3.50	HH	RLL 1	AUTO	N/A	N/A	150	9	1512	0	16	1021	51
PRODRIVE 425IS	SCSI	426.57	3.50	HH	RLL 1	AUTO	N/A	N/A	150	9	1512	0			
PRODRIVE 425S	SCSI	426.57	3.50	HH	RLL 1	AUTO	N/A	N/A	150	9	1512	0			
PRODRIVE 525S REV. 1	SCSI-2	525	3.50	3H	RLL 1	AUTO	N/A	N/A	350	6	2448	0			

QUANTUM CORPORATION

MODEL	INT	CAP	FF	HGT	CODE	LZ	WP	RWC	MTBF	PHDS	PCYL	PSPT	LHDS	LCYL	LSPT
PRODRIVE 525S REV. 2	SCSI-2	525	3.50	3H	RLL 1	AUTO	N/A	N/A	350	6	2448	0			
PRODRIVE 52S	SCSI	52.42	3.50	HH	RLL 2	AUTO	N/A	N/A	60	2	1219	0			
PRODRIVE 700S	SCSI	703.14	3.50	HH	RLL 1	AUTO	N/A	N/A	150	8	2448	0			
PRODRIVE 80AT	IDE(AT)	84.54	3.50	HH	RLL 2	AUTO	N/A	N/A	50	6	834	0	10	965	17
PRODRIVE 80S	SCSI	84.54	3.50	HH	RLL 2	AUTO	N/A	N/A	50	6	834	0			
PRODRIVE ELS 127AT	IDE(AT)	127.98	3.50	3H	RLL 1	AUTO	N/A	N/A	250	2	1536	0	16	919	17
PRODRIVE ELS 127S	SCSI	127.98	3.50	3H	RLL 1	AUTO	N/A	N/A	250	2	1536	0			
PRODRIVE ELS 170AT	IDE(AT)	170.81	3.50	3H	RLL 1	AUTO	N/A	N/A	250	4	1536	0	15	1011	22
PRODRIVE ELS 170S	SCSI	170.81	3.50	3H	RLL 1	AUTO	N/A	N/A	250	4	1536	0			
PRODRIVE ELS 42AT	IDE(AT)	42.12	3.50	3H	RLL 2	AUTO	N/A	N/A	250	1	1536	0	5	968	17
PRODRIVE ELS 42S	SCSI	42.12	3.50	3H	RLL 2	AUTO	N/A	N/A	250	1	1536	0			
PRODRIVE ELS 85AT	IDE(AT)	85.03	3.50	3H	RLL 2	AUTO	N/A	N/A	250	2	1536	0	10	977	17
PRODRIVE ELS 85S	SCSI	85.03	3.50	3H	RLL 2	AUTO	N/A	N/A	250	2	1536	0			
PRODRIVE LPS 105AT	IDE(AT)	104.85	3.50	3H	RLL 2	AUTO	N/A	N/A	60	4	1219	0	16	755	17
PRODRIVE LPS 105S	SCSI	104.85	3.50	3H	RLL 2	AUTO	N/A	N/A	60	4	1219	0			
PRODRIVE LPS 120AT	IDE(AT)	122.24	3.50	3H	RLL 2	AUTO	N/A	N/A	250	2	1800	0	5	901	53
PRODRIVE LPS 120S	SCSI	122	3.50	3H	RLL 1	AUTO	N/A	N/A	250	2	1800	0			
PRODRIVE LPS 127AT	IDE(AT)	127.9	3.50	3H	RLL	AUTO	N/A	N/A	300	0	0	0	16	919	17
PRODRIVE LPS 127S	SCSI-2	128.31	3.50	3H	RLL 1	AUTO	N/A	N/A	300	2	1745	0			
PRODRIVE LPS 170AT	IDE(AT)	426.57	3.50	HH	RLL 1	AUTO	N/A	N/A	150	9	1520	0	16	1021	51
PRODRIVE LPS 170S	SCSI-2	171.65	3.50	3H	RLL 1	AUTO	N/A	N/A	300	2	2337	0			
PRODRIVE LPS 210AT	IDE(AT)	211	3.50	3H	RLL 1	AUTO	N/A	N/A	300	2	0	0	15	723	38
PRODRIVE LPS 240AT	IDE(AT)	245.42	3.50	3H	RLL 1	AUTO	N/A	N/A	250	4	1800	0	13	723	51
PRODRIVE LPS 240S	SCSI	245	3.50	3H	RLL 1	AUTO	N/A	N/A	250	4	1800	0			
PRODRIVE LPS 270AT	IDE(AT)	270.66	3.50	3H	RLL 1	944	N/A	N/A	300	2	2740	0	14	944	40
PRODRIVE LPS 270S	SCSI-2	270.66	3.50	3H	RLL 1	AUTO	N/A	N/A	300	2	2740	0			
PRODRIVE LPS 340AT	IDE(AT)	340	3.50	3H	RLL 1	AUTO	N/A	N/A	250	4	1800	0	15	1011	44
PRODRIVE LPS 340S	SCSI-2	342.63	3.50	3H	RLL 1	AUTO	N/A	N/A	300	4	2337	0			
PRODRIVE LPS 420AT	IDE(AT)	421.9	3.50	3H	RLL	AUTO	N/A	N/A	300	0	0	0	16	1021	51
PRODRIVE LPS 525AT	IDE(AT)	524.86	3.50	HH	RLL 1	AUTO	N/A	N/A	250	6	2448	0	16	1017	63
PRODRIVE LPS 525S	SCSI-2	525	3.50	HH	RLL 1	AUTO	N/A	N/A	250	6	0	0			
PRODRIVE LPS 52AT	IDE(AT)	52.42	3.50	3H	RLL 2	AUTO	N/A	N/A	60	2	1219	0	8	751	17
PRODRIVE LPS 52S	SCSI	52.42	3.50	3H	RLL 2	AUTO	N/A	N/A	60	2	1219	0			
PRODRIVE LPS 540AT	IDE(AT)	541.32	3.50	3H	RLL 1	1120	N/A	N/A	300	4	2740	0	16	1048	63
PRODRIVE LPS 540S	SCSI-2	541.33	3.50	3H	RLL 1	AUTO	N/A	N/A	300	4	2740	0			
PRODRIVE LPS 80AT	IDE(AT)	85	3.50	3H	RLL 2	AUTO	N/A	N/A	60	0	0	0	16	611	17
PRODRIVE LPS 80S	SCSI	85	3.50	3H	RLL 2	AUTO	N/A	N/A	60	4	0	0			
Q-160	SCSI	200	5.25	HH	RLL 1	AUTO	N/A	N/A	0	12	1800	0			
Q-2010	SA1000	16.77	8.00	FH	MFM		256	256	12	2	512	32			
Q-2020	SA1000	33.55	8.00	FH	MFM		256	256	12	4	512	32			
Q-2030	SA1000	50.33	8.00	FH	MFM		256	256	12	6	512	32			
Q-2040	SA1000	67.1	8.00	FH	MFM		256	256	12	8	512	32			
Q-2080	SA1000	134.41	8.00	FH	MFM		256	256	8	7	1172	32			
Q-250	SCSI	53.93	5.25	HH	RLL 1	AUTO	256	256	25	4	823	32			

QUANTUM CORPORATION

MODEL	INT	CAP	FF	HGT	CODE	LZ	WP	RWC	MTBF	PHDS	PCYL	PSPT	LHDS	LCYL	LSPT
Q-280	SCSI	80.9	5.25	HH	RLL 1	AUTO	256	256	25	6	823	32			
Q-510	ST-506/412	8.91	5.25	HH	MFM		256	256	0	2	512	17			
Q-520	ST-506/412	17.82	5.25	HH	MFM		256	256	0	4	512	17			
Q-540	ST-506/412	35.65	5.25	FH	MFM		256	256	15	8	512	17			
SATURN VP31080	SCSI-2 F	1085	3.50	3H	UNIDEN	N/A	N/A	N/A	800	0	0	0		3432	
SATURN VP31080S	SCSI-2	1080	3.50	3H	RLL 1					5	0	0			
SATURN VP32170	SCSI-2 F	2170	3.50	3H	UNIDEN	N/A	N/A	N/A	800	0	0	0		3432	
SATURN VP32170S	SCSI-2	2170	3.50	3H	RLL 1	AUTO	N/A	N/A		0	0	0			
SIROCCO 1700AT	IDE(AT)	1700	3.50	3H	UNIDEN	AUTO	N/A	N/A	400	4	0	0	16	3309	63
SIROCCO 2550AT	IDE(AT)	2250	3.50	3H	UNIDEN	AUTO	N/A	N/A	400	6	0	0	16	4969	63
TRAILBLAZER 420AT	IDE(AT)	421.97	3.50	3H	RLL 1	AUTO	N/A	N/A	300	2	3653	0	16	1010	51
TRAILBLAZER 420S	SCSI-2	425	3.50	3H	RLL 1	AUTO	N/A	N/A	300	2	3653	0	16	1010	51
TRAILBLAZER 635AT	IDE(AT)	636.86	3.50	3H	RLL 1	AUTO	N/A	N/A	300	3	3653	0	16	1234	63
TRAILBLAZER 840S	SCSI-2	852	3.50	3H	RLL 1	AUTO	N/A	N/A	300	3	3653	0	16	1647	63
TRAILBLAZER 850AT	IDE(AT)	850	3.50	3H	RLL 1	AUTO	N/A	N/A	300	4	3653	0	16	1647	63
TRAILBLAZER 850S	SCSI-2	852	3.50	3H	RLL 1	AUTO	N/A	N/A	300	4	3653	0	16	1647	63
VIKING 2.2S	U-SCSI	2275	3.50	6						0	0	0			
VIKING 2.2SCA	U-SCSI W	2275	3.50	6						0	0	0			
VIKING 4.5S	U-SCSI	2275	3.50	6						0	0	0			
VIKING 4.5SCA	U-SCSI W	4550	3.50	6						0	0	0			

RICOH CORPORATION

MODEL	INT	CAP	FF	HGT	CODE	LZ	WP	RWC	MTBF	PHDS	PCYL	PSPT	LHDS	LCYL	LSPT
RH-5130	ST-506/412	10.65	5.25	HH	MFM		400	613	0	2	612	17			
RH-5260	ST-506/412	10.7	5.25	HH	MFM				0	2	615	17			
RH-5261	SCSI	10.65	5.25	HH	RLL	AUTO	N/A	N/A	0	2	612	17			
RH-5500	SCSI	100	5.25	HH	RLL 2	AUTO	N/A	N/A	20	2	1285	76			
RS-9150AR	SCSI	100	5.25	HH	RLL 2	AUTO	N/A	N/A	20	2	1285	76			

RMS

MODEL	INT	CAP	FF	HGT	CODE	LZ	WP	RWC	MTBF	PHDS	PCYL	PSPT	LHDS	LCYL	LSPT
RMS 506	ST-506/412	5.32	5.25	FH	MFM		77	77	0	4	153	17			
RMS 509	ST-506/412	7.99	5.25	FH	MFM		77	77	0	6	153	17			
RMS 512	ST-506/412	10.65	5.25	FH	MFM		77	77	0	8	153	17			

RODIME, INC.

MODEL	INT	CAP	FF	HGT	CODE	LZ	WP	RWC	MTBF	PHDS	PCYL	PSPT	LHDS	LCYL	LSPT
COBRA 110AT	IDE(AT)	110	3.50	HH	RLL 2	AUTO	N/A	N/A	40	4	0	0			
COBRA 210AT	IDE(AT)	210	3.50	HH	RLL 2	AUTO	N/A	N/A	40	5	0	0			
COBRA 40AT	IDE(AT)	40	3.50	HH	RLL 2	AUTO	N/A	N/A	40	0	0	0			
COBRA 80AT	IDE(AT)	80	3.50	HH	RLL 2	AUTO	N/A	N/A	40	0	0	0			
RO101	ST-506/412	3.34	5.25	FH	MFM		192	96	0	2	192	17			
RO102	ST-506/412	6.68	5.25	FH	MFM		192	96	0	4	192	17			

RODIME, INC.

MODEL	INT	CAP	FF	HGT	CODE	LZ	WP	RWC	MTBF	PHDS	PCYL	PSPT	LHDS	LCYL	LSPT
R0103	ST-506/412	10.02	5.25	FH	MFM		192	96	0	6	192	17			
R0104	ST-506/412	13.36	5.25	FH	MFM		192	96	0	8	192	17			
R0200	ST-506/412	11.14	5.25	FH	MFM		132		0	4	320	17			
R0201	ST-506/412	5.57	5.25	FH	MFM		0	132	0	2	320	17			
R0201E	ST-506/412	11.14	5.25	FH	MFM		300	264	0	2	640	17			
R0202	ST-506/412	11.14	5.25	HH	MFM		300	132	0	4	320	17			
R0202E	ST-506/412	22.28	5.25	FH	MFM		300	264	0	4	640	17			
R0203	ST-506/412	16.71	5.25	HH	MFM		300	132	0	6	320	17			
R0203E	ST-506/412	33.42	5.25	FH	MFM		300	264	0	6	640	17			
R0204	ST-506/412	22.28	5.25	FH	MFM		300	132	0	8	320	17			
R0204E	ST-506/412	44.56	5.25	FH	MFM		300	264	0	8	640	17			
R0251	ST-506/412	5.32	5.25	HH	MFM		307	307	0	2	306	17			
R0252	ST-506/412	10.65	5.25	HH	MFM		128	64	0	4	306	17			
R03045	ST-506/412	37.94	3.50	HH	MFM			873	0	5	872	17			
R03055	ST-506/412	45.53	3.50	HH	MFM			873	0	6	872	17			
R03055T	SCSI	45.28	3.50	HH	RLL	AUTO	N/A	N/A	0	3	1053	28			
R03057S	SCSI	62.04	3.50	HH	RLL 2	AUTO	N/A	N/A	0	3	1224	33			
R03058A	IDE(AT)	45.33	3.50	HH	RLL 2	AUTO	N/A	N/A	20	3	868	34	3	868	34
R03058T	SCSI	45.33	3.50	HH	RLL 2	AUTO	N/A	N/A	20	3	868	34			
R03059A	IDE(AT)	46	3.50	HH	RLL 2	AUTO	N/A	N/A	20	2	1216	0			
R03059T	SCSI	42.33	3.50	HH	RLL 2	AUTO	N/A	N/A	20	2	1216	34			
R03060R	ST-506/412	49.92	3.50	HH	RLL 2		600		20	5	750	26			
R03065	ST-506/412	53.12	3.50	HH	MFM		650		20	7	872	17			
R03070S	SCSI	71	3.50	HH	RLL 2	AUTO	N/A	N/A	0	0	0	0			
R03075R	ST-506/412	59.9	3.50	HH	RLL 2		650		20	6	750	26			
R03085R	ST-506/412	69.88	3.50	HH	RLL 2		650		20	7	750	26			
R03085S	SCSI	69.88	3.50	HH	RLL 2	AUTO	N/A	N/A	0	7	750	26			
R03088A	IDE(AT)	75.55	3.50	HH	RLL 2	AUTO	N/A	N/A	20	5	868	34	5	868	34
R03088T	SCSI	75.55	3.50	HH	RLL 2	AUTO	N/A	N/A	20	5	868	34			
R03089A	IDE(AT)	63.5	3.50	HH	RLL 2	AUTO	N/A	N/A	20	3	1216	34			
R03089T	SCSI	63.5	3.50	HH	RLL 2	AUTO	N/A	N/A	20	3	1216	34			
R03090T	SCSI	75.47	3.50	HH	RLL 2	AUTO	N/A	N/A	0	5	1053	28			
R03095A	IDE(AT)	80.33	3.50	HH	RLL 2	AUTO	N/A	N/A	20	5	923	34	5	923	34
R03099A	IDE(AT)	80.2	3.50	HH	RLL 2	AUTO	N/A	N/A	0	4	1030	0	15	614	17
R03099AP	IDE(AT)	80.2	3.50	HH	RLL 2	AUTO	N/A	N/A	0	4	1030	0	15	614	17
R03128A	IDE(AT)	105.77	3.50	HH	RLL 2	AUTO	N/A	N/A	20	7	868	34			
R03128T	SCSI	105.77	3.50	HH	RLL 2	AUTO	N/A	N/A	20	7	868	34			
R03129A	IDE(AT)	47.43	3.50	HH	RLL 2	AUTO	N/A	N/A	20	5	1090	17			
R03129T	SCSI	47.43	3.50	HH	RLL 2	AUTO	N/A	N/A	20	5	1090	17			
R03130S	SCSI	112.57	5.25	HH	RLL 2	AUTO	N/A	N/A	20	7	1047	30			
R03130T	SCSI	105.67	3.50	HH	RLL 2	AUTO	N/A	N/A	0	7	1053	28			
R03135A	IDE(AT)	112.47	3.50	HH	RLL 2	AUTO	N/A	N/A	20	7	923	34	7	923	34
R03139A	IDE(AT)	50.83	3.50	HH	RLL 2	AUTO	N/A	N/A	20	5	1168	0	15	861	17
R03139AP	IDE(AT)	112.46	3.50	HH	RLL 2	AUTO	N/A	N/A	0	5	1168	0	15	861	17

RODIME, INC.

MODEL	INT	CAP	FF	HGT	CODE	LZ	WP	RWC	MTBF	PHDS	PCYL	PSPT	LHDS	LCYL	LSPT
RO3139TP	SCSI	112.85	3.50	HH	RLL 2	AUTO	N/A	N/A	0	5	1148	0			
RO3258T	SCSI	210	3.50	HH	RLL 2	AUTO	N/A	N/A	32	0	0	0			
RO3259A	IDE(AT)	209.87	3.50	HH	RLL 2	AUTO	N/A	N/A	0	9	1235	0	15	976	28
RO3259AP	IDE(AT)	212.88	3.50	HH	RLL 2	AUTO	N/A	N/A	0	9	1235	0	15	990	28
RO3259T	SCSI	210	3.50	HH	RLL 2	AUTO	N/A	N/A	0	9	1216	0			
RO3259TP	SCSI	210.54	3.50	HH	RLL 2	AUTO	N/A	N/A	0	9	1148	0			
RO351	ST-506/412	5.32	3.50	HH	MFM		307	307	0	2	306	17			
RO352	ST-506/412	10.65	3.50	HH	MFM		128	64	0	4	306	17			
RO365	ST-506/412	21.3	3.50	HH	MFM		613	613	0	4	612	17			
RO5040S	SCSI	38	5.25	HH	RLL	AUTO	N/A	N/A	0	3	0	17			
RO5065	ST-506/412	53.26	5.25	HH	MFM		N/A	N/A	0	5	1224	17			
RO5075E	ESDI-10MHZ	65.8	5.25	HH	RLL 2		N/A	N/A	0	3	1224	35			
RO5075S	SCSI	62.04	5.25	HH	RLL 2	AUTO	N/A	N/A	0	3	1224	33			
RO5078S	SCSI	62.04	5.25	HH	RLL 2	AUTO	N/A	N/A	0	3	1224	33			
RO5090	ST-506/412	74.57	5.25	HH	MFM		600		0	7	1224	17			
RO5095R	ST-506/412	81.46	5.25	HH	RLL 2	AUTO	N/A	N/A	0	5	1224	26			
RO5125-1F2	SCSI	106.1	5.25	HH	RLL 2	AUTO	N/A	N/A	20	5	1219	34			
RO5125E	ESDI-10MHZ	109.67	5.25	HH	RLL 2	AUTO	N/A	N/A	25	5	1224	35			
RO5125S	SCSI	103.4	5.25	HH	RLL 2	AUTO	N/A	N/A	0	5	1224	33			
RO5128S	SCSI	103.4	5.25	HH	RLL 2	AUTO	N/A	N/A	0	5	1224	33			
RO5130R	ST-506/412	114.05	5.25	HH	RLL 2		N/A	N/A	20	7	1224	26			
RO5178S	SCSI	144.76	5.25	HH	RLL 2	AUTO	N/A	N/A	0	7	1224	33			
RO5180-1F2	SCSI	148.54	5.25	HH	RLL 2	AUTO	N/A	N/A	20	7	1219	34			
RO5180E	ESDI-10MHZ	153.53	5.25	HH	RLL 2	AUTO	N/A	N/A	25	7	1224	35			
RO5180S	SCSI	144.76	5.25	HH	RLL 2	AUTO	N/A	N/A	0	7	1224	33			
RO652	SCSI	20.68	3.50	HH	RLL 2	AUTO	N/A	N/A	0	4	306	33			
RO752	SCSI	20.68	5.25	HH	RLL 2	AUTO	N/A	N/A	0	4	306	33			

SAMSUNG ELECTRONICS, INC.

MODEL	INT	CAP	FF	HGT	CODE	LZ	WP	RWC	MTBF	PHDS	PCYL	PSPT	LHDS	LCYL	LSPT
PLS-30544A	IDE(AT)	540	3.50	3H	RLL 1				300	2	3952	0	16	1047	63
PLS-30854A	IDE(AT)	850	3.50	3H	RLL 1	1647	0	N/A	300	4	3858	0	16	1647	63
PLS-30854S	SCSI-2 F	850	3.50	3H	RLL 1	1647	0	N/A	300	4	3858	0	16	1647	63
PLS-31084A	IDE(AT)	1080	3.50	3H	RLL 1	2055	0	N/A	300	5	3840	0	16	2055	63
PLS-31084S	SCSI-2	1080	3.50	3H	RLL 1	2055	0	N/A	300	5	3840	0	16	2055	63
PLS-31274A	IDE(AT)	1272.6	3.50	3H	RLL 1	2466	0	N/A	300	5	3844	0	16	2466	63
PLS-31274S	SCSI-2	1272.6	3.50	3H	RLL 1	2466	0	N/A	300	5	3844	0	16	2466	63
SHD-2040N	ST-506/412	43.66	3.50	HH	RLL 2		544		0	4	820	26			
SHD-2041B	IDE(AT)	47.02	3.50	HH	RLL 2	AUTO	N/A	N/A	0	4	820	28	4	820	28
SHD-30280A	IDE(AT)	280.3	3.50	3H	RLL 1	AUTO	N/A	N/A	250	2	2668	0	10	869	63
SHD-30420A	IDE(AT)	420.7	3.50	3H	RLL 1	AUTO	N/A	N/A	250	3	2668	0	16	856	63
SHD-30560A	IDE(AT)	560.9	3.50	3H	RLL 1	AUTO	N/A	N/A	250	4	2668	0	16	1086	63
SHD-3061A	IDE(AT)	60.53	3.50	3H	RLL 1	AUTO	N/A	N/A	200	2	1478	40	7	993	17
SHD-3062A	IDE(AT)	121.07	3.50	3H	RLL 1	AUTO	N/A	N/A	200	4	1478	40	15	927	17

SAMSUNG ELECTRONICS, INC.

MODEL	INT	CAP	FF	HGT	CODE	LZ	WP	RWC	MTBF	PHDS	PCYL	PSPT	LHDS	LCYL	LSPT
SHD-3101A	IDE(AT)	105.02	3.50	3H	RLL 1	AUTO	N/A	N/A	200	4	1282	40	16	754	17
SHD-3121A	IDE(AT)	125.9	3.50	3H	RLL 2	AUTO	N/A	N/A	250	2	1956	0	16	615	25
SHD-3122A	IDE(AT)	251.9	3.50	3H	RLL 2	AUTO	N/A	N/A	250	4	1956	0	15	937	35
SHD-3171A	IDE(AT)	178.42	3.50	3H	RLL	AUTO	N/A	N/A	250	2	2227	0	8	968	45
SHD-3171A1	IDE(AT)	178	3.50	3H	RLL 1	AUTO	N/A	N/A	250	2	2210	0	8	968	VAR
SHD-3172A	IDE(AT)	356.84	3.50	3H	RLL	AUTO	N/A	N/A	250	4	2227	0	16	968	45
SHD-3172A1	IDE(AT)	356	3.50	3H	RLL 1	AUTO	N/A	N/A	250	4	2210	0	16	968	VAR
SHD-3202S	SCSI	212.05	3.50	HH	RLL 1	AUTO	N/A	N/A	50	7	1376	43			
SHD-3210S	SCSI	212.05	3.50	HH	RLL 1	AUTO	N/A	N/A	50	7	1376	43			
SHD-3211A	IDE(AT)	213.4	3.50	3H	RLL 1	AUTO	N/A	N/A	250	2	2570	0	8	1002	52
SHD-3212A	IDE(AT)	426.8	3.50	3H	RLL 1	AUTO	N/A	N/A	250	4	2570	0	16	1002	52
STG-31271A	IDE(AT)	1281	3.50	3H	UNIDEN	AUTO	N/A	N/A	300	4	0	0	16	2483	63
STG-31601A	IDE(AT)	1601	3.50	3H	UNIDEN	AUTO	N/A	N/A	300	4	0	0	16	3104	63
TBR-31081A	IDE(AT)	1080	3.50	3H	RLL 1	AUTO	N/A	N/A	500	4	0	0	16	2092	63
VG31702A	IDE(AT)	1700	3.50	3H	UNIDEN	AUTO	N/A	N/A	250	0	0	0	16	3295	63
VG32163A	IDE(AT)	2164	3.50	3H	UNIDEN	AUTO	N/A	N/A	500	0	0	0	9	7456	63
VG32502A	IDE(AT)	2551	3.50	3H	UNIDEN					0	0	0	16	4943	63
VG33402A	IDE(AT)	3401	3.50	3H	UNIDEN					0	0	0	16	6591	63
VG34202A	IDE(AT)	4251	3.50	3H	UNIDEN					0	0	0	16	8238	63
VG34323A	IDE(AT)	4324	3.50	3H	UNIDEN				500	0	0	0	9	14896	63
VG35102A	IDE(AT)	5102	3.50	3H	UNIDEN					0	0	0	16	9886	63
VG36483A	IDE(AT)	6480	3.50	3H	UNIDEN					0	0	0	15	13424	63
WN-310820A	IDE(AT)	1080	3.50	3H	UNIDEN				500	0	0	0	16	2093	63
WN-312021A	IDE(AT)	1207.6	3.50	3H	UNIDEN				500	0	0	0			
WN-31273A	IDE(AT)	1271.6	3.50	3H	UNIDEN				500	0	0	0			
WN-316025A	IDE(AT)	3121	3.50	3H	UNIDEN				500	0	0	0			
WN-321620A	IDE(AT)	2160	3.50	3H	UNIDEN				500	0	0	0			
WN-32543A	IDE(AT)	2541.2	3.50	3H	UNIDEN				500	0	0	0			
WN-33203A	IDE(AT)	3170.8	3.50	3H	UNIDEN				500	0	0	0			
WNR-31271A		0								0	0	0			
WNR-31601A		0								0	0	0			
WNR-32100A		0								0	0	0			
WNR-32101A		0								0	0	0			
WNR-32501A		0								0	0	0			

SEAGATE TECHNOLOGY, INC.

MODEL	INT	CAP	FF	HGT	CODE	LZ	WP	RWC	MTBF	PHDS	PCYL	PSPT	LHDS	LCYL	LSPT
BARRACUDA 1 (VER. 1)	SCSI-2	1690	3.50	HH	RLL 1	AUTO	N/A	N/A	500	15	2706	0			
BARRACUDA 1 (VER. 2)	SCSI-2 W	0	3.50	HH	RLL 1	AUTO	N/A	N/A	500	15	2706	0			
BARRACUDA 2 (VER. 1)		0	3.50	HH	RLL 1	AUTO	N/A	N/A	500	0	0	0			
BARRACUDA 2 (VER. 2)	SCSI-2	0	3.50	HH	RLL 1	AUTO	N/A	N/A	500	0	0	0			
BARRACUDA 2 (VER. 3)	SCSI-2 W	2139	3.50	HH	RLL 1	AUTO	N/A	N/A	500	19	2707	0			
BARRACUDA 2LP (VER. 1)	SCSI-2 F	0	3.50	3H	RLL 1	AUTO	N/A	N/A	800	5	3711	0			
BARRACUDA 2LP (VER. 2)	SCSI-2 F	0	3.50	3H	RLL 1	AUTO	N/A	N/A	800	5	3711	0			

SEAGATE TECHNOLOGY, INC.

MODEL	INT	CAP	FF	HGT	CODE	LZ	WP	RWC	MTBF	PHDS	PCYL	PSPT	LHDS	LCYL	LSPT
BARRACUDA 2LP (VER. 3)	SCSI-2 F	0	3.50	3H	RLL 1	AUTO	N/A	N/A	800	5	3711	0			
BARRACUDA 2LP (VER. 4)	SCSI-2 F	0	3.50	3H	RLL 1	AUTO	N/A	N/A	800	5	3711	0			
BARRACUDA 4 (VER. 1)		0	3.50	HH	RLL 1	AUTO	N/A	N/A	800	0	0	0			
BARRACUDA 4 (VER. 2)	SCSI-2 W	4297	3.50	HH	RLL 1	AUTO	N/A	N/A	800	21	3711	0			
BARRACUDA 4LP (VER. 1)	U-SCSI	0	3.50	3H	RLL	AUTO	N/A	N/A	1000	5	5167	0			
BARRACUDA 4LP (VER. 2)	U-SCSI	0	3.50	3H	RLL	AUTO	N/A	N/A	1000	5	5167	0			
BARRACUDA 9 (VER. 1)	U-SCSI	9100	3.50	HH	PRML	AUTO	N/A	N/A	1000	20	5273	0			
BARRACUDA 9 (VER. 2)	U-SCSI W	9100	3.50	HH	PRML	AUTO	N/A	N/A	1000	20	5273	0			
CHEETAH 4LP (VER. 1)	U-SCSI	4550	3.50	3H	UNIDEN	AUTO	N/A	N/A	300	6	0	0	16	1016	63
CHEETAH 4LP (VER. 2)	U-SCSI W	4550	3.50	3H	UNIDEN	AUTO	N/A	N/A		8	6526	0			
CHEETAH 9 (VER. 1)	U-SCSI	9100	3.50	HH		AUTO	N/A	N/A		16	6526	0			
CHEETAH 9 (VER. 2)	U-SCSI W	9100	3.50	HH		AUTO	N/A	N/A		16	6526	0			
DECATHALON 1080	IDE(AT)	1083.8	3.50	3H	RLL 1	AUTO	N/A	N/A	300	4	4834	0	16	2100	63
DECATHALON 540	IDE(AT)	540.1	3.50	3H	RLL 1	AUTO	N/A	N/A	300	2	4834	0	16	1050	63
DECATHALON 545 (VER. 1)	SCSI-2	545.29	3.50	3H	RLL 1	AUTO	N/A	N/A	0	4	3420	77			
DECATHALON 545 (VER. 2)	IDE(AT)	545.51	3.50	3H	RLL 1	AUTO	N/A	N/A	300	4	3420	0	16	1057	63
DECATHALON 850 (VER. 1)	IDE(AT)	854.7	3.50	3H	RLL 1	AUTO	N/A	N/A	500	4	0	0	32	828	63
DECATHALON 850 (VER. 2)	IDE(AT)	854.7	3.50	4H	RLL 1	AUTO	N/A	N/A	500	4	0	0	32	828	63
ELITE 23 (VER. 1)	U-SCSI	23400	5.25	FH	UNIDEN	AUTO	N/A	N/A		28	6880	0			
ELITE 23 (VER. 2)	U-SCSI W	23400	5.25	FH	UNIDEN	AUTO	N/A	N/A		28	6880	0			
ELITE 9 (VER. 1)		9080	5.25	FH	RLL 1	AUTO	N/A	N/A	500	27	4925	0			
ELITE 9 (VER. 2)	SCSI-2 W	9090	5.25	FH	RLL 1	AUTO			500	27	4925	0			
HAWK 1		0	3.50	HH	RLL 1	AUTO	N/A	N/A	200	0	0	0			
HAWK 1LP (VER. 1)	SCSI-2 FW	0	3.50	3H	RLL 1	AUTO	N/A	N/A	500	9	2700	84			
HAWK 1LP (VER. 2)	SCSI-2	1052.4	3.50	3H	RLL 1	AUTO	N/A	N/A	200	9	2700	0			
HAWK 1LP (VER. 3)	SCSI-2 F	0	3.50	3H	RLL 1	AUTO	N/A	N/A	500	5	2700	78			
HAWK 2 (VER. 1)		0	3.50	HH	RLL 1	AUTO	N/A	N/A	500	0	0	0			
HAWK 2 (VER. 2)	SCSI-2 W	0	3.50	HH	RLL 1	AUTO	N/A	N/A	500	19	2626	0			
HAWK 2 (VER. 3)	SCSI-2 FW	1430	3.50	HH	RLL 1	AUTO	N/A	N/A	500	13	2626	0			
HAWK 2 (VER. 4)	SCSI-2 FW	1430	3.50	HH	RLL 1	AUTO	N/A	N/A	500	13	2626	0			
HAWK 2LP (VER. 1)	SCSI-2 F	0	3.50	3H	RLL 1	AUTO	N/A	N/A	800	5	3992	103			
HAWK 2LP (VER. 2)	SCSI-2 W	0	3.50	3H	RLL	AUTO	N/A	N/A	800	5	3992	0			
HAWK 2LP (VER. 3)	SCSI-2	1700	3.50	3H	RLL	AUTO	N/A	N/A	0	0	0	0			
HAWK 2LP (VER. 4)	SCSI-2	1700	3.50	3H	RLL	AUTO	N/A	N/A	0	0	0	0			
HAWK 2XL (VER. 1)	SCSI-2 F	0	3.50	3H	RLL	AUTO	N/A	N/A	800	4	4569	0			
HAWK 2XL (VER. 2)	SCSI-2 FW	0	3.50	3H	RLL	AUTO	N/A	N/A	800	4	4569	0			
HAWK 4 (VER. 1)	SCSI-2 FW	0	3.50	3H	RLL 1	AUTO	N/A	N/A	800	19	3992	110			
HAWK 4 (VER. 2)	SCSI-2 F	0	3.50	HH	RLL 1	AUTO	N/A	N/A	800	19	3992	110			
MARATHON 130SL	IDE(44PIN)	131.07	2.50	4H	RLL 1	AUTO	N/A	N/A	300	2	0	0	13	419	47
MARATHON 170SL	IDE(44PIN)	171.63	2.50	4H	RLL 1	AUTO	N/A	N/A	300	4	0	0	16	873	24
MARATHON 210SL	IDE(44PIN)	210.43	2.50	4H	RLL 1	AUTO	N/A	N/A	300	4	0	0	8	988	52
MARATHON 260SL	IDE(44PIN)	262.19	2.50	4H	RLL 1	AUTO	N/A	N/A	300	4	0	0	15	569	60
MARATHON 340	IDE(AT)	341.44	2.50	4H	RLL 1	AUTO	N/A	N/A	300	6	0	0	14	934	51
MARATHON 420SL	IDE(44PIN)	420.8	2.50	5H	RLL 1	AUTO	N/A	N/A	300	4	0	0	16	988	53

SEAGATE TECHNOLOGY, INC.

MODEL	INT	CAP	FF	HGT	CODE	LZ	WP	RWC	MTBF	PHDS	PCYL	PSPT	LHDS	LCYL	LSPT
MARATHON 455	IDE(AT)	455.29	2.50	4H	RLL 1	AUTO	N/A	N/A	300	6	0	0	16	942	59
MARATHON 520	IDE(44PIN)	524.35	2.50	4H	RLL 1	AUTO	N/A	N/A	300	6	0	0	16	1016	63
MARATHON 810	IDE(44PIN)	810	2.50	5H	RLL 1	AUTO	N/A	N/A	300	8	0	0	16	1571	63
MEDALIST 1012	IDE(AT)	1012	3.50	3H		AUTO	N/A	N/A		0	0	0	16	2098	63
MEDALIST 1080 (VER. 1)	IDE(AT)	1083.2	3.50	3H	RLL 1	AUTO	N/A	N/A	300	6	0	0	16	2099	63
MEDALIST 1080 (VER. 2)	IDE(AT)	1088	3.50	3H	RLL 1	AUTO	N/A	N/A	300	6	0	0	16	2099	63
MEDALIST 1080SL	IDE(AT)	1083.8	3.50	3H	RLL 1	AUTO	N/A	N/A	300	4	4834	0	16	2100	63
MEDALIST 1270	IDE(AT)	1282.8	3.50	3H	RLL 1	AUTO	N/A	N/A	300	0	0	0	16	2485	63
MEDALIST 1270SL	IDE(AT)	1282.4	3.50	4H	RLL 1	AUTO	N/A	AUTO	300	4	5414	0	16	2485	63
MEDALIST 1276	IDE(AT)	1275	3.50			AUTO	N/A	N/A	300	0	0	0	16	2482	63
MEDALIST 1640	IDE(AT)	1625.7	3.50	3H	RLL 1	AUTO	N/A	N/A	500	6	4834	0	16	3150	63
MEDALIST 1720	IDE(AT)	1705	3.50			AUTO	N/A	N/A		0	0	0	16	3305	68
MEDALIST 1722	IDE(AT)	1722	3.50	3H		AUTO	N/A	N/A		0	0	0	16	3303	63
MEDALIST 210XE	IDE(AT)	213.9	3.50	3H	RLL 1	AUTO	N/A	N/A	300	2	0	0	12	1024	34
MEDALIST 2122	IDE(AT)	212.2	3.50	2		AUTO	N/A	N/A		0	0	0	64	1023	63
MEDALIST 2140	IDE(AT)	2113.4	3.50	3H	RLL 1	AUTO	N/A	N/A	500	8	4726	0	16	4095	63
MEDALIST 270XE (VER. 1)	IDE(AT)	272.7	3.50	3H	RLL 1	AUTO	N/A	N/A	300	2	0	0	14	761	50
MEDALIST 270XE (VER. 2)	IDE(AT)	272.74	3.50	3H	RLL 1	AUTO	N/A	N/A	300	4	0	0	14	761	50
MEDALIST 3232	IDE(AT)	3232	3.50	3H		AUTO	N/A	N/A		0	0	0	16	6253	63
MEDALIST 340XE	IDE(AT)	341.31	3.50	3H	RLL 1	AUTO	N/A	N/A	300	4	0	0	14	768	62
MEDALIST 425XE	IDE(AT)	428.1	3.50	3H	RLL 1	AUTO	N/A	N/A	300	4	0	0	15	899	62
MEDALIST 4342	IDE(AT)	4342	3.50	3H		AUTO	N/A	N/A		0	0	0	15	8894	63
MEDALIST 455	IDE(AT)	452.41	3.50	3H	RLL 2	AUTO	N/A	N/A	150	5	1691	0	14	1018	62
MEDALIST 540	IDE(AT)	528.48	3.50	3H	RLL 2	AUTO	N/A	N/A	200	5	0	0	16	1024	63
MEDALIST 540SL	IDE(AT)	540.1	3.50	3H	RLL 1	AUTO	N/A	N/A	300	2	4834	0	16	1050	63
MEDALIST 545XE	IDE(AT)	545	3.50	3H	RLL 1	AUTO	N/A	N/A	300	0	0	0	16	1057	63
MEDALIST 630XE	IDE(AT)	631.1	3.50	3H	RLL 1	AUTO	N/A	N/A	300	4	0	0	16	1223	63
MEDALIST 636	IDE(AT)	635	3.50		UNIDEN	AUTO	N/A	N/A	300	0	0	0	16	1241	63
MEDALIST 720	IDE(AT)	722.02	3.50	3H	RLL 1	AUTO	N/A	N/A	300	4	0	0	16	1399	63
MEDALIST 850	IDE(AT)	852	3.50			AUTO	N/A	N/A		0	0	0	16	1652	68
MEDALIST 850SL	IDE(AT)	854.7	3.50	4H	RLL 1	AUTO	N/A	AUTO	300	4	4834	0	16	1656	63
MEDALIST PRO 2160	IDE(AT)	810	3.50		UNIDEN	AUTO	N/A	N/A		8	0	0	16	4095	63
MEDALIST PRO 2520	IDE(AT)	810	3.50		UNIDEN	AUTO	N/A	N/A		8	0	0	16	4970	63
MEDALIST PRO 6450	IDE(AT)	6400	3.50	3H		AUTO	N/A	N/A		0	0	0	15	13328	63
MEDALIST PRO 6451	IDE(AT)	6400	3.50	3H		AUTO	N/A	N/A		0	0	0	15	13328	63
ST1057A	IDE(AT)	53.47	3.50	HH	RLL	AUTO	N/A	N/A	150	6	1024	17	6	1024	17
ST1057N	SCSI-2	49.09	3.50	HH	RLL 2	AUTO	N/A	N/A	150	3	940	34			
ST1090A	IDE(AT)	79.58	3.50	HH	RLL 2	AUTO	N/A	N/A	150	5	1072	29	16	335	29
ST1090N	SCSI	79.28	3.50	HH	RLL 2	AUTO	N/A	N/A	150	5	1068	29			
ST1096N	SCSI	84.42	3.50	HH	RLL 2	AUTO	N/A	N/A	150	7	906	26			
ST1100	ST-506/412	83.97	3.50	HH	MFM	AUTO	1,073	1,073	150	9	1072	17			
ST1102A	IDE(AT)	89.12	3.50	HH	RLL	AUTO	N/A	N/A	50	10	1024	17	10	1024	17
ST1102N	SCSI-2	83.99	3.50	HH	RLL	AUTO	N/A	N/A	50	5	965	0			
ST1106R	ST-506/412	91.04	3.50	HH	RLL 2		978	978	50	7	977	26			

SEAGATE TECHNOLOGY, INC.

MODEL	INT	CAP	FF	HGT	CODE	LZ	WP	RWC	MTBF	PHDS	PCYL	PSPT	LHDS	LCYL	LSPT
ST1111A	IDE(AT)	98.79	3.50	HH	RLL 2	AUTO	N/A	N/A	70	5	1072	36	10	536	36
ST1111E	ESDI-10MHZ	98.79	3.50	HH	RLL 2	AUTO	N/A	N/A	150	5	1072	36			
ST1111N	SCSI	98.42	3.50	HH	RLL 2	AUTO	N/A	N/A	70	5	1068	36			
ST11200N	SCSI-2	1054	3.50	HH	RLL 1	AUTO	N/A	N/A	200	15	1872	0			
ST11200NC	SCSI SCA	1050	3.50	HH	RLL 1	AUTO	N/A	N/A	200	15	1877	0			
ST11200ND	SCSI-2 D	1050	3.50	HH	RLL 1	AUTO	N/A	N/A	200	15	1877	0			
ST11201N	SCSI-2	1050	3.50	HH	RLL 1	AUTO	N/A	N/A	200	15	1877	0			
ST11201ND	SCSI-2 D	1050	3.50	HH	RLL 1	AUTO	N/A	N/A	500	15	1877	0			
ST1126A	IDE(AT)	111.41	3.50	HH	RLL 2	AUTO	N/A	N/A	150	7	1072	29	16	469	29
ST1126N	SCSI	111	3.50	HH	RLL 2	AUTO	N/A	N/A	150	7	1068	29			
ST1133A	IDE(AT)	117.22	3.50	HH	RLL 2	AUTO	N/A	N/A	150	5	1272	36	10	636	36
ST1133N	SCSI-2	116.85	3.50	HH	RLL 2	AUTO	N/A	N/A	150	5	1268	36			
ST1144A	IDE(AT)	130.69	3.50	HH	RLL	AUTO	N/A	N/A	150	7	0	0	15	1001	17
ST1144N	SCSI-2	125	3.50	HH	RLL	AUTO	N/A	N/A	50	7	0	0			
ST1150R	ST-506/412	128.43	3.50	HH	RLL 2	AUTO	300	1,073	150	9	1072	26			
ST1156A	IDE(AT)	138.31	3.50	HH	RLL 2	AUTO	N/A	N/A	150	7	1072	36	14	536	36
ST1156E	ESDI-10MHZ	138.31	3.50	HH	RLL 2	AUTO	N/A	N/A	70	7	1072	36			
ST1156N	SCSI	137.79	3.50	HH	RLL 2	AUTO	N/A	N/A	70	7	1068	36			
ST1156NS	SCSI-2	137.79	3.50	HH	RLL 2	AUTO	N/A	N/A	70	7	1068	36			
ST1162A	IDE(AT)	143.25	3.50	HH	RLL 2	AUTO	N/A	N/A	150	9	1072	29	16	603	29
ST1162N	SCSI	142.71	3.50	HH	RLL 2	AUTO	N/A	N/A	70	9	1068	29			
ST11700N	SCSI-2	1430	3.50	HH	RLL 1	AUTO	N/A	N/A	500	13	2626	0			
ST11700ND	SCSI-2 D	1430	3.50	HH	RLL 1	AUTO	N/A	N/A	500	13	2626	0			
ST1186A	IDE(AT)	164.11	3.50	HH	RLL 2	AUTO	N/A	N/A	150	7	1272	36	14	636	36
ST1186N	SCSI-2	163.6	3.50	HH	RLL 2	AUTO	N/A	N/A	150	7	1268	36			
ST11900N	SCSI-2	1700	3.50	HH	RLL 1	AUTO	N/A	N/A	500	15	2621	0			
ST11900NC	SCSI SCA	1700	3.50	HH	RLL 1	AUTO	N/A	N/A	500	15	2621	0			
ST11900ND	SCSI-2	1700	3.50	HH	RLL 1	AUTO	N/A	N/A	500	15	2621	0			
ST11900W	SCSI-2 FW	1430	3.50	HH	RLL 1	AUTO	N/A	N/A	500	13	2626	0			
ST11900WC	SCSI-2 FW	1700	3.50	HH	RLL 1	AUTO	N/A	N/A	500	15	2621	83			
ST11900WD	SCSI-2 FW	1430	3.50	HH	RLL 1	AUTO	N/A	N/A	500	13	2626	0			
ST11950N	SCSI-2	1690	3.50	HH	RLL 1	AUTO	N/A	N/A	500	15	2706	0			
ST11950ND	SCSI-2	1690	3.50	HH	RLL 1	AUTO	N/A	N/A	500	15	2706	0			
ST11950W	SCSI-2 W	1690	3.50	HH	RLL 1	AUTO	N/A	N/A	500	15	2706	0			
ST11950WD	SCSI-2 WD	1690	3.50	HH	RLL 1	AUTO	N/A	N/A	500	15	2706	0			
ST1201A	IDE(AT)	177.83	3.50	HH	RLL 2	AUTO	N/A	N/A	150	9	1072	36	12	804	36
ST1201E	ESDI-10MHZ	177.83	3.50	HH	RLL 2	AUTO	N/A	N/A	150	9	1072	36			
ST1201N	SCSI	177.16	3.50	HH	RLL 2	AUTO	N/A	N/A	150	9	1068	36			
ST1201NS	SCSI-2	177.16	3.50	HH	RLL 2	AUTO	N/A	N/A	150	9	1068	36			
ST1239A	IDE(AT)	211	3.50	HH	RLL 2	AUTO	N/A	N/A	150	9	1272	0	14	814	36
ST1239A (SHROUDED)	IDE(AT)	211	3.50	HH	RLL 2	AUTO	N/A	N/A	150	9	1272	0	12	954	36
ST1239N	SCSI-2	210.34	3.50	HH	RLL 2	AUTO	N/A	N/A	150	9	1268	36			
ST124	ST-506/412	21.41	3.50	HH	MFM	670	616	616	150	4	615	17			
ST12400N	SCSI-2	2148	3.50	HH	RLL 1	AUTO	N/A	N/A	500	19	2621	0			

SEAGATE TECHNOLOGY, INC.

MODEL	INT	CAP	FF	HGT	CODE	LZ	WP	RWC	MTBF	PHDS	PCYL	PSPT	LHDS	LCYL	LSPT
ST12400NC	SCSI SCA	2100	3.50	HH	RLL 1	AUTO	N/A	N/A	500	19	2626	0			
ST12400ND	SCSI-2 D	2148	3.50	HH	RLL 1	AUTO	N/A	N/A	500	19	2621	0			
ST12400W	SCSI-2 W	21	3.50	HH	RLL 1	AUTO	N/A	N/A	500	19	2626	0			
ST12400WC	SCSI-2 W	2148	3.50	HH	RLL 1	AUTO	N/A	N/A	500	19	2621	0			
ST12400WD	SCSI-2 WD	21	3.50	HH	RLL 1	AUTO	N/A	N/A	500	19	2626	0			
ST12401N	SCSI-2	21	3.50	HH	RLL 1	AUTO	N/A	N/A	500	19	2626	0			
ST12401ND	SCSI-2 D	21	3.50	HH	RLL 1	AUTO	N/A	N/A	500	19	2626	0			
ST12450W	SCSI-2 FW	2134	3.50	HH	RLL 1	AUTO	N/A	N/A	500	18	2710	0			
ST12450WD	SCSI-2 FW	2134	3.50	HH	RLL 1	AUTO	N/A	N/A	500	18	2710	0			
ST125	ST-506/412	21.41	3.50	HH	MFM	AUTO	616	616	150	4	615	17			
ST125-1	ST-506/412	21.41	3.50	HH	MFM	AUTO	616	616	150	4	615	17			
ST12550N	SCSI-2	2139	3.50	HH	RLL 1	AUTO	N/A	N/A	500	19	2707	0			
ST12550ND	SCSI-2	2139	3.50	HH	RLL 1	AUTO	N/A	N/A	500	19	2707	0			
ST12550W	SCSI-2 W	2139	3.50	HH	RLL 1	AUTO	N/A	N/A	500	19	2707	0			
ST12550WD	SCSI-2 WD	2139	3.50	HH	RLL 1	AUTO	N/A	N/A	500	19	2707	0			
ST12551N	SCSI-2	2100	3.50	HH	RLL 1	AUTO	N/A	N/A	500	19	2756	0			
ST12551ND	SCSI-2 D	2100	3.50	HH	RLL 1	AUTO	N/A	N/A	500	19	2756	0			
ST125A (10-PIN VER.)	IDE(AT)	21.51	3.50	HH	RLL 2	AUTO	N/A	N/A	70	4	404	26	4	404	26
ST125A (6-PIN VER.)	IDE(AT)	21.51	3.50	HH	RLL 2	AUTO	N/A	N/A	70	4	404	26	4	404	26
ST125A-1 (10-PIN VER.)	IDE(AT)	21.51	3.50	HH	RLL 2	AUTO	N/A	N/A	70	4	404	26	4	404	26
ST125A-1 (6-PIN VER.)	IDE(AT)	21.51	3.50	HH	RLL 2	AUTO	N/A	N/A	70	4	404	26	4	404	26
ST125N	SCSI	21.67	3.50	HH	RLL 2	AUTO	N/A	N/A	70	4	407	26			
ST125N-1	SCSI	21.67	3.50	HH	RLL 2	AUTO	N/A	N/A	70	4	407	26			
ST125R	ST-506/412	21.51	3.50	HH	RLL 2	404	N/A	N/A	150	4	404	26			
ST1274A	IDE(AT)	21.67	3.50	HH	RLL 2	AUTO	N/A	N/A	70	4	407	26	4	407	26
ST137R	ST-506/412	32.74	3.50	HH	RLL 2	AUTO			70	4	615	26			
ST138	ST-506/412	32.11	3.50	HH	MFM	AUTO	616	616	150	6	615	17			
ST138-1	ST-506/412	32.11	3.50	HH	MFM	AUTO	616	616	150	6	615	17			
ST138A (10-PIN VER.)	IDE(AT)	32.16	3.50	HH	RLL 2	AUTO	N/A	N/A	150	4	604	26	4	604	26
ST138A (6-PIN VER.)	IDE(AT)	32.16	3.50	HH	RLL 2	AUTO	N/A	N/A	150	4	604	26	4	604	26
ST138A-1 (10-PIN VER.)	IDE(AT)	32.16	3.50	HH	RLL 2	AUTO	N/A	N/A	150	4	604	26	4	604	26
ST138A-1 (6-PIN VER.)	IDE(AT)	32.16	3.50	HH	RLL 2	AUTO	N/A	N/A	150	4	604	26	4	604	26
ST138N	SCSI	32.74	3.50	HH	RLL 2	AUTO	N/A	N/A	150	4	615	26			
ST138N-1	SCSI	32.74	3.50	HH	RLL 2	AUTO	N/A	N/A	150	4	615	26			
ST138R	ST-506/412	32.74	3.50	HH	RLL 2	AUTO	616	616	150	4	615	26			
ST138R-1	ST-506/412	32.74	3.50	HH	RLL 2	AUTO	616	616	150	4	615	26			
ST1400A (J6-4 PINS)	IDE(AT)	331	3.50	HH	RLL	AUTO	N/A	N/A	150	7	1478	0	12	1018	53
ST1400A (J6-6 PINS)	IDE(AT)	331	3.50	HH	RLL	AUTO	N/A	N/A	150	7	1478	0	12	1018	53
ST1400N	SCSI-2	331	3.50	HH	RLL	AUTO	N/A	N/A	150	7	1476	0			
ST1401A (J6-4 PINS)	IDE(AT)	340	3.50	HH	RLL	AUTO	N/A	N/A	150	9	1100	0	15	726	61
ST1401A (J6-6 PINS)	IDE(AT)	340	3.50	HH	RLL	AUTO	N/A	N/A	150	9	1100	0	15	726	61
ST1401N	SCSI-2	340	3.50	HH	RLL	AUTO	N/A	N/A	150	9	1100	0			
ST14207N	SCSI-2 FW	4295	3.50	3H	RLL 1	AUTO	N/A	N/A	999	20	3924	96			
ST14207W	SCSI-2 FW	4295	3.50	3H	RLL 1	AUTO	N/A	N/A	999	20	3924	96			

SEAGATE TECHNOLOGY, INC.

MODEL	INT	CAP	FF	HGT	CODE	LZ	WP	RWC	MTBF	PHDS	PCYL	PSPT	LHDS	LCYL	LSPT
ST14207WC	SCSI-2 FW	4295	3.50	3H	RLL 1	AUTO	N/A	N/A	999	20	3924	96			
ST1480A (J6-4 PINS)	IDE(AT)	426	3.50	HH	RLL	AUTO	N/A	N/A	150	9	1478	0	15	895	62
ST1480A (J6-6 PINS)	IDE(AT)	426	3.50	HH	RLL	AUTO	N/A	N/A	150	9	1478	0	15	895	62
ST1480N	SCSI-2	426	3.50	HH	RLL	AUTO	N/A	N/A	150	9	1476	0			
ST1480NV	SCSI-2	426	3.50	HH	RLL 1	AUTO	N/A	N/A	150	9	1478	0			
ST1481N	SCSI-2	426	3.50	HH	RLL	AUTO	N/A	N/A	150	9	1476	0			
ST151	ST-506/412	42.51	3.50	HH	MFM	AUTO	978	978	150	5	977	17			
ST15150DC	SCSI-2 FW	4294	3.50	HH	RLL 1	AUTO	N/A	N/A	800	21	3711	0			
ST15150N	SCSI-2 W	4297	3.50	HH	RLL 1	AUTO	N/A	N/A	800	21	3711	0			
ST15150ND	SCSI-2 W	4294	3.50	HH	RLL 1	AUTO	N/A	N/A	800	21	3711	0			
ST15150W	SCSI-2 FW	4294	3.50	HH	RLL 1	AUTO	N/A	N/A	800	21	3711	0			
ST15150WC	SCSI-2 FW	4294	3.50	HH	RLL 1	AUTO	N/A	N/A	800	21	3711	0			
ST15150WD	SCSI-2 FW	4294	3.50	HH	RLL 1	AUTO	N/A	N/A	800	21	3711	0			
ST15230DC	SCSI SCA	4	3.50	HH	UNIDEN	AUTO	N/A	N/A	800	19	3992	0			
ST15230N	SCSI-2 F	4294	3.50	HH	RLL 1	AUTO	N/A	N/A	800	19	3992	110			
ST15230NC	SCSI-2 F	4294	3.50	HH	RLL 1	AUTO	N/A	N/A	800	19	3992	110			
ST15230ND	SCSI-2 F	4596	3.50	HH	RLL 1	AUTO	N/A	N/A	800	19	3992	110			
ST15230W	SCSI-2 FW	4294	3.50	3H	RLL 1	AUTO	N/A	N/A	800	19	3992	110			
ST15230WC	SCSI-2 FW	4294	3.50	3H	RLL 1	AUTO	N/A	N/A	800	19	3992	110			
ST15230WD	SCSI-2 FW	4294	3.50	HH	RLL 1	AUTO	N/A	N/A	800	19	3992	0			
ST157A (10-PIN VER.)	IDE(AT)	44.72	3.50	HH	RLL 2	AUTO	N/A	N/A	150	6	560	26	6	560	26
ST157A (6-PIN VER.)	IDE(AT)	44.72	3.50	HH	RLL 2	AUTO	N/A	N/A	150	6	560	26	6	560	26
ST157A-1 (10-PIN VER.)	IDE(AT)	44.72	3.50	HH	RLL 2	AUTO	N/A	N/A	150	6	560	26	6	560	26
ST157A-1 (6-PIN VER.)	IDE(AT)	44.72	3.50	HH	RLL 2	AUTO	N/A	N/A	150	6	560	26	6	560	26
ST157N	SCSI	48.96	3.50	HH	RLL 2	AUTO	N/A	N/A	150	6	613	26			
ST157N-1	SCSI	48.96	3.50	HH	RLL 2	AUTO	N/A	N/A	150	6	613	26			
ST157R	ST-506/412	49.12	3.50	HH	RLL 2	AUTO	616	616	150	6	615	26			
ST157R-1	ST-506/412	49.12	3.50	HH	RLL 2	AUTO	616	616	150	6	615	26			
ST1581N	SCSI-2	525	3.50	HH	RLL	AUTO	N/A	N/A	150	9	1476	0			
ST177N	SCSI	61.3	3.50	HH	RLL 2	AUTO	N/A	N/A	150	5	921	26			
ST1830N	SCSI-2 F	702	3.50	HH	RLL	AUTO	N/A	N/A	0	0	0	0			
ST19101DC	U-SCSI D	9100	3.50	HH		AUTO	N/A	N/A		16	6526	0			
ST19101N	U-SCSI	9100	3.50	HH		AUTO	N/A	N/A		16	6526	0			
ST19101W	U-SCSI W	9100	3.50	HH		AUTO	N/A	N/A		16	6526	0			
ST19101WC	U-SCSI W	9100	3.50	HH		AUTO	N/A	N/A		16	6526	0			
ST19101WD	U-SCSI WD	9100	3.50	HH		AUTO	N/A	N/A		16	6526	0			
ST19171DC	U-SCSI D	9100	3.50	HH	PRML	AUTO	N/A	N/A	1000	20	5273	0			
ST19171N	U-SCSI	9100	3.50	HH	PRML	AUTO	N/A	N/A	1000	20	5273	0			
ST19171W	U-SCSI W	9100	3.50	HH	PRML	AUTO	N/A	N/A	1000	20	5273	0			
ST19171WC	U-SCSI W	9100	3.50	HH	PRML	AUTO	N/A	N/A	1000	20	5273	0			
ST19171WD	U-SCSI WD	9100	3.50	HH	PRML	AUTO	N/A	N/A	1000	20	5273	0			
ST1950N	SCSI-2 F	803	3.50	HH	RLL	AUTO	N/A	N/A	0	13	1575	0			
ST1980N	SCSI-2	860	3.50	HH	RLL 1	AUTO	N/A	N/A	200	13	1730	0			
ST1980NC	SCSI SCA	860	3.50	HH	RLL 1	AUTO	N/A	N/A	200	13	1730	0			

SEAGATE TECHNOLOGY, INC.

MODEL	INT	CAP	FF	HGT	CODE	LZ	WP	RWC	MTBF	PHDS	PCYL	PSPT	LHDS	LCYL	LSPT
ST1980ND	SCSI-2 D	860	3.50	HH	RLL 1	AUTO	N/A	N/A	200	13	1730	0			
ST206	ST-506/412	5.32	5.25	HH	MFM		128	307	0	2	306	17			
ST2106E	ESDI-10MHZ	94.37	5.25	HH	RLL 2	AUTO	N/A	N/A	100	5	1024	36			
ST2106N	SCSI	94.18	5.25	HH	RLL 2	AUTO	N/A	N/A	100	5	1022	36			
ST2106NM	SCSI	94.18	5.25	HH	RLL 2	AUTO	N/A	N/A	100	5	1022	36			
ST212 (VER. 1)	ST-506/412	10.65	5.25	HH	MFM	319	128	307	11	4	306	17			
ST212 (VER. 2)	ST-506/412	10.65	5.25	HH	MFM	319	128	307	11	4	306	17			
ST212 (VER. 3)	ST-506/412	10.65	5.25	HH	MFM	319	128	307	11	4	306	17			
ST2125N	SCSI	107	5.25	HH	RLL	AUTO	N/A	N/A	100	3	1544	0			
ST2125NM	SCSI	107	5.25	HH	RLL	AUTO	N/A	N/A	100	3	1544	0			
ST2125NV	SCSI	107	5.25	HH	RLL	AUTO	N/A	N/A	100	3	1544	0			
ST213	ST-506/412	10.7	5.25	HH	MFM	670	300	N/A	20	2	615	17			
ST2182E	ESDI-15MHZ	160.69	5.25	HH	RLL 2	AUTO	N/A	N/A	100	4	1453	54			
ST2209N	SCSI	183	5.25	HH	RLL	AUTO	N/A	N/A	100	5	1544	0			
ST2209NM	SCSI	183	5.25	HH	RLL	AUTO	N/A	N/A	100	5	1544	0			
ST2209NV	SCSI	183	5.25	HH	RLL	AUTO	N/A	N/A	100	5	1544	0			
ST224N	SCSI	21.41	5.25	HH	RLL 2	AUTO	N/A	N/A	100	4	615	17			
ST225	ST-506/412	21.41	5.25	HH	MFM	670	300	N/A	100	4	615	17			
ST225N	SCSI	21.41	5.25	HH	RLL	670	N/A	N/A	100	4	615	17			
ST225R	ST-506/412	21.17	5.25	HH	RLL 2	667			100	2	667	31			
ST2274A	IDE(AT)	241.5	5.25	HH	RLL 2	AUTO	N/A	N/A	100	5	1747	54	10	873	54
ST238	ST-506/412	32.74	5.25	HH	RLL 2	615			100	4	615	26			
ST2383A	IDE(AT)	338.1	5.25	HH	RLL 2	AUTO	N/A	N/A	100	7	1747	54	14	873	54
ST2383E	ESDI-15MHZ	338.1	5.25	HH	RLL 2	AUTO	N/A	N/A	100	7	1747	54			
ST2383N	SCSI	332	5.25	HH	RLL	AUTO	N/A	N/A	100	7	1261	0			
ST2383NM	SCSI	332	5.25	HH	RLL	AUTO	N/A	N/A	100	7	1261	0			
ST238R	ST-506/412	32.74	5.25	HH	RLL 2	615			100	4	615	26			
ST2502N	SCSI	442	5.25	HH	RLL	AUTO	N/A	N/A	100	7	1755	0			
ST2502NM	SCSI	442	5.25	HH	RLL	AUTO	N/A	N/A	100	7	1755	0			
ST2502NV	SCSI	442	5.25	HH	RLL	AUTO	N/A	N/A	100	7	1755	0			
ST250N	SCSI	42.34	5.25	HH	RLL 2	AUTO	N/A	N/A	100	4	667	31			
ST250R	ST-506/412	42.34	5.25	HH	RLL 2	670			100	4	667	31			
ST251	ST-506/412	42.82	5.25	HH	MFM	AUTO			100	6	820	17			
ST251-1	ST-506/412	42.82	5.25	HH	MFM	AUTO	821	821	100	6	820	17			
ST251N	SCSI	43.66	5.25	HH	RLL 2	AUTO	N/A	N/A	70	4	820	26			
ST251N-1	SCSI	43.86	5.25	HH	RLL 2	AUTO	N/A	N/A	70	4	630	34			
ST251R	ST-506/412	43.66	5.25	HH	RLL 2				100	4	820	26			
ST252	ST-506/412	42.82	5.25	HH	MFM	AUTO			100	6	820	17			
ST253	ST-506/412	43.04	5.25	HH	MFM	AUTO			40	5	989	17			
ST274A	IDE(AT)	63.09	5.25	HH	RLL 2	AUTO	N/A	N/A	40	5	948	26	5	948	26
ST277N	SCSI	65.49	5.25	HH	RLL 2	AUTO	N/A	N/A	70	6	820	26			
ST277N-1	SCSI	65.59	5.25	HH	RLL 2	AUTO	N/A	N/A	70	6	628	34			
ST277R	ST-506/412	65.49	5.25	HH	RLL 2	AUTO			70	6	820	26			
ST277R-1	ST-506/412	65.49	5.25	HH	RLL 2	AUTO			70	6	820	26			

SEAGATE TECHNOLOGY, INC.

MODEL	INT	CAP	FF	HGT	CODE	LZ	WP	RWC	MTBF	PHDS	PCYL	PSPT	LHDS	LCYL	LSPT
ST278R	ST-506/412	65.49	5.25	HH	RLL 2	AUTO			70	6	820	26			
ST279R	ST-506/412	65.82	5.25	HH	RLL 2	AUTO			40	5	989	26			
ST280A	IDE(AT)	71.33	5.25	HH	RLL 2	AUTO	N/A	N/A	40	5	1032	27	10	516	27
ST296N	SCSI	85.64	5.25	HH	RLL 2	AUTO	N/A	N/A	70	6	820	34			
ST3025A	IDE(AT)	21.51	3.50	3H	RLL 2	AUTO	N/A	N/A	50	1	1616	26	2	808	26
ST3025N	SCSI-2	21.51	3.50	3H	RLL 2	AUTO	N/A	N/A	50	1	1616	26			
ST3051A	IDE(AT)	43.1	3.50	3H	RLL 2	AUTO	N/A	N/A	150	0	0	0	6	820	17
ST3057N	SCSI-2	49.09	3.50	3H	RLL 2	AUTO	N/A	N/A	50	3	940	34			
ST3096A	IDE(AT)	89.12	3.50	3H	RLL 2	AUTO	N/A	N/A	50	10	1024	17	10	1024	17
ST3096N	SCSI-2	89.12	3.50	3H	RLL 2	AUTO	N/A	N/A	50	10	1024	17			
ST31012A	IDE(AT)	1012	3.50	3H		AUTO	N/A	N/A		0	0	0	16	2098	63
ST31051N	SCSI-2 F	1050	3.50	3H	RLL	AUTO	N/A	N/A	800	4	4569	0			
ST31051W	SCSI-2 FW	1050	3.50	3H	RLL	AUTO	N/A	N/A	800	4	4569	0			
ST31051WC	SCSI SCA	1050	3.50	4H	RLL	AUTO	N/A	N/A	800	4	4569	0			
ST31055N	U-SCSI	1060	3.50	3H	RLL	AUTO	N/A	N/A	800	4	4176	0			
ST31055W	U-SCSI	1060	3.50	3H	RLL	AUTO	N/A	N/A	800	4	4176	123			
ST31055WC	U-SCSI	1060	3.50	3H	RLL	AUTO	N/A	N/A	800	4	4176	0			
ST31080N	SCSI-2 F	0	3.50	3H	RLL 1	AUTO	N/A	N/A	500	0	0	0			
ST31081A	IDE(AT)	1080	3.50	3H	RLL 1	2097	N/A	N/A	300	4	3924	0	16	2097	63
ST31200N	SCSI-2	0	3.50	3H	RLL 1	AUTO	N/A	N/A	200	9	2700	0			
ST31200NC	SCSI SCA	1052	3.50	3H	RLL 1	AUTO	N/A	N/A	500	9	2700	84			
ST31200ND	SCSI-2	1052.4	3.50	3H	RLL 1	AUTO	N/A	N/A	200	9	2700	0			
ST31200W	SCSI-2 FW	1052	3.50	3H	RLL 1	AUTO	N/A	N/A	500	9	2700	84			
ST31200WC	SCSI-2 FW	1052	3.50	3H	RLL 1	AUTO	N/A	N/A	500	9	2700	84			
ST31200WD	SCSI-2 FW	1052	3.50	3H	RLL 1	AUTO	N/A	N/A	500	9	2700	84			
ST3120A	IDE(AT)	106.9	3.50	3H	RLL 2	AUTO	N/A	N/A	150	0	0	0	12	1024	17
ST31220A	IDE(AT)	1083.2	3.50	3H	RLL 1	AUTO	N/A	N/A	300	6	0	0	16	2099	63
ST31220A (VER. 2)	IDE(AT)	1088	3.50	3H	RLL 1	AUTO	N/A	N/A	300	6	0	0	16	2099	63
ST31230DC	SCSI-2 WD	1060	3.50	3H	UNIDEN	AUTO	N/A	N/A	800	5	3992	0			
ST31230N	SCSI-2 F	1060	3.50	3H	RLL 1	AUTO	N/A	N/A	800	5	3992	103			
ST31230NC	SCSI SCA	1060	3.50	3H	RLL 1	AUTO	N/A	N/A	800	5	3992	103	5	3992	103
ST31230ND	SCSI-2 D	1060	3.50	3H	RLL 1	AUTO	N/A	N/A	800	5	3992	103			
ST31230W	SCSI-2 W	1060	3.50	3H	RLL	AUTO	N/A	N/A	800	5	3992	0			
ST31230WC	SCSI-2 FW	1060	3.50	3H	RLL 1	AUTO	N/A	N/A	800	5	3992	0			
ST31230WD	SCSI-2 FW	1060	3.50	3H	RLL	AUTO	N/A	N/A	800	5	3992	0			
ST31231N	SCSI-2 F	1060	3.50	3H	RLL 1	AUTO	N/A	N/A	800	5	3992	103			
ST3123A	IDE(AT)	106.6	3.50	3H	RLL 1	AUTO	N/A	N/A	250	0	0	0	12	1024	17
ST31250DC	SCSI SCA	1020.9	3.50	3H	RLL 1	AUTO	N/A	N/A	800	5	3711	0			
ST31250N	SCSI-2 F	1020.9	3.50	3H	RLL 1	AUTO	N/A	N/A	800	5	3711	0			
ST31250ND	SCSI-2 F	1020.9	3.50	3H	RLL 1	AUTO	N/A	N/A	800	5	3711	0			
ST31250W	SCSI-2 F	1020.9	3.50	3H	RLL 1	AUTO	N/A	N/A	800	5	3711	0			
ST31250WC	SCSI SCA	1020.9	3.50	3H	RLL 1	AUTO	N/A	N/A	800	5	3711	0			
ST31250WD	SCSI-2 F	1020.9	3.50	3H	RLL 1	AUTO	N/A	N/A	800	5	3711	0			
ST31270A	IDE(AT)	1282.8	3.50	3H	RLL 1	AUTO	N/A	N/A	300	0	0	0	16	2485	63

SEAGATE TECHNOLOGY, INC.

MODEL	INT	CAP	FF	HGT	CODE	LZ	WP	RWC	MTBF	PHDS	PCYL	PSPT	LHDS	LCYL	LSPT
ST31274A	IDE(AT)	1278	3.50	3H	RLL 1	2480	N/A	N/A	300	6	2479	0	16	2479	63
ST31275A	IDE(AT)	1278	3.50	3H	RLL 1	2479	N/A	N/A	250	0	0	0	16	2479	63
ST31276A	IDE(AT)	1275	3.50			AUTO	N/A	N/A	300	0	0	0	16	2482	63
ST3144A	IDE(AT)	130.69	3.50	3H	RLL 2	AUTO	N/A	N/A	150	3	0	0	15	1001	17
ST3145A	IDE(AT)	130.2	3.50	3H	RLL 1	AUTO	N/A	N/A	250	2	0	0	15	1001	17
ST31621A	IDE(AT)	1620	3.50	3H		3146	N/A	N/A	300	6	3924	0	16	3146	63
ST31640A	IDE(AT)	1625.7	3.50	3H	RLL 1	AUTO	N/A	N/A	500	6	4834	0	16	3150	63
ST31720A	IDE(AT)	1705	3.50			AUTO	N/A	N/A		0	0	0	16	3305	68
ST31722A	IDE(AT)	1722	3.50	3H		AUTO	N/A	N/A		0	0	0	16	3303	63
ST31930N	SCSI-2	1700	3.50	3H	RLL	AUTO	N/A	N/A	0	0	0	0			
ST31930ND	SCSI-2	1700	3.50	3H	RLL	AUTO	N/A	N/A	0	0	0	0			
ST3195A	IDE(AT)	170.77	3.50	3H	RLL 2	AUTO	N/A	N/A	150	4	0	0	10	981	34
ST32105N	SCSI-2 F	2147	3.50	3H	UNIDEN	N/A	N/A	AUTO	999	10	3892	0			
ST32105W	SCSI-2 FW	2147	3.50	3H	UNIDEN	N/A	N/A	AUTO	999	10	3892	0			
ST32107N	SCSI-2 FW	2110	3.50	3H	RLL 1	AUTO	N/A	N/A	999	10	3924	91			
ST32107W	SCSI-2 FW	2110	3.50	3H	RLL 1	AUTO	N/A	N/A	999	10	3924	91			
ST32107WC	SCSI-2 FW	2110	3.50	3H	RLL 1	AUTO	N/A	N/A	999	10	3924	91			
ST32122A	IDE(AT)	2122	3.50	3H		AUTO	N/A	N/A		0	0	0	16	4092	63
ST32140A	IDE(AT)	2113.4	3.50	3H	RLL 1	AUTO	N/A	N/A	500	8	4726	0	16	4095	63
ST32151N	SCSI-2 F	2147	3.50	3H	RLL	AUTO	N/A	N/A	800	8	4569	0			
ST32151W	SCSI-2 FW	2147	3.50	3H	RLL	AUTO	N/A	N/A	800	8	4569	0			
ST32151WC	SCSI SCA	2147	3.50	4H	RLL	AUTO	N/A	N/A	800	8	4569	0			
ST32155N	U-SCSI	2148	3.50	3H	RLL	AUTO	N/A	N/A	800	8	4176	0			
ST32155W	U-SCSI	2148	3.50	3H	RLL	AUTO	N/A	N/A	800	8	4176	125			
ST32155WC	SCSI-2 W	2148	3.50	3H	RLL	AUTO	N/A	N/A	800	8	4176	0			
ST32161A	IDE(AT)	2147	3.50	3H	RLL 1	AUTO	N/A	N/A	500	8	4474	0	16	4095	63
ST32171DC	U-SCSI D	2150	3.50	3H	RLL	AUTO	N/A	N/A	1000	5	5167	0			
ST32171N	U-SCSI	2150	3.50	3H	RLL	AUTO	N/A	N/A	1000	5	5167	0			
ST32171ND	U-SCSI D	2150	3.50	3H	RLL	AUTO	N/A	N/A	1000	5	5167	0			
ST32171W	U-SCSI	2150	3.50	3H	RLL	AUTO	N/A	N/A	1000	5	5167	0			
ST32171WC	U-SCSI W	2150	3.50	3H	RLL	AUTO	N/A	N/A	1000	5	5167	0			
ST32171WD	U-SCSI	2150	3.50	3H	RLL	AUTO	N/A	N/A	1000	5	5167	0			
ST32430DC	SCSI SCA	2140	3.50	3H	UNIDEN	AUTO	N/A	N/A	800	9	3992	0			
ST32430N	SCSI-2 F	2140	3.50	3H	RLL 1	AUTO	N/A	N/A	800	9	3992	116			
ST32430NC	SCSI SCA	2140	3.50	3H	RLL 1	AUTO	N/A	N/A	800	9	3992	116			
ST32430ND	SCSI-2 D	2140	3.50	3H	RLL 1	AUTO	N/A	N/A	800	9	3992	116			
ST32430W	SCSI-2 FW	2140	3.50	3H	RLL 1	AUTO	N/A	N/A	800	9	3892	0			
ST32430WC	SCSI-2 W	2140	3.50	3H	RLL 1	AUTO	N/A	N/A	800	9	3992	116			
ST32430WD	SCSI-2 FW	2140	3.50	3H	RLL 1	AUTO	N/A	N/A	800	9	3992	0			
ST3243A	IDE(AT)	213.9	3.50	3H	RLL 2	AUTO	N/A	N/A	150	4	0	0	12	1024	34
ST3250A	IDE(AT)	213.9	3.50	3H	RLL 1	AUTO	N/A	N/A	300	2	0	0	12	1024	34
ST32550DC	SCSI SCA	2147.8	3.50	3H	RLL 1	AUTO	N/A	N/A	800	11	3510	0			
ST32550N	SCSI-2 F	2147.8	3.50	3H	RLL 1	AUTO	N/A	N/A	800	11	3510	0			
ST32550ND	SCSI-2 F	2147.8	3.50	3H	RLL 1	AUTO	N/A	N/A	800	11	3510	0			

SEAGATE TECHNOLOGY, INC.

MODEL	INT	CAP	FF	HGT	CODE	LZ	WP	RWC	MTBF	PHDS	PCYL	PSPT	LHDS	LCYL	LSPT
ST32550W	SCSI-2 F	2147.8	3.50	3H	RLL 1	AUTO	N/A	N/A	800	11	3510	0			
ST32550WC	SCSI SCA	2147.8	3.50	3H	RLL 1	AUTO	N/A	N/A	800	11	3510	0			
ST32550WD	SCSI-2 F	2147.8	3.50	3H	RLL 1	AUTO	N/A	N/A	800	11	3510	0			
ST325A\X (12-PIN VER.)	IDE(XT/AT)	21.41	3.50	3H	RLL 2	AUTO	N/A	N/A	150	2	697	30	2	697	30
ST325A\X (18-PIN VER.)	IDE(XT/AT)	21.41	3.50	3H	RLL 2	AUTO	N/A	N/A	150	2	697	30	2	697	30
ST325N	SCSI	21.41	3.50	3H	RLL 2	AUTO	N/A	N/A	50	2	697	30			
ST3270A	IDE(AT)	270.9	3.50	3H	RLL 1	AUTO	N/A	N/A	250	4	0	0	16	525	63
ST3283A	IDE(AT)	245.36	3.50	3H	RLL 2	AUTO	N/A	N/A	150	5	1691	0	14	978	35
ST3283N PCB <260035	SCSI	245.36	3.50	3H	RLL 2	AUTO	N/A	N/A	150	5	1691	0			
ST3283N PCB >260035	SCSI	245.36	3.50	3H	RLL 2	AUTO	N/A	N/A	150	5	1691	0			
ST3285N	SCSI-2 F	248.62	3.50	3H	RLL 1	AUTO	N/A	N/A	250	3	1689	0			
ST3290A	IDE(AT)	261.38	3.50	3H	RLL 1	AUTO	N/A	N/A	150	4	1691	0	15	1001	34
ST3291A	IDE(AT)	272.74	3.50	3H	RLL 1	AUTO	N/A	N/A	300	4	0	0	14	761	50
ST3295A	IDE(AT)	272.7	3.50	3H	RLL 1	AUTO	N/A	N/A	300	2	0	0	14	761	50
ST33232A	IDE(AT)	3232	3.50	3H		AUTO	N/A	N/A		0	0	0	16	6253	63
ST3385A	IDE(AT)	340	3.50	3H	RLL 2	AUTO	N/A	N/A	150	5	1691	0	14	768	62
ST3390A	IDE(AT)	341.3	3.50	3H	RLL 2	AUTO	N/A	N/A	250	3	2676	0	14	768	62
ST3390N	SCSI-2	344.3	3.50	3H	RLL 2	AUTO	N/A	N/A	250	3	2676	0			
ST3391A	IDE(AT)	341.31	3.50	3H	RLL 1	AUTO	N/A	N/A	300	4	0	0	14	768	62
ST3423A	IDE(AT)	426	3.50	3H	RLL 1	863	N/A	N/A	300	2	862	0	16	826	63
ST3425A	IDE(AT)	425	3.50	3H	RLL 1	AUTO	N/A	N/A	250	2	0	0	16	826	63
ST34342A	IDE(AT)	4342	3.50	3H		AUTO	N/A	N/A		0	0	0	15	8894	63
ST34371DC	U-SCSI D	4350	3.50	3H	RLL	AUTO	N/A	N/A	1000	10	5167	0			
ST34371N	U-SCSI	4350	3.50	3H	RLL	AUTO	N/A	N/A	1000	10	5167	0			
ST34371ND	U-SCSI D	4350	3.50	3H	RLL	AUTO	N/A	N/A	1000	10	5167	0			
ST34371W	U-SCSI	4350	3.50	3H	RLL	AUTO	N/A	N/A	1000	10	5167	0			
ST34371WC	U-SCSI W	4350	3.50	3H	RLL	AUTO	N/A	N/A	1000	10	5167	0			
ST34371WD	U-SCSI	4350	3.50	3H	RLL	AUTO	N/A	N/A	1000	10	5167	0			
ST34501DC	U-SCSI D	4550	3.50	3H	UNIDEN	AUTO	N/A	N/A		8	6526	0			
ST34501N	U-SCSI	4550	3.50	3H	UNIDEN	AUTO	N/A	N/A	300	6	0	0	16	1016	63
ST34501W	U-SCSI W	4550	3.50	3H	UNIDEN	AUTO	N/A	N/A		8	6526	0			
ST34501WC	U-SCSI W	4550	3.50	3H	UNIDEN	AUTO	N/A	N/A		8	6526	0			
ST34501WD	U-SCSI WD	4550	3.50	3H	UNIDEN	AUTO	N/A	N/A		8	6526	0			
ST3491A	IDE(AT)	428.1	3.50	3H	RLL 1	AUTO	N/A	N/A	300	4	0	0	15	899	62
ST3500A	IDE(AT)	426.16	3.50	3H	RLL 1	AUTO	N/A	N/A	200	7	1874	0	15	895	62
ST3500N	SCSI-2	426	3.50	3H	RLL 1	AUTO	N/A	N/A	200	7	1872	0			
ST351A\X (12-PIN VER.)	IDE(XT/AT)	42.82	3.50	3H	RLL 2	AUTO	N/A	N/A	150	6	820	17	6	820	17
ST351A\X (18-PIN VER.)	IDE(XT/AT)	42.82	3.50	3H	RLL 2	AUTO	N/A	N/A	150	6	820	17	6	820	17
ST352	IDE(XT/AT)	42.82	3.50	3H	RLL 2	AUTO	N/A	N/A	150	6	820	17			
ST3541A	IDE(AT)	540	3.50	3H	RLL 1	1045	N/A	N/A	300	2	3924	0	16	1048	63
ST3543A	IDE(AT)	540	3.50	3H	RLL 1	AUTO	N/A	N/A	250	4	0	0	16	1050	63
ST3550A	IDE(AT)	452.41	3.50	3H	RLL 2	AUTO	N/A	N/A	150	5	1691	0	14	1018	62
ST3550N	SCSI-2	456.48	3.50	3H	RLL 1	AUTO	N/A	N/A	200	5	2128	0			
ST3600A (VER. A)	IDE(AT)	528.48	3.50	3H	RLL 1	AUTO	N/A	N/A	200	7	1874	0	16	1024	63

SEAGATE TECHNOLOGY, INC.

MODEL	INT	CAP	FF	HGT	CODE	LZ	WP	RWC	MTBF	PHDS	PCYL	PSPT	LHDS	LCYL	LSPT
ST3600A (VER. B)	IDE(AT)	528.48	3.50	3H	RLL 1	AUTO	N/A	N/A	200	7	1874	0	16	1024	63
ST3600N	SCSI-2	525	3.50	3H	RLL 1	AUTO	N/A	N/A	200	7	1872	0			
ST3600ND	SCSI-2 D	525	3.50	3H	RLL 1	AUTO	N/A	N/A	200	7	1872	0			
ST3610N	SCSI-2	535	3.50	3H	RLL 1	AUTO	N/A	N/A	200	7	1872	0			
ST3610NC	SCSI SCA	570	3.50	3H	RLL 1	AUTO	N/A	N/A	0	7	1872	0			
ST3610ND	SCSI-2 D	535	3.50	3H	RLL 1	AUTO	N/A	N/A	200	7	1872	0			
ST3620N	SCSI-2 F	545.9	3.50	3H	RLL 1	AUTO	N/A	N/A	500	5	2700	78			
ST3620NC	SCSI SCA	545.9	3.50	3H	RLL 1	AUTO	N/A	N/A	500	5	2700	78			
ST3620ND	SCSI-2 D	545.9	3.50	3H	RLL 1	AUTO	N/A	N/A	500	5	2700	78			
ST3620W	SCSI-2 W	545.9	3.50	3H	RLL 1	AUTO	NA	N/A	400	0	0	0			
ST3630A	IDE(AT)	631.1	3.50	3H	RLL 1	AUTO	N/A	N/A	300	4	0	0	16	1223	63
ST3636A	IDE(AT)	635	3.50		UNIDEN	AUTO	N/A	N/A	300	0	0	0	16	1241	63
ST36450A	IDE(AT)	6400	3.50	3H		AUTO	N/A	N/A		0	0	0	15	13328	63
ST36451A	IDE(AT)	6400	3.50	3H		AUTO	N/A	N/A		0	0	0	15	13328	63
ST3655A	IDE(AT)	528.48	3.50	3H	RLL 2	AUTO	N/A	N/A	200	5	0	0	16	1024	63
ST3655N	SCSI-2	545.48	3.50	3H	RLL 1	AUTO	N/A	N/A	200	5	2676	79			
ST3660A	IDE(AT)	545	3.50	3H	RLL 1	AUTO	N/A	N/A	300	0	0	0	16	1057	63
ST3780A	IDE(AT)	722.02	3.50	3H	RLL 1	AUTO	N/A	N/A	300	4	0	0	16	1399	63
ST3852A	IDE(AT)	852	3.50			AUTO	N/A	N/A		0	0	0	16	1652	68
ST3853A	IDE(AT)	850	3.50	3H	RLL 1	N/A	N/A	N/A	250	4	3640	0	16	1652	63
ST4026	ST-506/412	21.41	5.25	FH	MFM	AUTO	307	616	15	4	615	17			
ST4038	ST-506/412	31.9	5.25	FH	MFM	AUTO	734	734	25	5	733	17			
ST4038M	ST-506/412	31.9	5.25	FH	MFM	AUTO	734	734	25	5	733	17			
ST4051	ST-506/412	42.51	5.25	FH	MFM	AUTO	978	978	40	5	977	17			
ST4053	ST-506/412	44.56	5.25	FH	MFM	AUTO	1,024	1,024	40	5	1024	17			
ST406	ST-506/412	5.32	5.25	FH	MFM	319	128	307	11	2	306	17			
ST4077N	SCSI	68.15	5.25	FH	RLL 2	AUTO	N/A	N/A	0	5	1024	26			
ST4077R	ST-506/412	68.15	5.25	FH	RLL 2		1,025	1,025	0	5	1024	26			
ST4085	ST-506/412	71.3	5.25	FH	MFM	AUTO	1,025	1,025	40	8	1024	17			
ST4086	ST-506/412	72.46	5.25	FH	MFM	AUTO	926	926	40	9	925	17			
ST4096	ST-506/412	80.21	5.25	FH	MFM	AUTO	1,025	1,025	40	9	1024	17			
ST4096N	SCSI	80.21	5.25	FH	RLL	AUTO	N/A	N/A	40	9	1024	17			
ST4097	ST-506/412	80.21	5.25	FH	MFM	AUTO	1,024	1,024	40	9	1024	17			
ST410800N	SCSI-2 W	9090	5.25	FH	RLL 1	AUTO			500	27	4925	0			
ST410800ND	SCSI-2	9080	5.25	FH	RLL 1	AUTO	N/A	N/A	500	27	4925	0			
ST410800W	SCSI-2 FW	9090	5.25	FH	RLL 1	AUTO	N/A	N/A	500	27	4925	0			
ST410800WD	SCSI-2 FW	9090	5.25	FH	RLL 1	AUTO	N/A	N/A	500	27	4925	0			
ST412	ST-506/412	10.65	5.25	FH	MFM	319	128	307	11	4	306	17			
ST41200N	SCSI	1037	5.25	FH	RLL 1	AUTO	N/A	N/A	150	15	1931	0			
ST41200ND	SCSI	1037	5.25	FH	RLL 1	AUTO	N/A	N/A	150	15	1931	0			
ST41200NM	SCSI	1037	5.25	FH	RLL 1	AUTO	N/A	N/A	150	15	1931	0			
ST41200NV	SCSI	1037	5.25	FH	RLL 1	AUTO	N/A	N/A	150	15	1931	0			
ST4135R	ST-506/412	115.01	5.25	FH	RLL 2	AUTO	128	1,025	40	9	960	26			
ST4144N	SCSI	122.68	5.25	FH	RLL 2	1023	N/A	N/A	0	9	1024	26			

SEAGATE TECHNOLOGY, INC.

MODEL	INT	CAP	FF	HGT	CODE	LZ	WP	RWC	MTBF	PHDS	PCYL	PSPT	LHDS	LCYL	LSPT
ST4144R	ST-506/412	122.68	5.25	FH	RLL 2	AUTO	1,025	1,025	40	9	1024	26			
ST41520N	SCSI-2	370	5.25	FH	RLL 2	AUTO	N/A	N/A	150	17	2101	0			
ST41520ND	SCSI-2 D	1370	5.25	FH	RLL 2	AUTO	N/A	N/A	150	17	2101	0			
ST41600N	SCSI-2	1370	5.25	FH	RLL 2	AUTO	N/A	N/A	150	17	2101	0			
ST41600ND	SCSI-2 D	1370	5.25	FH	RLL 2	AUTO	N/A	N/A	150	17	2101	0			
ST41601N	SCSI-2	1370	5.25	FH	RLL 2	AUTO	N/A	N/A	150	17	2101	0			
ST41601ND	SCSI-2	1370	5.25	FH	RLL 2	AUTO	N/A	N/A	150	17	2101	0			
ST41650N	SCSI-2	1415	5.25	FH	RLL 1	AUTO	N/A	N/A	150	15	2107	0			
ST41650ND	SCSI-2 D	1415	5.25	FH	RLL 1	AUTO	N/A	N/A	150	15	2107	0			
ST41651N	SCSI-2	1415	5.25	FH	RLL 1	AUTO	N/A	N/A	150	15	2107	0			
ST41651ND	SCSI-2 D	1415	5.25	FH	RLL 1	AUTO	N/A	N/A	150	15	2107	0			
ST4182E	ESDI-10MHZ	151.81	5.25	FH	RLL 2	AUTO	N/A	N/A	100	9	969	34			
ST4182N (2X7 PIN VER.)	SCSI	160.74	5.25	FH	RLL 2	AUTO	N/A	N/A	100	9	969	36			
ST4182N (3X5 PIN VER.)	SCSI	160.74	5.25	FH	RLL 2	AUTO	N/A	N/A	100	9	969	36			
ST4182NM (2X7 PIN VER.)	SCSI	160.74	5.25	FH	RLL 2	AUTO	N/A	N/A	100	9	969	36			
ST419	ST-506/412	15.98	5.25	FH	MFM		128	307	11	6	306	17			
ST4192E	ESDI-10MHZ	169.13	5.25	FH	RLL 2	AUTO	N/A	N/A	20	8	1147	36			
ST4192N	SCSI	169.13	5.25	FH	RLL 2	AUTO	N/A	N/A	20	8	1147	36			
ST42000N	SCSI-2	1792	5.25	FH	RLL 2	AUTO	N/A	N/A	150	16	2627	0			
ST42000ND	SCSI-2 D	1792	5.25	FH	RLL 2	AUTO	N/A	N/A	150	16	2627	0			
ST42100N	SCSI-2	1900	5.25	FH	RLL 2	AUTO	N/A	N/A	150	16	2627	0			
ST42101N	SCSI-2	1900	5.25	FH	RLL 2	AUTO	N/A	N/A	150	16	2627	0			
ST423451N	U-SCSI	23400	5.25	FH	UNIDEN	AUTO	N/A	N/A		28	6880	0			
ST423451W	U-SCSI W	23400	5.25	FH	UNIDEN	AUTO	N/A	N/A		28	6880	0			
ST423451WD	U-SCSI WD	23400	5.25	FH	UNIDEN	AUTO	N/A	N/A		28	6880	0			
ST42400N	SCSI-2	2129	5.25	FH	RLL 2	AUTO	N/A	N/A	150	19	2627	0			
ST42400ND	SCSI-2	2129	5.25	FH	RLL 2	AUTO	N/A	N/A	150	19	2627	0			
ST425	ST-506/412	21.3	5.25	FH	MFM		128	307	0	8	306	17			
ST43400N	SCSI-2	2912	5.25	FH	RLL 1	AUTO	N/A	N/A	150	21	2738	0			
ST43400ND	SCSI-2 D	2912	5.25	FH	RLL 1	AUTO	N/A	N/A	150	21	2738	0			
ST43401N	SCSI-2 FW	2912	5.25	FH	RLL 1	AUTO	N/A	N/A	150	21	2738	0			
ST43401ND	SCSI-2 FW	2912	5.25	FH	RLL 1	AUTO	N/A	N/A	150	21	2738	0			
ST43402ND	SCSI-2 FWD	2912	5.25	FH	RLL 1	AUTO	N/A	N/A	150	21	2738	0			
ST4350N	SCSI	307	5.25	FH	RLL	AUTO	N/A	N/A	100	9	1412	0			
ST4350NM	SCSI	307	5.25	FH	RLL	AUTO	N/A	N/A	100	9	1412	0			
ST4376N	SCSI	330	5.25	FH	RLL	AUTO	N/A	N/A	100	9	1541	0			
ST4376NM	SCSI	330	5.25	FH	RLL	AUTO	N/A	N/A	100	9	1541	0			
ST4376NV	SCSI	330	5.25	FH	RLL	AUTO	N/A	N/A	100	9	1541	0			
ST4383E	ESDI-10MHZ	319.54	5.25	FH	RLL 2	AUTO	N/A	N/A	100	13	1412	34			
ST4384E	ESDI-10MHZ	319.61	5.25	FH	RLL 2	AUTO	N/A	N/A	100	15	1224	34			
ST4385N	SCSI	337	5.25	FH	RLL	AUTO	N/A	N/A	100	15	791	0			
ST4385NM	SCSI	337	5.25	FH	RLL	AUTO	N/A	N/A	100	15	791	0			
ST4385NV	SCSI	337	5.25	FH	RLL	AUTO	N/A	N/A	100	15	791	0			
ST4442E	ESDI-10MHZ	368.7	5.25	FH	RLL 2	AUTO	N/A	N/A	100	15	1412	34			

SEAGATE TECHNOLOGY, INC.

MODEL	INT	CAP	FF	HGT	CODE	LZ	WP	RWC	MTBF	PHDS	PCYL	PSPT	LHDS	LCYL	LSPT
ST4702N	SCSI	613	5.25	FH	RLL	AUTO	N/A	N/A	100	15	1546	0			
ST4702NM	SCSI	613	5.25	FH	RLL	AUTO	N/A	N/A	100	15	1546	0			
ST4766E	ST-506/412	664.28	5.25	FH	RLL 1	AUTO	N/A	N/A	150	15	1632	53			
ST4766N	SCSI	676.82	5.25	FH	RLL 2	AUTO	N/A	N/A	150	15	1632	54			
ST4766NM	SCSI	676.82	5.25	FH	RLL 2	AUTO	N/A	N/A	150	15	1632	54			
ST4766NV	SCSI	676.82	5.25	FH	RLL 2	AUTO	N/A	N/A	150	15	1632	54			
ST4767ES	ESDI-15MHZ	676.89	5.25	FH	RLL 1	AUTO	N/A	N/A	150	15	1399	63			
ST4767N	SCSI-2	666.5	5.25	FH	RLL 1	AUTO	N/A	N/A	150	15	1356	64			
ST4767ND	SCSI-2	666.5	5.25	FH	RLL 1	AUTO	N/A	N/A	150	15	1356	64			
ST4767NM	SCSI-2	666.5	5.25	FH	RLL 1	AUTO	N/A	N/A	150	15	1356	64			
ST4767NV	SCSI-2	666.5	5.25	FH	RLL 1	AUTO	N/A	N/A	150	15	1356	64			
ST4769ES	ESDI-15MHZ	631.72	5.25	FH	RLL 1	AUTO	N/A	N/A	150	15	1552	53			
ST506	ST-506/412	5.32	5.25	FH	MFM	157	128	128	11	4	153	17			
ST51080A	IDE(AT)	1083.8	3.50	3H	RLL 1	AUTO	N/A	N/A	300	4	4834	0	16	2100	63
ST51080N	SCSI-2 F	1080.23	3.50	4H	RLL 1	AUTO	N/A	N/A	300	4	4826	0			
ST51270A	IDE(AT)	1282.4	3.50	4H	RLL 1	AUTO	N/A	AUTO	300	4	5414	0	16	2485	63
ST52160A	IDE(AT)	810	3.50		UNIDEN	AUTO	N/A	N/A		8	0	0	16	4095	63
ST52520A	IDE(AT)	810	3.50		UNIDEN	AUTO	N/A	N/A		8	0	0	16	4970	63
ST5540A	IDE(AT)	540.1	3.50	3H	RLL 1	AUTO	N/A	N/A	300	2	4834	0	16	1050	63
ST5660A	IDE(AT)	545.51	3.50	3H	RLL 1	AUTO	N/A	N/A	300	4	3420	0	16	1057	63
ST5660N	SCSI-2	545.29	3.50	3H	RLL 1	AUTO	N/A	N/A	0	4	3420	77			
ST5660NC	SCSI SCA	545.29	3.50	3H	RLL 1	AUTO	N/A	N/A	300	4	3002	0			
ST5850A (VER. 1)	IDE(AT)	854.7	3.50	3H	RLL 1	AUTO	N/A	N/A	500	4	0	0	32	828	63
ST5850A (VER. 2)	IDE(AT)	854.7	3.50	4H	RLL 1	AUTO	N/A	N/A	500	4	0	0	32	828	63
ST5851A	IDE(AT)	854.7	3.50	4H	RLL 1	AUTO	N/A	AUTO	300	4	4834	0	16	1656	63
ST7050P	PCM	42	1.80	4H	RLL 1	AUTO	N/A	N/A	250	2	0	0			
ST706	ST-506/412	5.32	5.25	FH	MFM		128	307	0	2	306	17			
ST81236N	SCSI	1056	8.00	FH	RLL 2	AUTO	N/A	N/A	150	15	1635	0			
ST81236N VER. 1	SCSI	1000	8.00	FH	RLL 2				0	15	1371	0		1381	
ST81236N VER. 2	SCSI	1000	8.00	FH	RLL 2				0	15	1371	0		1381	
ST81236N VER. 3	SCSI	1000	8.00	FH	RLL 2				0	15	1371	0		1381	
ST81236ND VER. 1	SCSI D	1000	8.00	FH	RLL 2				0	15	1371	0		1381	
ST81236ND VER. 2	SCSI D	1000	8.00	FH	RLL 2				0	15	1371	0		1381	
ST81236ND VER. 3	SCSI D	1000	8.00	FH	RLL 2				0	15	1371	0		1381	
ST82500N	SCSI	2140	8.00	FH	RLL 2	AUTO	N/A	N/A	150	19	2611	0			
ST8741N VER. 1	SCSI	608	8.00	FH	RLL 2				0	15	1371	0		1381	
ST8741N VER. 2	SCSI	608	8.00	FH	RLL 2				0	15	1371	0		1381	
ST8741N VER. 3	SCSI	608	8.00	FH	RLL 2				0	15	1371	0		1381	
ST8741ND VER. 1	SCSI D	608	8.00	FH	RLL 2				0	15	1371	0		1381	
ST8741ND VER. 2	SCSI D	608	8.00	FH	RLL 2				0	15	1371	0		1381	
ST8741ND VER. 3	SCSI D	608	8.00	FH	RLL 2				0	15	1371	0		1381	
ST8851N	SCSI	727	8.00	FH	RLL 2	AUTO	N/A	N/A	100	15	1381	0			
ST8851N VER. 1	SCSI	693	8.00	FH	RLL 2				0	15	1371	0		1381	
ST8851N VER. 2	SCSI	693	8.00	FH	RLL 2				0	15	1371	0		1381	

SEAGATE TECHNOLOGY, INC.

MODEL	INT	CAP	FF	HGT	CODE	LZ	WP	RWC	MTBF	PHDS	PCYL	PSPT	LHDS	LCYL	LSPT
ST8851N VER. 3	SCSI	693	8.00	FH	RLL 2				0	15	1371	0		1381	
ST8851ND VER. 1	SCSI D	693	8.00	FH	RLL 2				0	15	1371	0		1381	
ST8851ND VER. 2	SCSI D	693	8.00	FH	RLL 2				0	15	1371	0		1381	
ST8851ND VER. 3	SCSI D	693	8.00	FH	RLL 2				0	15	1371	0		1381	
ST9051A	IDE(44PIN)	42.86	2.50	4H	RLL 2	AUTO	N/A	N/A	150	4	654	32	6	820	17
ST9052A	IDE(44PIN)	42.6	2.50	4H	RLL 2	AUTO	N/A	N/A	150	0	0	0	5	980	17
ST9077A	IDE(44PIN)	64.05	2.50	4H	RLL 2	AUTO	N/A	N/A	150	4	802	39	11	669	17
ST9080A	IDE(44PIN)	64	2.50	4H	RLL 2	AUTO	N/A	N/A	150	0	0	0	4	823	38
ST9096A	IDE(44PIN)	85.3	2.50	4H	RLL 2	AUTO	N/A	N/A	50	0	0	0	10	980	17
ST9100A	IDE(44PIN)	85.8	2.50	5H	RLL 1	AUTO	N/A	N/A	300	0	0	0	14	748	16
ST9100AG	IDE(44PIN)	85.3	2.50	3H	RLL 1	AUTO	N/A	N/A	150	0	0	0	14	748	16
ST9140AG	IDE(44PIN)	127.94	2.50	4H	RLL 1	AUTO	N/A	N/A	300	4	0	0	15	980	17
ST9144A	IDE(44PIN)	130.69	2.50	4H	RLL 2	AUTO	N/A	N/A	150	3	0	0	15	980	17
ST9145A	IDE(44PIN)	127.94	2.50	4H	RLL 1	AUTO	N/A	N/A	300	4	1463	0	15	980	17
ST9145AG	IDE(44PIN)	127.94	2.50	4H	RLL 1	AUTO	N/A	N/A	300	4	1463	0	15	980	17
ST9150AG	IDE(44PIN)	131.07	2.50	4H	RLL 1	AUTO	N/A	N/A	300	2	0	0	13	419	47
ST9190AG	IDE(44PIN)	171.63	2.50	4H	RLL 1	AUTO	N/A	N/A	300	4	0	0	16	873	24
ST9235A	IDE(AT)	209.7	3.50	4H	RLL 2	AUTO	N/A	N/A	150	0	0	0	13	985	32
ST9235AG	IDE(44PIN)	209.7	2.50	4H	RLL 2	AUTO	N/A	N/A	150	0	0	0	13	985	32
ST9235N	SCSI	209	2.50	4H	RLL 2	AUTO	N/A	N/A	150	0	0	0			
ST9240A	IDE(44PIN)	210.43	2.50	4H	RLL 1	AUTO	N/A	N/A	300	4	0	0	8	988	52
ST9240AG	IDE(44PIN)	210.43	2.50	4H	RLL 1	AUTO	N/A	N/A	300	4	0	0	8	988	52
ST9295AG	IDE(44PIN)	261	2.50	4H	RLL 2	AUTO	N/A	N/A	150	0	0	0			
ST9295N	SCSI	250.6	2.50	4H	RLL 2	AUTO	N/A	N/A	150	0	0	0			
ST9300AG	IDE(44PIN)	262.19	2.50	4H	RLL 1	AUTO	N/A	N/A	300	4	0	0	15	569	60
ST9352A	IDE(AT)	350	2.50	FH	RLL 1	905	N/A	N/A	300	4	2225	0	12	905	63
ST9385AG	IDE(AT)	341.44	2.50	4H	RLL 1	AUTO	N/A	N/A	300	6	0	0	14	934	51
ST9420A	IDE(44PIN)	420.8	2.50	5H	RLL 1	AUTO	N/A	N/A	300	4	0	0	16	988	52
ST9420AG	IDE(44PIN)	420.8	2.50	5H	RLL 1	AUTO	N/A	N/A	300	4	0	0	16	988	53
ST9422A	IDE(AT)	422	2.50	5H	RLL 1	818	N/A	N/A	300	4	2393	0	16	818	63
ST9550AG	IDE(AT)	455.29	2.50	4H	RLL 1	AUTO	N/A	N/A	300	6	0	0	16	942	59
ST9655AG	IDE(44PIN)	524.35	2.50	4H	RLL 1	AUTO	N/A	N/A	300	6	0	0	16	1016	63
ST9816AG	IDE(44PIN)	810	2.50	5H	RLL 1	AUTO	N/A	N/A	300	8	0	0	16	1571	63

SHUGART CORPORATION

MODEL	INT	CAP	FF	HGT	CODE	LZ	WP	RWC	MTBF	PHDS	PCYL	PSPT	LHDS	LCYL	LSPT
SA 1002	ST-506/412	5	8.00	FH	MFM				0	0	0	17			
SA 1004	ST-506/412	10	8.00	FH	MFM				0	0	0	17			
SA 1106	ST-506/412	30	8.00	FH	MFM				0	0	0	17			
SA 4004	ST-506/412	14	14.0	FH	MFM				0	0	0	17			
SA 4008	ST-506/412	29	14.0	FH	MFM				0	0	0	17			
SA 4100	ST-506/412	56	14.0	FH	MFM				0	0	0	17			
SA 604	ST-506/412	5.57	5.25	FH	MFM		128	128	0	4	160	17			
SA 606	ST-506/412	8.35	5.25	FH	MFM		128	128	0	6	160	17			

SHUGART CORPORATION

MODEL	INT	CAP	FF	HGT	CODE	LZ	WP	RWC	MTBF	PHDS	PCYL	PSPT	LHDS	LCYL	LSPT
SA 607	ST-506/412	5.32	5.25	FH	MFM				0	2	306	17			
SA 612	ST-506/412	10.65	5.25	FH	MFM		128	307	0	4	306	17			
SA 706	ST-506/412	5.57	5.25	FH	MFM		128	321	0	2	320	17			
SA 712	ST-506/412	11.14	5.25	HH	MFM		128	321	0	4	320	17			

SIEMENS NIXDORF INFORMATIONSSYSTEME AG

MODEL	INT	CAP	FF	HGT	CODE	LZ	WP	RWC	MTBF	PHDS	PCYL	PSPT	LHDS	LCYL	LSPT
1200	ESDI-10MHZ	174.32	5.25	FH	RLL 2	AUTO	N/A	N/A	0	8	1216	35			
1300	ESDI-10MHZ	261.48	5.25	FH	RLL 2	AUTO	N/A	N/A	0	12	1216	35			
2200	ESDI-10MHZ	174	5.25	FH	RLL 2	AUTO	N/A	N/A	0	8	1216	35			
2300	SCSI	261.48	5.25	FH	RLL 2	AUTO	N/A	N/A	0	12	1216	35			
4410	ESDI-15MHZ	322.15	5.25	FH	RLL 2	AUTO	N/A	N/A	30	11	1100	52			
4420	SCSI	334.54	5.25	FH	RLL 2	AUTO	N/A	N/A	40	11	1100	54			
5710	ESDI-15MHZ	655	5.25	FH	RLL 2	AUTO	N/A	N/A	0	15	1224	48			
5720	SCSI	655	5.25	FH	RLL	AUTO	N/A	N/A	0	15	1224	48			
5810	ESDI-15MHZ	733.45	5.25	FH	RLL 2	AUTO	N/A	N/A	0	16	1658	54			
5820	SCSI	733.45	5.25	FH	RLL 2	AUTO	N/A	N/A	0	16	1658	54			
6200	SCSI	1200	5.25	–	RLL 2	AUTO	N/A	N/A	0	0	0	0			
7520	SCSI	655	5.25	FH	RLL 2	AUTO	N/A	N/A	0	15	0	0			

SIMPLE TECHNOLOGY, INC.

MODEL	INT	CAP	FF	HGT	CODE	LZ	WP	RWC	MTBF	PHDS	PCYL	PSPT	LHDS	LCYL	LSPT
STI-105HD	PCM	105	CARD	NO	UNIDEN	AUTO	N/A	N/A	0	0	0	0	8	828	31
STI-130HD	PCM	130	CARD	NO	UNIDEN	AUTO	N/A	N/A	0	0	0	0	4	1010	31
STI-170HD	PCM	170	CARD	NO	UNIDEN	AUTO	N/A	N/A	0	0	0	0	8	923	45

SONY CORPORATION

MODEL	INT	CAP	FF	HGT	CODE	LZ	WP	RWC	MTBF	PHDS	PCYL	PSPT	LHDS	LCYL	LSPT
SRD 2040Z	SCSI	41.92	3.50	HH					0	4	622	33			

STORAGE DIMENSIONS

MODEL	INT	CAP	FF	HGT	CODE	LZ	WP	RWC	MTBF	PHDS	PCYL	PSPT	LHDS	LCYL	LSPT
AT-1000S	SCSI	1000	5.25	–	RLL	AUTO	N/A	N/A	100	15	0	0			
AT-100S	SCSI	105	3.50	HH	RLL 2	AUTO	N/A	N/A	150	3	0	0			
AT-120	ST-506/412	119.85	5.25	FH	MFM				100	15	918	17			
AT-155E	ESDI-10MHZ	157.92	5.25	FH	RLL 2	AUTO	N/A	N/A	100	7	1224	36			
AT-155S	SCSI	157.92	5.25	FH	RLL 2	AUTO	N/A	N/A	100	7	1224	36			
AT-160	ST-506/412	159.8	5.25	FH	MFM				100	15	1224	17			
AT-200S	SCSI	204	3.50	HH	RLL 2	AUTO	N/A	N/A	150	7	0	0			
AT-320S	SCSI	338.41	5.25	FH	RLL 2	AUTO	N/A	N/A	100	15	1224	36			
AT-335E	ESDI-10MHZ	338.41	5.25	FH	RLL 2	AUTO	N/A	N/A	100	15	1224	36			
AT-650E	ESDI-15MHZ	651.75	5.25	FH	RLL 2	AUTO	N/A	N/A	70	15	1632	52			
AT-650S	SCSI	651.75	5.25	FH	RLL 2	AUTO	N/A	N/A	100	15	1632	52			

STORAGE DIMENSIONS

MODEL	INT	CAP	FF	HGT	CODE	LZ	WP	RWC	MTBF	PHDS	PCYL	PSPT	LHDS	LCYL	LSPT
MAC-195	SCSI	195	3.50	HH	RLL 2	AUTO	N/A	N/A	150	7	0	0			
PS-155E	ESDI-15MHZ	203.04	5.25	FH	RLL 2	AUTO	N/A	N/A	70	9	1224	36			
PS-155S	SCSI	203.04	5.25	FH	RLL 2	AUTO	N/A	N/A	70	9	1224	36			
PS-320S	SCSI	338.41	5.25	FH	RLL 2	AUTO	N/A	N/A	150	15	1224	36			
PS-335E	ESDI-10MHZ	338.41	5.25	FH	RLL 2	AUTO	N/A	N/A	70	15	1224	36			
PS-650S	SCSI	676.82	5.25	FH	RLL 2	AUTO	N/A	N/A	100	15	1632	54			

SYQUEST TECHNOLOGY

MODEL	INT	CAP	FF	HGT	CODE	LZ	WP	RWC	MTBF	PHDS	PCYL	PSPT	LHDS	LCYL	LSPT
EZ135 IDE	IDE(AT)	135			UNIDEN	AUTO	N/A	N/A		0	0	0	16	512	32
EZ135 SCSI	SCSI	135								0	0	0			
EZFLYER EIDE	IDE(AT)	230							20000	0	0	0			
SQ2542	IDE(AT)	42			UNIDEN	AUTO	N/A	N/A		0	0	0	5	985	17
SQ306F	ST-506/412	5	5.25	HH	MFM				0	0	0	17			
SQ306R	ST-506/412	5.32	5.25	HH	RLL				0	2	306	17			
SQ306RD	ST-506/412	5.32	5.25	HH	MFM		307	307	0	2	306	17			
SQ3105A	IDE(AT)	105	3.50	3H	UNIDEN	AUTO	N/A	N/A	100	0	0	0	16	841	16
SQ3105S	SCSI	105	3.50	3H	UNIDEN	AUTO	N/A	N/A	100	0	0	0			
SQ312	ST-506/412	10.7	5.25	HH	MFM				0	2	615	17			
SQ312RD	ST-506/412	10.7	5.25	HH	MFM		616	616	0	2	615	17			
SQ319	ST-506/412	10.65	5.25	HH	RLL				0	2	612	17			
SQ325	ST-506/412	21.3	5.25	HH	MFM				0	4	612	17			
SQ325F	ST-506/412	21.41	5.25	HH	MFM		616	616	0	4	615	17			
SQ3270A	IDE(AT)	256	3.50	3H	UNIDEN	AUTO	N/A	N/A	100	0	0	0	16	1024	16
SQ3270S	SCSI	256	3.50	3H	UNIDEN	AUTO	N/A	N/A	100	0	0	0			
SQ338F	ST-506/412	32.11	5.25	HH	MFM		616	616	0	6	615	17			
SQ340AF	ST-506/412	33.42	5.25	HH	MFM		616	616	0	6	640	17			
SQ5110	SCSI	88	5.25	HH	RLL 2	AUTO	N/A	N/A	60	0	0	0			
SQ5110C	SCSI	88	5.25	HH	RLL 2	AUTO	N/A	N/A	60	0	0	0			
SQ5200C	SCSI	200	5.25	HH	RLL 1	AUTO	N/A	N/A	100	0	0	0			
SQ555	SCSI	40	5.25	HH	RLL 2	AUTO	N/A	N/A	60	0	0	17			
SYJET EIDE	IDE(AT)	1.5								0	0	0			
SYJET SCSI MAC	SCSI	1.5							25000	0	0	0			
SYJET SCSI PC	IDE(AT)	1.5							20000	0	0	0			
SYJET SCSI PORTABLE	SCSI	1.5							25000	0	0	0			
SYSDOS MARLIN 105I	IDE(AT)	105	5.25							0	0	0			

TALON

MODEL	INT	CAP	FF	HGT	CODE	LZ	WP	RWC	MTBF	PHDS	PCYL	PSPT	LHDS	LCYL	LSPT
SHD-3061A	IDE(AT)	0	3.50	3H		AUTO	N/A	N/A	200	0	0	0			
SHD-3062A	IDE(AT)	0	3.50	3H		AUTO	N/A	N/A	200	0	0	0			
SHD-3101A	IDE(AT)	0	3.50	3H		AUTO	N/A	N/A	200	0	0	0			
TA3020A	IDE(AT)	121.07	3.50	3H	RLL 1	AUTO	N/A	N/A	200	4	1478	40	8	739	40
TA3101A	IDE(AT)	105.02	3.50	3H	RLL 1	AUTO	N/A	N/A	200	4	1282	40	8	641	40

TANDON CORPORATION

MODEL	INT	CAP	FF	HGT	CODE	LZ	WP	RWC	MTBF	PHDS	PCYL	PSPT	LHDS	LCYL	LSPT
TM 2128	SCSI	120.28	5.25	FH	RLL 2	AUTO	N/A	N/A	0	9	1004	26			
TM 2170	SCSI	161.02	5.25	FH	RLL 2	AUTO	N/A	N/A	0	9	1344	26			
TM 244	ST-506/412	41.63	3.50	HH	RLL 2		783	783	0	4	782	26			
TM 246	ST-506/412	62.45	3.50	HH	RLL 2		783	783	0	6	782	26			
TM 251	ST-506/412	5.32	5.25	–	MFM				0	2	306	17			
TM 252	ST-506/412	10.65	5.25	HH	MFM		307	307	0	4	306	17			
TM 261	ST-506/412	10.7	5.25	–	MFM		616	616	0	2	615	17			
TM 262	ST-506/412	21.41	3.50	HH	MFM		616	616	0	4	615	17			
TM 262R	ST-506/412	20.81	3.50	HH	RLL 2		783	783	0	2	782	26			
TM 264	ST-506/412	41.63	3.50	HH	RLL 2		783	783	0	4	782	26			
TM 3085	ST-506/412	71.3	5.25	–	MFM		1,024	1,024	0	8	1024	17			
TM 3085-R	ST-506/412	109.05	5.25	–	RLL				0	8	1024	26			
TM 344	ST-506/412	41.63	3.50	HH	RLL 2		783	783	0	4	782	26			
TM 346	ST-506/412	62.45	3.50	HH	RLL 2		783	783	0	6	782	26			
TM 361	ST-506/412	10.7	5.25	–	MFM		616	616	0	2	615	17			
TM 362	ST-506/412	21.41	5.25	–	MFM		616	616	0	4	615	17			
TM 362R	ST-506/412	20.81	3.50	HH	RLL 2		783	783	0	2	782	26			
TM 364	ST-506/412	41.63	3.50	HH	RLL 2		783	783	0	4	782	26			
TM 501	ST-506/412	5.32	5.25	FH	MFM		153	128	0	2	306	17			
TM 502	ST-506/412	10.65	5.25	FH	MFM		153	128	0	4	306	17			
TM 503	ST-506/412	15.98	5.25	FH	MFM		153	128	0	6	306	17			
TM 602S	ST-506/412	5.32	5.25	FH	MFM		128	128	0	4	153	17			
TM 602SE	ST-506/412	12	5.25	FH	MFM				0	0	0	17			
TM 603S	ST-506/412	7.99	5.25	FH	MFM		128	128	0	6	153	17			
TM 603SE	ST-506/412	12.01	5.25	FH	MFM		128	128	0	6	230	17			
TM 702	ST-506/412	32.74	5.25	FH	RLL 2		616	616	0	4	615	26			
TM 702AT	ST-506/412	21.41	5.25	FH	MFM	615	615	615	0	4	615	17			
TM 703	ST-506/412	30.24	5.25	FH	MFM	AUTO	695	695	0	5	695	17			
TM 703-C	ST-506/412	25	5.25	FH	MFM	AUTO	576	576	0	5	576	17			
TM 703AT	ST-506/412	31.9	5.25	FH	MFM	733	733	733	0	5	733	17			
TM 705	ST-506/412	41.86	5.25	FH	MFM	962			0	5	962	17			
TM 755	ST-506/412	42.69	5.25	HH	MFM		982	982	0	5	981	17			

TANDY/RADIO SHACK

MODEL	INT	CAP	FF	HGT	CODE	LZ	WP	RWC	MTBF	PHDS	PCYL	PSPT	LHDS	LCYL	LSPT
25-1045	IDE(XT)	28	3.50	HH	RLL	AUTO			0	0	0	0			
25-1046	IDE(XT)	43.24	3.50	HH	RLL 2	AUTO	N/A	N/A	40	4	782	27			
25-4130	IDE(AT)	100	3.50	HH	RLL 2	AUTO	N/A	N/A	0	4	1219	0			

TEAC CORPORATION

MODEL	INT	CAP	FF	HGT	CODE	LZ	WP	RWC	MTBF	PHDS	PCYL	PSPT	LHDS	LCYL	LSPT
SD-150	ST-506/412	10.65	5.25	–	MFM				0	4	306	17			
SD-3105	IDE(AT)	105.02	3.50	3H	RLL 2	AUTO	N/A	N/A	30	4	1282	40	8	641	40

TEAC CORPORATION

MODEL	INT	CAP	FF	HGT	CODE	LZ	WP	RWC	MTBF	PHDS	PCYL	PSPT	LHDS	LCYL	LSPT
SD-3105(S)	SCSI	105.02	3.50	3H	RLL 2	AUTO	N/A	N/A	30	4	1282	40			
SD-3210	IDE(AT)	209.92	3.50	3H	RLL	AUTO	N/A	N/A	0	0	0	0	8	847	62
SD-3210(S)	SCSI	215.22	3.50	3H	RLL 1	AUTO	N/A	N/A	100	4	1695	62			
SD-3240	IDE(AT)	245.06	3.50	3H	RLL	AUTO	N/A	N/A	100	4	1930	0	8	965	62
SD-340	IDE(AT)	43	3.50	3H	RLL 2	AUTO	N/A	N/A	30	2	1050	40	4	525	40
SD-340S	SCSI	43	3.50	3H	RLL 2	AUTO	N/A	N/A	30	2	1050	40			
SD-380	IDE(AT)	86.01	3.50	3H	RLL 2	AUTO	N/A	N/A	30	4	1050	40	8	965	62
SD-380H(A)	IDE(AT)	86.01	3.50	3H	RLL 2	AUTO	N/A	N/A	30	4	1050	40			
SD-380H(S)	SCSI	86.01	3.50	3H	RLL 2	AUTO	N/A	N/A	30	4	1050	40			
SD-380S	SCSI	86.01	3.50	3H	RLL 2	AUTO	N/A	N/A	30	4	1050	40			
SD-510	ST-506/412	10.65	5.25	FH	MFM		128	128	0	4	306	17			
SD-520	ST-506/412	21.41	5.25	HH	MFM		128	128	0	4	615	17			
SD-540	ST-506/412	42.82	5.25	HH	MFM	615	125	N/A	0	8	615	17			

TEXAS INSTRUMENTS

MODEL	INT	CAP	FF	HGT	CODE	LZ	WP	RWC	MTBF	PHDS	PCYL	PSPT	LHDS	LCYL	LSPT
DB260	SCSI-2	212.6	3.50	HH	RLL	AUTO	N/A	N/A	0	8	0	0			
DB380	SCSI	333.43	5.25	FH	RLL	AUTO	N/A	N/A	0	15	0	0			
DB760	SCSI	669.77	5.25	FH	RLL	AUTO	N/A	N/A	0	15	0	0			
TI-5	ST-506/412	5.32	5.25	FH	MFM		64	64	0	4	153	17			

TOKICO

MODEL	INT	CAP	FF	HGT	CODE	LZ	WP	RWC	MTBF	PHDS	PCYL	PSPT	LHDS	LCYL	LSPT
DK 503-2	ST-506/412	10.65	5.25	FH	MFM				0	4	306	17			

TOSHIBA

MODEL	INT	CAP	FF	HGT	CODE	LZ	WP	RWC	MTBF	PHDS	PCYL	PSPT	LHDS	LCYL	LSPT
HDD 2214	IDE(44PIN)	86.03	2.50	4H	RLL 2	AUTO	N/A	N/A	80	0	0	0	10	988	17
HDD 2216	IDE(AT)	130.09	2.50	4H	RLL 2	934	N/A	N/A		0	0	0			
HDD 2238	IDE(44PIN)	213	2.50	4H	RLL	AUTO	N/A	N/A	150	4	1560	0	16	684	38
HDD 2324	IDE(AT)	86	2.50	3H	RLL	AUTO	N/A	N/A	150	2	1501	56	10	988	17
HDD 2326	IDE(AT)	126	2.50		RLL	AUTO	N/A	N/A		2	0	0	8	812	38
HDD 2336	IDE(AT)	131	2.50		RLL	AUTO	N/A	N/A		2	0	0	16	842	38
HDD 2339	IDE(AT)	262	2.50			AUTO	N/A	N/A		0	0	0	16	842	38
HDD 2412	IDE(44PIN)	340	2.50	4H	RLL 1	AUTO	N/A	N/A	150	6	1830	0	14	969	49
HDD 2414	IDE(44PIN)	524	2.50	4H	RLL 1	AUTO	N/A	N/A	150	8	1920	0	16	1016	63
HDD 2416		704	2.50	5H	RLL 1	1365	N/A	N/A	300	0	0	0	16	1365	63
HDD 2417		810	2.50	5H	RLL 1	1571	N/A	N/A	300	0	0	0	16	1571	63
HDD 2434		528	2.50	5H	RLL 1	1023	N/A	N/A	300	0	0	0	16	1023	63
HDD 2512	IDE(44PIN)	352	2.50	5H	RLL 1	AUTO	N/A	N/A	300	0	0	0	16	682	63
HDD 2514	IDE(44PIN)	543	2.50	5H	RLL 1	AUTO	N/A	N/A	300	0	0	0			
HDD 2517	IDE(44PIN)	810	2.50	4H	RLL 1	1579	N/A	N/A	300	0	0	0	16	1579	63
HDD 2612		1.35	2.50	5H	RLL 1	AUTO	N/A	N/A	300	10	2920	0			

TOSHIBA

MODEL	INT	CAP	FF	HGT	CODE	LZ	WP	RWC	MTBF	PHDS	PCYL	PSPT	LHDS	LCYL	LSPT
HDD 2616	IDE(AT)	2.16	2.50	4H	RLL	AUTO	N/A	N/A	300	10	3650	0			
HDD 2619	IDE(44PIN)	3300	2.50	3					300	0	0	0	16	6409	63
HDD 2632	IDE(AT)	1080	2.50	4H		AUTO	N/A	N/A	300	8	0	0	16	2633	63
HDD 2634	IDE(AT)	1080	2.50	4H		AUTO	N/A	N/A	300	10	0	0	16	3294	63
HDD 2710	IDE(AT)	1080	2.50	4H		AUTO	N/A	N/A	300	3	0	0	16	2100	63
HDD 2712	IDE(44PIN)	1350	2.50	3					300	0	0	0	16	2800	63
HDD 2714	IDE(44PIN)	1440	2.50	3					300	0	0	0	16	2800	63
HDD 2716	IDE(AT)	2160	2.50	4H		AUTO	N/A	N/A	300	3	0	0	16	4200	63
HDD 2731	IDE(AT)	1080	2.50	4H		AUTO	N/A	N/A	300	3	0	0	16	2100	63
MK-1002	IDE(AT)	1080	2.50	4H		AUTO	N/A	N/A	300	6	0	0	16	2100	63
MK-1034FC	IDE(AT)	107.42	3.50	3H	RLL 2	AUTO	N/A	N/A	40	4	1345	39	8	664	39
MK-1122FC	IDE(AT)	43.01	2.50	4H	RLL 2	AUTO	N/A	N/A	80	2	977	43	5	988	17
MK-1301MAV	IDE(AT)	1.35	2.50	5H	UNIDEN	AUTO		128	300	6	3650	0			
MK-1302	IDE(AT)	1700	2.50	4H		AUTO	N/A	N/A	300	4	0	0	16	2633	63
MK-134FA	ST-506/412	44.66	3.50	HH	MFM		512	N/A	30	7	733	17			
MK-134FA(R)	ST-506/412	68.3	3.50	HH	RLL 2				30	7	733	26			
MK-1422FCV	IDE(AT)	86	2.50	3H	RLL	AUTO	N/A	N/A	150	2	1501	56	10	988	17
MK-153FA	ESDI-10MHZ	74.36	5.25	FH	RLL 2	AUTO	N/A	N/A	30	5	830	35			
MK-153FA-I	ESDI-10MHZ	74.36	5.25	FH	RLL 2	AUTO	N/A	N/A	30	5	830	35			
MK-154FA	ESDI-10MHZ	104.11	5.25	FH	RLL 2	AUTO	N/A	N/A	30	7	830	35			
MK-154FA-I	ESDI-10MHZ	104.11	5.25	FH	RLL 2	AUTO	N/A	N/A	30	7	830	35			
MK-154FB	ESDI-10MHZ	104.11	5.25	FH	RLL 2	AUTO	N/A	N/A	30	7	830	35			
MK-156FA	ESDI-10MHZ	148.73	5.25	FH	RLL 2	AUTO	N/A	N/A	30	10	830	35			
MK-156FB	ESDI-10MHZ	148.73	5.25	FH	RLL 2	AUTO	N/A	N/A	30	10	830	35			
MK-1824FCV	IDE(44PIN)	352	2.50	5H	RLL 1	682	N/A	N/A	300	0	0	0	16	682	63
MK-1924FCV	IDE(44PIN)	543	2.50	5H	RLL 1	1053	N/A	N/A	300	0	0	0	16	1053	63
MK-1926FCV	IDE(44PIN)	810	2.50	5H	RLL 1	1579	N/A	N/A	300	0	0	0	16	1579	63
MK-2024FC	IDE(AT)	86.03	2.50	4H	RLL 2	AUTO	N/A	N/A	80	4	977	43	10	988	17
MK-2124FC	IDE(AT)	130.09	2.50	4H	RLL 2	934	N/A	N/A	80	4	1155	55	16	934	17
MK-2224FB	SCSI	213	2.50	4H	RLL 1		N/A	N/A		4	1560	0	16	684	38
MK-2224FC	IDE(AT)	213	2.50	3H	RLL	AUTO	N/A	N/A	150	4	1560	0	16	684	38
MK-2326FB	SCSI	340	2.50	4H	RLL 1		N/A	N/A		6	1830	0	14	969	49
MK-2326FC	IDE(44PIN)	340.34	2.50	4H	RLL 1	AUTO	N/A	N/A	150	6	1830	0	14	969	49
MK-232FB	SCSI	45.42	3.50	HH	RLL 2	AUTO	N/A	N/A	30	3	845	35			
MK-232FBS	SCSI	45.42	3.50	HH	RLL 2	AUTO	N/A	N/A	30	3	845	35			
MK-232FC	IDE(AT)	45.42	3.50	HH	RLL 2	AUTO	N/A	N/A	30	3	845	35			
MK-232FCH	IDE(AT)	0	2.50	4H	RLL 1	AUTO	N/A	N/A	150	14	969	49			
MK-233FB	SCSI	75.71	3.50	HH	RLL 2	AUTO	N/A	N/A	30	5	845	35			
MK-234FB	SCSI	105.99	3.50	HH	RLL 2	AUTO	N/A	N/A	30	7	845	35			
MK-234FBS	SCSI	105.99	3.50	HH	RLL 2	AUTO	N/A	N/A	30	7	845	35			
MK-234FC	IDE(AT)	110.44	3.50	HH	RLL 2	AUTO	N/A	N/A	30	7	856	36	7	845	35
MK-234FCF	IDE(AT)	0	2.50	4H	RLL 1				150	0	0	0	14	969	49
MK-234FCH	IDE(AT)	107.37	3.50	HH	RLL 2	AUTO	N/A	N/A	30	7	856	35			
MK-234FCH-I	IDE(AT)	107.37	3.50	HH	RLL 2	AUTO	N/A	N/A	30	7	856	35			

TOSHIBA

MODEL	INT	CAP	FF	HGT	CODE	LZ	WP	RWC	MTBF	PHDS	PCYL	PSPT	LHDS	LCYL	LSPT
MK-2428FB	SCSI	524	2.50	4H	RLL 1		N/A	N/A		8	1920	0	16	1016	63
MK-2428FC	IDE(44PIN)	524.35	2.50	4H	RLL 1	AUTO	N/A	N/A	150	8	1920	0	16	1016	63
MK-250FB	SCSI	219.34	5.25	FH	RLL 2	AUTO	N/A	N/A	30	10	1224	35			
MK-2526FB	SCSI-2	528	2.50	5H						6	2050	0			104
MK-2526FC	IDE(44PIN)	528	2.50	5H	RLL 1	1023	N/A	N/A	300	0	0	0	16	1023	63
MK-2528FC	IDE(44PIN)	704	2.50	5H	RLL 1	1365	N/A	N/A	300	0	0	0	16	1365	63
MK-253FA	ESDI-15MHZ	162.8	5.25	FH	RLL 1	AUTO	N/A	N/A	30	5	1223	52			
MK-254FA	ESDI-15MHZ	227.92	5.25	FH	RLL 1	AUTO	N/A	N/A	30	7	1223	52			
MK-256FA	ESDI-15MHZ	325.61	5.25	FH	RLL 1	AUTO	N/A	N/A	30	10	1223	52			
MK-2628FB	SCSI-2	811	2.50	5H						8	2360	0			104
MK-2628FC	IDE(44PIN)	810	2.50	5H	RLL 1	1571	N/A	N/A	300	0	0	0	16	1571	63
MK-2720	IDE(AT)	1035	2.50		UNIDEN	AUTO	N/A	N/A	250	0	0	0			
MK-355FA	ESDI-15MHZ	405.65	5.25	FH	RLL 2	AUTO	N/A	N/A	30	9	1661	53			
MK-355FB	SCSI-2	405.65	5.25	FH	RLL 2	AUTO	N/A	N/A	30	9	1661	53			
MK-358FA	ESDI-15MHZ	676.09	5.25	FH	RLL 2	AUTO	N/A	N/A	30	9	1661	53			
MK-358FB	SCSI-2	676.09	5.25	FH	RLL 2	AUTO	N/A	N/A	30	15	1661	53			
MK-438FB ASSY. 0605	SCSI-2	877	3.50	HH	RLL 1	AUTO	N/A	N/A	200	15	1692	0			
MK-438FB ASSY. 0817	SCSI-2	900	3.50	HH	RLL 1	AUTO	N/A	N/A	200	11	1980	0			
MK-438FB ASSY. 0834	SCSI-2	877	5.25	FH	RLL 1	AUTO	N/A	N/A	200	11	1980	35			
MK-537FB	SCSI-2	1064	3.50	HH	RLL 1	AUTO	N/A	N/A	200	13	1980	0			
MK-538FB	SCSI-2	1228	3.50	HH	RLL 1	AUTO	N/A	N/A	200	15	1980	0			
MK-53FA	ST-506/412	36.12	5.25	FH	MFM	830	512		20	5	830	17			
MK-53FB	ST-506/412	36.12	5.25	FH	MFM		512	830	20	5	830	17			
MK-53FB(M)	ST-506/412	36.12	5.25	FH	MFM		512	830	20	5	830	17			
MK-53FB(R)	ST-506/412	36.12	5.25	FH	RLL		512	830	20	5	830	17			
MK-53FB-I	ST-506/412	36.12	5.25	FH	MFM		512	830	20	5	830	17			
MK-54FA(M)	ST-506/412	50.57	5.25	FH	MFM	830	512		20	7	830	17			
MK-54FB(M)	ST-506/412	50.57	5.25	FH	MFM	830	512	830	20	7	830	17			
MK-54FB-I	ST-506/412	50.57	5.25	FH	MFM	830	512	830	20	7	830	17			
MK-56FA(M)	ST-506/412	72.24	5.25	FH	MFM		831	831	20	10	830	17			
MK-56FA(R)	ST-506/412	110.48	5.25	FH	RLL 2	830	512		20	10	830	26			
MK-56FB(M)	ST-506/412	72.24	5.25	FH	MFM	830	512	830	20	10	830	17			
MK-56FB(R)	ST-506/412	110.48	5.25	FH	RLL 2		831	831	20	10	830	26			
MK-56FB-I	ST-506/412	72.24	5.25	FH	MFM	830	512	830	20	10	830	17			
MK-72PC	ST-506/412	72.24	5.25	FH	MFM				20	10	830	17			
MK-72PCR	ST-506/412	110.48	5.25	FH	RLL 2				20	10	830	26			
MKM-0351E	ST-506/412	36.12	5.25	FH	MFM	830	512	830	20	5	830	17			
MKM-0351J	ST-506/412	36.12	5.25	FH	MFM	830	512	830	20	5	830	17			
MKM-0352E	ST-506/412	50.57	5.25	FH	MFM	830	512		20	7	830	17			
MKM-0352J	ST-506/412	50.57	5.25	FH	MFM	830	512		20	7	830	17			
MKM-0353E	ST-506/412	72.24	5.25	FH	MFM	830	512	830	20	10	830	17			
MKM-0353J	ST-506/412	72.24	5.25	FH	MFM	830	512	830	20	10	830	17			
MKM-0363A	ESDI-10MHZ	74.36	5.25	FH	RLL 2	AUTO	N/A	N/A	30	5	830	35			
MKM-0363J	ESDI-10MHZ	74.36	5.25	FH	RLL 2	AUTO	N/A	N/A	30	5	830	35			

TOSHIBA

MODEL	INT	CAP	FF	HGT	CODE	LZ	WP	RWC	MTBF	PHDS	PCYL	PSPT	LHDS	LCYL	LSPT
MKM-0364A	ESDI-10MHZ	104.11	5.25	FH	RLL 2	AUTO	N/A	N/A	30	7	830	35			
MKM-0364J	ESDI-10MHZ	104.11	5.25	FH	RLL 2	AUTO	N/A	N/A	30	7	830	35			
MKM-0381E	ST-506/412	36.12	5.25	FH	MFM		512	830	20	5	830	17			
MKM-0381J	ST-506/412	36.12	5.25	FH	MFM	830	512	830	20	5	830	17			
MKM-0382E	ST-506/412	50.57	5.25	FH	MFM	830	512		20	7	830	17			
MKM-0382J	ST-506/412	50.57	5.25	FH	MFM	830	512		20	7	830	17			
MKM-0383E	ST-506/412	72.24	5.25	FH	MFM	830	512	830	20	10	830	17			
MKM-0383J	ST-506/412	72.24	5.25	FH	MFM	830	512	830	20	10	830	17			

TULIN

MODEL	INT	CAP	FF	HGT	CODE	LZ	WP	RWC	MTBF	PHDS	PCYL	PSPT	LHDS	LCYL	LSPT
TL 213	ST-506/412	11.14	5.25	HH	MFM		656	656	0	2	640	17			
TL 226	ST-506/412	22.28	5.25	HH	MFM		656	656	0	4	640	17			
TL 238	ST-506/412	22.28	5.25	HH	MFM	640			0	4	640	17			
TL 240	ST-506/412	33.42	5.25	HH	MFM		656	656	0	6	640	17			
TL 258	ST-506/412	33.42	5.25	HH	MFM	640			0	6	640	17			
TL 326	ST-506/412	22.28	5.25	HH	MFM		641	641	0	4	640	17			
TL 340	ST-506/412	33.42	5.25	HH	MFM		641	641	0	6	640	17			

UNIDENTIFIED

MODEL	INT	CAP	FF	HGT	CODE	LZ	WP	RWC	MTBF	PHDS	PCYL	PSPT	LHDS	LCYL	LSPT
HIPER 380	ESDI-15MHZ	388	5.25	–	RLL	AUTO	N/A	N/A	100	8	0	0			
M2637T	IDE(AT)	240	3.50	3H	RLL 1	AUTO	N/A	N/A	0	12	698	56			
MT-4115E	ESDI-10MHZ	114.4	5.25	HH	RLL 2	AUTO	N/A	N/A	100	4	1597	34			
MT-4115S	SCSI	114.47	5.25	HH	RLL 2	AUTO	N/A	N/A	100	4	1597	35			
MT-4140E	SCSI	143.09	5.25	HH	RLL 2	AUTO	N/A	N/A	100	5	1597	35			
MT-4140S	SCSI	143.09	5.25	HH	RLL 2	AUTO	N/A	N/A	100	5	1597	35			
MT-4170E	ESDI-10MHZ	143.09	5.25	HH	RLL 2	AUTO	N/A	N/A	100	5	1597	35			
MT-4170S	SCSI	171.7	5.25	HH	RLL 2	AUTO	N/A	N/A	100	6	1597	35			
MT-5760	SCSI	673.09	5.25	FH	RLL 1	AUTO	N/A	N/A	150	15	1623	54			
MT-5760E	ESDI-15MHZ	676.82	5.25	FH	RLL 1	AUTO	N/A	N/A	150	15	1225	54			
MT-6120	SCSI	1050.75	5.25	FH	RLL 1	AUTO	N/A	N/A	100	15	1927	71			
PROPAQ 185-15	IDE(AT)	189	3.50	HH	RLL	AUTO	N/A	N/A	70	5	0	0			
SI 1000/S5	SCSI	1037	5.25	–	RLL	AUTO	N/A	N/A	40	8	0	0			
SI 200/PS3	SCSI	209	3.50	HH	RLL 2	AUTO	N/A	N/A	70	4	0	0			
SI 585/S5	SCSI	601	5.25	–	RLL	AUTO	N/A	N/A	100	8	0	0			

VERTEX

MODEL	INT	CAP	FF	HGT	CODE	LZ	WP	RWC	MTBF	PHDS	PCYL	PSPT	LHDS	LCYL	LSPT
V130	ST-506/412	25.77	5.25	FH	MFM		988	988	0	3	987	0			
V150	ST-506/412	42.95	5.25	FH	MFM		988	988	0	5	987	0			
V160	ST-506/412	50.74	5.25	FH	MFM		1,167	1,167	0	5	1166	0			
V160A	ST-506/412	50.74	5.25	FH	MFM		1,167	1,167	0	5	1166	17			

VERTEX

MODEL	INT	CAP	FF	HGT	CODE	LZ	WP	RWC	MTBF	PHDS	PCYL	PSPT	LHDS	LCYL	LSPT
V170	ST-506/412	60.13	5.25	FH	MFM		988	988	0	7	987	0			
V170R	ST-506/412	91.97	5.25	FH	RLL 2				0	7	987	0			
V185	ST-506/412	71.04	5.25	FH	MFM		1,167	1,167	0	7	1166	0			
V519	ST-506/412	159.8	5.25	FH	MFM	1223			0	15	1224	0			

WESTERN DIGITAL CORPORATION

MODEL	INT	CAP	FF	HGT	CODE	LZ	WP	RWC	MTBF	PHDS	PCYL	PSPT	LHDS	LCYL	LSPT
PHD1000	IDE(AT)	1083	3.50	5H					300	0	0	0			
WD-262	ST-506/412	20	5.25	HH	MFM	663			20	4	615	17			
WD-344R	ST-506/412	41.63	3.50	HH	RLL		783	783	0	4	782	26			
WD-362	ST-506/412	20	3.50	HH	MFM	663			20	4	615	17			
WD-382R	ST-506/412	20.81	3.50	HH	RLL 2		783	783	0	2	782	26			
WD-383R	ST-506/412	32.74	3.50	HH	RLL 2		616	616	0	4	615	26			
WD-384R	ST-506/412	41.63	3.50	HH	RLL 2		783	783	0	4	782	26			
WD-544R	ST-506/412	41.63	3.50	HH	RLL 2		783	783	0	4	782	26			
WD-562-5	ST-506/412	21.41	3.50	HH	MFM				40	4	615	17			
WD-582R	ST-506/412	20.81	3.50	HH	RLL 2		783	783	0	2	782	26			
WD-583R	ST-506/412	32.74	3.50	HH	RLL 2		616	616	0	4	615	26			
WD-584R	ST-506/412	41.63	3.50	HH	RLL 2		783	783	0	4	782	26			
WD-MI130-44	MCA	31.08	3.50	3H	RLL 2	AUTO	N/A	N/A	45	2	920	33			
WD-MI130-72	MCA	30.4	3.50	3H	RLL 2	AUTO	N/A	N/A	45	2	928	32			
WD-MI260-72	MCA	63.61	3.50	3H	RLL 2	AUTO	N/A	N/A	45	6	767	27			
WD-MI4120-72	MCA	125.03	3.50	3H	RLL 2	AUTO	N/A	N/A	45	8	925	33			
WD-TM262R	ST-506/412	20.81	3.50	HH	RLL 2		783	783	0	2	782	26			
WD-TM364	ST-506/412	41.63	3.50	HH	RLL 2		783	783	0	4	782	26			
WD93024-A	IDE(AT)	21.62	3.50	HH	RLL 2	862	N/A	N/A	40	2	782	27	4	615	17
WD93024-X	IDE(XT)	21.62	3.50	HH	RLL 2	862	N/A	N/A	50	2	782	27			
WD93028-AD	IDE(AT)	21.62	3.50	HH	RLL 2	862	N/A	N/A	40	2	782	27	4	615	17
WD93028-X	IDE(XT)	21.62	3.50	HH	RLL 2	862	N/A	N/A	40	2	782	27			
WD93034-X	IDE(XT)	32.43	3.50	HH	RLL 2	862	N/A	N/A	50	3	782	27			
WD93038-X	IDE(XT)	32.43	3.50	HH	RLL 2	862	N/A	N/A	40	3	782	27			
WD93044-A	IDE(AT)	43.24	3.50	HH	RLL 2	862	N/A	N/A	40	4	782	27	5	977	17
WD93044-X	IDE(XT)	43.24	3.50	HH	RLL 2	862	N/A	N/A	50	4	782	27			
WD93048-AD	IDE(AT)	43.24	3.50	HH	RLL 2	862	N/A	N/A	40	4	782	27			
WD93048-X	IDE(XT)	43.24	3.50	HH	RLL 2	862	N/A	N/A	40	4	782	27			
WD95024-A	IDE(AT)	21.62	3.50	HH	RLL 2	862	N/A	N/A	40	2	782	27	4	615	17
WD95028-AD	IDE(AT)	21.62	3.50	HH	RLL 2	862	N/A	N/A	40	2	782	27			
WD95028-X	IDE(XT)	21.62	3.50	HH	RLL 2	862	N/A	N/A	40	2	782	27			
WD95038-X	IDE(XT)	32.43	3.50	HH	RLL 2	862	N/A	N/A	40	3	782	27			
WD95044-A	IDE(AT)	43.24	3.50	HH	RLL 2	862	N/A	N/A	40	4	782	27	5	977	17
WD95048-AD	IDE(AT)	43.24	3.50	HH	RLL 2	862	N/A	N/A	40	4	782	27	5	977	17
WD95048-X	IDE(XT)	43.24	3.50	HH	RLL 2	862	N/A	N/A	40	4	782	27			
WDAB130	IDE(44PIN)	31.9	2.50	4H	RLL 2	AUTO	N/A	N/A	50	2	1020	0	4	916	17
WDAB140	IDE(44PIN)	42.7	2.50	5H	RLL 2	AUTO	N/A	N/A	0	2	1390	0	5	980	17

WESTERN DIGITAL CORPORATION

MODEL	INT	CAP	FF	HGT	CODE	LZ	WP	RWC	MTBF	PHDS	PCYL	PSPT	LHDS	LCYL	LSPT
WDAC11200	IDE(AT)	1281.9	3.50	3H	UNIDEN					0	0	0			
WDAC11200 (10-PIN)	IDE(AT)	1281.9	3.50	3H	UNIDEN					0	0	0			
WDAC1170	IDE(AT)	170.64	3.50	3H	RLL 2	AUTO	N/A	N/A	100	2	2233	0	6	1010	55
WDAC1210	IDE(AT)	212	3.50	3H	RLL 2	AUTO	N/A	N/A	100	0	0	0	12	989	35
WDAC1270	IDE(AT)	270.4	3.50	3H	RLL 2	AUTO	N/A	N/A	250	2	2233	0	12	917	48
WDAC1365	IDE(AT)	365.4	3.50	3H	RLL 1	AUTO	N/A	N/A	250	0	0	0	16	708	63
WDAC140	IDE(AT)	42.65	3.50	3H	RLL 2	AUTO	N/A	N/A	50	2	1082	39	5	980	17
WDAC1425	IDE(AT)	426.8	3.50	3H	RLL 1	AUTO	N/A	N/A	250	0	0	0	16	827	63
WDAC160	IDE(AT)	63.2	3.50	3H	RLL 2	AUTO	N/A	N/A	100	2	1349	0	7	1024	17
WDAC21000	IDE(AT)	1083.8	3.50	3H	UNIDEN	AUTO	N/A	N/A	300	4	0	0	16	2100	63
WDAC21000 (10-PIN)	IDE(AT)	1083.8	3.50	3H	UNIDEN					0	0	0			
WDAC2120	IDE(AT)	126.4	3.50	3H	RLL 2	AUTO	N/A	N/A	100	4	1349	0	8	872	35
WDAC21200	IDE(AT)	1281.9	3.50	3H	PRML	AUTO	N/A	N/A	300	4	0	0	16	2484	63
WDAC21200 (10-PIN)	IDE(AT)	1281.9	3.50	3H	UNIDEN					0	0	0			
WDAC21600	IDE(AT)	1624.6	3.50	3H	PRML	AUTO	N/A	N/A	300	4	0	0	16	3148	63
WDAC21600 (10-PIN)	IDE(AT)	1624.6	3.50	3H	UNIDEN					0	0	0			
WDAC2170	IDE(AT)	170.6	3.50	3H	RLL 2	AUTO	N/A	N/A	100	4	1584	0	6	1010	55
WDAC2200	IDE(AT)	212.67	3.50	3H	RLL 2	AUTO	N/A	N/A	100	4	1971	0	12	989	35
WDAC22000	IDE(AT)	2000.3	3.50	3H	UNIDEN					0	0	0			
WDAC22000 (10-PIN)	IDE(AT)	2000.3	3.50	3H	UNIDEN					0	0	0			
WDAC22100	IDE(AT)	2111.8	3.50	3H	UNIDEN					0	0	0			
WDAC22100 (10-PIN)	IDE(AT)	2111.8	3.50	3H	UNIDEN					0	0	0			
WDAC2250	IDE(AT)	256	3.50	3H	RLL 2	AUTO	N/A	N/A	100	3	2233	0	9	1010	55
WDAC22500	IDE(AT)	2559.8	3.50	3H	UNIDEN					0	0	0			
WDAC22500 (10-PIN)	IDE(AT)	2559.8	3.50	3H	UNIDEN					0	0	0			
WDAC2340	IDE(AT)	341.29	3.50	3H	RLL 2	AUTO	N/A	N/A	100	4	2233	0	12	1010	55
WDAC2420	IDE(AT)	425.3	3.50	3H	RLL 2	AUTO	N/A	N/A	100	0	0	0	15	989	56
WDAC2540	IDE(AT)	540.86	3.50	3H	RLL 2	AUTO	N/A	N/A	100	0	0	0	16	1048	63
WDAC2540 (10-PIN)	IDE(AT)	540.8	3.50	3H	UNIDEN	AUTO	N/A	N/A		0	0	0			
WDAC2635	IDE(AT)	639.9	3.50	3H	RLL 1	AUTO	N/A	N/A	250	0	0	0	16	1240	63
WDAC2635 (10-PIN)	IDE(AT)	639.9	3.50	3H	UNIDEN	AUTO	N/A	N/A		0	0	0			
WDAC2700	IDE(AT)	696.6	3.50	3H	RLL 2	AUTO	N/A	N/A	100	0	0	0	16	1416	63
WDAC2700 (10-PIN)	IDE(AT)	730.8	3.50	3H	UNIDEN					0	0	0			
WDAC280	IDE(AT)	85.29	3.50	3H	RLL 2	AUTO	N/A	N/A	50	4	1082	39	10	980	17
WDAC2850	IDE(AT)	853.6	3.50	3H	RLL 1	AUTO	N/A	N/A	250	0	0	0	16	1654	63
WDAC2850 (10-PIN)	IDE(AT)	853.6	3.50	3H	UNIDEN					0	0	0			
WDAC31000	IDE(AT)	1084	3.50	3H	RLL 2	AUTO	N/A	N/A	250	6	0	0	16	2100	63
WDAC31000 (10-PIN)	IDE(AT)	1083.8	3.50	3H	UNIDEN					0	0	0			
WDAC31200	IDE(AT)	1222.6	3.50	3H	RLL 2	AUTO	N/A	N/A	0	0	0	0	16	2484	63
WDAC31200 (10-PIN)	IDE(AT)	1281.9	3.50	3H	UNIDEN					0	0	0			
WDAC31600	IDE(AT)	1549	3.50	3H	RLL 2	AUTO	N/A	N/A	250	0	0	0	16	3148	63
WDAC31600 (10-PIN)	IDE(AT)	1624.6	3.50	3H	UNIDEN					0	0	0			
WDAC32100	IDE(AT)	2111.8	3.50	3H	PRML	AUTO	N/A	N/A	300	5	0	0	16	4092	63
WDAC32100 (10-PIN)	IDE(AT)	2111.8	3.50	3H	UNIDEN					0	0	0			

WESTERN DIGITAL CORPORATION

MODEL	INT	CAP	FF	HGT	CODE	LZ	WP	RWC	MTBF	PHDS	PCYL	PSPT	LHDS	LCYL	LSPT
WDAC32500	IDE(XT)	2559.8	3.50	3H	PRML	AUTO	N/A	N/A	300	6	0	0	16	4960	63
WDAC32500 (10-PIN)	IDE(AT)	2559.8	3.50	3H	UNIDEN					0	0	0			
WDAC33100	IDE(AT)	3166	3.50	3H	UNIDEN	AUTO	N/A	N/A	300	6	0	0	16	6136	63
WDAC33100 (10-PIN)	IDE(AT)	3166.7	3.50	3H	UNIDEN					0	0	0			
WDAC33200	IDE(AT)	3249.3	3.50	3H	UNIDEN					0	0	0			
WDAC33200 (10-PIN)	IDE(AT)	3249.3	3.50	3H	UNIDEN					0	0	0			
WDAC34000	IDE(AT)	1083	3.50	3H		AUTO	N/A	N/A	350	0	0	0	16	7752	63
WDAC34000 (10-PIN)	IDE(AT)	4000.7	3.50	3H	UNIDEN					0	0	0			
WDAH260	IDE(44PIN)	63.25	2.50	4H	RLL 2	AUTO	N/A	N/A	50	4	1020	0	7	1024	17
WDAH280	IDE(44PIN)	85.57	2.50	4H	RLL 2	AUTO	N/A	N/A	50	4	1390	0	10	980	17
WDAL1100	IDE(44PIN)	100.06	2.50	4H	RLL 1	AUTO	N/A	N/A	100	2	1900	0	6	958	34
WDAL2120	IDE(AT)	126.4	2.50	3H	RLL 2	AUTO	N/A	N/A	100	4	1349	0	8	872	35
WDAL2170	IDE(AT)	170.6	2.50	3H	RLL 2	AUTO	N/A	N/A	100	4	1584	0	6	1010	55
WDAL2200	IDE(44PIN)	200.12	2.50	4H	RLL 1	AUTO	N/A	N/A	100	2	1900	0	12	958	34
WDAL2540	IDE(44PIN)	540.8	2.50	5H	UNIDEN	AUTO	N/A	N/A	300	4	0	0	16	1048	63
WDAP105 PIRANHA	IDE(AT)	104	3.50	HH	RLL 2	AUTO	N/A	N/A	50	4	0	0			
WDAP4200 PIRANHA	IDE(AT)	214.95	3.50	HH	RLL 2	AUTO	N/A	N/A	86	8	1280	41	12	987	35
WDCU140	PCM	42.65	2.50	4H	RLL 1	AUTO	N/A	N/A	255	2	1050	0	5	980	17
WDE2170 (50-PIN)	U-SCSI	2170	3.50	3H		AUTO	N/A	N/A	1000	0	0	0			
WDE2170 (68-PIN)	U-SCSI W	2170	3.50	3H		AUTO	N/A	N/A	1000	0	0	0			
WDE2170 (80-PIN)	U-SCSI W	2170	3.50	3H		AUTO	N/A	N/A	1000	0	0	0			
WDE4360 (50-PIN)	U-SCSI	4360	3.50	3H		AUTO	N/A	N/A	1000	0	0	0			
WDE4360 (68-PIN)	U-SCSI W	4360	3.50	3H		AUTO	N/A	N/A	1000	0	0	0			
WDE4360 (80-PIN)	U-SCSI W	4360	3.50	3H		AUTO	N/A	N/A	1000	0	0	0			
WDSC8320	SCSI-2	326.51	3.50	HH	RLL 1	AUTO	N/A	N/A	150	14	949	48			
WDSC8400	SCSI-2	412.53	3.50	HH	RLL 1	AUTO	N/A	N/A	150	14	1199	48			
WDSP105 PIRANHA	SCSI	104	3.50	HH	RLL 2	AUTO	N/A	N/A	50	4	0	0			
WDSP2100 PIRANHA	SCSI-2	106.21	3.50	HH	RLL 2	AUTO	N/A	N/A	50	4	1265	41			
WDSP4200 REV. 1	SCSI	209.71	3.50	HH	RLL 2	AUTO	N/A	N/A	86	8	1280	40			
WDSP4200 REV. 2	SCSI	209.71	3.50	3H	RLL 2	AUTO	N/A	N/A	86	8	1280	40			
WDUC140	PCM	42.65	5.25	FH	RLL 2	AUTO	N/A	N/A	50	2	1050	0			

XEBEC AMERICA

MODEL	INT	CAP	FF	HGT	CODE	LZ	WP	RWC	MTBF	PHDS	PCYL	PSPT	LHDS	LCYL	LSPT
OWL I	ST-506/412	25	5.25	HH	MFM	AUTO	N/A	N/A	0	4	0	17			
OWL II	ST-506/412	38	5.25	HH	MFM				0	4	0	17			
OWL III	SCSI	52	5.25	HH	RLL	AUTO	N/A	N/A	0	4	0	17			

Y-E DATA AMERICA, INC.

MODEL	INT	CAP	FF	HGT	CODE	LZ	WP	RWC	MTBF	PHDS	PCYL	PSPT	LHDS	LCYL	LSPT
YD-3042	SCSI	43.57	3.50	HH	RLL	AUTO	N/A	N/A	40	4	788	27			
YD-3081B	SCSI	45.45	3.50	HH	RLL 2	AUTO	N/A	N/A	30	2	1057	42			
YD-3082	SCSI	87.14	3.50	3H	RLL	AUTO	N/A	N/A	40	8	788	27			

Y-E DATA AMERICA, INC.

MODEL	INT	CAP	FF	HGT	CODE	LZ	WP	RWC	MTBF	PHDS	PCYL	PSPT	LHDS	LCYL	LSPT
YD-3082B	SCSI	90.91	3.50	HH	RLL 2	AUTO	N/A	N/A	30	4	1057	42			
YD-3083B	SCSI	136.37	3.50	HH	RLL 2	AUTO	N/A	N/A	30	6	1057	42			
YD-3084B	SCSI	181.83	3.50	HH	RLL 2	AUTO	N/A	N/A	30	8	1057	42			
YD-3161B	IDE(AT)	45.45	3.50	3H	RLL 2	AUTO	N/A	N/A	40	2	1057	42			
YD-3162B	IDE(AT)	90.91	3.50	3H	RLL 2	AUTO	N/A	N/A	40	4	1057	42			
YD-3181B	SCSI	45.45	3.50	3H	RLL 2	AUTO	N/A	N/A	40	2	1057	42			
YD-3182B	SCSI	90.91	3.50	3H	RLL 2	AUTO	N/A	N/A	40	4	1057	42			
YD-3530	ST-506/412	31.81		–	MFM		732	732	0	5	731	0			
YD-3540	ST-506/412	44.53	5.25	HH	MFM		732	732	0	7	731	17			

ZENTEK STORAGE, INC.

MODEL	INT	CAP	FF	HGT	CODE	LZ	WP	RWC	MTBF	PHDS	PCYL	PSPT	LHDS	LCYL	LSPT
ZH3100(A)	IDE(AT)	86	3.50	HH	RLL	AUTO	N/A	N/A	50	0	0	0			
ZH3100(S)	SCSI	86	3.50	HH	RLL	AUTO	N/A	N/A	50	0	0	0			
ZH3140(A)	IDE(AT)	121	3.50	HH	RLL	AUTO	N/A	N/A	50	0	0	0			
ZH3140(S)	SCSI	121	3.50	HH	RLL	AUTO	N/A	N/A	50	0	0	0			

Chapter 7

IDE Quick Reference Chart

About IDE Drive Types

Most newer IDE systems can automatically configure for a BIOS Drive Type. However, there are still many drives and BIOSes in use that do not support the IDE Identify Drive command (see "The IDE Interface" in Chapter Two for more information). If this is the case and you do not have a drive that fits the parameters of one of the defined drive types for your particular BIOS, then you will have to enter a user-definable drive type. This chapter contains a table that allows you to simply look up the capacity of your particular drive and enter the corresponding parameters into the BIOS User Definable Drive Type.

The User Definable Drive Type does not have to use the physical parameters of a particular drive or even match the drive's capacity exactly. This is because most IDE drives do a translation so that they may be used in systems that do not support the drive type needed.

There are two types of specifications that you will encounter with this kind of hard drive. The first type of specifications are the *physical* (actual) specifications. These are the hard drive's real values. If you opened up the hard drive, looked inside and saw four heads, you would be looking at the physical (actual) number of heads. These physical specifications are supplied in Chapter 6 and indicated by PHDS, PCYL, and PSPT in the tables there. In some cases, the physical specifications are not available to us; we then supply you with the logical values.

The second type of specifications are the *logical* (translation) specifications. These are the heads, cylinders, and sectors that both the AT BIOS table and DOS can agree on. Most AT BIOSes have a limit of 16 heads and 1,024 cylinders. And DOS can never use more than 1,024 cylinders (unless special software is used). Most IDE hard drives exceed these limits. The IDE hard drive will do a mathematical conversion from the logical (translation) specifications to the physical (actual) specifications. These specifications are normally supplied by the hard drive manufacturer. The manufacturer may supply you with one set of logical (translation) specifications but by no means are you limited to those. Just remember to choose a drive type that is equal to or less than, but never more than the size of the drive in megabytes. The IDE drive will handle the translation from there.

If you have the physical (actual) specifications and a user-definable drive type in your BIOS, you may enter these in (but don't go over the DOS 1,024 cylinder limit). IDE Quick Reference Chart

On the table that follows, look up your drive capacity in column one and enter the specifications into the host system's user-definable drive type section.

This table is only valid when used with IDE-type hard drives.

IDE QUICK REFERENCE TABLE

CAP	HDS	SPT	CYL
20	4	17	574
21	4	17	603
22	4	17	631
23	4	17	660
24	4	17	689
25	4	17	718
26	4	17	746
27	4	17	775
28	4	17	804
29	4	17	832
30	4	17	861
31	4	17	890
32	4	17	919
33	4	17	947
34	4	17	976
35	4	17	1005
36	5	17	827
37	5	17	850
38	5	17	873
39	5	17	896
40	5	17	919
41	5	17	942
42	5	17	965
43	5	17	988
44	5	17	1011
45	6	17	861
46	6	17	880
47	6	17	899
48	6	17	919
49	6	17	938
50	6	17	957
51	6	17	976
52	6	17	995
53	6	17	1014
54	7	17	886
55	7	17	902
56	7	17	919
57	7	17	935
58	7	17	951
59	7	17	968
60	7	17	984
61	7	17	1001
62	7	17	1017
63	8	17	904
64	8	17	919
65	8	17	933
66	8	17	947
67	8	17	962
68	8	17	976
69	8	17	990
70	8	17	1005
71	8	17	1019
72	9	17	919
73	9	17	931
74	9	17	944
75	9	17	957
76	9	17	970
77	9	17	982
78	9	17	995
79	9	17	1008
80	9	17	1021
81	10	17	930
82	10	17	942
83	10	17	953
84	10	17	965
85	10	17	976
86	10	17	988
87	10	17	999
88	10	17	1011
89	10	17	1022
90	11	17	940
91	11	17	950
92	11	17	960
93	11	17	971
94	11	17	981
95	11	17	992
96	11	17	1002
97	11	17	1013
98	11	17	1023
99	12	17	947
100	12	17	957
101	12	17	966
102	12	17	976
103	12	17	986
104	12	17	995
105	12	17	1005
106	12	17	1014
107	12	17	1024
108	13	17	954
109	13	17	963
110	13	17	972
111	13	17	980
112	13	17	989
113	13	17	998
114	13	17	1007
115	13	17	1016
116	14	17	951
117	14	17	960
118	14	17	968
119	14	17	976
120	14	17	984
121	14	17	992
122	14	17	1001
123	14	17	1009
124	14	17	1017
125	15	17	957
126	15	17	965
127	15	17	972
128	15	17	980
129	15	17	988
130	15	17	995
131	15	17	1003
132	15	17	1011
133	15	17	1018
134	16	17	962
135	16	17	969
136	16	17	976
137	16	17	983
138	16	17	990
139	16	17	998
140	16	17	1005
141	16	17	1012
142	16	17	1019
143	9	33	940
144	9	33	946
145	9	33	953
146	9	33	960
147	9	33	966
148	9	33	973
149	9	33	979
150	9	33	986
151	9	33	993
152	9	33	999
153	9	33	1006
154	9	33	1012
155	9	33	1019
156	10	33	923
157	10	33	929
158	10	33	935
159	10	33	941
160	10	33	946
161	10	33	952
162	10	33	958
163	10	33	964
164	10	33	970

CAP	HDS	SPT	CYL	CAP	HDS	SPT	CYL	CAP	HDS	SPT	CYL
166	10	33	982	217	13	33	987	268	16	33	991
167	10	33	988	218	13	33	992	269	16	33	995
168	10	33	994	219	13	33	997	270	16	33	998
169	10	33	1000	220	13	33	1001	271	16	33	1002
170	10	33	1006	221	13	33	1006	272	16	33	1006
171	10	33	1012	222	13	33	1010	273	16	33	1009
172	10	33	1017	223	13	33	1015	274	16	33	1013
173	10	33	1023	224	13	33	1019	275	16	33	1017
174	11	33	936	225	13	33	1024	276	16	33	1020
175	11	33	941	226	14	33	955	277	16	33	1024
176	11	33	946	227	14	33	959	278	9	63	957
177	11	33	952	228	14	33	963	279	9	63	961
178	11	33	957	229	14	33	968	280	9	63	964
179	11	33	963	230	14	33	972	281	9	63	967
180	11	33	968	231	14	33	976	282	9	63	971
181	11	33	973	232	14	33	980	283	9	63	974
182	11	33	979	233	14	33	985	284	9	63	978
183	11	33	984	234	14	33	989	285	9	63	981
184	11	33	990	235	14	33	993	286	9	63	985
185	11	33	995	236	14	33	997	287	9	63	988
186	11	33	1000	237	14	33	1001	288	9	63	992
187	11	33	1006	238	14	33	1006	289	9	63	995
188	11	33	1011	239	14	33	1010	290	9	63	998
189	11	33	1016	240	14	33	1014	291	9	63	1002
190	11	33	1022	241	14	33	1018	292	9	63	1005
191	12	33	942	242	14	33	1023	293	9	63	1009
192	12	33	946	243	15	33	958	294	9	63	1012
193	12	33	951	244	15	33	962	295	9	63	1016
194	12	33	956	245	15	33	966	296	9	63	1019
195	12	33	961	246	15	33	970	297	9	63	1023
196	12	33	966	247	15	33	974	298	10	63	923
197	12	33	971	248	15	33	978	299	10	63	926
198	12	33	976	249	15	33	982	300	10	63	930
199	12	33	981	250	15	33	986	301	10	63	933
200	12	33	986	251	15	33	990	302	10	63	936
201	12	33	991	252	15	33	994	303	10	63	939
202	12	33	996	253	15	33	998	304	10	63	942
203	12	33	1001	254	15	33	1002	305	10	63	945
204	12	33	1006	255	15	33	1006	306	10	63	948
205	12	33	1011	256	15	33	1010	307	10	63	951
206	12	33	1016	257	15	33	1014	308	10	63	954
207	12	33	1020	258	15	33	1017	309	10	63	957
208	13	33	946	259	15	33	1021	310	10	63	961
209	13	33	951	260	16	33	961	311	10	63	964
210	13	33	956	261	16	33	965	312	10	63	967
211	13	33	960	262	16	33	969	313	10	63	970
212	13	33	965	263	16	33	972	314	10	63	973
213	13	33	969	264	16	33	976	315	10	63	976
214	13	33	974	265	16	33	980	316	10	63	979
215	13	33	978	266	16	33	983	317	10	63	982
216	13	33	983	267	16	33	987	318	10	63	985

CAP	HDS	SPT	CYL	CAP	HDS	SPT	CYL	CAP	HDS	SPT	CYL
362	11	63	1020	414	13	63	987	464	15	63	958
363	11	63	1023	415	13	63	989	465	15	63	961
364	12	63	940	416	13	63	992	466	15	63	963
365	12	63	942	417	13	63	994	467	15	63	965
366	12	63	945	418	13	63	996	468	15	63	967
367	12	63	948	419	13	63	999	469	15	63	969
368	12	63	950	420	13	63	1001	470	15	63	971
369	12	63	953	421	13	63	1003	471	15	63	973
370	12	63	955	422	13	63	1006	472	15	63	975
371	12	63	958	423	13	63	1008	473	15	63	977
372	12	63	961	424	13	63	1011	474	15	63	979
373	12	63	963	425	13	63	1013	475	15	63	981
374	12	63	966	427	13	63	1018	476	15	63	983
375	12	63	968	428	13	63	1020	477	15	63	985
376	12	63	971	427	13	63	1018	478	15	63	987
377	12	63	973	427	13	63	1018	479	15	63	989
378	12	63	976	428	13	63	1020	480	15	63	992
379	12	63	979	429	13	63	1023	481	15	63	994
380	12	63	981	430	14	63	952	482	15	63	996
381	12	63	984	431	14	63	954	483	15	63	998
382	12	63	986	432	14	63	956	484	15	63	1000
383	12	63	989	433	14	63	958	485	15	63	1002
384	12	63	992	434	14	63	961	486	15	63	1004
385	12	63	994	435	14	63	963	487	15	63	1006
386	12	63	997	436	14	63	965	488	15	63	1008
387	12	63	999	437	14	63	967	489	15	63	1010
388	12	63	1002	438	14	63	969	490	15	63	1012
389	12	63	1004	439	14	63	972	491	15	63	1014
390	12	63	1007	440	14	63	974	492	15	63	1016
391	12	63	1010	441	14	63	976	493	15	63	1018
392	12	63	1012	442	14	63	978	494	15	63	1020
393	12	63	1015	443	14	63	980	495	15	63	1023
394	12	63	1017	444	14	63	983	496	16	63	960
395	12	63	1020	445	14	63	985	497	16	63	962
396	12	63	1023	446	14	63	987	498	16	63	964
397	13	63	946	447	14	63	989	499	16	63	966
398	13	63	949	448	14	63	992	500	16	63	968
399	13	63	951	449	14	63	994	501	16	63	970
400	13	63	953	450	14	63	996	502	16	63	972
401	13	63	956	451	14	63	998	503	16	63	974
402	13	63	958	452	14	63	1000	504	16	63	976
403	13	63	961	453	14	63	1003	505	16	63	978
404	13	63	963	454	14	63	1005	506	16	63	980
405	13	63	965	455	14	63	1007	507	16	63	982
406	13	63	968	456	14	63	1009	508	16	63	984
407	13	63	970	457	14	63	1011	509	16	63	986
408	13	63	972	458	14	63	1014	510	16	63	988
409	13	63	975	459	14	63	1016	511	16	63	990
410	13	63	977	460	14	63	1018	512	16	63	992
411	13	63	980	461	14	63	1020	513	16	63	994
412	13	63	982	462	14	63	1023	514	16	63	995
413	13	63	984	463	15	63	956	515	16	63	997

CAP	HDS	SPT	CYL
516	16	63	999
517	16	63	1001
518	16	63	1003
519	16	63	1005
520	16	63	1007
521	16	63	1009
522	16	63	1011
523	16	63	1013
524	16	63	1015
525	16	63	1017
526	16	63	1019
527	16	63	1021
528	16	63	1023

The parameters provided are calculated *not to exceed* the formatted capacity specified. For example:

The parameter's list for 130MBs, 15HDs x 995CYL x 17SPT x 512BPS = 129,907,200 bytes.

If you know the exact capacity of the drive, you may increase the cylinder count to achieve maximum capacity. For example:

If the drive's capacity is 130.76MB, you could increase the cylinder count to 1001 for a capacity of 130,690,560 bytes, thus only forfeiting about 7KB.

You'll notice that the largest drive available with 17 sectors is 142MB. You can still use a drive with a higher capacity by increasing the sectors, such as the 143 to 528MB drives above.

In addition to this table, a list of manufacturers' recommended parameters is provided in Chapter 6 and indicated by LHDS, LCYL, and LSPT in the tables there.

Appendix A

Manufacturer Contact Information

Appendix A lists contact information for most companies involved with hard drives, drive controllers, and other technologies related to drives including software and removable media. In addition to mailing addresses and phone numbers for most companies, technical support numbers, fax numbers, fax-back services, BBSs, Web addresses, and e-mail addresses are provided where available.

ABIT COMPUTER CORPORATION

Address	47889 FREMONT BLVD. FREMONT, CA 94538
Phone	(510) 623-0500
Fax	(510) 623-1092
Web	http://www.abit.com.tw

ACCULOGIC, INC.

Address	7 WHATNEY IRVINE, CA 92718
Phone	(714) 454-2441
Fax	(714) 454-8527
Toll Free	(800) 234-7811
BBS	(714) 470-1759
Web	http://www.acctechnology.com
E-mail	techsupport@acctechnology.com

ACER, INC.

Address	2641 ORCHARD PARKWAY SAN JOSE, CA 95134
Phone	(408) 432-6200
Fax	(408) 922-2933
BBS	(408) 428-0140

Fax-Back	(800) 554-2494
Web	http://www.acer.com

ACHME COMPUTER, INC.

Address	4059 CLIPPER COURT FREMONT, CA 94538
Phone	(510) 623-8818
Fax	(510) 623-8585
BBS	(510) 623-7398
Web	http://www.achme.com
E-mail	support@achme.com

ACTIONTEC ELECTRONICS, INC.

Address	750 NORTH MARY AVENUE SUNNYVALE, CA 94086
Phone	(408) 739-7000
Fax	(408) 739-7001
Toll Free	(800) 797-7001
BBS	(408) 739-7071
Web	http://www.actiontec.com
E-mail	techsupp@actiontec.com

ADAPTEC, INC.

Address	691 SOUTH MILPITAS BLVD. MILPITAS, CA 95035
Phone	(408) 945-8600
Fax	(408) 945-6776
Toll Free	(800) 959-7274
BBS	(408) 945-7727
Fax-Back	(408) 957-7150
Web	http://www.adaptec.com

ADS INTERNATIONAL, INC.

Address	418-B CLOVERLEAF DRIVE BALDWIN PARK, CA 91706-6511
Phone	(818) 369-2332
Fax	(818) 855-5059
Web	http://www.ads.i1.com

ADVANCED COMPUTER TECHNOLOGY, LTD.

Address	12/F, CHUANGS FINANCE CENTER 81 85 LOCKHART ROAD WANCHAI, HONG KONG

ADVANCED DIGITAL INFORMATION CORPORATION

Address	14737 NE 87TH STREET REDMOND, WA 98073
Phone	(206) 881-8004
Fax	(206) 881-2296
Toll Free	(800) 336-1233
Web	http://www.adic.com
E-mail	support@adic.com

ADVANCED INTEGRATION RESEARCH, INC.

Address	2188 DEL FRANCO STREET SAN JOSE, CA 95131
Phone	(408) 428-0800
Fax	(408) 428-0950
Toll Free	(800) 866-1945
BBS	(408) 428-1735
Web	http://www.airwebs.com
E-mail	air@ix.netcom.com

ADVANCED LOGIC RESEARCH, INC.

Address	9401 JERONIMO ROAD IRVINE, CA 92618
Fax	(714) 458-0532
Toll Free	(800) 257-1230
BBS	(714) 458-6834
Fax-Back	(714) 581-3332
Web	http://www.alr.com

ADVANCED STORAGE CONCEPTS

Address	10713 RANCH ROAD, 620 NORTH, SUITE 305 AUSTIN, TX 78726
Phone	(512) 335-1077
Fax	(512) 335-1078
BBS	(512) 335-3499
Web	http://www.eden.com/~asc
E-mail	asc@eden.com

ADVANCED SYSTEM PRODUCTS, INC.

Address	1150 RINGWOOD COURT SAN JOSE, CA 95131
Phone	(408) 383-5714
Fax	(408) 383-9612
Toll Free	(800) 525-7443
BBS	(408) 383-9540
Fax-Back	(408) 383-9753

Web	http://www.advansys.com
E-mail	support@advansys.com

ADVANTAGE MEMORY CORPORATION

Address	25-A TECHNOLOGY DRIVE, BLDG. 1
	IRVINE, CA 92618
Phone	(714) 453-8111
Fax	(714) 453-8158
Toll Free	(800) 245-5299
Web	http://www.advantagememory.com
E-mail	advantagemem@applelink.apple.c

ALPHA OMEGA COMPUTER CORPORATION

Address	3041 E. LA JOLLA ST.
	ANAHEIM, CA 92806
Phone	(714) 632-0388

ALPS ELECTRIC (USA), INC.

Address	3553 NORTH FIRST STREET
	SAN JOSE, CA 95134
Phone	(408) 432-6000
Fax	(408) 432-6035
BBS	(408) 432-6424
Web	http://www.alps.com

AMERICAN DIGICOM CORPORATION

Address	1233 MIDAS WAY
	SUNNYVALE, CA 94086
Phone	(408) 245-1580
Fax	(408) 245-1584
Web	http://www.digicomgroup.com

AMERICAN MEGATRENDS, INC.

Address	6145-F NORTHBELT PARKWAY
	NORCROSS, GA 30071
Phone	(770) 263-8181
Fax	(770) 246-8791
Toll Free	(800) 828-9264
BBS	(770) 246-8780
Fax-Back	(770) 246-8787
Web	http://www.megatrends.com

AMPEX CORPORATION

Address	198 NORTH NASH STREET
	EL SEGUNDO, CA 90245
Phone	(310) 640-0150

Fax	(310) 640--0450
Web	http://www.ampex.com

AMPRO COMPUTERS, INC.

Address	990 ALMANOR AVENUE SUNNYVALE, CA 94086
Phone	(408) 522-2100
Toll Free	(800) 966-5200
Web	http://www.ampro.com

ANDROMEDA SYSTEMS, INC.

Address	9000 ETON AVENUE CANOGA PARK, CA 91304
Phone	(818) 709-7600
Fax	(818) 709-7407
Web	http://www.andromedasystems.co

AQUARIUS SYSTEMS, INC.

Address	47381 BAYSIDE PARKWAY FREMONT, CA 94538
Phone	(510) 656-9800
Fax	(510) 656-9876

ARCHIVE CORPORATION

Address	1650 SUNFLOWER AVENUE COSTA MESA, CA 92626
Phone	(714) 641-1230
Toll Free	(800) 526-6637

ARCO COMPUTER PRODUCTS

Address	2750 N. 29TH AVENUE SUITE 316 HOLLYWOOD, FL 33020
Phone	(305) 925-2688
Fax	(305) 925-2889
Toll Free	(800) 458-1666
BBS	(954) 925-2791
Fax-Back	(954) 925-2889
Web	http://www.arcoide.com
E-mail	arco@arcoide.com

AREAL TECHNOLOGY, INC.

Address	2146 BERING DRIVE SAN JOSE, CA 95131
Phone	(408) 241-8290
Fax	(408) 248-2960

Web	http://www.areal.com

ASUS COMPUTER INTERNATIONAL

Address	721 CHARCOT AVENUE SAN JOSE, CA 95131
Phone	(408) 474-0567
Fax	(408) 474-0568
BBS	(408) 474-0555
Web	http://asustek.asus.com.tw

ATC/UNITRON COMPUTERS & COMPUTER PARTS

Address	PO BOX 2140 5700 DA HELMOND, NETHERLANDS
Phone	492 528 899
Fax	492 55 42 45

ATRONICS INTERNATIONAL, INC.

Address	44700-B INDUSTRIAL DRIVE FREMONT, CA 94538
Phone	(510) 656-8400
Fax	(510) 656-8560
BBS	(510) 226-2671
Web	http://www.atronics.com
E-mail	info@ati1.com

ATTO TECHNOLOGY, INC.

Address	40 HAZELWOOD DRIVE, UNIT 106 AMHERST, NY 14228
Phone	(716) 691-1999
Fax	(716) 691-9353
BBS	(716) 691-9403
Web	http://www.attotech.com

AUVA COMPUTER, INC.

Address	NO. 18, PEI YUAN RD. CHUNG LI IND., PARK TAOYUAN, TAIWAN, REP OF CHINA
Phone	3 452 8200
Fax	3 452 9155

BASF

Address	CARL-BOSCH-STRABE 38 67056 LUDWIGSHAFEN, GERMANY
Fax	621 604 2525

Web http://www.basf-magnetics.be

BCM ADVANCED RESEARCH, INC.

Address 1 HUGHES
IRVINE, CA 92718
Phone (714) 470-1888
Fax (714) 470-0883
Web http://www.bcmcom.com

BEHAVIOR TECH COMPUTER CORPORATION

Address 4180 BUSINESS CENTER DRIVE
FREMONT, CA 94538
Phone (510) 657-3956
Fax (510) 657-1859
BBS (510) 249-0342
Web http://www.btc.com.tw

BELFORT MEMORY INTERNATIONAL

Address 170 KNOWLES
SUITE 1
LOS GATOS, CA 95030
Phone (408) 370-5300
Fax (408) 370-3464

BIOTEQ

Address EKKERSRIJT 7316
5692 HH SON, POSTBUS 204
5690 AE SON,
NETHERLANDS
Phone 49 907 1200
Fax 49 907 1876
Web http://www.bioteq.com

BOCA RESEARCH, INC.

Address 1377 CLINT MOORE ROAD
BOCA RATON, FL 33487-2722
Phone (407) 997-6227
Fax (407) 994-5848
Web http://www.bocaresearch.com

BULL HN INFORMATION SYSTEMS, INC.

Address TECHNOLOGY PARK
BILLERICA, MA 01821-9903
Phone (508) 294-6000
Fax (508) 294-7411
Toll Free (800) 285-5727

Web	http://www.bull.com

BUSLOGIC, INC.

Address	4151 BURTON DRIVE SANTA CLARA, CA 95054
Phone	(408) 492-9090
Fax	(408) 492-1542
BBS	(408) 492-1984
Web	http://www.mylex.com

C.ITOH ELECTRONICS

Address	2701 DOW AVENUE TUSTIN, CA 92780
Fax	(714) 757-4488
Toll Free	(800) 347-2484

CALLUNA TECHNOLOGY, INC.

Address	1762 TECHNOLOGY DRIVE SUITE 223 SAN JOSE, CA 95110
Phone	(408) 453-4753
Fax	(408) 453-0427
Web	http://www.calluna1.demon.co.u

CHAINTECH COMPUTER COMPANY, LTD.

Address	12880 LAKELAND ROAD SANTA FE SPRINGS, CA 90670
Phone	(310) 906-1698
Fax	(310) 906-1699
Web	http://www.chaintech.com.tw

CHICONY, INC.

Address	53 PARKER IRVINE, CA 92718
Phone	(714) 380-0928
Fax	(714) 380-9204
Web	http://www.chicony.com

CIPRICO, INC.

Address	2800 CAMPUS DRIVE SUITE 60 PLYMOUTH, MN 55441
Phone	(612) 551-4000
Fax	(612) 551-4002
Web	http://www.ciprico.com

CMD TECHNOLOGY, INC.

Address	1 VANDERBILT IRVINE, CA 92618
Phone	(714) 454-0800
Fax	(714) 455-1656
Toll Free	(800) 426-3832
Web	http://www.cmd.com

CMS ENHANCEMENTS, INC.

Address	THREE IMPERIAL PROMENADE SANTA ANA, CA 92707
Phone	(714) 437-0099
Fax	(714) 513-2465
Toll Free	(800) 555-1671
Fax-Back	(714) 445-5327
Web	http://www.cmsenh.com

CNET TECHNOLOGY, INC.

Address	2199 ZANKER ROAD SAN JOSE, CA 95131
Phone	(408) 954-8000
Fax	(408) 954-8866
Toll Free	(800) 486-2638
BBS	(408) 954-1787
Web	http://www.cnet.com.tw

COGENT DATA TECHNOLOGIES, INC.

Address	640 MULLIS STREET FRIDAY HARBOR, WA 98250
Phone	(360) 378-2929
Fax	(360) 378-2882
Toll Free	(800) 426-4368
BBS	(360) 378-5405
Web	http://www.cogentdata.com

COMPAQ COMPUTER CORPORATION

Address	20555 STATE HIGHWAY 249 PO BOX 692000 HOUSTON, TX 77269-2000
Phone	(713) 370-0670
Fax	(713) 374-1740
Toll Free	(800) 345-1518
BBS	(713) 378-1418
Fax-Back	(800) 345-1518
Web	http://www.compaq.com

CONNER PERIPHERALS, INC.

Address	3081 ZANKER ROAD SAN JOSE, CA 95134-2128
Phone	(408) 456-4500
Fax	(408) 456-4501
Toll Free	(800) 426-6637
BBS	(800) 456-4415
Fax-Back	(408) 456-4903
Web	http://www.conner.com

CONTROL CONCEPTS, INC.

Address	8500 EXECUTIVE PARK AVENUE FAIRFAX, VA 22031-2218
Phone	(703) 876-6400
Fax	(703) 876-6416
Toll Free	(800) 922-9259
Web	http://www.controlconceptsinc.
E-mail	techsupp@presearch.com

CORE INTERNATIONAL, INC.

Address	6500 EAST ROGERS CIRCLE BOCA RATON, FL 33487
Phone	(561) 997-6033
Fax	(561) 997-6202
BBS	(561) 241-2929

COREL CORPORATION

Address	1600 CARLING AVENUE OTTAWA, ONTARIO, K1Z 8R7 CANADA
Phone	(613) 728-8200
Fax	(613) 761-9177
Toll Free	(800) 394-3729
Web	http://www.corel.com

CSS LABORATORIES, INC.

Address	1641 MCGAW AVENUE IRVINE, CA 92714
Phone	(714) 852-8161
Fax	(714) 852-0410
BBS	(714) 852-9231
Web	http://www.csslabs.com

DATA TECHNOLOGY CORPORATION

Address	1515 CENTRE POINTE DRIVE

	MILPITAS, CA 95035
Phone	(408) 942-4000
Fax	(408) 942-4027
BBS	(408) 942-4010
Fax-Back	(408) 942-4005
Web	http://www.datatechnology.com

DATASTOR TECHNOLOGY CO., INC.

Address	39833 PASEO PADRE PKWY. SUITE 1 FREMONT, CA 94538
Phone	(510) 770-9707
Fax	(510) 797-1840
Web	http://www.datastortech.com
E-mail	support@datastortech.com

DELL COMPUTER CORPORATION

Address	2214 W BREAKER LANE SUITE D AUSTIN, TX 78758
Phone	(512) 338-4400
Fax	(512) 728-3653
Toll Free	(800) 624-9897
BBS	(512) 728-8528
Fax-Back	(800) 950-1329
Web	http://www.dell.com

DIAMOND FLOWER, INC.

Address	135 MAIN AVENUE SACRAMENTO, CA 95838
Phone	(916) 568-1234
Fax	(916) 568-1233
Toll Free	(800) 808-4334
BBS	(908) 390-4820
Web	http://www.dfiusa.com
E-mail	info@dfiusa.com

DIGITAL EQUIPMENT CORPORATION

Address	111 POWDERMILL ROAD MAYNARD, MA 01754
Phone	(508) 493-2211
Fax	(508) 493-8780
Toll Free	(800) DIGITAL
BBS	(508) 264-7227
Fax-Back	(800) 344-4825

Web	http://www.digital.com

DIGITAL STORAGE SOLUTIONS

Address	5612 REGINA AVE. SAN DIEGO, CA 92117-1420

DISCTEC

Address	925 S. SEMORAN BLVD. SUITE 114 WINTER PARK, FL 32792
Phone	(407) 671-5500
Fax	(407) 671-6606
Toll Free	(800) 553-0337
BBS	(407) 671-6099
Web	http://www.disctec.com

DISTRIBUTED PROCESSING TECHNOLOGY

Address	140 CANDACE DRIVE MAITLAND, FL 32751
Phone	(407) 830-5522
Fax	(407) 260-5366
BBS	(407) 830-1070
Web	http://www.dpt.com

DIVERSE LOGISTICS, INC.

Address	2862 MCGAW AVENUE IRVINE, CA 92614
Phone	(714) 476-0303
Fax	(714) 476-0633
Toll Free	(800) 345-6432
Web	http://www.dilog.com
E-mail	info@dilog.com

DTK COMPUTER, INC.

Address	18645 EAST GALE AVENUE SUITE 233 CITY OF INDUSTRY, CA 91748
Phone	(626) 810-8880
Fax	(626) 810-5233
Toll Free	(800) 289-2385
BBS	(626) 854-0797
Fax-Back	(800) 806-1385
Web	http://www.dtkcomputer.com

EASTERN ELECTRONICS COMPANY

Address	NO. 4, SHIN LONG ROAD, KWEISHAN IND. AREA

	TAOYUAN,
	TAIWAN, REP OF CHINA
Phone	3 361 1141
Fax	3 362 5407

ELITEGROUP COMPUTER SYSTEMS, INC.

Address	45401 RESEARCH AVENUE
	FREMONT, CA 94539
Phone	(510) 226-7333
Fax	(510) 226-7350
Toll Free	(800) 829-8890
Web	http://www.ecsusa.com

EMULEX CORPORATION

Address	3535 HARBOR BLVD.
	COSTA MESA, CA 92626
Phone	(714) 662-5600
Fax	(714) 241-0792
Toll Free	(800) 854-7112
Web	http://www.emulex.com

EPSON, INC.

Address	20770 MADRONA AVENUE
	TORRANCE, CA 90503
Phone	(310) 782-0770
Fax	(310) 782-5220
Web	http://www.epson.com

EVEREX SYSTEMS, INC.

Address	5020 BRANDIN COURT
	FREMONT, CA 94538
Phone	(510) 498-1111
Fax	(510) 683-2186
Toll Free	(800) 821-0806
BBS	(510) 226-9694
Fax-Back	(510) 683-2800
Web	http://www.everex.com

EVERGREEN SYSTEMS, INC.

Address	120 LANDING COURT
	NOVATO, CA 94945
Phone	(415) 897-8888
Fax	(415) 897-6158
Toll Free	(800) EVERSYS
BBS	(415) 898-9398
Web	http://www.eversys.com

FIRST INTERNATIONAL COMPUTER, INC.

Address	980A MISSION COURT FREMONT, CA 94539
Phone	(510) 252-7777
Fax	(510) 252-8888
Toll Free	(800) 342-2636
BBS	(510) 252-7750
Fax-Back	(510) 252-8844
Web	http://www.fica.com
E-mail	sales@fica.com

FUJI ELECTRIC CORPORATION

Address	47520 WESTINGHOUSE DRIVE FREMONT, CA 94538
Phone	(510) 438-9700
Fax	(510) 428-9753
Web	http://www.fujielectric.co.jp

FUJITSU CORPORATION

Address	2904 ORCHARD PARKWAY SAN JOSE, CA 95134
Phone	(408) 432-6333
Fax	(408) 894-1709
Toll Free	(800) 626-4686
Fax-Back	(408) 428-0456
Web	http://www.fujitsu.com

FUTURE DOMAIN CORPORATION

Address	PO BOX 57020 IRVINE, CA 92619-7020
Phone	(714) 455-8100
Fax	(714) 253-0913
BBS	(714) 253-0432
Web	http://www.adaptec.com

GIGA-BYTE TECHNOLOGY CO., LTD.

Address	18325 VALLEY BLVD. SUITE E LA PUENTE, CA 91748
Phone	(818) 854-9338
Fax	(818) 854-9339
BBS	(818) 369-3985
Web	http://www.giga-byte.com
E-mail	gigabyte@tpts1.seed.net.tw

GOLDSTAR ELECTRONICS INTERNATIONAL, INC.

Address	1000 SYLVAN AVENUE ENGLEWOOD CLIFFS, NJ 07632
Phone	(201) 816-2000
Fax	(201) 816-0636
Toll Free	(800) 243-0000

GSI, INC.

Address	17951-H SKYPARK CIRCLE IRVINE, CA 92614
Phone	(714) 261-7949
Fax	(714) 757-1778
Toll Free	(800) 486-7800
Web	http://www.gsi-inc.com

HEWLETT-PACKARD COMPANY

Address	3000 HANNOVER STREET PALO ALTO, CA 94304
Phone	(800) 752-0900
Fax	(415) 857-5518
Toll Free	(800) 637-7740
BBS	(408) 553-3500
Fax-Back	(800) 231-9300
Web	http://www.hp.com

HITACHI CORPORATION

Address	2520 JUNCTION AVENUE BUILDING D SAN JOSE, CA 95134
Phone	(408) 321-5000
Toll Free	(800) 314-4743
Web	http://www.hitachi.com

IBM CORPORATION

Toll Free	(800) 772-2227
BBS	(919) 517-0001
Web	http://www.ibm.com

ICL

Address	WILLSTAETTER STRASSE 13 40549 DUESSELDORF, GERMANY
BBS	346 560 748
Web	http://www.icl.com

INFOMATIC POWER SYSTEMS CORPORATION (IPX)

Address	9945 SOUTH PIONEER BLVD. SANTA FE SPRINGS, CA 90670
Phone	(562) 948-2217
Fax	(562) 948-5264
Web	http://www.infomatic.com

INFORTREND TECHNOLOGY, INC.

Address	149 STONY CIRCLE SUITE 210 SANTA ROSA, CA 95401
Phone	(707) 541-3400
Fax	(707) 541-3409
Web	http://www.infortrend.com

INITIO CORPORATION

Address	2188 DEL FRANCO STREET SUITE B SAN JOSE, CA 95131
Phone	(408) 577-1919
Fax	(408) 577-0640
BBS	(408) 577-0431
Web	http://www.initio.com

INTEGRAL PERIPHERALS

Address	5775 FLATIRON PARKWAY SUITE 100 BOULDER, CO 80301-5730
Phone	(303) 449-8009
Fax	(303) 449-8089
Web	http://www.integralnet.com

INTEL CORPORATION

Address	P.O. BOX 58119 SANTA CLARA, CA 95052-8119
Phone	(408) 765-8080
Fax-Back	(800) 525-3019
Web	http://www.intel.com

INTELLIGENT COMPUTER PERIPHERALS

Address	4857 W. VAN BUREN PHOENIX, AZ 85043
Phone	(602) 353-0303
Fax	(602) 353-0051
Web	http://www.icp-vortex.com

E-mail	info@vortex.de

IOMEGA CORPORATION

Address	1821 WEST IOMEGA WAY ROY, UT 84067
Phone	(801) 778-1000
Fax	(801) 778-3748
Toll Free	(800) 778-0056
BBS	(801) 392-9819
Web	http://www.iomega.com

J&A COMPUTER CO., LTD.

Address	6FL-1, NO. 1, KUANG FU S. ROAD TAIPEI, TAIWAN, REP OF CHINA
Phone	2 749 1850
Fax	2 749 1858

JAMECO ELECTRONIC COMPONENTS

Address	1355 SHOREWAY ROAD BELMONT, CA 94002
Phone	(415) 592-8097
Fax	(415) 592-2503
Toll Free	(800) 831-4242
BBS	(415) 637-9025
Web	http://www.jameco.com
E-mail	tech@jameco.com

JC INFORMATION SYSTEMS CORPORATION

Address	4487 TECHNOLOGY DRIVE FREMONT, CA 94538
Phone	(510) 659-8440
Fax	(510) 659-8449
Web	http://www.jcis.com
E-mail	info@jcis.com

JPN CORPORATION

Address	48006 FREMONT BLVD. FREMONT, CA 94538
Phone	(510) 770-3990
Fax	(510) 770-3994
BBS	(510) 226-0632

JTS CORPORATION

Address	166 BAYPOINTE PARKWAY SAN JOSE, CA 95134

Phone	(408) 468-1800
Fax	(408) 468-1801
BBS	(408) 435-1741
E-mail	tech@jtscorp.com

JUKO LABORATORIES, LTD.

Address	UNIT B8, 6/F, BLK. B, HOPETITE IND. CENTRE 3-5 WANG TAI ROAD KOWLOON BAY, HONG KONG
Phone	2 953 0300
Fax	2 953 0363

JVC

Address	17811 MITCHELL AVENUE IRVINE, CA 92614
Phone	(714) 261-1292
Fax	(714) 261-9690
Web	http://www.jvc-us.com

KENTECH COMPUTER (HK), LTD.

Address	UNIT B, 21/F, ROXY INDUSTRIAL CENTRE 58-66 TAI LIN PAI ROAD, KWAI CHUNG KOWLOON, HONG KONG
Phone	2 489 0168
Fax	2 428 2770

KIMPSION INTERNATIONAL

Address	4701 PATRICK HENRY DRIVE SUITE 401 SANTA CLARA, CA 95054
Phone	(408) 988-8808
Fax	(408) 988-8809

KINGSTON TECHNOLOGY CORPORATION

Address	17600 NEWHOPE STREET FOUNTAIN VALLEY, CA 92708
Phone	(714) 435-2600
Fax	(714) 438-2699
Toll Free	(800) 435-2620
BBS	(714) 435-2636
Fax-Back	(714) 435-2677
Web	http://www.kingston.com

KOUWELL ELECTRONIC CORPORATION

Address	7F, NO. 99, NAN KANG ROAD, SEC. 3 TAIPEI, TAIWAN, REP OF CHINA
Phone	2 783 1166
Fax	2 783 5500
Web	http://www.kouwell.com.tw

KYOCERA ELECTRONICS, INC.

Address	8611 BALBOA AVENUE SAN DIEGO, CA 92123
Phone	(619) 576-2600
Fax	(619) 268-3035
Web	http://www.kyocera.com

LAVA COMPUTER MANUFACTURING, INC.

Address	28-A DANSK COURT REXDALE, ON M9W 5V8 CANADA
Phone	(416) 674-5942
Fax	(416) 674-8262
Toll Free	(800) 241-5282
Web	http://www.lavalink.com
E-mail	e-mail@lavalink.com

LEADINGSPECT CORPORATION

Address	1025 SEGOVIA CIRCLE PLACENTIA, CA 92870
Phone	(714) 666-2626
Fax	(714) 666-2900
Toll Free	(800) 234-0688
E-mail	leadspect@ix.netcom.com

LINKSYS

Address	17401 ARMSTRONG AVENUE IRVINE, CA 92614
Phone	(714) 261-1288
Fax	(714) 261-8868
Toll Free	(800) 546-5797
BBS	(714) 261-2888
Web	http://www.linksys.com

LOMAS DATA PRODUCTS, INC.

Address	420 MAPLE STREET SUITE 2

	MARLBORO, MA 01752
Phone	(508) 460-0333
Fax	(508) 460-0616
Web	http://ourworld.compuserve.com

LONGSHINE MICROSYSTEM, INC.

Address	10400-9 PIONEER BLVD. SANTA FE SPRINGS, CA 90670
Phone	(562) 903-0899
Fax	(562) 944-2201
BBS	(562) 903-4590
Web	http://www.longshin.com.tw

MAGICRAM, INC.

Address	1850 BEVERLY BLVD. LOS ANGELES, CA 90057
Phone	(213) 413-9999
Fax	(213) 413-0828
Toll Free	(800) 272-6242
Web	http://www.magicram.com

MAGITRONICS

Address	10 HUB DRIVE MELVILLE, NJ 11747
Web	http://www.magitronic.com

MAXTOR CORPORATION

Address	510 COTTONWOOD DRIVE MILPITAS, CA 95035
Phone	(408) 432-1700
Fax	(408) 432-4510
Toll Free	(800) 262-9867
BBS	(303) 678-2222
Fax-Back	(303) 678-2618
Web	http://www.maxtor.com

MECER CORPORATION

Address	3217 WHIPPLE ROAD UNION CITY, CA 94587
Phone	(510) 475-5730
Fax	(510) 475-0982
Web	http://www.posiflexusa.com

MEDIA VISION TECHNOLOGY, INC.

Address	KELTENRING 12 82041 MUNCHEN,

	GERMANY
Phone	89 613 813 000
Web	http://www.mediavis.com

MEGA DRIVE SYSTEMS

Address	489 S. ROBERTSON BLVD. BEVERLY HILLS, CA 90211
Phone	(310) 247-0006
Fax	(310) 247-8118
Web	http://www.megadrive.com

MEMOREX TELEX CORPORATION

Address	545 E. JOHN CARPENTER FREEWAY SUITE 200 IRVING, TX 75062
Phone	(972) 444-3500
Toll Free	(800) 980-6767

MEMTECH TECHNOLOGY

Address	1259 BIRCHWOOD DR. SUNNYVALE, CA 94089
Phone	(408) 745-1600
Fax	(408) 745-1733
Web	http://www.memtech.com
E-mail	info@memtech.com

MICRO EQUIPMENT CORPORATION

Address	2900 JONES MILL ROAD NORCROSS, GA 30071
Phone	(770) 447-1726
Fax	(770) 449-6103
Toll Free	(800) 226-8088
Web	http://www.mecworld.com

MICROPOLIS CORPORATION

Address	21211 NORDHOFF STREET CHATSWORTH, CA 91311
Phone	(818) 709-3300
Fax	(818) 718-5312
Toll Free	(800) 395-3748

MICRO-STAR INTERNATIONAL CO., LTD.

Address	4059 CLIPPER COURT FREMONT, CA 94538
Phone	(510) 623-8818
Fax	(510) 623-8585

Web	http://www.achme.com

MICROTECH INTERNATIONAL, INC.

Address	158 COMMERCE STREET EAST HAVEN, CT 06512
Phone	(203) 468-6223
Fax	(203) 468-6466
Toll Free	(800) 626-4276
BBS	(203) 469-6430
Web	http://www.microtechint.com

MITSUBISHI ELECTRONICS

Address	5665 PLAZA DRIVE CYPRESS, CA 90630
Phone	(800) 843-2515
Fax	(714) 236-6425
BBS	(714) 236-6286
Fax-Back	(714) 236-6453
Web	http://www.mitsubishi-display.

MODULAR CIRCUIT TECHNOLOGY

Web	http://www.jdr.com

MONOLITHIC SYSTEMS, INC. (COLORADO MSI)

Address	8150 S. ACKRON STREET UNIT 402 ENGLEWOOD, CO 80112
Phone	(303) 790-0180
Fax	(303) 790-0182

MOUNTAIN NETWORK SOLUTIONS, INC.

Web	http://www.ncegroup.com

MULTIBEST INDUSTRIAL & MANUFACTURING, INC.

Address	560 DENISON STREET UNIT 2 MARKHAM, ONTARIO, L3R 2M8 CANADA
Phone	(905) 889-0968

MYLEX CORPORATION

Address	34551 ARDENWOOD BLVD. FREMONT, CA 94555-3607
Phone	(510) 796-6100
Fax	(510) 745-7653
Toll Free	(800) 776-9539

BBS	(510) 793-3491
Web	http://www.mylex.com

NCR CORPORATION

Address	1700 SOUTH PATTERSON BLVD. DAYTON, OH 45479
Phone	(513) 445-5000
Toll Free	(800) 225-5627
Web	http://www.ncr.com

NEC TECHNOLOGIES, INC.

Address	8 CORPORATE CENTER DRIVE MELVILLE, NY 11747
Phone	(516) 753-7000
Fax	(516) 753-7434
Toll Free	(800) 338-7549
Web	http://www.nec.com

NEW MEDIA CORPORATION

Address	1 TECHNOLOGY, BUILDING A IRVINE, CA 92618
Phone	(714) 453-0100
Fax	(714) 453-0114
Toll Free	(800) 227-3748
BBS	(714) 453-0214
Fax-Back	(714) 789-5212
Web	http://www.newmediacorp.com

NEXSTOR, INC.

Address	631 SOUTH MILPITAS BLVD. MILPITAS, CA 95035
Phone	(408) 262-1056
Fax	(408) 262-1082
BBS	(408) 946-6932
Web	http://www.mycomp-tmc.com

NORTHGATE COMPUTER SYSTEMS, INC.

Address	6840 HAVENHURST AVENUE VAN NUIS, CA 91406
Toll Free	(800) 446-5037
Web	http://www.northgate.net

OCEAN INFORMATION SYSTEMS, INC.

Address	688 ARROW GRAND CIRLCE COVINA, CA 91722
Phone	(818) 339-8888

Fax	(818) 859-7668
Toll Free	(818) 339-8888
BBS	(818) 859-7639
Web	http://www.ocean-usa.com
E-mail	Webmaster@ocean-usa.com

OKIDATA

Address	532 FELLOWSHIP ROAD MT. LAUREL, NJ 08054
Phone	(609) 235-2600
Fax	(609) 778-4184
Toll Free	(800) OKI-DATA
Fax-Back	(800) 654-3282
Web	http://www.okidata.com

OLIVETTI

Address	22425 E. APPLEWAY AVENUE LIBERTY LAKE, WA 99019-9534
Phone	(509) 927-5600
Fax	(509) 927-5774
Toll Free	(800) 255-4319
Web	http://www.olivetti.com

OPTIMA TECHNOLOGY

Address	17062 MURPHY AVENUE IRVINE, CA 92714
Phone	(714) 476-0515
Fax	(714) 476-0613
Web	http://www.optimatech.com
E-mail	sales@optimatech.com

ORCHID TECHNOLOGY

Address	221 WARREN AVENUE FREMONT, CA 94539
Phone	(510) 683-0300
Fax	(510) 490-9312
BBS	(510) 683-0555
Web	http://www.orchid.com

OTARI

Address	378 VINTAGE PARK DRIVE FOSTER CITY, CA 94404
Phone	(415) 341-5900
Fax	(415) 341-7200
Web	http://www.otari.com

PANASONIC INDUSTRIAL COMPANY

Address	TWO PANASONIC WAY SECAUCUS, NJ 07094
Phone	(201) 392-4222
Fax	(201) 392-4482
Toll Free	(800) 726-2797
Web	http://www.panasonic.com

PC WARE INTERNATIONAL, INC.

Address	4372 10TH STREET LONG ISLAND CITY, NY 11101
Phone	(718) 706-7770
Fax	(718) 706-7864
Toll Free	(800) 572-9273
Web	http://pcware.com

PERCEPTIVE SOLUTIONS, INC.

Address	2700 FLORA STREET DALLAS, TX 75201
Phone	(214) 954-1774
Fax	(214) 953-1774
BBS	(214) 954-1856
Web	http://www.psidisk.com

PERIPHERALS ENTERPRISE CO., LTD.

Address	3F.NO. 10, ALLEY 6, LANE 45, PAO SHIN ROAD HSIN TIEN TAIPEI HSIEN, TAIWAN, REP OF CHINA
Phone	2 917 7322
Fax	2 915 4672
E-mail	peripherals@twinner.com.tw

PINE TECHNOLOGY

Address	777 NEW DURHAM ROAD EDISON, NJ 08817
Phone	(908) 321-3700
Fax	(908) 321-3636
Web	http://www.pinegroup.com

PRACTICAL ENHANCED LOGIC CORPORATION

Address	22695 OLD CANAL ROAD YORBA LINDA, CA 92687
Phone	(714) 282-6188
Fax	(714) 282-6199

Web	http://www.pelogic.com

PROLINK COMPUTER, INC.

Address	15336 EAST VALLEY BLVD. CITY OF INDUSTRY, CA 91746
Phone	(818) 369-3833
Fax	(818) 369-4883
Web	http://www.prolink-usa.com

PROMISE TECHNOLOGY, INC.

Address	1460 KOLL CIRCLE SAN JOSE, CA 95112
Phone	(408) 452-0948
Fax	(408) 452-1534
Toll Free	(800) 888-0245
Web	http://www.promise.com

Q LOGIC CORPORATION

Address	3545 HARBOR BLVD. COSTA MESA, CA 92626
Fax	(714) 668-5090
Toll Free	(800) 662-4471
BBS	(714) 708-3170
Web	http://www.qlc.com

QDI COMPUTER, INC.

Address	11552 EAST WASHINGTON BLVD., UNIT D WHITTIER, CA 90606
Phone	(310) 908-1029
Fax	(310) 908-1033
Web	http://www.qdigrp.com

QUANTUM CORPORATION

Address	500 MCCARTHY BLVD. MILPITAS, CA 95035
Phone	(408) 894-4000
Fax	(408) 894-3218
Toll Free	(800) 624-5525
Web	http://www.quantum.com

QUATECH, INC.

Address	662 WOLF LEDGES PARKWAY AKRON, OH 44311
Phone	(330) 434-3154
Fax	(330) 434-1409
Toll Free	(800) 553-1170

BBS	(330) 434-2481
Web	http://www.quatech.com
E-mail	trier@quatech.com

QUICK TECHNOLOGY, INC.

Address	1542 EDINGER AVENUE, #B TUSTIN, CA 92680
Phone	(714) 258-4500
Fax	(714) 258-4508
Toll Free	(800) 950-8999
Web	http://www.iwillusa.com

QUICKPATH SYSTEMS, INC.

Address	46723 FREMONT BLVD. FREMONT, CA 94538
Phone	(510) 440-7288
Fax	(510) 440-7289
Toll Free	(800) 995-8828
BBS	(510) 440-7284
Web	http://www.quickpath.com
E-mail	qpinfo@quickpath.com

RANCHO TECHNOLOGY, INC.

Address	10783 BELL COURT RANCHO CUCAMONGA, CA 91730
Phone	(909) 987-3966
Fax	(909) 989-2365
BBS	(909) 980-7699
Web	http://www.rancho.com

RELIALOGIC CORPORATION PRIVATE, LTD.

Address	48006 FREMONT BLVD. FREMONT, CA 94538
Phone	(510) 770-3990
Fax	(510) 770-3994
Toll Free	(800) 998-3966
BBS	(510) 226-0632
E-mail	rlogic@aol.com

RICOH CORPORATION

Address	1231 WARNER AVENUE TUSTIN, CA 92680
Phone	(714) 259-1310
Fax	(714) 566-3664
Toll Free	(800) 955-3453
Web	http://www.ricohcorp.com

RMS

Address	TANJONG PAGAR P.O. BOX 0322 910811 REP OF SINGAPORE

SAMSUNG ELECTRONICS, INC.

Address	105 CHALLENGER BLVD. RIDGEFIELD PARK, NJ 07660
Phone	(201) 229-7000
Fax	(201) 229-4110
Toll Free	(800) 446-0262
BBS	(201) 691-6238
Fax-Back	(201) 229-4053
Web	http://www.sosimple.com

SANYO ELECTRIC CO., LTD.

Address	52 7 25, EDOBOR, NISHI KU OSAKA, 550 OSAKA, JAPAN
Phone	64 432 949
Fax	64 432 849
Web	http://www.sanyo.co.jp

SBE, INC.

Address	4550 NORRIS CANYON ROAD SAN RAMON, CA 94583-1369
Phone	(510) 355-2000
Fax	(510) 355-2020
Toll Free	(800) 925-2666
BBS	(510) 355-2048
Fax-Back	(800) 214-4723
Web	http://www.sbei.com
E-mail	netxpand@sbei.com

SCIENTIFIC MICRO SYSTEMS

Address	2345 HARRIS WAY SAN JOSE, CA 95131
Phone	(408) 954-8333
Fax	(408) 954-0622
BBS	(408) 954-8231
Web	http://www.televideo.com

SEAGATE TECHNOLOGY, INC.

Address	PO BOX 66360

	SCOTTS VALLEY, CA 95067-0360
Phone	(408) 438-6550
Fax	(408) 438-9120
Toll Free	(800) SEAGATE
BBS	(408) 438-8771
Fax-Back	(408) 438-2620
Web	http://www.seagate.com

SEANIX TECHNOLOGY, INC.

Address	6651 ELMBRIDGE WAY
	SUITE 150
	RICHMOND, B. V7C 4N1
	CANADA
Phone	(604) 273-3692
Fax	(604) 273-0768
Toll Free	(800) 663-5656
BBS	(604) 278-7418
Web	http://www.seanix.com

SIEMENS NIXDORF INFORMATIONSSYSTEME AG

Address	DYBENDALSVAENGET 3
	2630 TASTRUP,
	DENMARK
Phone	44 774 600
Fax	44 774 610

SIIG, INC.

Address	6078 STEWART AVENUE
	FREMONT, CA 94538
Phone	(510) 657-8688
Fax	(510) 657-5962
BBS	(510) 353-7532
Web	http://www.siig.com

SILICON STAR INTERNATIONAL, INC.

Address	3F-7, NO. 79, SEC. 1, HSIN TAI WU ROAD
	HSI-CHI
	TAIPEI HSIEN,
	TAIWAN, REP OF CHINA
Phone	2 698 1888
Fax	2 698 1811
Web	http://www.abit.com.tw

SIMPLE TECHNOLOGY, INC.

Address	3001 DAIMLER STREET
	SANTA ANA, CA 92705

Phone	(714) 476-1180
Fax	(714) 476-1209
Toll Free	(800) 367-7330
Web	http://www.simpletech.com

SONY CORPORATION

Address	1 SONY DRIVE PARK RIDGE, NJ 07656
Phone	(201) 930-1000
Fax	(201) 573-8608
Web	http://www.sony.com

STANDARD MICROSYSTEMS CORPORATION

Address	80 ARKAY DRIVE HAUPPAUGE, NY 11788
Phone	(516) 435-6000
Fax	(516) 273-1803
Toll Free	(800) SMC-4YOU
BBS	(516) 434-3162
Fax-Back	(800) 762-8329
Web	http://www.smc.com
E-mail	techsupt@ccmail.west.smc.com

STB SYSTEMS, INC.

Address	1651 N. GLENVILLE DRIVE SUITE 210 RICHARDSON, TX 75081
Phone	(214) 234-8750
Fax	(214) 234-1306
BBS	(214) 437-9615
Fax-Back	(214) 669-1326
Web	http://www.stb.com

STORAGE DIMENSIONS

Address	1656 MC CARTHY BLVD. MILPITAS, CA 95035
Phone	(408) 954-0710
Fax	(408) 944-1203
Web	http://www.storagedimensions.c

SUNIX CO., LTD.

Address	791 FOAM STREET SUITE 205 MONTEREY, CA 93940
Phone	(408) 665-1977
Fax	(408) 665-1919

Web	http://www.dolphinfast.com

SYMBIOS LOGIC

Address	4420 ARROWS WEST DRIVE COLORADO SPRINGS, CO 80907
Phone	(719) 533-7000
Fax	(719) 533-7244
Web	http://www.symbios.com

SYQUEST TECHNOLOGY

Address	47071 BAYSIDE PARKWAY FREMONT, CA 94538
Phone	(510) 226-4000
Fax	(510) 226-4102
Toll Free	(800) 245-2278
BBS	(510) 656-0473
Fax-Back	(510) 226-4120
Web	http://www.syquest.com

TANDON CORPORATION

Phone	(414) 638-1851
Fax	(414) 638-1852
Web	http://www.sierrainc.com

TANDY/RADIO SHACK

Address	1500 ONE TANDY CENTER FORT WORTH, TX 76102
Phone	(817) 390-3011
Toll Free	(800) THE-SHACK
Fax-Back	(800) 323-6586
Web	http://www.tandy.com

TEAC CORPORATION

Address	7733 TELEGRAPH ROAD MONTEBELLO, CA 90640
Phone	(213) 726-0303
Fax	(213) 727-7672
BBS	(213) 727-7660
Fax-Back	(213) 727-7629
Web	http://www.teac.com

TEKRAM TECHNOLOGY CO., LTD.

Address	11500 METRIC BLVD. SUITE 190 AUSTIN, TX 78758
Phone	(512) 833-6550

Fax	(512) 833-7276
BBS	(512) 833-7985
Web	http://www.tekram.com
E-mail	sales@tekram.com.tw

TEXAS INSTRUMENTS

Address	12201 SOUTHWEST FREEWAY STAFFORD, TX 77477
Phone	(713) 274-3388
Toll Free	(800) TI-TEXAS
Web	http://www.ti.com
E-mail	odin@msg.ti.com

TMC RESEARCH CORPORATION

Address	631 SOUTH MILPITAS BLVD. MILPITAS, CA 95035
Phone	(408) 262-0888
Fax	(408) 262-1082
BBS	(408) 946-6932
Web	http://www.mycomp-tmc.com

TOSHIBA

Address	9740 IRVINE BLVD. PO BOX 19724 IRVINE, CA 92713
Phone	(714) 455-0407
Fax	(714) 583-3140
Toll Free	(800) 999-4273
BBS	(714) 837-8864
Fax-Back	(714) 455-0407
Web	http://www.toshiba.com

TRANTOR SYSTEMS, LTD.

Address	691 SOUTH MILPITAS BLVD. MILPITAS, CA 95035
Phone	(408) 945-8600
Toll Free	(800) 872-6867
Web	http://www.adaptec.com

TULIN

Address	2156 O'TOOLE AVENUE UNIT H SAN JOSE, CA 95131
Phone	(408) 432-9057
Fax	(408) 943-0782

TYAN COMPUTER CORPORATION

Address	1753 S MAIN STREET MILPITAS, CA 95035
Phone	(408) 956-8000
Fax	(408) 956-8044
BBS	(408) 956-8171
Web	http://www.tyan.com

UNITED MICROELECTRONICS CORPORATION

Address	NO. 13, INNOVATION ROAD 1 SCIENCE BASED INDUSTRIAL PARK HSINCHU, TAIWAN, REP OF CHINA
Phone	35 782 258
Fax	35 774 767
Web	http://www.umc.com.tw

VELA RESEARCH, INC.

Address	PO BOX 9090 CLEARWATER, FL 34618
Phone	(813) 572-1230
Fax	(813) 573-2508
Web	http://www.vela.com

VISIONTEK

Address	1175 LAKESIDE DRIVE GURNEE, IL 60031
Phone	(847) 360-7500
Fax	(847) 360-7144
Toll Free	(800) 726-9695
BBS	(847) 360-7432
Fax-Back	(800) 360-7170
Web	http://www.visiontek.com

VIVA COMPUTER COMPANY

Address	48346 MILMONT DRIVE FREMONT, CA 94538
Phone	(510) 490-4988
Fax	(510) 490-4999

VTECH INDUSTRIES, INC.

Address	800 CHURCH STREET LAKE ZURICH, IL 60047
Phone	(708) 215-9806
Web	http://www.vtechinc.com

WESTERN DIGITAL CORPORATION

Address	8105 IRVINE CENTER DRIVE IRVINE, CA 92718
Phone	(714) 932-5000
Fax	(714) 932-6294
Toll Free	(800) 832-4778
BBS	(714) 753-1234
Fax-Back	(714) 932-4300
Web	http://www.wdc.com

WYSE TECHNOLOGY, INC.

Address	3471 NORTH FIRST STREET SAN JOSE, CA 95134-1803
Phone	(408) 473-1200
Fax	(408) 473-2788
Toll Free	(800) 438-9973
BBS	(408) 922-4400
Fax-Back	(800) 800-9973
Web	http://www.wyse.com

XIRLINK, INC.

Address	2210 O'TOOLE AVENUE SAN JOSE, CA 95131
Phone	(408) 324-2100
Fax	(408) 324-2101
Web	http://www.xirlink.com

Y-E DATA AMERICA, INC.

Address	2942 MACARTHUR BLVD. NORTHBROOK, IL 60062
Phone	(847) 291-2340
Fax	(847) 291-4203
Web	http://www.yaskawa.com

ZENTEK STORAGE, INC.

Address	NO. 1, INNOVATION ROAD 1 SCIENCE BASE INDUSTRIAL PARK HSINCHU, TAIWAN, REP OF CHINA
Phone	35 783 300
Fax	35 774 734

ZOBEX

Address	9020 JUNCTION DRIVE ANNAPOLIS JUNCTION, MD 20701

Phone (301) 317-7710
Fax (301) 317-7220
Toll Free (800) 359-7710

Manufacturer Index

A

E

F-G

H

I

J

K

L

M

N

O

P

Q

R

S

X-Z

Index

Symbols

A

B

D

E

F

I

L

M

N

O

P

R

T

U

V

W

X

Z